THERMO-FLUID DYNAMICS OF TWO-PHASE FLOW

discussed in detail. The macroscopic formulations are presented in terms of the two-fluid model and drift-flux model which are two of the most accurate and useful formulations for practical engineering problems.

The change of the interfacial structures in two-phase flow is dynamically modeled through the interfacial area transport equation. This is a new approach which can replace the static and inaccurate approach based on the flow regime transition criteria. The interfacial momentum transfer models are discussed in great detail, because for most two-phase flow, thermo-fluid dynamics are dominated by the interfacial structures and interfacial momentum transfer. Some other necessary constitutive relations such as the turbulence modeling, transient forces and lift forces are also discussed.

Mamoru Ishii, Ph.D.
School of Nuclear Engineering
Purdue University
West Lafayette, IN, USA

Takashi Hibiki, Ph.D.
Research Reactor Institute
Kyoto University
Kumatori, Osaka, Japan

September 2005

THERMO-FLUID DYNAMICS OF TWO-PHASE FLOW

Authored by

MAMORU ISHII
Purdue University

TAKASHI HIBIKI
Kyoto University

Springer

Mamoru Ishii
School of Nuclear Engineering
Purdue University
1290 Nuclear Engineering Building
West Lafayette, IN 47906
U.S.A.

Takashi Hibiki
Research Reactor Institute
Kyoto University
Noda, Kumatori, Sennan
Osaka 590-0494
Japan

Thermo-fluid Dynamics of Two-phase Flow

Library of Congress Control Number: 20055934802

ISBN-10: 0-387-28321-8
ISBN-13: 9780387283210
ISBN-10: 0-387-29187-3 (e-book)
ISBN-13: 9780387291871 (e-book)

Printed on acid-free paper.

Printed in the United States of America.

9 8 7 6 5 4 3 2 1 SPIN 11429425

springer.com

Dedication

This book is dedicated to our parents.

Table of Contents

Preface

This book is intended to be an introduction to the theory of thermo-fluid dynamics of two-phase flow for graduate students, scientists and practicing engineers seriously involved in the subject. It can be used as a text book at the graduate level courses focused on the two-phase flow in Nuclear Engineering, Mechanical Engineering and Chemical Engineering, as well as a basic reference book for two-phase flow formulations for researchers and engineers involved in solving multiphase flow problems in various technological fields.

The principles of single-phase flow fluid dynamics and heat transfer are relatively well understood, however two-phase flow thermo-fluid dynamics is an order of magnitude more complicated subject than that of the single-phase flow due to the existence of moving and deformable interface and its interactions with the two phases. However, in view of the practical importance of two-phase flow in various modern engineering technologies related to nuclear energy, chemical engineering processes and advanced heat transfer systems, significant efforts have been made in recent years to develop accurate general two-phase formulations, mechanistic models for interfacial transfer and interfacial structures, and computational methods to solve these predictive models.

A strong emphasis has been put on the rational approach to the derivation of the two-phase flow formulations which represent the fundamental physical principles such as the conservations laws and constitutive modeling for various transfer mechanisms both in bulk fluids and at interface. Several models such as the local instant formulation based on the single-phase flow model with explicit treatment of interface and the macroscopic continuum formulations based on various averaging methods are presented and

Foreword

Thermo-Fluid Dynamics of Two-Phase Flow takes a major step forward in our quest for understanding fluids as they metamorphose through change of phase, properties and structure. Like Janus, the mythical Roman God with two faces, fluids separating into liquid and gas, each state sufficiently understood on its own, present a major challenge to the most astute and insightful scientific minds when it comes to deciphering their dynamic entanglement.

The challenge stems in part from the vastness of scale where two phase phenomena can be encountered. Between the microscopic *nano*-scale of molecular dynamics and deeply submerged modeling assumptions and the *macro*-scale of measurements, there is a *meso*-scale as broad as it is nebulous and elusive. This is the scale where everything is in a permanent state of exchange, a Heraclitean state of flux, where nothing ever stays the same and where knowledge can only be achieved by firmly grasping the underlying principles of things.

The subject matter has sprung from the authors' own firm grasp of fundamentals. Their bibliographical contributions on two-phase principles reflect a scientific tradition that considers theory and experiment a duality as fundamental as that of appearance and reality. In this it differs from other topical works in the science of fluids. For example, the leading notion that runs through two-phase flow is that of interfacial velocity. It is a concept that requires, amongst other things, continuous improvements in both modeling and measurement. In the *meso*-scale, this gives rise to new science of the interface which, besides the complexity of its problems and the fuzziness of its structure, affords ample scope for the creation of elegant, parsimonious formulations, as well as promising engineering applications.

The two-phase flow theoretical discourse and experimental inquiry are closely linked. The synthesis that arises from this connection generates immense technological potential for measurements informing and validating dynamic models and conversely. The resulting technology finds growing utility in a broad spectrum of applications, ranging from next generation nuclear machinery and space engines to pharmaceutical manufacturing, food technology, energy and environmental remediation.

This is an intriguing subject and its proper understanding calls for exercising the rigorous tools of advanced mathematics. The authors, with enormous care and intellectual affection for the subject reach out and invite an inclusive audience of scientists, engineers, technologists, professors and students.

It is a great privilege to include the *Thermo-Fluid Dynamics of Two-Phase Flow* in the series ***Smart Energy Systems: Nanowatts to Terawatts***. This is work that will stand the test of time for its scientific value as well as its elegance and aesthetic character.

Lefteri H. Tsoukalas, Ph.D.
School of Nuclear Engineering
Purdue University
West Lafayette, IN, USA

September 2005

Acknowledgments

The authors would like to express their sincere appreciation to those persons who have contributed in preparing this book. Professors N. Zuber and J. M. Delhaye are acknowledged for their early input and discussions on the development of the fundamental approach for the theory of thermo-fluid dynamics of multiphase flow. We would like to thank Dr. F. Eltawila of the U.S. Nuclear Regulatory Commission for long standing support of our research focused on the fundamental physics of two-phase flow. This research led to some of the important results included in the book. Many of our former students such as Professors Qiao Wu, Seungjin Kim, Xiaodong Sun and Dr. X.Y. Fu contributed significantly through their Ph.D. thesis research. Current Ph.D. students S. Paranjape and B. Ozar deserve many thanks for checking the equations and taking the two-phase flow images, respectively. The authors thank Professor Lefteri Tsoukalas for inviting us to write this book under the new series, "Smart Energy Systems: Nanowatts to Terawatts".

Chapter 1

INTRODUCTION

1.1 Relevance of the problem

This book is intended to be a basic reference on the thermo-fluid dynamic theory of two-phase flow. The subject of two or multiphase flow has become increasingly important in a wide variety of engineering systems for their optimum design and safe operations. It is, however, by no means limited to today's modern industrial technology, and multiphase flow phenomena can be observed in a number of biological systems and natural phenomena which require better understandings. Some of the important applications are listed below.

Power Systems
Boiling water and pressurized water nuclear reactors; liquid metal fast breeder nuclear reactors; conventional power plants with boilers and evaporators; Rankine cycle liquid metal space power plants; MHD generators; geothermal energy plants; internal combustion engines; jet engines; liquid or solid propellant rockets; two-phase propulsors, etc.

Heat Transfer Systems
Heat exchangers; evaporators; condensers; spray cooling towers; dryers, refrigerators, and electronic cooling systems; cryogenic heat exchangers; film cooling systems; heat pipes; direct contact heat exchangers; heat storage by heat of fusion, etc.

Process Systems
Extraction and distillation units; fluidized beds; chemical reactors; desalination systems; emulsifiers; phase separators; atomizers; scrubbers; absorbers; homogenizers; stirred reactors; porous media, etc.

Transport Systems
Air-lift pump; ejectors; pipeline transport of gas and oil mixtures, of slurries, of fibers, of wheat, and of pulverized solid particles; pumps and hydrofoils with cavitations; pneumatic conveyors; highway traffic flows and controls, etc.

Information Systems
Superfluidity of liquid helium; conducting or charged liquid film; liquid crystals, etc.

Lubrication Systems
Two-phase flow lubrication; bearing cooling by cryogenics, etc.

Environmental Control
Air conditioners; refrigerators and coolers; dust collectors; sewage treatment plants; pollutant separators; air pollution controls; life support systems for space application, etc.

Geo-Meteorological Phenomena
Sedimentation; soil erosion and transport by wind; ocean waves; snow drifts; sand dune formations; formation and motion of rain droplets; ice formations; river floodings, landslides, and snowslides; physics of clouds, rivers or seas covered by drift ice; fallout, etc.

Biological Systems
Cardiovascular system; respiratory system; gastrointestinal tract; blood flow; bronchus flow and nasal cavity flow; capillary transport; body temperature control by perspiration, etc.

It can be said that all systems and components listed above are governed by essentially the same physical laws of transport of mass, momentum and energy. It is evident that with our rapid advances in engineering technology, the demands for progressively accurate predictions of the systems in interest have increased. As the size of engineering systems becomes larger and the operational conditions are being pushed to new limits, the precise understanding of the physics governing these multiphase flow systems is indispensable for safe as well as economically sound operations. This means a shift of design methods from the ones exclusively based on static experimental correlations to the ones based on mathematical models that can predict dynamical behaviors of systems such as transient responses and stabilities. It is clear that the subject of multiphase flow has immense importance in various engineering technology. The optimum design, the prediction of operational limits and, very often, the safe control of a great number of important systems depend upon the availability of realistic and accurate mathematical models of two-phase flow.

1.2 Characteristic of multiphase flow

Many examples of multiphase flow systems are noted above. At first glance it may appear that various two or multiphase flow systems and their physical phenomena have very little in common. Because of this, the tendency has been to analyze the problems of a particular system, component or process and develop system specific models and correlations of limited generality and applicability. Consequently, a broad understanding of the thermo-fluid dynamics of two-phase flow has been only slowly developed and, therefore, the predictive capability has not attained the level available for single-phase flow analyses.

The design of engineering systems and the ability to predict their performance depend upon both the availability of experimental data and of conceptual mathematical models that can be used to describe the physical processes with a required degree of accuracy. It is essential that the various characteristics and physics of two-phase flow should be modeled and formulated on a rational basis and supported by detailed scientific experiments. It is well established in continuum mechanics that the conceptual model for single-phase flow is formulated in terms of field equations describing the conservation laws of mass, momentum, energy, charge, etc. These field equations are then complemented by appropriate constitutive equations for thermodynamic state, stress, energy transfer, chemical reactions, etc. These constitutive equations specify the thermodynamic, transport and chemical properties of a specific constituent material.

It is to be expected, therefore, that the conceptual models for multiphase flow should also be formulated in terms of the appropriate field and constitutive relations. However, the derivation of such equations for multiphase flow is considerably more complicated than for single-phase flow. The complex nature of two or multiphase flow originates from the existence of multiple, deformable and moving interfaces and attendant significant discontinuities of fluid properties and complicated flow field near the interface. By focusing on the interfacial structure and transfer, it is noticed that many of two-phase systems have a common geometrical structure. It is recalled that single-phase flow can be classified according to the structure of flow into laminar, transitional and turbulent flow. In contrast, two-phase flow can be classified according to the structure of interface into several major groups which can be called flow regimes or patterns such as separated flow, transitional or mixed flow and dispersed flow. It can be expected that many of two-phase flow systems should exhibit certain degree of physical similarity when the flow regimes are same. However, in general, the concept of two-phase flow regimes is defined based on a macroscopic volume or length scale which is often comparative to the system length scale. This implies that the concept of two-phase flow regimes and regime-dependent model require an introduction of a large length scale and associated limitations. Therefore, regime-dependent models may lead to an analysis that cannot mechanistically address the physics and phenomena occurring below the reference length scale.

For most two-phase flow problems, the local instant formulation based on the single-phase flow formulation with explicit moving interfaces encounters insurmountable mathematical and numerical difficulties, and therefore it is not a realistic or practical approach. This leads to the need of a macroscopic formulation based on proper averaging which gives a two-phase flow continuum formulation by effectively eliminating the interfacial discontinuities. The essence of the formulation is to take into account for the various multi-scale physics by a cascading modeling approach, bringing the micro and meso-scale physics into the macroscopic continuum formulation.

The above discussion indicates the origin of the difficulties encountered in developing broad understanding of multiphase flow and the generalized method for analyzing such flow. The two-phase flow physics are fundamentally multi-scale in nature. It is necessary to take into account these cascading effects of various physics at different scales in the two-phase flow formulation and closure relations. At least four different scales can be important in multiphase flow. These are 1) system scale, 2) macroscopic scale required for continuum assumption, 3) mesoscale related to local structures, and 4) microscopic scale related to fine structures and molecular transport. At the highest level, the scale is the system where system transients and component interactions are the primary focus. For example, nuclear reactor accidents and transient analysis requires specialized system analysis codes. At the next level, macro physics such as the structure of interface and the transport of mass, momentum and energy are addressed. However, the multiphase flow field equations describing the conservation principles require additional constitutive relations for bulk transfer. This encompasses the turbulence effects for momentum and energy as well as for interfacial exchanges for mass, momentum and energy transfer. These are meso-scale physical phenomena that require concentrated research efforts. Since the interfacial transfer rates can be considered as the product of the interfacial flux and the available interfacial area, the modeling of the interfacial area concentration is essential. In two-phase flow analysis, the void fraction and the interfacial area concentration represent the two fundamental first-order geometrical parameters and, therefore, they are closely related to two-phase flow regimes. However, the concept of the two-phase flow regimes is difficult to quantify mathematically at the local point because it is often defined at the scale close to the system scale.

This may indicate that the modeling of the changes of the interfacial area concentration directly by a transport equation is a better approach than the conventional method using the flow regime transitions criteria and regime-dependent constitutive relations for interfacial area concentration. This is particularly true for a three-dimensional formulation of two-phase flow. The next lower level of physics in multiphase flow is related to the local microscopic phenomena, such as: the wall nucleation or condensation; bubble coalescence and break-up; and entrainment and deposition.

1.3 Classification of two-phase flow

There are a variety of two-phase flows depending on combinations of two phases as well as on interface structures. Two-phase mixtures are characterized by the existence of one or several interfaces and discontinuities at the interface. It is easy to classify two-phase mixtures according to the combinations of two phases, since in standard conditions we have only three states of matters and at most four, namely, solid, liquid, and gas phases and possibly plasma (Pai, 1972). Here, we consider only the first three phases, therefore we have:

1. Gas-solid mixture;
2. Gas-liquid mixture;
3. Liquid-solid mixture;
4. Immiscible-liquid mixture.

It is evident that the fourth group is not a two-phase flow, however, for all practical purposes it can be treated as if it is a two-phase mixture.

The second classification based on the interface structures and the topographical distribution of each phase is far more difficult to make, since these interface structure changes occur continuously. Here we follow the standard flow regimes reviewed by Wallis (1969), Hewitt and Hall Taylor (1970), Collier (1972), Govier and Aziz (1972) and the major classification of Zuber (1971), Ishii (1971) and Kocamustafaogullari (1971). The two-phase flow can be classified according to the geometry of the interfaces into three main classes, namely, separated flow, transitional or mixed flow and dispersed flow as shown in Table 1-1.

Table 1-1. Classification of two-phase flow (Ishii, 1975)

Class	Typical regimes	Geometry	Configuration	Examples
Separated flows	Film flow	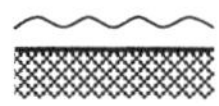	Liquid film in gas Gas film in liquid	Film condensation Film boiling
	Annular flow	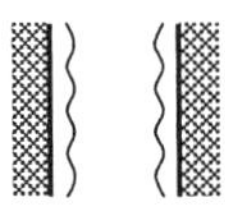	Liquid core and gas film Gas core and liquid film	Film boiling Boilers
	Jet flow	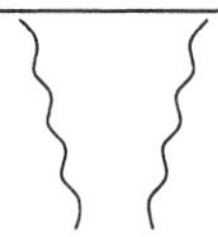	Liquid jet in gas Gas jet in liquid	Atomization Jet condenser
Mixed or Transitional flows	Cap, Slug or Churn-turbulent flow		Gas pocket in liquid	Sodium boiling in forced convection
	Bubbly annular flow	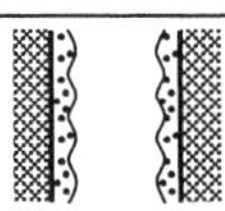	Gas bubbles in liquid film with gas core	Evaporators with wall nucleation
	Droplet annular flow	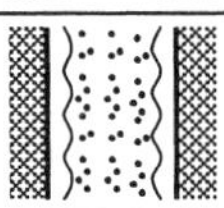	Gas core with droplets and liquid film	Steam generator
	Bubbly droplet annular flow	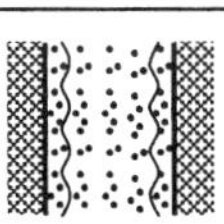	Gas core with droplets and liquid film with gas bubbles	Boiling nuclear reactor channel
Dispersed flows	Bubbly flow	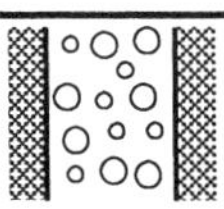	Gas bubbles in liquid	Chemical reactors
	Droplet flow	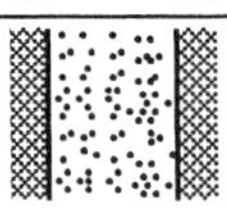	Liquid droplets in gas	Spray cooling
	Particulate flow	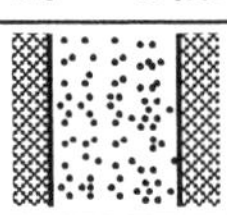	Solid particles in gas or liquid	Transportation of powder

Depending upon the type of the interface, the class of separated flow can be divided into plane flow and quasi-axisymmetric flow each of which can be subdivided into two regimes. Thus, the plane flow includes film and stratified flow, whereas the quasi-axisymmetric flow consists of the annular and the jet-flow regimes. The various configurations of the two phases and of the immiscible liquids are shown in Table 1-1.

The class of dispersed flow can also be divided into several types. Depending upon the geometry of the interface, one can consider spherical, elliptical, granular particles, etc. However, it is more convenient to subdivide the class of dispersed flows by considering the phase of the dispersion. Accordingly, we can distinguish three regimes: bubbly, droplet or mist, and particulate flow. In each regime the geometry of the dispersion can be spherical, spheroidal, distorted, etc. The various configurations between the phases and mixture component are shown in Table 1-1.

As it has been noted above, the change of interfacial structures occurs gradually, thus we have the third class which is characterized by the presence of both separated and dispersed flow. The transition happens frequently for liquid-vapor mixtures as a phase change progresses along a channel. Here too, it is more convenient to subdivide the class of mixed flow according to the phase of dispersion. Consequently, we can distinguish five regimes, i.e., cap, slug or churn-turbulent flow, bubbly-annular flow, bubbly annular-droplet flow and film flow with entrainment. The various configurations between the phases and mixtures components are shown in Table 1-1.

Figures 1-1 and 1-2 show typical air-water flow regimes observed in vertical 25.4 mm and 50.8 mm diameter pipes, respectively. The flow regimes in the first, second, third, fourth, and fifth figures from the left are bubbly, cap-bubbly, slug, churn-turbulent, and annular flows, respectively. Figure 1-3 also shows typical air-water flow regimes observed in a vertical rectangular channel with the gap of 10 mm and the width of 200 mm. The flow regimes in the first, second, third, and fourth figures from the left are bubbly, cap-bubbly, churn-turbulent, and annular flows, respectively. Figure 1-4 shows inverted annular flow simulated adiabatically with turbulent water jets, issuing downward from large aspect ratio nozzles, enclosed in gas annuli (De Jarlais et al., 1986). The first, second, third and fourth images from the left indicate symmetric jet instability, sinuous jet instability, large surface waves and skirt formation, and highly turbulent jet instability, respectively. Figure 1-5 shows typical images of inverted annular flow at inlet liquid velocity 10.5 cm/s, inlet gas velocity 43.7 cm/s (nitrogen gas) and inlet Freon-113 temperature 23 °C with wall temperature of near 200 °C (Ishii and De Jarlais, 1987). Inverted annular flow was formed by introducing the test fluid into the test section core through thin-walled, tubular nozzles coaxially centered within the heater quartz tubing, while vapor or gas is introduced in the annular gap between the liquid nozzle and the heated quartz tubing. The absolute vertical size of each image is 12.5 cm. The visualized elevation is higher from the left figure to the right figure.

Figure 1-1. Typical air-water flow images observed in a vertical 25.4 mm diameter pipe

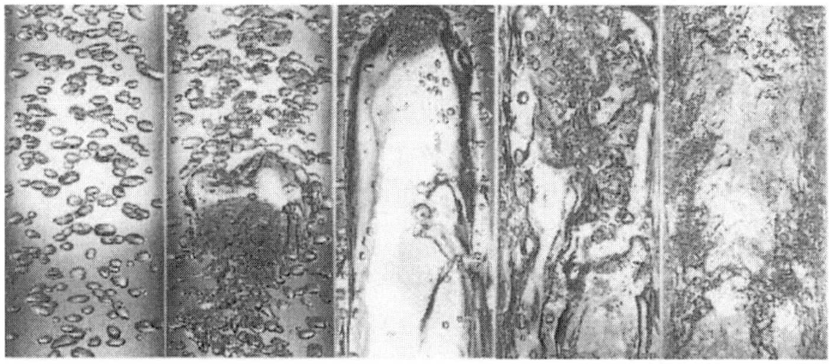

Figure 1-2. Typical air-water flow images observed in a vertical 50.8 mm diameter pipe

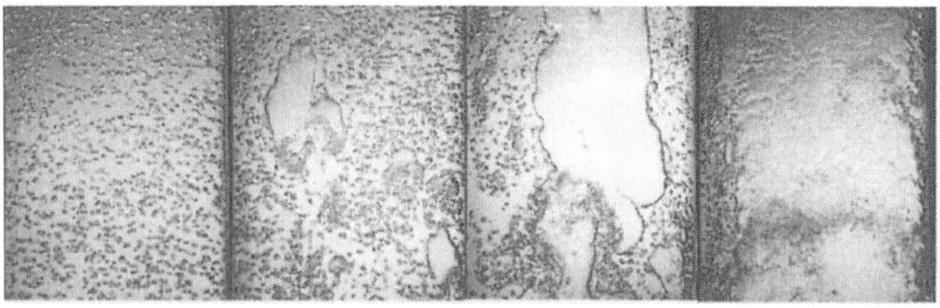

Figure 1-3. Typical air-water flow images observed in a rectangular channel of 200 mm × 10 mm

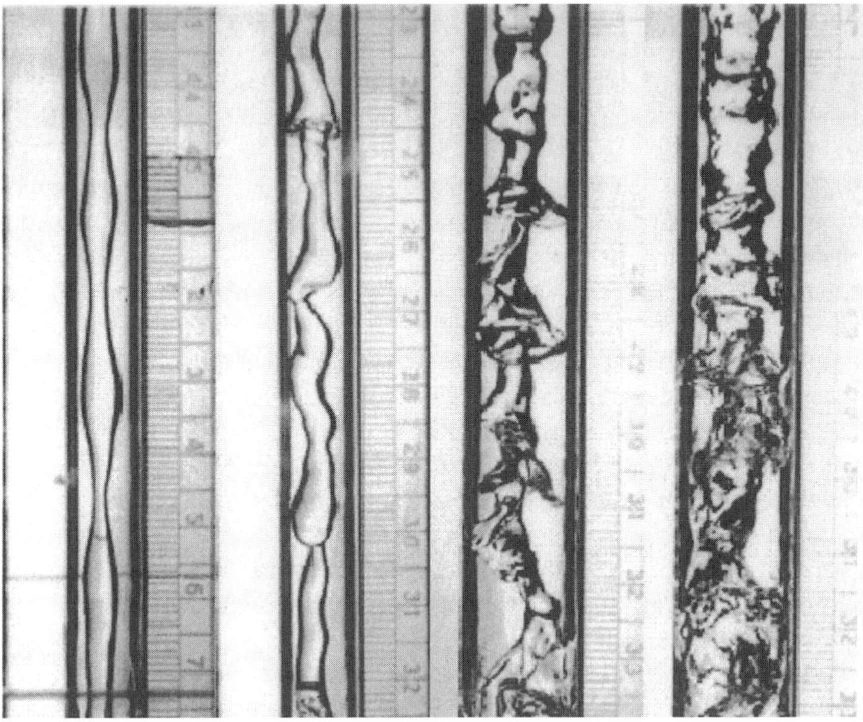

Figure 1-4. Typical images of simulated air-water inverted annular flow (It is cocurrent down flow)

Figure 1-5. Axial development of Inverted annular flow (It is cocurrent up flow)

1.4 Outline of the book

The purpose of this book is to present a detailed two-phase flow formulation that is rationally derived and developed using mechanistic modeling. This book is an extension of the earlier work by the author (Ishii, 1975) with special emphasis on the modeling of the interfacial structure with the interfacial area transport equation and modeling of the hydrodynamic constitutive relations. However, special efforts are made such that the formulation and mathematical models for complex two-phase flow physics and phenomena are realistic and practical to use for engineering analyses. It is focused on the detailed discussion of the general formulation of various mathematical models of two-phase flow based on the conservation laws of mass, momentum, and energy. In Part I, the foundation of the two-phase flow formulation is given as the local instant formulation of the two-phase flow based on the single-phase flow continuum formulation and explicit existence of the interface dividing the phases. The conservation equations, constitutive laws, jump conditions at the interface and special thermo-mechanical relations at the interface to close the mathematical system of equations are discussed.

Based on this local instant formulation, in Part II, macroscopic two-phase continuum formulations are developed using various averaging techniques which are essentially an integral transformation. The application of time averaging leads to general three-dimensional formulation, effectively eliminating the interfacial discontinuities and making both phases co-existing continua. The interfacial discontinuities are replaced by the interfacial transfer source and sink terms in the averaged differential balance equations

Details of the three-dimensional two-phase flow models are presented in Part III. The two-fluid model, drift-flux model, interfacial area transport, and interfacial momentum transfer are major topics discussed. In Part IV, more practical one-dimensional formulation of two-phase flow is given in terms of the two-fluid model and drift-flux model. It is planned that a second book will be written for many practical two-phase flow models and correlations that are necessary for solving actual engineering problems and the experimental base for these models.

Chapter 2

LOCAL INSTANT FORMULATION

The singular characteristic of two-phase or of two immiscible mixtures is the presence of one or several interfaces separating the phases or components. Examples of such flow systems can be found in a large number of engineering systems as well as in a wide variety of natural phenomena. The understanding of the flow and heat transfer processes of two-phase systems has become increasingly important in nuclear, mechanical and chemical engineering, as well as in environmental and medical science.

In analyzing two-phase flow, it is evident that we first follow the standard method of continuum mechanics. Thus, a two-phase flow is considered as a field that is subdivided into single-phase regions with moving boundaries between phases. The standard differential balance equations hold for each subregion with appropriate jump and boundary conditions to match the solutions of these differential equations at the interfaces. Hence, in theory, it is possible to formulate a two-phase flow problem in terms of the local instant variable, namely, $F = F(\boldsymbol{x},t)$. This formulation is called a *local instant formulation* in order to distinguish it from formulations based on various methods of averaging.

Such a formulation would result in a multiboundary problem with the positions of the interface being unknown due to the coupling of the fields and the boundary conditions. Indeed, mathematical difficulties encountered by using this local instant formulation can be considerable and, in many cases, they may be insurmountable. However, there are two fundamental importances in the local instant formulation. The first importance is the *direct application* to study the separated flows such as film, stratified, annular and jet flow, see Table 1-1. The formulation can be used there to study pressure drops, heat transfer, phase changes, the dynamic and stability of an interface, and the critical heat flux. In addition to the above applications, important examples of when this formulation can be used include: the problems of single or several bubble dynamics, the growth or collapse of a single bubble or a droplet, and ice formation and melting.

The second importance of the local instant formulation is as a *fundamental base of the macroscopic two-phase flow models* using various averaging. When each subregion bounded by interfaces can be considered as a continuum, the local instant formulation is mathematically rigorous. Consequently, two-phase flow models should be derived from this formulation by proper averaging methods. In the following, the general formulation of two-phase flow systems based on the local instant variables is presented and discussed. It should be noted here that the balance equations for a single-phase one component flow were firmly established for some time (Truesdell and Toupin, 1960; Bird et al, 1960). However, the axiomatic construction of the general constitutive laws including the equations of state was put into mathematical rigor by specialists (Coleman, 1964; Bowen, 1973; Truesdell, 1969). A similar approach was also used for a single-phase diffusive mixture by Muller (1968).

Before going into the detailed derivation and discussion of the local instant formulation, we review the method of mathematical physics in connection with the continuum mechanics. The next diagram shows the basic procedures used to obtain a mathematical model for a physical system.

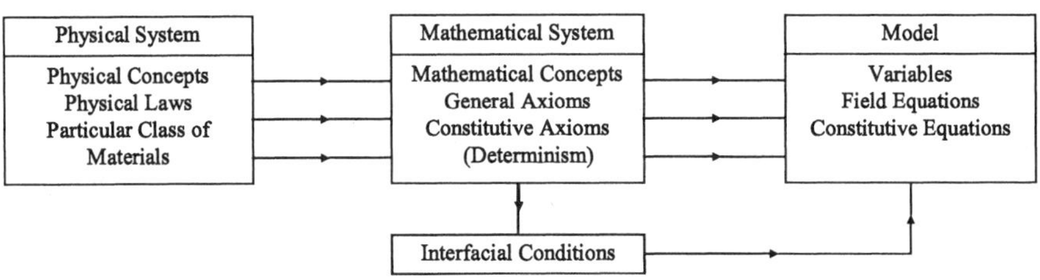

As it can be seen from the diagram, a physical system is first replaced by a mathematical system by introducing mathematical concepts, general axioms and constitutive axioms. In the continuum mechanics they correspond to variables, field equations and constitutive equations, whereas at the singular surface the mathematical system requires the interfacial conditions. The latter can be applied not only at the interface between two phases, but also at the outer boundaries which limit the system. It is clear from the diagram that the continuum formulation consists of three essential parts, namely: the derivations of field equations, constitutive equations, and interfacial conditions.

Now let us examine the basic procedure used to solve a particular problem. The following diagram summarizes the standard method. Using the continuum formulation, the physical problem is represented by idealized boundary geometries, boundary conditions, initial conditions, field and

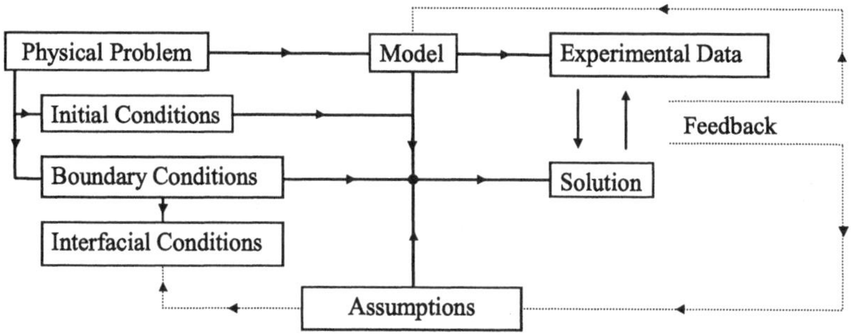

constitutive equations. It is evident that in two-phase flow systems, we have interfaces within the system that can be represented by general interfacial conditions. The solutions can be obtained by solving these sets of differential equations together with some idealizing or simplifying assumptions. For most problems of practical importance, experimental data also play a key role. First, experimental data can be taken by accepting the model, indicating the possibility of measurements. The comparison of a solution to experimental data gives feedback to the model itself and to the various assumptions. This feedback will improve both the methods of the experiment and the solution. The validity of the model is shown in general by solving a number of simple physical problems.

The continuum approach in single-phase thermo-fluid dynamics is widely accepted and its validity is well proved. Thus, if each subregion bounded by interfaces in two-phase systems can be considered as continuum, the validity of local instant formulation is evident. By accepting this assumption, we derive and discuss the field equations, the constitutive laws, and the interfacial conditions. Since an interface is a singular case of the continuous field, we have two different conditions at the interface. The balance at an interface that corresponds to the field equation is called a jump condition. Any additional information corresponding to the constitutive laws in space, which are also necessary at interface, is called an interfacial boundary condition.

1.1 Single-phase flow conservation equations

1.1.1 General balance equations

The derivation of the differential balance equation is shown in the following diagram. The general integral balance can be written by introducing the fluid density ρ_k, the efflux $\mathcal{J}_k$ and the body source ϕ_k of any quantity ψ_k defined for a unit mass. Thus we have

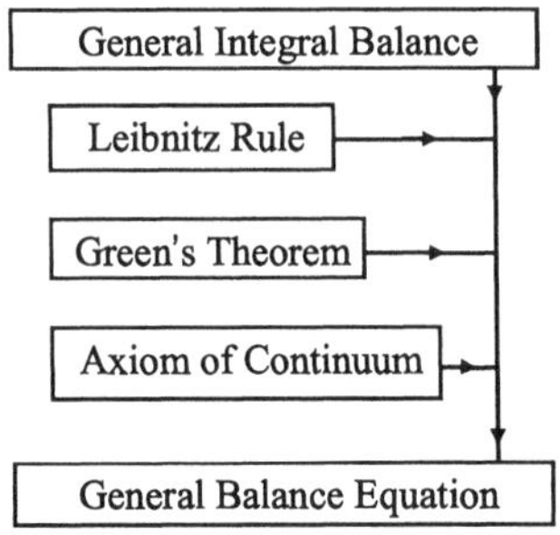

$$\frac{d}{dt}\int_{V_m}\rho_k\psi_k dV = -\oint_{A_m}\boldsymbol{n}_k\cdot\mathcal{J}_k dA + \int_{V_m}\rho_k\phi_k dV \tag{2-1}$$

where V_m is a material volume with a material surface A_m. It states that the time rate of change of $\rho_k\psi_k$ in V_m is equal to the influx through A_m plus the body source. The subscript k denotes the kth-phase. If the functions appearing in the Eq.(2-1) are sufficiently smooth such that the Jacobian transformation between material and spatial coordinates exists, then the familiar differential form of the balance equation can be obtained. This is done by using the Reynolds transport theorem (Aris, 1962) expressed as

$$\frac{d}{dt}\int_{V_m}\mathcal{F}_k dV = \int_{V_m}\frac{\partial \mathcal{F}_k}{\partial t}dV + \oint_{A_m}\mathcal{F}_k\boldsymbol{v}_k\cdot\boldsymbol{n}\ dA \tag{2-2}$$

where $\boldsymbol{v}_k$ denotes the velocity of a fluid particle. The Green's theorem gives a transformation between a certain volume and surface integral, thus

$$\int_V \nabla\cdot\mathcal{F}_k dV = \oint_A \boldsymbol{n}\cdot\mathcal{F}_k dA. \tag{2-3}$$

Hence, from Eqs.(2-2) and (2-3) we obtain

$$\frac{d}{dt}\int_{V_m}\mathcal{F}_k dV = \int_{V_m}\left[\frac{\partial \mathcal{F}_k}{\partial t} + \nabla\cdot\left(\boldsymbol{v}_k\mathcal{F}_k\right)\right]dV. \tag{2-4}$$

Furthermore, we note that the Reynolds transport theorem is a special case of Leibnitz rule given by

$$\frac{d}{dt}\int_V \mathcal{F}_k dV = \int_V \frac{\partial \mathcal{F}_k}{\partial t}dV + \int_A \mathcal{F}_k\boldsymbol{u}\cdot\boldsymbol{n}dA \tag{2-5}$$

where $V(t)$ is an arbitrary volume bounded by $A(t)$ and $\boldsymbol{u}\cdot\boldsymbol{n}$ is the surface displacement velocity of $A(t)$.

In view of Eqs.(2-1), (2-3) and (2-4) we obtain a differential balance equation

$$\frac{\partial \rho_k \psi_k}{\partial t} + \nabla\cdot\left(\boldsymbol{v}_k \rho_k \psi_k\right) = -\nabla\cdot \mathcal{J}_k + \rho_k \phi_k. \tag{2-6}$$

The first term of the above equation is the time rate of change of the quantity per unit volume, whereas the second term is the rate of convection per unit volume. The right-hand side terms represent the surface flux and the volume source.

1.1.2 Conservation equation

Continuity Equation

The conservation of mass can be expressed in a differential form by setting

$$\psi_k = 1,\quad \phi_k = 0,\quad \mathcal{J}_k = 0 \tag{2-7}$$

since there is no surface and volume sources of mass with respect to a fixed mass volume. Hence from the general balance equation we obtain

$$\frac{\partial \rho_k}{\partial t} + \nabla\cdot\left(\rho_k \boldsymbol{v}_k\right) = 0. \tag{2-8}$$

Momentum Equation

The conservation of momentum can be obtained from Eq.(2-6) by introducing the surface stress tensor $\mathcal{T}_k$ and the body force $\boldsymbol{g}_k$, thus we set

$$\begin{aligned} \psi_k &= \boldsymbol{v}_k \\ \mathcal{J}_k &= -\mathcal{T}_k = p_k \mathcal{I} - \mathfrak{T}_k \\ \phi_k &= \boldsymbol{g}_k \end{aligned} \tag{2-9}$$

where $\mathcal{I}$ is the unit tensor. Here we have split the stress tensor into the pressure term and the viscous stress $\mathfrak{T}_k$. In view of Eq.(2-6) we have

$$\frac{\partial \rho_k \boldsymbol{v}_k}{\partial t} + \nabla\cdot\left(\rho_k \boldsymbol{v}_k \boldsymbol{v}_k\right) = -\nabla p_k + \nabla\cdot \mathfrak{T}_k + \rho_k \boldsymbol{g}_k. \tag{2-10}$$

Conservation of Angular Momentum

If we assume that there is no body torque or couple stress, then all torques arise from the surface stress and the body force. In this case, the conservation of angular momentum reduces to

$$T_k = T_k^+ \tag{2-11}$$

where T_k^+ denotes the transposed stress tensor. The above result is correct for a non-polar fluid, however, for a polar fluid we should introduce an intrinsic angular momentum. In that case, we have a differential angular momentum equation (Aris, 1962).

Conservation of Energy

The balance of energy can be written by considering the total energy of the fluid. Thus we set

$$\begin{aligned} \psi_k &= u_k + \frac{v_k^2}{2} \\ J_k &= \boldsymbol{q}_k - T_k \cdot \boldsymbol{v}_k \\ \phi_k &= \boldsymbol{g}_k \cdot \boldsymbol{v}_k + \frac{\dot{q}_k}{\rho_k} \end{aligned} \tag{2-12}$$

where u_k, $\boldsymbol{q}_k$ and $\dot{q}_k$ represent the internal energy, heat flux and the body heating, respectively. It can be seen here that both the flux and the body source consist of the thermal effect and the mechanical effect. By substituting Eq.(2-12) into Eq.(2-6) we have the total energy equation

$$\begin{aligned} &\frac{\partial \rho_k \left(u_k + \dfrac{v_k^2}{2} \right)}{\partial t} + \nabla \cdot \left[\rho_k \left(u_k + \frac{v_k^2}{2} \right) \boldsymbol{v}_k \right] \\ &= -\nabla \cdot \boldsymbol{q}_k + \nabla \cdot \left(T_k \cdot \boldsymbol{v}_k \right) + \rho_k \boldsymbol{g}_k \cdot \boldsymbol{v}_k + \dot{q}_k . \end{aligned} \tag{2-13}$$

These four local equations, namely, Eqs.(2-8), (2-10), (2-11) and (2-13), express the four basic physical laws of the conservation of mass, momentum, angular momentum and energy. In order to solve these equations, it is necessary to specify the fluxes and the body sources as well as the fundamental equation of state. These are discussed under the constitutive laws. Apart from these constitutive laws, we note that there are several important transformations of above equations. A good review of transformed equations can be found in Bird et al. (1960). The important ones are given below.

The Transformation on Material Derivative

In view of the continuity equation we have

$$\frac{\partial \rho_k \psi_k}{\partial t} + \nabla \cdot \left(\rho_k \psi_k \boldsymbol{v}_k\right) = \rho_k \left(\frac{\partial \psi_k}{\partial t} + \boldsymbol{v}_k \cdot \nabla \psi_k\right) \equiv \rho_k \frac{D_k \psi_k}{Dt}. \quad \text{(2-14)}$$

This special time derivative is called the material or substantial derivative, since it expresses the rate of change with respect to time when an observer moves with the fluid.

Equation of Motion

By using the above transformation the momentum equation becomes the equation of motion

$$\rho_k \frac{D_k \boldsymbol{v}_k}{Dt} = -\nabla p_k + \nabla \cdot \mathcal{T}_k + \rho_k \boldsymbol{g}_k. \quad \text{(2-15)}$$

Here it is noted that $D_k \boldsymbol{v}_k / Dt$ is the fluid acceleration, thus the equation of motion expresses Newton's Second Law of Motion.

Mechanical Energy Equation

By dotting the equation of motion by the velocity we obtain

$$\begin{aligned} &\frac{\partial}{\partial t}\left(\rho_k \frac{v_k^2}{2}\right) + \nabla \cdot \left(\rho_k \frac{v_k^2}{2} \boldsymbol{v}_k\right) \\ &= -\boldsymbol{v}_k \cdot \nabla p_k + \boldsymbol{v}_k \cdot \left(\nabla \cdot \mathcal{T}_k\right) + \rho_k \boldsymbol{v}_k \cdot \boldsymbol{g}_k. \end{aligned} \quad \text{(2-16)}$$

For a symmetrical stress tensor

$$\mathcal{T}_k : \nabla \boldsymbol{v}_k \equiv \left(\mathcal{T}_k \cdot \nabla\right) \cdot \boldsymbol{v}_k = \nabla \cdot \left(\mathcal{T}_k \cdot \boldsymbol{v}_k\right) - \boldsymbol{v}_k \cdot \left(\nabla \cdot \mathcal{T}_k\right). \quad \text{(2-17)}$$

Thus, Eq.(2-16) may be written as

$$\frac{\partial}{\partial t}\left(\rho_k \frac{v_k^2}{2}\right) + \nabla \cdot \left(\rho_k \frac{v_k^2}{2} \boldsymbol{v}_k\right) = -\boldsymbol{v}_k \cdot \nabla p_k + \nabla \cdot \left(\mathcal{T}_k \cdot \boldsymbol{v}_k\right) - \mathcal{T}_k : \nabla \boldsymbol{v}_k + \rho_k \boldsymbol{v}_k \cdot \boldsymbol{g}_k. \tag{2-18}$$

This mechanical energy equation is a scalar equation, therefore it represents only some part of the physical law concerning the fluid motion governed by the momentum equation.

Internal Energy Equation

By subtracting the mechanical energy equation from the total energy equation, we obtain the internal energy equation

$$\frac{\partial \rho_k u_k}{\partial t} + \nabla \cdot \left(\rho_k u_k \boldsymbol{v}_k\right) = -\nabla \cdot \boldsymbol{q}_k - p_k \nabla \cdot \boldsymbol{v}_k + \mathcal{T}_k : \nabla \boldsymbol{v}_k + \dot{q}_k. \tag{2-19}$$

Enthalpy Equation

By introducing the enthalpy defined by

$$i_k \equiv u_k + \frac{p_k}{\rho_k} \tag{2-20}$$

the enthalpy energy equation can be obtained as

$$\frac{\partial \rho_k i_k}{\partial t} + \nabla \cdot \left(\rho_k i_k \boldsymbol{v}_k\right) = -\nabla \cdot \boldsymbol{q}_k + \frac{D_k p_k}{Dt} + \mathcal{T}_k : \nabla \boldsymbol{v}_k + \dot{q}_k. \tag{2-21}$$

1.1.3 Entropy inequality and principle of constitutive law

The constitutive laws are constructed on three different bases. The *entropy inequality* can be considered as a restriction on the constitutive laws, and it should be satisfied by the proper constitutive equations regardless of the material responses. Apart from the entropy inequality there is an important group of *constitutive axioms* that idealize in general terms the responses and behaviors of all the materials included in the theory. The principles of determinism and local action are frequently used in the continuum mechanics.

The above two bases of the constitutive laws define the general forms of the constitutive equations permitted in the theory. The third base of the constitutive laws is the *mathematical modeling* of material responses of a certain group of fluids based on the experimental observations. Using these three bases, we obtain specific constitutive equations that can be used to solve the field equations. It is evident that the balance equations and the proper constitutive equations should form a mathematically closed set of equations.

Now we proceed to the discussion of the entropy inequality. In order to state the second law of thermodynamics, it is necessary to introduce the concept of a temperature T_k and of the specific entropy s_k. With these variables the second law can be written as an inequality

$$\frac{d}{dt}\int_{V_m} \rho_k s_k dV + \oint_{A_m} \frac{\boldsymbol{q}_k}{T_k}\cdot \boldsymbol{n}_k dA - \int_{V_m} \frac{\dot{q}_k}{T_k} dV \geq 0. \tag{2-22}$$

Assuming the sufficient smoothness on the variables we obtain

$$\frac{\partial}{\partial t}(\rho_k s_k) + \nabla\cdot(\rho_k s_k \boldsymbol{v}_k) + \nabla\cdot\left(\frac{\boldsymbol{q}_k}{T_k}\right) - \frac{\dot{q}_k}{T_k} \equiv \Delta_k \geq 0 \tag{2-23}$$

where Δ_k is the rate of entropy production per unit volume. In this form it appears that Eq.(2-23) yields no clear physical or mathematical meanings in relation to the conservation equations, since the relations of s_k and T_k to the other dependent variables are not specified. In other words, the constitutive equations are not given yet. The inequality thus can be considered as a restriction on the constitutive laws rather than on the process itself.

As it is evident from the previous section, the number of dependent variables exceed that of the field equations, thus the balance equations of mass, momentum, angular momentum and total energy with proper boundary conditions are insufficient to yield any specific answers. Consequently, it is necessary to supplement them with various constitutive equations that define a certain type of ideal materials. Constitutive equations, thus, can be considered as a mathematical model of a particular group of materials. They are formulated on experimental data characterizing specific behaviors of materials together with postulated principles governing them.

From their physical significances, it is possible to classify various constitutive equations into three groups:

1. Mechanical constitutive equations;
2. Energetic constitutive equations;
3. Constitutive equation of state.

The first group specifies the stress tensor and the body force, whereas the second group supplies the heat flux and the body heating. The last equation gives a relation between the thermodynamic properties such as the entropy, internal energy and density of the fluid with the particle coordinates as a parameter. If it does not depend on the particle, it is called thermodynamically homogenous. It implies that the field consists of same material.

As it has been explained, the derivation of a general form of constitutive laws follows the postulated principles such as the entropy inequality, determinism, frame indifference and local action. The most important of them all is the principle of determinism that roughly states the predictability of a present state from a past history. The principle of material frame-indifference is the realization of the idea that the response of a material is independent of the frame or the observer. And the entropy inequality requires that the constitutive equations should satisfy inequality (2-23) unconditionally. Further restrictions such as the equipresence of the variables are frequently introduced into the constitutive equations for flux, namely, $\mathcal{T}_k$ and $\boldsymbol{q}_k$.

1.1.4 Constitutive equation

We restrict our attention to particular type of materials and constitutive equations which are most important and widely used in the fluid mechanics.

Fundamental Equation of State

The standard form of the fundamental equation of state for thermodynamically homogeneous fluid is given by a function relating the internal energy to the entropy and the density, hence we have

$$u_k = u_k\left(s_k, \rho_k\right). \tag{2-24}$$

And the temperature and the thermodynamic pressure are given by

$$T_k \equiv \frac{\partial u_k}{\partial s_k}, \quad -p_k \equiv \frac{\partial u_k}{\partial\left(1/\rho_k\right)}. \tag{2-25}$$

Thus in a differential form, the fundamental equation of state becomes

$$du_k = T_k ds_k - p_k d\left(\frac{1}{\rho_k}\right). \tag{2-26}$$

The Gibbs free energy, enthalpy and Helmholtz free energy function are defined by

$$g_k \equiv u_k - T_k s_k + \frac{p_k}{\rho_k} \tag{2-27}$$

$$i_k \equiv u_k + \frac{p_k}{\rho_k} \tag{2-28}$$

$$f_k \equiv u_k - T_k s_k \tag{2-29}$$

respectively. These can be considered as a Legendre transformation* (Callen, 1960) which changes independent variables from the original ones to the first derivatives. Thus in our case we have

$$g_k = g_k\left(T_k, p_k\right) \tag{2-30}$$

---

* If we have

$$y = y\left(x_1, x_2, \cdots, x_n\right);\ P_i \equiv \frac{\partial y}{\partial x_i}$$

then the Legendre transformation is given by

$$Z = y - \sum_{i=1}^{j} P_i x_i$$

and in this case Z becomes

$$Z = Z\ \left(P_1,\ P_2, \cdots, P_j,\ x_{j+1}, \cdots, x_n\right).$$

Thus, we have

$$dZ = -\sum_{i=1}^{j} x_i dP_i + \sum_{i=j+1}^{n} P_i dx_i.$$

$$i_k = i_k\left(s_k, p_k\right) \tag{2-31}$$

$$f_k = f_k\left(T_k, \rho_k\right) \tag{2-32}$$

which are also a fundamental equation of state.

Since the temperature and the pressure are the first order derivatives of u_k of the fundamental equation of state, Eq.(2-24) can be replaced by a combination of *thermal and caloric equations of state* (Bird et al., 1960; Callen, 1960) given by

$$p_k = p_k\left(\rho_k, T_k\right) \tag{2-33}$$

$$u_k = u_k\left(\rho_k, T_k\right). \tag{2-34}$$

The temperature and pressure are easily measurable quantities; therefore, it is more practical to obtain these two equations of state from experiments as well as to use them in the formulation. A simple example of these equations of state is for an incompressible fluid

$$\begin{aligned} \rho_k &= \text{constant} \\ u_k &= u_k\left(T_k\right). \end{aligned} \tag{2-35}$$

And in this case the pressure cannot be defined thermodynamically, thus we use the hydrodynamic pressure which is the average of the normal stress. Furthermore, an ideal gas has the equations of state

$$\begin{aligned} p_k &= R_M T_k \rho_k \\ u_k &= u_k\left(T_k\right) \end{aligned} \tag{2-36}$$

where R_M is the ideal gas constant divided by a molecular weight.

Mechanical Constitutive Equation

The simplest rheological constitutive equation is the one for an inviscid fluid expressed as

$$\mathcal{T}_k = 0. \tag{2-37}$$

For most fluid, Newton's Law of Viscosity apples. The generalized linearly viscous fluid of Navier-Stokes has a constitutive equation (Bird et al., 1960)

$$\mathcal{T}_k = \mu_k\left[\nabla \boldsymbol{v}_k + (\nabla \boldsymbol{v}_k)^+\right] - \left(\frac{2}{3}\mu_k - \lambda_k\right)(\nabla\cdot\boldsymbol{v}_k)\mathcal{I} \tag{2-38}$$

where μ_k and λ_k are the viscosity and the bulk viscosity of the kth-phase, respectively.

The body forces arise from external force fields and from mutual interaction forces with surrounding bodies or fluid particles. The origins of the forces are Newtonian gravitational, electrostatic, and electromagnetic forces. If the mutual interaction forces are important the body forces may not be considered as a function only of the independent variables $\boldsymbol{x}$ and t. In such a case, the principle of local actions cannot be applied. For most problems, however, these mutual interaction forces can be neglected in comparison with the gravitational field force $\boldsymbol{g}$. Thus we have

$$\boldsymbol{g}_k = \boldsymbol{g}. \tag{2-39}$$

Energetic Constitutive Equation

The contact heat transfer is expressed by the heat flux vector $\boldsymbol{q}_k$, and its constitutive equation specifies the nature and mechanism of the contact energy transfer. Most fluids obey the generalized Fourier's Law of Heat Conduction having the form

$$\boldsymbol{q}_k = -\mathcal{K}_k\cdot\nabla T_k. \tag{2-40}$$

The second order tensor $\mathcal{K}_k$ is the conductivity tensor which takes account for the anisotropy of the material. For an isotropic fluid the constitutive law can be expressed by a single coefficient as

$$\boldsymbol{q}_k = -K_k(T_k)\nabla T_k. \tag{2-41}$$

This is the standard form of Fourier's Law of Heat Conduction and the scalar K_k is called the thermal conductivity.

The body heating $\dot{q}_k$ arises from external energy sources and from mutual interactions. Energy can be generated by nuclear fission and can be transferred from distance by radiation, electric conduction and magnetic induction. The mutual interaction or transfer of energy is best exemplified by the mutual radiation between two parts of the fluid. In most cases these interaction terms are negligibly small in comparison with the contact heating. The radiation heat transfer becomes increasingly important at elevated temperature and in that case the effects are not local. If the radiation effects are negligible and the nuclear, electric or magnetic heating are absent, then the constitutive law for body heating is simply

$$\dot{q}_k = 0 \tag{2-42}$$

which can be used in a wide range of practical problems.

Finally, we note that the entropy inequality requires the transport coefficients μ_k, λ_k and K_k to be non-negative. Thus, viscous stress works as a resistance of fluid motions and it does not give out work. Furthermore, the heat flows only in the direction of higher to lower temperatures.

1.2 Interfacial balance and boundary condition

1.2.1 Interfacial balance (Jump condition)

The standard differential balance equations derived in the previous sections can be applied to each phase up to an interface, but not across it. A particular form of the balance equation should be used at an interface in order to take into account the singular characteristics, namely, the sharp changes (or discontinuities) in various variables. By considering the interface as a singular surface across which the fluid density, energy and velocity suffer jump discontinuities, the so-called jump conditions have been developed. These conditions specify the exchanges of mass, momentum, and energy through the interface and stand as matching conditions between two phases, thus they are indispensable in two-phase flow analyses. Furthermore since a solid boundary in a single-phase flow problem also constitutes an interface, various simplified forms of the jump conditions are in frequent use without much notice. Because of its importances, we discuss in detail the derivation and physical significance of the jump conditions.

The interfacial jump conditions without any surface properties were first put into general form by Kotchine (1926) as the dynamical compatibility condition at shock discontinuities, though special cases had been developed earlier by various authors. It can be derived from the integral balance equation by assuming that it holds for a material volume with a surface of discontinuity. Various authors (Scriven, 1960; Slattery, 1964; Standart, 1964; Delhaye, 1968; Kelly, 1964) have attempted to extend the Kotchine's theorem. These include the introduction of interfacial line fluxes such as the surface tension, viscous stress and heat flux or of surface material properties. There are several approaches to the problem and the results of the above authors are not in complete agreement. The detailed discussion on this subject as well as a comprehensive analysis which shows the origins of various discrepancies among previous studies have been presented by Delhaye (1974). A particular emphasis is directed there to the correct form of the energy jump condition and of the interfacial entropy production.

Since it will be convenient to consider a finite thickness interface in applying time average to two-phase flow fields, we derive a general interfacial balance equation based on the control volume analyses. Suppose the position of an interface is given by a mathematical surface $f(\boldsymbol{x},t)=0$. The effect of the interface on the physical variables is limited only to the neighborhood of the surface, and the domain of influence is given by a thin layer of thickness δ with δ_1 and δ_2 at each side of the surface. Let's denote the simple connected region on the surface by A_i, then the control volume is bounded by a surface Σ_i which is normal to A_i and the intersection of A_i and Σ_i is a closed curve C_i. Thus Σ_i forms a ring with a width δ, whereas the boundaries of the interfacial region at each side are denoted by A_1 and A_2. Our control volume V_i is formed by Σ_i, A_1 and A_2.

Since the magnitude of δ is assumed to be much smaller than the characteristic dimension along the surface A_i, we put

$$\boldsymbol{n}_1=-\boldsymbol{n}_2 \tag{2-43}$$

where $\boldsymbol{n}_1$ and $\boldsymbol{n}_2$ are the outward unit normal vectors from the bulk fluid of phase 1 and 2, respectively. The outward unit vector normal to Σ_i is denoted by $\boldsymbol{N}$, then the extended general integral balance equation for the control volume V_i is given by

$$\begin{aligned}\frac{d}{dt}\int_{V_i}\rho\psi dV=&\sum_{k=1}^{2}\int_{A_k}\boldsymbol{n}_k\cdot\left[(\boldsymbol{v}_k-\boldsymbol{v}_i)\rho_k\psi_k+\mathcal{J}_k\right]dA\\&-\int_{C_i}\int_{-\delta_2}^{\delta_1}\boldsymbol{N}\cdot\left[(\boldsymbol{v}-\boldsymbol{v}_i)\rho\psi+\mathcal{J}\right]d\delta dC+\int_{V_i}\rho\phi dV.\end{aligned} \tag{2-44}$$

The first two integrals on the right-hand side take account for the fluxes from the surface A_1, A_2 and Σ_i. In order to reduce the volume integral balance to a surface integral balance over A_i, we should introduce surface properties defined below.

The surface mean particle velocity $\boldsymbol{v}_s$ is given by

$$\rho_s\boldsymbol{v}_s\delta\equiv\int_{-\delta_2}^{\delta_1}\rho\,\boldsymbol{v}\,d\delta \tag{2-45}$$

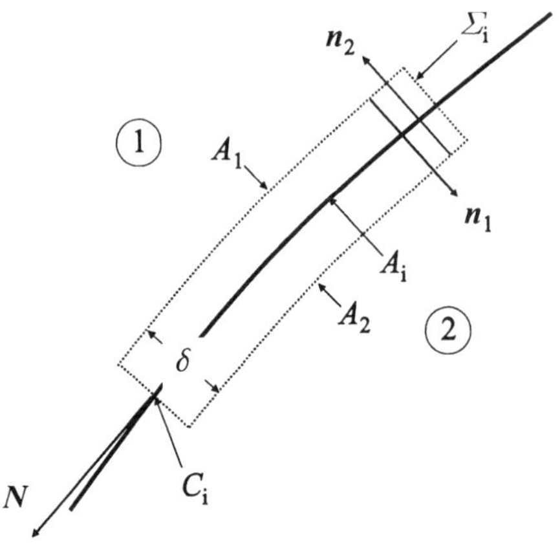

Figure 2-1. Interface (Ishii, 1975)

where the mean density ρ_s and the mean density per unit surface area ρ_a are defined as

$$\rho_a = \rho_s \delta \equiv \int_{-\delta_2}^{\delta_1} \rho \, d\delta. \tag{2-46}$$

Then the weighted mean values of ψ and ϕ are given by

$$\rho_a \psi_s \equiv \int_{-\delta_2}^{\delta_1} \rho\psi \, d\delta \tag{2-47}$$

and

$$\rho_a \phi_s \equiv \int_{-\delta_2}^{\delta_1} \rho\phi d\delta. \tag{2-48}$$

The notation here is such that a quantity per unit interface mass and per unit surface area is denoted by the subscript *s* and *a*, respectively.

The control surface velocity can be split into the tangential and normal components, thus

$$\boldsymbol{v}_i = \boldsymbol{v}_{ti} + \boldsymbol{v}_{ni} \tag{2-49}$$

where

$$\boldsymbol{v}_{ti} = \boldsymbol{v}_{ts}$$
$$\boldsymbol{v}_i \cdot \boldsymbol{n} = -\frac{\dfrac{\partial f}{\partial t}}{|\nabla f|}. \tag{2-50}$$

Hence the normal component is the surface displacement velocity and the tangential component is given by the mean tangential particle velocity $\boldsymbol{v}_{ts}$. Since the unit vector $\boldsymbol{N}$ is in the tangential plane and normal to C_i, we have

$$\boldsymbol{N} \cdot \boldsymbol{v}_i = \boldsymbol{N} \cdot \boldsymbol{v}_s. \tag{2-51}$$

Thus, from Eqs.(2-45) and (2-51) we obtain

$$\int_{-\delta_2}^{\delta_1} \rho \boldsymbol{N} \cdot (\boldsymbol{v}_i - \boldsymbol{v}) d\delta = 0 \tag{2-52}$$

and

$$\int_{-\delta_2}^{\delta_1} \rho\psi \boldsymbol{N} \cdot (\boldsymbol{v}_i - \boldsymbol{v}) d\delta = \int_{-\delta_2}^{\delta_1} \rho\psi \boldsymbol{N} \cdot (\boldsymbol{v}_s - \boldsymbol{v}) d\delta. \tag{2-53}$$

In view of Eqs.(2-44) and (2-53) we define the average line efflux along C_i by

$$\mathcal{J}_a \equiv \int_{-\delta_2}^{\delta_1} \left\{\mathcal{J} - (\boldsymbol{v}_s - \boldsymbol{v})\rho\psi\right\} d\delta. \tag{2-54}$$

Using the above definitions the integral balance at the interfacial region becomes

$$\begin{aligned}
&\frac{d}{dt}\int_{A_i} \rho_a \psi_s \, dA \\
&= \sum_{k=1}^{2} \int_{A_k} \boldsymbol{n}_k \cdot \left[(\boldsymbol{v}_k - \boldsymbol{v}_i)\rho_k \psi_k + \mathcal{J}_k\right] dA - \int_{C_i} \boldsymbol{N} \cdot \mathcal{J}_a \, dC \\
&+ \int_{A_i} \rho_a \phi_s \, dA.
\end{aligned} \tag{2-55}$$

As in the case for the derivation of the field equation, here we need two mathematical transformations, namely, the surface transport theorem and the surface Green's theorem (Weatherburn, 1927; McConnell, 1957; Aris 1962). The surface transport theorem is given by

$$\frac{d}{dt}\int_{A_i} F\, dA = \int_{A_i}\left\{\frac{d_s}{dt}(F) + F\nabla_s \cdot \boldsymbol{v}_i\right\} dA \tag{2-56}$$

where d_s/dt denotes the convective derivative with the surface velocity $\boldsymbol{v}_i$ defined by Eq.(2-50), and ∇_s denotes the surface divergence operator. The surface Green's theorem is given by

$$\int_{C_i} \boldsymbol{N}\cdot J_a dC = \int_{A_i} A^{\alpha\beta} g_{\ln}\left(t_\alpha^n J_a^{l\cdot}\right)_{,\beta} dA. \tag{2-57}$$

Here, $A^{\alpha\beta}$, $g_{\ln}$, t_α^n and $(\quad)_{,\beta}$ denote the surface metric tensor, the space metric tensor, the hybrid tensor, and the surface covariant derivative, respectively (Aris, 1962).

The surface flux, J_a in space coordinates is expressed by $J_a^{l\cdot}$ which represents the space vector for mass and energy balance and the space tensor for momentum balance. The essential concepts of the tensor symbols are given below. First the Cartesian space coordinates are denoted by (y_1, y_2, y_3) and a general coordinates by (x_1, x_2, x_3), then the space metric tensor is defined by

$$g_{\ln} \equiv \sum_{k=1}^{3} \frac{\partial y^k}{\partial x^l}\frac{\partial y^k}{\partial x^n} \tag{2-58}$$

which relates the distance of the infinitesimal coordinate element between these two systems. As shown in Fig.2-2, if the Cartesian coordinates y^k give a point of a surface with the surface coordinates of (u^1, u^2) as $y^k = y^k(u^1, u^2)$, then the surface metric tensor is defined by

$$A^{\alpha\beta} = \sum_{k=1}^{3} \frac{\partial y^k}{\partial u^\alpha}\frac{\partial y^k}{\partial u^\beta} \tag{2-59}$$

and the small distance ds is given by

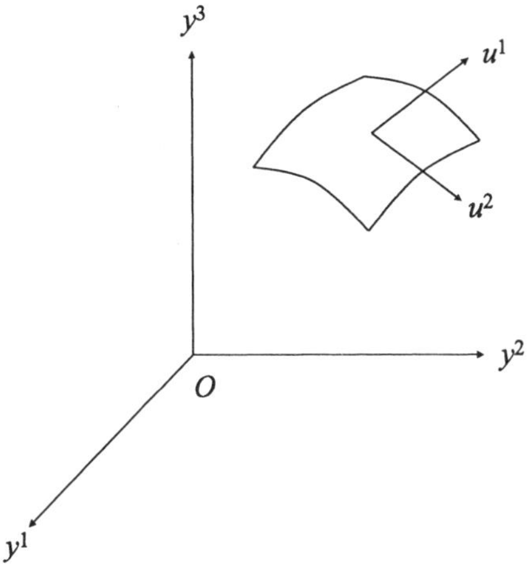

Figure 2-2. Relationship between Cartesian coordinates and surface coordinates

$$(ds)^2 = (dy^1)^2 + (dy^2)^2 + (dy^3)^2 = A^{\alpha\beta} du^\alpha du^\beta. \tag{2-60}$$

By introducing the general space coordinates, the surface position is given by $x^i = x^i(u^1, u^2)$. The hybrid tensor is then defined by

$$t_\alpha^i = \frac{\partial x^i}{\partial u^\alpha}. \tag{2-61}$$

The covariance surface derivative $(\quad)_{,\beta}$ is similar to the space derivative but it also takes into account for the curved coordinate effects. Furthermore, if $\boldsymbol{N} \cdot \underset{a}{J}$ has only a tangential component as in the case of surface tension force, $A^{\alpha\beta} g_{\ln} t_\alpha^n \underset{a}{J}^{lm} = t_\alpha^m \underset{a}{J}^{\alpha\beta}$. Hence, the surface flux contribution can be written as $\left(t_\alpha^m \underset{a}{J}^{\alpha\beta}\right)_{,\beta}$ or $\left(\boldsymbol{t}_\alpha \underset{a}{J}^{\alpha\beta}\right)_{,\beta}$ where $\boldsymbol{t}_\alpha$ denotes the hybrid tensor in vector notation. It is noted that for the momentum transfer, the dominant interfacial momentum flux is the isotropic surface tension σ. Then, $\underset{a}{J}^{\alpha\beta} = \sigma A^{\alpha\beta}$. In this case, the surface flux contribution becomes as follows

$$\left(\boldsymbol{t}_\alpha \sigma A^{\alpha\beta}\right)_{,\beta} = 2H\sigma\boldsymbol{n} + \boldsymbol{t}_\alpha A^{\alpha\beta}(\sigma)_{,\beta}. \tag{2-62}$$

The first term represents the net effect of the curved surface and gives the normal component force with the mean curvature H, whereas the second term represents the tangential force due to surface tension gradient.

Since we assumed that δ is sufficiently small, the surface A_1 and A_2 coincide with A_i geometrically. Thus, Eq.(2-55) reduces to

$$\begin{aligned}&\int_{A_i}\left\{\frac{d_s}{dt}(\rho_a\psi_s)+\rho_a\psi_s\nabla_s\cdot\boldsymbol{v}_i\right\}dA\\&=\int_{A_i}\left\{\sum_{k=1}^{2}\left[\rho_k\psi_k\boldsymbol{n}_k\cdot(\boldsymbol{v}_k-\boldsymbol{v}_i)+\boldsymbol{n}_k\cdot\mathcal{J}_k\right]\right.\\&\left.-A^{\alpha\beta}g_{\text{ln}}\left(t_\alpha^n\mathcal{J}_a^{l\cdot}\right)_{,\beta}+\rho_a\phi_s\right\}dA.\end{aligned} \tag{2-63}$$

This balance equation holds for any arbitrary portion of an interface with $A_i >> \delta^2$, thus we obtain a differential balance equation

$$\begin{aligned}&\frac{d_s}{dt}(\rho_a\psi_s)+\rho_a\psi_s\nabla_s\cdot\boldsymbol{v}_i\\&=\sum_{k=1}^{2}\left\{\rho_k\psi_k\boldsymbol{n}_k\cdot(\boldsymbol{v}_k-\boldsymbol{v}_i)+\boldsymbol{n}_k\cdot\mathcal{J}_k\right\}\\&-A^{\alpha\beta}g_{\text{ln}}\left(t_\alpha^n\mathcal{J}_a^{l\cdot}\right)_{,\beta}+\rho_a\phi_s.\end{aligned} \tag{2-64}$$

We note here this result has exactly the same form as the one derived by Delhaye (1974), although the method used and the definition of the surface velocity $\boldsymbol{v}_i$ is different. Let's define a surface quantity and a source per surface area as

$$\psi_a \equiv \rho_a\psi_s \tag{2-65}$$

and

$$\phi_a \equiv \rho_a\phi_s. \tag{2-66}$$

Then the surface balance equation becomes

$$\begin{aligned}&\frac{d_s}{dt}(\psi_a)+\psi_a\nabla_s\cdot\boldsymbol{v}_i=\sum_{k=1}^{2}\left\{\rho_k\psi_k\boldsymbol{n}_k\cdot(\boldsymbol{v}_k-\boldsymbol{v}_i)+\boldsymbol{n}_k\cdot\mathcal{J}_k\right\}\\&-A^{\alpha\beta}g_{\text{ln}}\left(t_\alpha^n\mathcal{J}_a^{l\cdot}\right)_{,\beta}+\phi_a.\end{aligned} \tag{2-67}$$

The left-hand side represents the time rate of change of ψ_a from the observer moving at $\boldsymbol{v}_i$, plus the effect of the surface dilatation. Whereas the three terms on the right-hand side are the fluxes from the bulk phases, the line flux along the surface, and the surface source respectively. We note that Eq.(2-6) and Eq.(2-67) govern the physical laws in the bulk phases and at an interface.

In order to obtain a simpler expression for interfacial jump of quantities, we make further assumptions which are consistent with our thin layer assumption given by

$$\delta^2 << A_i. \tag{2-68}$$

First the mass density of interface ρ_a is negligibly small so that its momentum and mechanical energy can also be neglected. Secondly, all the molecular diffusion fluxes along the line are neglected, namely, no surface viscous stress or surface heat flux. Furthermore all the surface sources are neglected, namely, no particular body force other than the gravity and no radiation effect.

The thermodynamic tension and hence the interfacial energy are *included* in the following analysis, consequently from the principle of determinism we should postulate the existence of the surface equation of state. Under these assumptions we obtain

Interfacial Mass Balance

$$\sum_{k=1}^{2} \rho_k \boldsymbol{n}_k \cdot (\boldsymbol{v}_k - \boldsymbol{v}_i) = 0. \tag{2-69}$$

By defining the interfacial mass efflux from the kth-phase as

$$\dot{m}_k \equiv \rho_k \boldsymbol{n}_k \cdot (\boldsymbol{v}_k - \boldsymbol{v}_i) \tag{2-70}$$

we have from Eq.(2-69)

$$\sum_{k=1}^{2} \dot{m}_k = 0. \tag{2-71}$$

This equation simply states that there is no capacity of mass at the interface, hence phase changes are pure exchanges of mass between the two phases.

Interfacial Momentum Balance

$$\sum_{k=1}^{2}\left\{\rho_k \boldsymbol{n}_k\cdot\left(\boldsymbol{v}_k-\boldsymbol{v}_i\right)\boldsymbol{v}_k-\boldsymbol{n}_k\cdot\mathcal{T}_k\right\}+\left(\boldsymbol{t}_\alpha A^{\alpha\beta}\sigma\right)_{,\beta}=0. \tag{2-72}$$

Equation (2-72) is a balance between the momentum fluxes from the bulk fluids and the interfacial tension.

Interfacial Energy Balance

Substituting the interfacial energy u_a per unit surface area for ψ_a, we obtain from Eq.(2-67)

$$\begin{aligned}&\frac{d_s u_a}{dt}+u_a\nabla_s\cdot\boldsymbol{v}_i\\&=\sum_{k=1}^{2}\left\{\rho_k\boldsymbol{n}_k\cdot\left(\boldsymbol{v}_k-\boldsymbol{v}_i\right)\left(u_k+\frac{v_k^2}{2}\right)+\boldsymbol{n}_k\cdot\left(-\mathcal{T}_k\cdot\boldsymbol{v}_k+\boldsymbol{q}_k\right)\right\}\\&+\left(\boldsymbol{t}_\alpha A^{\alpha\beta}\sigma\cdot\boldsymbol{v}_i\right)_{,\beta}.\end{aligned} \tag{2-73}$$

The left-hand side represents the rate change of the surface energy, whereas the right-hand side accounts for the energy transfer from the bulk at each side and for the work done by the surface tension.

1.2.2 Boundary conditions at interface

As in the case of the three-dimensional field equations the surface balance equations should be supplemented by various constitutive laws. In order to establish the principle of determinism, first we introduce a simple equation of state. Since the mass of interface is negligible, we have

$$u_a=u_a\left(s_a\right) \tag{2-74}$$

where u_a and s_a are the specific internal energy and the specific entropy per unit surface area, respectively.

The thermodynamic tension is given by

$$\sigma\equiv -T_i s_a+u_a \tag{2-75}$$

where the temperature T_i is defined by

$$T_i \equiv \frac{du_a}{ds_a}. \tag{2-76}$$

Thus, in a differential form, Eq.(2-74) becomes

$$du_a = T_i ds_a \tag{2-77}$$

and the Gibbs-Duhem relation is given by

$$s_a dT_i + d\sigma = 0. \tag{2-78}$$

The interfacial enthalpy is defined by

$$i_a = u_a - \sigma. \tag{2-79}$$

From Eq.(2-78) we have

$$\frac{d\sigma}{dT_i} = -s_a. \tag{2-80}$$

Hence, from Eqs.(2-77) and (2-80) we obtain

$$du_a = -T_i d\left(\frac{d\sigma}{dT_i}\right). \tag{2-81}$$

By combining Eqs.(2-75), (2-79) and (2-80) we get

$$u_a = -T_i\left(\frac{d\sigma}{dT_i}\right) + \sigma;\ i_a = -T_i\left(\frac{d\sigma}{dT_i}\right). \tag{2-82}$$

Thus the thermal equation of state

$$\sigma = \sigma\left(T_i\right) \tag{2-83}$$

supplies all the necessary information to interrelate the thermodynamic properties. By substituting Eq.(2-81) into Eq.(2-73) we obtain an energy jump condition in terms of the surface tension as

$$
\begin{aligned}
&-T_i\left\{\frac{d_s}{dt}\left(\frac{d\sigma}{dT_i}\right)+\left(\frac{d\sigma}{dT_i}\right)\nabla_s\cdot\boldsymbol{v}_i\right\}\\
&=\left(\boldsymbol{t}_\alpha A^{\alpha\beta}\sigma\right)_{,\beta}\cdot\boldsymbol{v}_i \qquad (2\text{-}84)\\
&+\sum_{k=1}^{2}\left\{\dot{m}_k\left(u_k+\frac{v_k^{\,2}}{2}\right)+\boldsymbol{n}_k\cdot\left(-\mathcal{T}_k\cdot\boldsymbol{v}_k+\boldsymbol{q}_k\right)\right\}.
\end{aligned}
$$

Interfacial Entropy Inequality

Following the above discussion, we assume the existence of the surface temperature T_i which enables us to write an entropy inequality at the interface. Thus, in the absence of surface heat flux and source terms, we have

$$\Delta_a=\frac{d_s s_a}{dt}+s_a\nabla_s\cdot\boldsymbol{v}_i-\sum_{k=1}^{2}\left(\dot{m}_k s_k+\frac{\boldsymbol{n}_k\cdot\boldsymbol{q}_k}{T_k}\right)\geq 0. \qquad (2\text{-}85)$$

The entropy s_a in above inequality can be eliminated by using the energy balance equation, Eq.(2-73), and the equation of state, Eq.(2-77), hence we obtain

$$
\begin{aligned}
T_i\Delta_a=\sum_{k=1}^{2}&\left\{\dot{m}_k\left[u_k-s_kT_i+\frac{\left|\boldsymbol{v}_k-\boldsymbol{v}_i\right|^2}{2}+\frac{p_k}{\rho_k}\right]\right. \qquad (2\text{-}86)\\
&\left.-\boldsymbol{n}_k\cdot\mathcal{T}_k\cdot\left(\boldsymbol{v}_k-\boldsymbol{v}_i\right)+\boldsymbol{n}_k\cdot\boldsymbol{q}_k\left(1-\frac{T_i}{T_k}\right)\right\}\geq 0.
\end{aligned}
$$

We note here that this expression has the same form as the one obtained by Delhaye (1974). Also a similar result was derived by Standart (1968) without considering the surface properties and the surface tension term, but including the effect of chemical reactions.

In general, the interfacial jump conditions, Eqs.(2-69), (2-72) and (2-84), do not constitute sufficient matching conditions which are necessary to define the problem uniquely. Consequently, they should be supplemented by various boundary conditions that restrict the kinematical, dynamical and thermal relations between two phases. These relations can also be considered as interfacial constitutive laws, satisfying the restriction imposed by the entropy inequality (2-86). They may be obtained from the standard argument of the irreversible thermodynamics. In order to do so, first suitable combinations of fluxes and potentials should be postulated in the inequality (2-86), and then the fluxes were expanded linearly in terms of the potentials. Here, the principle of equipresence and the symmetric relations between the expansion coefficients are normally used. The standard procedure for a general system is discussed in detail by De Groot and Mazur (1962) among others, and it has been applied to an interface by Standart (1968), and Bornhorst and Hatsopoulos (1967). Standart based his argument on the correct jump conditions and the entropy inequality and obtained the interfacial constitutive laws with great care, though he neglected from the beginning all the surface properties and the surface tension that are generally important in a two-phase system. The results of Bornhorst are limited to particular cases and the argument is based on the classical thermodynamic tools of piston, reservoir, homogeneous system, etc.

The analysis based on the constitutive laws of the interface may be important for a detailed study of a two-phase system. However, they are generally too complicated to apply as boundary conditions. Furthermore, the effects of the potentials, namely, the discontinuities of temperature, chemical potential, tangential velocity, etc., as driving forces of transfer of quantities, or resulting interfacial resistances to heat, momentum and mass transfer are relatively insignificant in the total system.

Consequently a much simpler theory for providing the necessary boundary conditions is desirable. As a limiting case, it is possible to consider the case when entropy production of the interface Δ_a becomes zero. This means that there are no resistances to interfacial transfer of quantities. Hence, the exchanges between two phases are governed by the conditions of the bulk fluid at each side, but not by the interface itself. Furthermore, from the classical thermodynamic point of view, the transfer at the interface is said to be reversible. This is not so for a shock discontinuity in a single-phase flow.

By setting the entropy production of Eq.(2-86) to be zero we obtain

$$\begin{aligned}&\sum_{k=1}^{2}\frac{\dot{m}_k}{T_i}\left(g_k+\frac{\left|\boldsymbol{v}_k-\boldsymbol{v}_i\right|^2}{2}-\frac{\tau_{nnk}}{\rho_k}\right)-\sum_k\frac{\boldsymbol{\tau}_{tk}}{T_i}\cdot\left(\boldsymbol{v}_{tk}-\boldsymbol{v}_{ti}\right)\\&+\sum_{k=1}^{2}\left(\boldsymbol{n}_k\cdot\boldsymbol{q}_k+\dot{m}_k s_k T_k\right)\left(\frac{1}{T_i}-\frac{1}{T_k}\right)=0.\end{aligned} \tag{2-87}$$

Moreover, we assume that the three terms in Eq.(2-87) are independently zero for all combinations of the mass flux, the tangential stresses and the heat fluxes.

Thermal Boundary Condition

Thus, from the last term of Eq.(2-87), we obtain a thermal equilibrium condition at the interface

$$T_{1i} = T_{2i} = T_i \tag{2-88}$$

that is consistent with the assumption of the existence of the equation of state at the interface, Eqs.(2-74) and (2-83). In view of Eqs.(2-82) and (2-84) this thermal boundary condition sets the energy level of the interface. In contrast to the above equation, the energy jump condition, Eq.(2-73), specifies the relation between the energy transfers to the interface. Furthermore, the thermal equilibrium condition, Eq.(2-88), eliminates a variable T_i, and it stands as a matching condition for the temperature of each phase at the interface. We note here that, in reality, the discontinuity of the temperature at the interface exists and can be estimated from the kinetic theories (Hirschfelder et al., 1954). However, its value in comparison with the absolute temperature is very small for most materials with few exceptions, such as for liquid metals (Brodkey, 1971). Thus, the influence on the interfacial transfer is negligible under the standard conditions.

No-Slip Condition

In view of the definition of the interfacial surface velocity $\boldsymbol{v}_i$, Eq.(2-50), the tangential velocity $\boldsymbol{v}_{ti}$ is an unknown parameter, whereas the normal component is directly related to the position of the interface. Furthermore, it appears in the dissipation term in the entropy inequality (2-86) and Eq.(2-87). Thus, it is natural to supply a constitutive relation between the tangential stress $\boldsymbol{\tau}_{tk}$ and the tangential relative velocity $\boldsymbol{v}_{tk} - \boldsymbol{v}_{ti}$, as it has been discussed previously. However, in the present analysis we have assumed that the interfacial entropy production is identically zero. By taking the second term of Eq.(2-87) to be zero independently, we obtain a no-slip condition

$$\boldsymbol{v}_{t1} = \boldsymbol{v}_{t2} = \boldsymbol{v}_{ti}. \tag{2-89}$$

The no-slip condition for a moving viscous fluid in contact with a solid wall is well established (Goldstein, 1938; Serrin, 1959). It is called a classical adherence condition and it has been verified experimentally and also analytically from kinetic theories. The relation given by Eq.(2-89) can be used to eliminate the interfacial tangential particle velocity and then it can be utilized as a velocity boundary condition at an interface.

However, it should be noted here that for an inviscid fluid the no-slip condition (2-89) is not necessary and cannot be satisfied generally, due to the tangential component of the momentum jump condition, Eq.(2-72). This is in complete agreement with our analysis, since the viscous dissipation term in Eq.(2-87) is identically zero for an inviscid fluid and does not appear in the entropy inequality. Consequently, Eq.(2-89) cannot be obtained. Furthermore, under the condition of no-slip, the momentum jump condition, Eq.(2-72), in the tangential and the normal directions becomes

$$\sum_{k=1}^{2} \boldsymbol{\tau}_{tk} = A^{\alpha\beta} \boldsymbol{t}_{\beta} (\sigma)_{,\alpha} \tag{2-90}$$

and

$$\sum_{k=1}^{2} \left(\boldsymbol{n}_k \frac{\dot{m}_k^2}{\rho_k} + \boldsymbol{n}_k p_k - \boldsymbol{\tau}_{nk} \right) = -2H_{21} \boldsymbol{n}_1 \sigma \tag{2-91}$$

where the normal and the tangential viscous stress is given by

$$\boldsymbol{n}_k \cdot \mathcal{T}_k = \boldsymbol{\tau}_{nk} + \boldsymbol{\tau}_{tk} = \boldsymbol{n}_k \tau_{nnk} + \boldsymbol{\tau}_{tk}. \tag{2-92}$$

And the mean curvature H_{21} is taken from phase 2 to 1, namely, $H_{21} > 0$ if the interface makes a convex surface in phase 1.

Chemical (Phase Change) Boundary Condition

In analogy with the preceding discussion, the chemical (or phase change) boundary condition can be obtained by setting the first term of Eq.(2-87) to be independently zero for all values of $\dot{m}_k$. This implies that the entropy production due to a phase transition is zero, and hence the phase change is considered not as a transfer due to non-equilibrium forces, but rather as an equilibrium transformation of state.

Substituting the thermal equilibrium condition, Eq.(2-88), into the first term of Eq.(2-87) and equating it to zero, we obtain

$$(g_1 - g_2) = \left(\frac{|\boldsymbol{v}_2 - \boldsymbol{v}_i|^2}{2} - \frac{|\boldsymbol{v}_1 - \boldsymbol{v}_i|^2}{2} \right) - \left(\frac{\tau_{nn2}}{\rho_2} - \frac{\tau_{nn1}}{\rho_1} \right). \tag{2-93}$$

The phase change condition given by the above equation shows that the difference in the chemical potential compensates for the mechanical effects of the relative kinetic energy difference and of the normal stresses. Here it should be noted that this phase change condition is only applicable to the case when the transfer of mass across the interface is possible. In other words, if the transfer of mass is identically zero for all conditions as in the case of two immiscible non-reacting liquids, the boundary condition should be

$$\dot{m}_k = 0 \tag{2-94}$$

which replaces the condition on the chemical potentials.

1.2.3 Simplified boundary condition

In the preceding sections the interfacial jump conditions and supplementary boundary conditions have been given. It is important to realize that the thermal equilibrium condition, Eq.(2-88), normal component of the momentum jump condition, Eq.(2-91), and the phase change boundary condition, Eq.(2-93), correspond to the standard thermal, mechanical and chemical equilibrium conditions of the thermostatics (Gibbs, 1948). The difference is that the present analysis takes into account the dynamic effects of mass transfer and of the normal stresses in the mechanical and phase change boundary conditions. These interesting properties between the results of dynamical analysis and of the thermostatic theory can be summarized in the following table.

It can be seen from the table that except the thermal condition these interfacial relations are still very complicated for many practical applications. This is mainly due to the terms arisen from the mass transfer and from the normal stresses. The former contributes as a thrust force due to the density change in the mechanical boundary condition and also as an impact kinetic energy change in the chemical (phase) boundary condition. The latter introduces complicated coupling effects of the flow fields with the thermodynamic properties at the interface. Under standard conditions, however, the normal stresses may be neglected with respect to the pressure terms, which greatly simplify the mechanical boundary condition, Eq.(2-91). The same argument can be applied to the chemical boundary condition, since the order of magnitude of the term $\rho_k g_k$ is p_k, thus the normal stress terms can be neglected also in Eq.(2-93). Similarly the mass transfer terms are negligibly small in most practical problems, though they can be important for problems with large mass transfer rate or with vapor film boiling.

Since in the standard formulation of field equations the Gibbs free energy g_k does not appear explicitly, it is desirable to transform the variable g_k in the chemical boundary condition, Eq.(2-93), into other variables which have

Table 2-1. Interfacial relations of thermodynamic potentials (Ishii, 1975)

Condition \ Analysis	Thermostatics	Present Dynamical Analysis
Thermal	$T_1 - T_2 = 0$	$T_1 - T_2 = 0$
Mechanical	$p_1 - p_2 = 0$	$p_1 - p_2 = -2H_{21}\sigma - \dot{m}_1^2\left(\frac{1}{\rho_1} - \frac{1}{\rho_2}\right) + \left(\tau_{nn1} - \tau_{nn2}\right)$
Chemical (phase change)	$g_1 - g_2 = 0$	$g_1 - g_2 = -\frac{\dot{m}_1^2}{2}\left(\frac{1}{\rho_1^2} - \frac{1}{\rho_2^2}\right) + \left(\frac{\tau_{nn1}}{\rho_1} - \frac{\tau_{nn2}}{\rho_2}\right)$

already been used in the field equations. For this purpose, we recall here that the Gibbs free energy expressed as a function of the temperature and pressure is a fundamental equation of state, Eq.(2-30), thus we have

$$g_k = g_k\left(T_k, p_k\right) \tag{2-95}$$

and

$$dg_k = -s_k\, dT_k + \frac{1}{\rho_k} dp_k. \tag{2-96}$$

The thermostatic phase equilibrium condition is then given by

$$T_1 = T_2 = T^{sat};\ p_1 = p_2 = p^{sat};\ \text{and } g_1 = g_2. \tag{2-97}$$

Hence from Eqs.(2-95) and (2-97) we obtain

$$g_1\left(T^{sat}, p^{sat}\right) = g_2\left(T^{sat}, p^{sat}\right) \tag{2-98}$$

which reduces to the classical saturation condition

$$p^{sat} = p^{sat}\left(T^{sat}\right). \tag{2-99}$$

This relation shows that the thermostatic equilibrium condition uniquely relates the thermodynamic potentials of each phase. Furthermore, the differential form of Eq.(2-99) known as the Clausius-Clapeyron equation can be obtained from Eqs.(2-27) and (2-28) and Eqs.(2-96) and (2-97)

$$\frac{dp^{sat}}{dT^{sat}} = \frac{i_1 - i_2}{T^{sat}\left(\dfrac{1}{\rho_1} - \dfrac{1}{\rho_2}\right)} \tag{2-100}$$

where all values of the right-hand side are calculated on the saturation line given by Eq.(2-99).

If we assume that the deviations of the interface pressures of each phase from the saturation pressure corresponding to the interfacial temperature T_i are sufficiently small in comparison with the pressure level, the Gibbs free energy function can be expanded around the static saturation point. Thus we have

$$g_k\left(p^{sat}, T_i\right) \doteq g_k\left(p_k, T_i\right) - \frac{\delta p_k}{\rho_k\left(p_k^{sat}, T_i\right)} \tag{2-101}$$

where δp_k is defined by

$$\delta p_k \equiv p_k - p^{sat}\left(T_i\right). \tag{2-102}$$

Since we have

$$g_1\left(p^{sat}\left(T_i\right), T_i\right) = g_2\left(p^{sat}\left(T_i\right), T_i\right). \tag{2-103}$$

Equation (2-93) can be reduced to

$$\frac{\delta p_1}{\rho_1} - \frac{\delta p_2}{\rho_2} \doteq -\frac{\dot{m}_1^2}{2}\left(\frac{1}{\rho_1^2} - \frac{1}{\rho_2^2}\right) + \left(\frac{\tau_{nn1}}{\rho_1} - \frac{\tau_{nn2}}{\rho_2}\right) \tag{2-104}$$

whereas the mechanical boundary condition, Eq.(2-91), with the definition of δp_k becomes

$$\delta p_1 - \delta p_2 = -2H_{21}\sigma - \dot{m}_1^2\left(\frac{1}{\rho_1} - \frac{1}{\rho_2}\right) + \left(\tau_{nn1} - \tau_{nn2}\right). \tag{2-105}$$

These above two equations can be solved for the pressure deviation from the saturation pressure as

$$\delta p_1 = -2H_{21}\sigma\left(\frac{\rho_1}{\rho_1-\rho_2}\right)+\frac{(\dot{m}_1)^2}{2}\left(\frac{1}{\rho_2}-\frac{1}{\rho_1}\right)+\tau_{nn1}$$

and (2-106)

$$\delta p_2 = -2H_{21}\sigma\left(\frac{\rho_2}{\rho_1-\rho_2}\right)+\frac{(\dot{m}_1)^2}{2}\left(\frac{1}{\rho_1}-\frac{1}{\rho_2}\right)+\tau_{nn2}.$$

This result shows that neither phase is in the saturation condition given by Eq.(2-99). The amount of deviation of pressure from p^{sat} depends on the mean curvature, the surface tension, the mass transfer rate and the normal stress. An interesting result follows if we take into account only the effect of the surface tension and drop the other terms which are generally negligibly small. In this case, we can approximate

$$\delta p_g = 2H_{fg}\sigma\left(\frac{\rho_g}{\rho_f-\rho_g}\right) \text{ and } \delta p_f = 2H_{fg}\sigma\left(\frac{\rho_f}{\rho_f-\rho_g}\right). \qquad (2\text{-}107)$$

Since the mean curvature of the liquid phase H_{fg} is positive for a droplet and negative for a bubble, the phase pressures at the interface are both over the saturation pressure for a droplet flow, and they are both under it for a bubbly flow.

Now we recall the existence of the limits on heating of liquid or cooling of vapor beyond the saturation condition in terms of the pressure deviation at fixed temperature, namely, the instability points of the equation of state in the thermostatics. Thus, we write

$$\begin{aligned} &\delta p_g \leqq \delta p_{g\max}(T_i) \\ &\delta p_f \geqq \delta p_{f\min}(T_i) \end{aligned} \qquad (2\text{-}108)$$

which are shown in Fig.2-3.

Figure 2-3 shows the saturation line corresponding to the Clausius-Clapeyron equation or Eq.(2-99) and the limits of the metastable liquid and vapor phases. These two limits can be obtained from the van der Waals equation of state given by

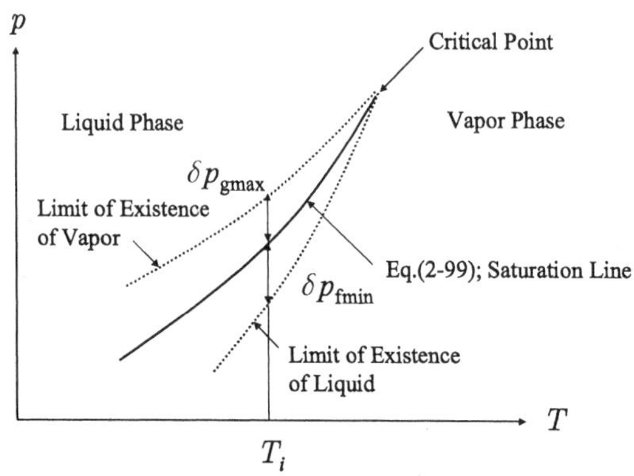

Figure 2-3. *p-T* diagram (Ishii, 1975)

$$\left\{p+\frac{a}{\left(M/\rho\right)^2}\right\}\left(\frac{M}{\rho}-b\right)=RT \tag{2-109}$$

where R and M are the gas constant and the molecular weight, respectively. a and b are empirical constants. The thermodynamic theory states that the intrinsic thermodynamic stability requires

$$\left.\frac{\partial p}{\partial\left(1/\rho\right)}\right|_T<0. \tag{2-110}$$

Therefore, by using the van der Waals equation, the loci of $\partial p/\partial\left(1/\rho\right)=0$ can be found. These loci actually represent two limits, namely the superheated liquid limit and subcooled vapor limit. These two loci are shown by the broken curves in Fig.2-3.

It is interesting to note that Eq.(2-107) with the limiting condition of Eq.(2-108) gives the smallest droplet and the bubble sizes. In other words, these sizes are the lowest natural level of the disturbances in the statistical sense. Beyond these limits the liquid or the vapor phase cannot stay without changing the phase, because the statistical fluctuations create a core which can grow to a bubble or a droplet.

The relations given by Eqs.(2-107) and (2-108) at a temperature T_i are exhibited in Fig.2-4. The widely used interfacial condition that the vapor

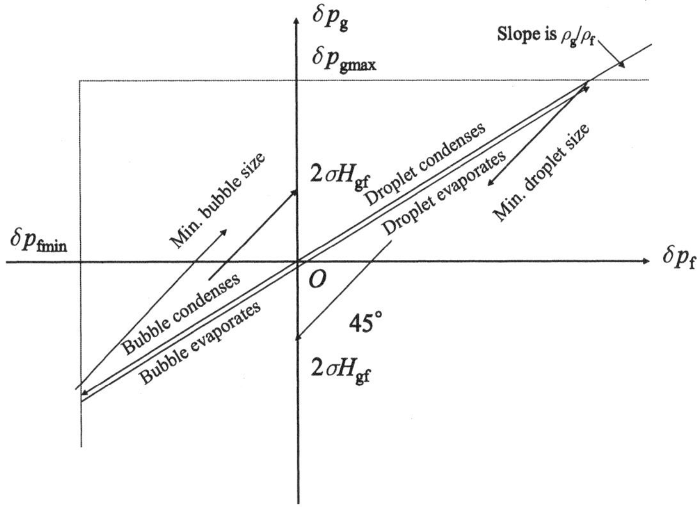

Figure 2-4. δp_g-δp_f relation (Ishii, 1975)

interfacial pressure equals the saturation pressure p^{sat} at a temperature T_i can be derived as a further approximation to Eq.(2-107). Since the density ratio between phases is very large at a small reduce pressure, namely, $p/p_c << 1$ where p_c is the critical pressure, Eq.(2-107) can be approximated by

$$\begin{aligned} &\delta p_g \approx 0, \qquad p_g \approx p^{sat}(T_i) \\ &\delta p_f \approx 2H_{fg}\sigma, \qquad p_f \approx p^{sat}(T_i) + 2H_{fg}\sigma. \end{aligned} \tag{2-111}$$

1.2.4 External boundary condition and contact angle

The external boundary condition is a special case of the jump and the supplemental interfacial boundary conditions which have been discussed in the previous section. For a standard single-phase flow problem, these conditions become particularly simple because the mass transfer rate $\dot{m}_k$, the effect of the surface tension and the velocity of the solid-wall interface are all set to be zero. Similar simplifications could also be applied to a two-phase flow system, however, two exceptional characteristics should be taken into account here. These are:

1. The wall microstructure effect on bubble nucleations;
2. The intersection of a phase interface with the external boundary.

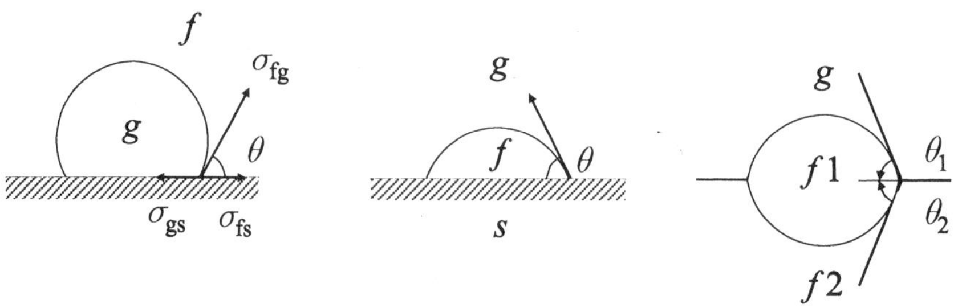

Figure 2-5. Contact angle (Ishii, 1975)

The first effect characterizes the necessity to consider the existence of surface nucleation sites which have irregular geometries deviating from the standard idealized wall boundary. These microstructures and the gas content in these sites often decide the bubble nucleation conditions and the degree of thermodynamic non-equilibrium. The second case is the singularity created by meeting of two different interfaces, see Fig.2-5. As a bubble or a droplet comes in contact with the external boundary, the vapor-liquid interface attaches to the wall and forms a singular curve at the intersection. When such a contact line is formed, the angle of contact θ measured though the liquid characterizes the condition along the curve. An analysis similar to the one for the interface can be developed also for this singular line. In this case, since the area of transport from the bulk fluids is the thickness of the interface δ, the effects of the mass transfers and of the fluxes of the fluids can be neglected. Hence, only the surface fluxes and possibly the properties associated with the curve, namely, energy of the contact line, are important. By considering only the surface fluxes, we have from the force balance in the normal plane to the singular curve

$$\cos\theta = \frac{\sigma_{gs} - \sigma_{fs}}{\sigma_{fg}} \tag{2-112}$$

where σ_{fg}, σ_{gs} and σ_{fs} denote the surface tension between vapor-liquid, vapor-solid, and liquid-solid respectively.

We note here that Eq.(2-112) is consistent with the jump conditions, if we neglect the tension tangent to the singular curve and thus the thermal energy of the curve. If these effects are neglected, Eq.(2-112) is the only condition obtainable in parallel with the jump conditions. Hence, as it has been mentioned, the contact angle θ characterizes the phenomenon and an appropriate constitutive law should be supplied if σ_{gs} and σ_{fs} are not available. The static contact angle θ is well measured and tabulated for various interfaces: in reality however it is greatly influenced by the surface roughness, the deposit of foreign materials and the purity of fluid itself.

Furthermore, the dynamic contact angle of a moving interface can be significantly different from the static values. However, in the absence of a well established constitutive law for θ under dynamic condition, the static values are frequently used in practical problems. We only note here that it is generally accepted that the apparent difference between the static and the dynamic contact angle is a function of a surface tension σ_{fg} and the normal slipping velocity of the singular curve (Schwartz and Tejada, 1972; Phillips and Raddiford, 1972).

In summarizing this section we list standard external boundary conditions at the solid wall:

The position of an external boundary

$$f_w(\boldsymbol{x}) = 0 \tag{2-113}$$

No-mass transfer condition

$$\boldsymbol{v}_{nk} = \boldsymbol{v}_{nw} = 0 \tag{2-114}$$

No-slip condition for a viscous fluid

$$\boldsymbol{v}_{tk} = \boldsymbol{v}_{tw} = 0 \tag{2-115}$$

The force balance from the momentum jump condition

$$\boldsymbol{n}_k \cdot \mathcal{T}_k + \boldsymbol{n}_w \cdot \mathcal{T}_w = 0 \tag{2-116}$$

The energy balance from the energy jump condition

$$\boldsymbol{n}_k \cdot \boldsymbol{q}_k + \boldsymbol{n}_w \cdot \boldsymbol{q}_w = 0 \tag{2-117}$$

The thermal equilibrium condition

$$T_k = T_w \tag{2-118}$$

These above conditions can be applied where a fluid is in contact with the wall. It cannot be applied however at an intersection of an interface with the solid boundary. On such a singular curve the constitutive equation for the contact angle θ should be given. Finally, we summarize the local instant formulation of a two-phase flow system in the following diagram.

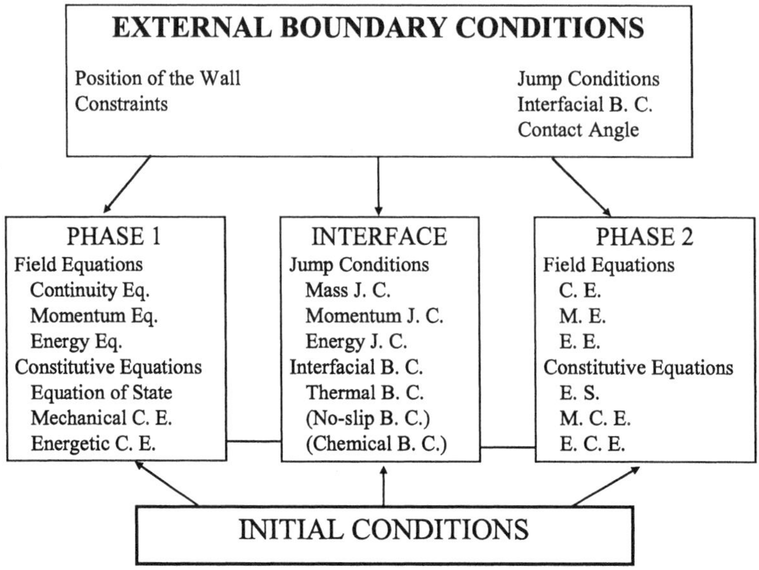

1.3 Application of local instant formulation to two-phase flow problems

1.3.1 Drag force acting on a spherical particle in a very slow stream

As an example of applying local instant formulation to two-phase flow problems, let us study the drag constitutive equation of a solid sphere of radius r_d in a very slow stream of speed U_0 (creeping flow) (Stokes, 1851; Schlichting, 1979). In order to analyze this problem analytically, we assume (1) Newtonian viscous fluid with constant viscosity, (2) incompressible flow (fluid density is constant), and (3) very small Reynolds number ($Re\left(\equiv 2r_d\rho_c U_0/\mu_c\right) << 1$) where viscous effects dominate the flow and the inertia term can be neglected in the momentum equation. Then, the continuity equation, Eq.(2-8), and the momentum equation, Eq.(2-10), can be linearized as

$$\nabla \cdot \boldsymbol{v}_c = 0 \tag{2-119}$$

$$\nabla p_c = \mu_c \nabla^2 \boldsymbol{v}_c . \tag{2-120}$$

The gravity term is dropped by considering the pressure field which excludes the hydrostatic effect. The velocity components and the pressure in spherical coordinates (r,θ) with $\theta = 0$ in the direction of U_0 can be derived under the boundary condition of no-slip on the solid sphere as

$$v_{rc} = U_0 \cos\theta \left(1 - \frac{3}{2}\frac{r_d}{r} + \frac{1}{2}\frac{r_d^3}{r^3}\right) \tag{2-121}$$

$$v_{\theta c} = -U_0 \cos\theta \left(1 - \frac{3}{4}\frac{r_d}{r} - \frac{1}{4}\frac{r_d^3}{r^3}\right) \tag{2-122}$$

$$p_c = p_\infty - \frac{3\mu_c r_d U_0}{2r^2}\cos\theta. \tag{2-123}$$

where p_∞ is the uniform freestream pressure. The shear stress acting on the solid sphere, $\tau_{r\theta c}$, is given by

$$\tau_{r\theta c}\Big|_{r=a} = \mu_c \left(\frac{1}{r}\frac{\partial v_{rc}}{\partial \theta} + \frac{\partial v_{rc}}{\partial r}\right)\Bigg|_{r=r_d} = \frac{3}{2}\frac{\mu_c U_0}{r_d}\sin\theta. \tag{2-124}$$

Thus, the total drag force, F_D, acting on the solid sphere is given by integrating the pressure and the shear stress around the surface as

$$\begin{aligned} F_D &= \int_0^\pi \tau_{r\theta c} \sin\theta dA - \int_0^\pi p_c \cos\theta dA \\ &= 4\pi r_d \mu_c U_0 + 2\pi r_d \mu_c U_0 = 6\pi r_d \mu_c U_0 \end{aligned} \tag{2-125}$$

where A is the surface area. This indicates that the drag consists of the pressure and shear forces even in this viscosity dominated flow. Then, we define the drag coefficient, C_D, by

$$C_D \equiv \frac{F_D}{\frac{1}{2}\rho_c U_0^2 A_p} \tag{2-126}$$

where A_p is the projected area of a particle. Thus, we have

$$C_D = \frac{24}{Re}. \tag{2-127}$$

This analysis was extended by Rybczynski (1911) and Hadamard (1911) to creeping motion of a spherical fluid particle in an infinite Navier-Stokes fluid (Brodkey, 1967; Soo, 1967). Thus, the total force acting on a fluid particle is given by

$$F_D = 6\pi r_d \mu_c \left(v_{c\infty} - v_d\right)\left\{\frac{2\mu_c + 3\mu_d}{3\left(\mu_c + \mu_d\right)}\right\}. \tag{2-128}$$

Then, we define the drag coefficient, $C_{D\infty}$, by

$$C_{D\infty} \equiv \frac{F_D}{\frac{1}{2}\rho_c\left(v_{c\infty} - v_d\right)A_p} \tag{2-129}$$

and the particle Reynolds number by

$$Re_d \equiv \frac{2r_d\rho_c\left(v_{c\infty} - v_d\right)}{\mu_c}. \tag{2-130}$$

It is evident here that $v_{c\infty}$ and v_d are the undisturbed flow velocity and the particle velocity. Thus, we have

$$C_{D\infty} = \frac{24}{Re_d}\left\{\frac{2\mu_c + 3\mu_d}{3\left(\mu_c + \mu_d\right)}\right\}; Re_d < 1. \tag{2-131}$$

The drag law given by Rybczynski and Hadamard is good up to a Reynolds number of about 1.

1.3.2 Kelvin-Helmholtz instability

As another example of application of local instant formulation to two-phase flow problems, let us study the Kelvin-Helmholtz instability (Helmholtz, 1868; Kelvin; 1871; Lamb, 1945). The Kelvin-Helmholtz instability arises at the interface of two fluid layers of different densities ρ_1 and ρ_2 flowing with average velocities v_1 and v_2 in a horizontal duct. In order to analyze this problem analytically, we assume: (1) inviscid flow (viscous force is negligible); (2) incompressible flow (fluid density is constant); and (3) irrotational flow. It is convenient to use rectangular coordinates (x,y) where x and y indicate the coordinate in the horizontal direction and the coordinate in the vertical direction measured from the average interface of the two fluid layers, respectively. Then, the velocity components are given in terms of the velocity potential, ϕ_k, as

$$v_{xk} = -\frac{\partial \phi_k}{\partial x}, v_{yk} = -\frac{\partial \phi_k}{\partial y}. \tag{2-132}$$

Thus, the continuity equation, Eq.(2-8), is given in terms of the velocity potential as

$$\frac{\partial^2 \phi_k}{\partial x^2} + \frac{\partial^2 \phi_k}{\partial y^2} = 0 \tag{2-133}$$

and the momentum equation, Eq.(2-10), is given by

$$\frac{p_k}{\rho_k} + \frac{1}{2} v_k^2 + gy = \frac{\partial \phi}{\partial t} + F(t) \tag{2-134}$$

where $F(t)$ is the function of t, respectively. The shape of the interface between two phases are approximated by a sinusoidal wave as

$$\eta = \eta_0 \sin\{k(x - Ct)\} \tag{2-135}$$

where η_0, k, and C are the amplitude, the wave number, and the wave velocity, respectively. Then, the velocity potentials of the upper fluid $(k = 1)$ and lower fluid $(k = 2)$ are derived under the boundary condition of no fluid penetration on the upper and lower duct surfaces and the assumption of small perturbation.

$$\phi_1 = -v_1 x + \eta_0 (v_1 - C) \frac{\cosh\{k(h_1 - y)\}}{\sinh(kh_1)} \cos\{k(x - Ct)\} \tag{2-136}$$

$$\phi_2 = -v_2 x - \eta_0 (v_2 - C) \frac{\cosh\{k(h_2 + y)\}}{\sinh(kh_2)} \cos\{k(x - Ct)\} \tag{2-137}$$

where h_1 and h_2 are the average thickness of the upper and lower fluid layers, respectively. Substituting Eqs.(2-136) and (2-137) into Eq.(2-134) and assuming $v_{yk} << v_{xk}$ yield the pressure of each phase at the interface as

$$p_{i1} = -\rho_1 \left\{ (v_1 - C)^2 k\coth(kh_1) + g \right\} \eta_0 \sin\{k(x - Ct)\} + p_i \tag{2-138}$$

$$p_{i2} = \rho_2 \left\{ (v_2 - C)^2 k\coth(kh_2) - g \right\} \eta_0 \sin\{k(x - Ct)\} + p_i \tag{2-139}$$

where p_i is the pressure at a smooth interface. The interfacial pressure difference between two fluid layers is due to the surface tension, and can be approximated by

$$p_{i2} - p_{i1} = -\sigma \frac{\partial^2 \eta}{\partial x^2}. \tag{2-140}$$

Then, the wave velocity can be obtained from Eq.(2-135) and Eqs.(2-138)-to-(2-140) as

$$C = \frac{\rho_1' v_1 + \rho_2' v_2}{\rho_1' + \rho_2'} \pm \sqrt{\frac{\sigma k + (\rho_2 - \rho_1) g/k}{\rho_1' + \rho_2'} - \rho_1' \rho_2' \left(\frac{v_1 - v_2}{\rho_1' + \rho_2'} \right)^2} \tag{2-141}$$

where $\rho_k' \equiv \rho_k \coth(kh_k)$. Under the deep water assumption of $h_1/(2\pi/k)$, $h_2/(2\pi/k) > 0.25$, ρ_k' can be approximated to be ρ_k. In this case, Eq.(2-141) can be simplified as

$$C = \frac{\rho_1 v_1 + \rho_2 v_2}{\rho_1 + \rho_2} \pm \sqrt{C_\infty^2 - \rho_1 \rho_2 \left(\frac{v_1 - v_2}{\rho_1 + \rho_2} \right)^2} \tag{2-142}$$

where

$$C_\infty^2 = \frac{g}{k} \frac{\rho_2 - \rho_1}{\rho_1 + \rho_2} + \frac{\sigma k}{\rho_1 + \rho_2}. \tag{2-143}$$

When the root in the expression for the wave velocity C has a nonzero imaginary part, then the interfacial disturbance can grow exponentially. Hence, the flow is unstable if

$$\frac{g}{k}\frac{\rho_2-\rho_1}{\rho_1+\rho_2}+\frac{\sigma k}{\rho_1+\rho_2}<\rho_1\rho_2\left(\frac{v_1-v_2}{\rho_1+\rho_2}\right)^2. \tag{2-144}$$

There are several important points to be recognized in this stability criterion. First, the viscous effects of the fluids are neglected; therefore, the Reynolds number plays no role in this type of interfacial instability. The stability of the system then is governed by three effects, namely, the gravity force, surface-tension force, and relative motion. The relative-motion term is always destabilizing due to the inertia force from Bernoulli effect. The surface-tension force is always stabilizing, since the flat interface has the minimum surface area, and the surface-tension force acts to resist any deformation from the equilibrium configuration. The gravity term is stabilizing only if the upper fluid is lighter than the lower fluid ($\rho_2>\rho_1$).

The propagation velocity C_∞ in the absence of the flows (or the left-hand side of the stability criterion) is a function of the wave number k. Therefore, as the wavelength $\lambda=2\pi/k$ changes from zero to infinite, the wave velocity decreases to the minimum value and then increases. This minimum value of C_∞^2 is given by $C_{\infty c}^2=2\left[\sigma g(\rho_2-\rho_1)/(\rho_2-\rho_1)^2\right]^{1/2}$, which occurs at $k_c^2=g(\rho_2-\rho_1)/\sigma$. This corresponds to the critical wavelength of $\lambda_c=2\pi/k_c$. This is known as Taylor wave length that is one of the most important internal length scales in two-phase flow. Then the system is stable for small disturbances of all wavelengths if the relative velocity is sufficiently small to satisfy

$$(v_1-v_2)^2<\frac{2(\rho_1+\rho_2)}{\rho_1\rho_2}\sqrt{\sigma g(\rho_2-\rho_1)}. \tag{2-145}$$

For a relative velocity larger than this limit, the system is only conditionally stable for a certain range of the wavelength. When the wavelength is large, the value of C_∞^2 in Eq.(2-143) is mainly determined by the gravity term. Conversely, if λ is sufficiently small, the capillary force governs the wave motion.

Furthermore, it is possible to develop a similar stability criterion based on the one-dimensional two-phase flow equations (Wallis, 1969; Kocamustafaogullari, 1971). It is noted (Miles, 1957) that the Kelvin-Helmholtz instability theory tends to overpredict the critical relative velocity for the initial generation of surface waves, except in the case of highly viscous fluids. However, the Kelvin-Helmholtz instability mechanism is important in wave-propagation phenomena, particularly for flows in a confined channel (Kordyban, 1977). Based on the analysis, Kelvin proposed the word "Ripples" to describe waves having a wavelength of less than $\lambda_c = 2\pi\sqrt{\sigma/g(\rho_2 - \rho_1)}$.

For a gravity dominated flow with a relatively large wave length $\lambda >> \lambda_c$, the surface tension effect can be neglected. By considering the finite channel flow, Eq.(2-141) can give a criterion for instability as

$$\frac{g}{k}\frac{\rho_2 - \rho_1}{\rho_1' + \rho_2'} < \rho_1'\rho_2'\left(\frac{v_1 - v_2}{\rho_1' + \rho_2'}\right)^2. \tag{2-146}$$

By taking a Taylor expansion and retaining only the first order term for the hyperbolic functions, a following simplified but useful criterion can be obtained.

$$\left(v_1 - v_2\right)^2 > \frac{g}{k}\frac{\left(\rho_2 - \rho_1\right)\left(\rho_1' + \rho_2'\right)}{\rho_1'\rho_2'} \approx \frac{g\left(\rho_2 - \rho_1\right)h_1}{\rho_1}. \tag{2-147}$$

When this criterion is compared to experimental data for slug formation in a channel, the critical relative velocity is overpredicted by a factor close to two. This discrepancy can be explained by a theoretical analysis introducing a finite amplitude or wave front propagation method (Mishima and Ishii, 1980; Wu and Ishii, 1996).

1.3.3 Rayleigh-Taylor instability

The Rayleigh-Taylor instability is the interfacial instability between two fluids of different densities that are stratified in the gravity field or accelerated normal to the interface. It is commonly observed that the boundary between two stratified fluid layers at rest is not stable if the upper-fluid density ρ_1 is larger than the lower-fluid density ρ_2. Since the Rayleigh-Taylor instability can lead to the destruction of the single common interface, it is important in the formation of bubbles or droplets. In particular, the critical wavelength predicted by the related stability analysis is one of the most significant length scales for two-phase flow.

The Rayleigh-Taylor instability can be considered as a special case of the Kelvin-Helmholtz instability with zero flows and $\rho_1 > \rho_2$. Hence, the propagation velocity can be obtained from Eq.(2-142) by setting $v_1 = v_2 = 0$

$$C^2 = \frac{g}{k}\frac{\rho_2 - \rho_1}{\rho_2 + \rho_1} + \frac{\sigma k}{\rho_2 + \rho_1}. \tag{2-148}$$

The system is unstable if the root of the propagation velocity has a nonzero imaginary part. Therefore, Eq.(2-148) shows that the gravitational force is destabilizing for $\rho_1 > \rho_2$, whereas the surface-tension force is stabilizing. There is a critical wavelength λ below which C^2 is always positive. This is given by $\lambda_c = 2\pi\sqrt{\sigma/g(\rho_2 - \rho_1)}$. If the wavelength of a disturbance is larger than the critical wave length ($\lambda > \lambda_c$), then C^2 becomes negative and the interface is unstable. For fluids that are unlimited laterally, the wavelength of the disturbance can be as large as desired; therefore such a system is always unstable. However, if the fluids are confined laterally, the maximum wavelength is limited to twice the system dimension. This implies that a system is stable if the lateral characteristic dimension is less than half the critical wavelength λ_c. For an air-water system, this characteristic dimension is 0.86 cm. A similar dimension can be obtained from fluids contained in a vertical cylinder by using polar coordinates in the stability analysis.

For an unstable system, any disturbance having a wavelength greater than λ_c can grow in time. However, the dominant waves are those having the maximum growth factor. Since the wave amplitude grows with $\exp(-ikCt)$, the predominant wavelength should be

$$\lambda_m = 2\pi\sqrt{\frac{3\sigma}{g(\rho_1 - \rho_2)}}. \tag{2-149}$$

These unstable waves can be observed as water droplets dripping from a wire in a rainy day, or condensed water droplets falling from a horizontal downward-facing surface. Quite regular waveforms and generation of bubbles due to the Rayleigh-Taylor instability can also be observed in film boiling. Note that this instability is not limited to the gravitational field. Any interface, and fluids that are accelerated normal to the interface, can exhibit the same instability. This can occur for example in nuclear explosion and inertia confinement of a fusion pellet. In such a case the acceleration should replace the gravity field g in the analysis.

Chapter 3

VARIOUS METHODS OF AVERAGING

1.1 Purpose of averaging

The design of engineering systems and the ability to predict their performance depend on the availability of experimental data and conceptual models that can be used to describe a physical process with a required degree of accuracy. From both a scientific and a practical point of view, it is essential that the various characteristics and properties of such conceptual models and processes are clearly formulated on rational bases and supported by experimental data. For this purpose, specially designed experiments are required which must be conducted in conjunction with and in support of analytical investigations. It is well established in continuum mechanics that the conceptual models for single-phase flow of a gas or of a liquid are formulated in terms of field equations describing the conservation laws of mass, momentum, energy, charge, etc. These field equations are then complemented by appropriate constitutive equations such as the constitutive equations of state, stress, chemical reactions, etc., which specify the thermodynamic, transport and chemical properties of a given constituent material, namely, of a specified solid, liquid or gas.

It is to be expected, therefore, that the conceptual models describing the steady state and dynamic characteristics of multiphase or multi-component media should also be formulated in terms of the appropriate field and constitutive equations. However, the derivation of such equations for the flow of structured media is considerably more complicated than for strictly continuous homogeneous media for single-phase flow. In order to appreciate the difficulties in deriving balance equations for structured, namely, inhomogeneous media with interfacial discontinuities, we recall that in continuum mechanics the field theories are constructed on integral balances of mass, momentum and energy. Thus, if the variables in the region of integration are continuously differentiable and the Jacobian transformation between material and spatial coordinates exists, then the Euler-type differential balance can be obtained by using the Leibnitz's rule; more specifically, however, the Reynolds's transport theorem allows us to interchange differential and integral operations.

In multi-phase or multi-component flows, the presence of interfacial surfaces introduces great difficulties in the mathematical and physical formulation of the problem. From the mathematical point of view, a multi-phase flow can be considered as a field that is subdivided into single-phase regions with moving boundaries separating the constituent phases. The differential balance holds for *each* sub-region. It cannot be applied, however, to the *set* of these sub-regions in the normal sense without violating the above conditions of continuity. From the point of view of physics, the difficulties encountered in deriving the field and constitutive equations appropriate to multi-phase flow systems stem from the presence of the interface. It also stems from the fact that both the steady and dynamic characteristics of multi-phase flows depend upon the interfacial structure of the flow. For example, the steady state and the dynamic characteristics of dispersed two-phase flow systems depend upon the collective dynamics of solid particles, bubbles or droplets interacting with each other and with the surrounding continuous phase; whereas, in the case of separated flows, these characteristics depend upon the structure and wave dynamics of the interface. In order to determine the collective interaction of particles and the dynamics of the interface, it is necessary to describe first the local properties of the flow and then to obtain a macroscopic description by means of appropriate averaging procedures. For dispersed flows, for example, it is necessary to determine the rates of nucleation, evaporation or condensation, motion and disintegration of single droplets (bubbles) as well as the collisions and coalescence processes of several droplets (or bubbles).

For separated flow, the structure and the dynamics of the interface greatly influence the rates of mass, heat and momentum transfer as well as the stability of the system. For example, the performance and flow stability of a condenser for space application depend upon the dynamics of the vapor interface. Similarly, the rate of droplet entrainment from a liquid film, and therefore, the effectiveness of film cooling, depend upon the stability of the vapor liquid interface.

It can be concluded from this discussion that in order to derive the field and constitutive equations appropriate to structured multiphase flow, it is necessary to describe the local characteristics of the flow. From that flow, the macroscopic properties should be obtained by means of an appropriate averaging procedure. It is evident also that the design, performance and, very often, the safe operation of a great number of important technological systems (which were enumerated in the preceding sections) depend upon the availability of realistic and accurate field and constitutive equations.

The formulation based on the local instant variables of Chapter 2 shows that, in general, it results in a multi-boundary problem with the positions of the interfaces being unknown. In such a case the mathematical difficulties encountered in obtaining solutions are prohibitively great and in many practical problems they are beyond our present computational capability. In order to appreciate these difficulties we recall that even in single-phase turbulent flow without moving interfaces, it has not been possible to obtain exact solutions expressing local instant fluctuations. It can be said that overwhelming difficulties encountered in the local instant formulations stem from:

1. Existence of the multiple deformable moving interfaces with their motions being unknown;
2. Existence of the fluctuations of variables due to turbulences and to the motions of the interfaces;
3. Significant discontinuities of properties at interface.

The first effect causes complicated coupling between the field equations of each phase and the interfacial conditions, whereas the second effect inevitably introduces a statistical characteristic originated from the instability of the Navier-Stokes equation and of the interfacial waves. The third effect introduces huge local jumps in various variables in space and time. Since these difficulties exist in almost all two-phase flow systems, an application of the local instant formulation to obtain a solution is severely limited. For a system with a simple interfacial geometry, however, as in the case of a single or several bubble problem or of a separated flow, it has been used extensively and very useful information have been obtained. As most two-phase flow observed in practical engineering systems have extremely complicated interfacial geometry and motions, it is not possible to solve for local instant motions of the fluid particles. Such microscopic details of the fluid motions and of other variables are rarely needed for an engineering problem, but rather macroscopic aspects of the flow are much more important.

By proper averaging, we can obtain the mean values of fluid motions and properties that effectively eliminate local instant fluctuations. The averaging procedure can be considered as low-pass filtering, excluding unwanted high frequency signals from local instant fluctuations. However, it is important to note that the statistical properties of these fluctuations influencing the macroscopic phenomena should be taken into account in a formulation based on averaging.

1.2 Classification of averaging

The importance and the necessity of averaging procedures in order to derive macroscopic field and constitutive equations for structured two-phase media have been discussed in the Section 1.1 of Chapter 3 In this section we study various methods of averaging that can be applied to thermo-fluid dynamics in general and to two-phase flow in particular. Depending on the basic physical concepts used to formulate thermal-hydraulic problems, averaging procedures can be classified into three main groups: the Eulerian averaging; the Lagrangian averaging; and the Boltzmann statistical averaging. They can be further divided into sub-groups based on a variable with which a mathematical operator of averaging is defined. The summary of the classifications and the definitions of various averaging are given below.

i) *Eulerian Average - Eulerian Mean Value*

Function: $F = F(t, \boldsymbol{x})$ (3-1)

Time (Temporal) mean value: $\dfrac{1}{\Delta t}\displaystyle\int_{\Delta t} F\left(t, \boldsymbol{x}\right) dt$ (3-2)

Spatial mean value: $\dfrac{1}{\Delta R}\displaystyle\int_{\Delta R} F\left(t, \boldsymbol{x}\right) dR\left(\boldsymbol{x}\right)$ (3-3)

Volume: $\dfrac{1}{\Delta V}\displaystyle\int_{\Delta V} F\left(t, \boldsymbol{x}\right) dV$ (3-4)

Area: $\dfrac{1}{\Delta A}\displaystyle\int_{\Delta A} F\left(t, \boldsymbol{x}\right) dA$ (3-5)

Line: $\dfrac{1}{\Delta C}\displaystyle\int_{\Delta C} F\left(t, \boldsymbol{x}\right) dC$ (3-6)

Statistical mean value: $$\frac{1}{N}\sum_{n=1}^{N} F_n\left(t,\boldsymbol{x}\right) \tag{3-7}$$

Mixed mean value: combination of above operations

ii) *Lagrangian Average - Lagrangian Mean Value*

Function: $$F = F\left(t,\boldsymbol{X}\right); \quad \boldsymbol{X} = \boldsymbol{X}\left(\boldsymbol{x},t\right) \tag{3-8}$$

Time (Temporal) mean value: $$\frac{1}{\Delta t}\int_{\Delta t} F\left(t,\boldsymbol{X}\right) dt \tag{3-9}$$

Statistical mean value: $$\frac{1}{N}\sum_{n=1}^{N} F_n\left(t,\boldsymbol{X}\right) \tag{3-10}$$

iii) *Boltzmann Statistical Average*

Particle density function: $$f = f\left(\boldsymbol{x},\xi,t\right) \tag{3-11}$$

Transport properties: $$\psi\left(t,\boldsymbol{x}\right) = \frac{\int \psi(\xi)\, f d\xi}{\int f d\xi} \tag{3-12}$$

Here we note that $\boldsymbol{x}$ and $\boldsymbol{X}$ are the spatial and the material coordinates, respectively, whereas ξ is the phase velocity or kinetic energy of particles. Furthermore, we point out that the true time or statistical averaging is defined by taking the limit $\Delta t \to \infty$ or $N \to \infty$, which is only possible in concept. The material coordinates can be considered as the initial positions of all the particles, thus if $\boldsymbol{X}$ is fixed it implies the value of a function following a particle.

The most important and widely used group of averaging in continuum mechanics is the Eulerian averaging, because it is closely related to human observations and most instrumentations. The basic concept underlining this method is the time-space description of physical phenomena. In a so-called Eulerian description, the time and space coordinates are taken as independent variables and various dependent variables express their changes with respect to these coordinates. Since the standard field equations of continuum mechanics developed in Chapter 2 adapt to this description, it is natural to consider averaging with respect to these independent variables, namely, the time and the space. Furthermore, these averaging processes are basically integral operators, therefore, they have an effect of smoothing out instant or local variations within a domain of integration.

The Lagrangian mean values are directly related to the Lagrangian description of mechanics. As the particle coordinate $\boldsymbol{X}$ displaces the spatial variable $\boldsymbol{x}$ of the Eulerian description, this averaging is naturally fitted to a study of the dynamics of a particle. If our interest is focused on a behavior of an individual particle rather than on the collective mechanics of a group of particles, the Lagrangian average is important and useful for analyses. The Lagrangian time average is taken by following a certain particle and observing it over some time interval. A simple example is the average speed of a particular vehicle such as a car, a train or an airplane. Furthermore, the Eulerian temporal mean values can be exemplified by an average velocity of all cars passing at a point on a road over some time interval.

In contrast to the mean values explained above, the Eulerian and the Lagrangian statistical mean values are based on a statistical assumption, since they involve a collection of N similar samples denoted by F_n with $n = 1, \cdots, N$. A fundamental question arises as we ask, "What are the similar samples for a system with fluctuating signals?" To visualize a group of similar samples, it is useful to consider a time averaging as a filtering process that eliminates unwanted fluctuations. The similar samples may then be considered as a group of samples which have time mean values of all the important variables within certain ranges of deviations. In this case, the time interval of the averaging and the ranges of deviations define the unwanted fluctuations, thus the statistical averaging depends on them. For a steady-state flow based on time averaging, random sampling over a time domain can constitute a proper set of samples as it is often done in experimental measurements. In this case, the time averaging and the statistical averaging are equivalent. There are many other factors to consider, however it is also possible to leave it as an abstract concept. The difficulties arise when the constitutive equations are studied in connection with experimental data. The true statistical averaging involving an infinite number of similar samples is only possible in concept, and it cannot be realized. Thus, if it is considered alone, the ensemble averaging faces two difficulties, namely, choosing a group of similar samples and connecting the experimental data to a model.

The Boltzmann statistical averaging with a concept of the particle number density is important when the collective mechanics of a large number of particles are in question. As the number of particles and their interactions between them increase, the behavior of any single particle becomes so complicated and diversified, it is not practical to solve for each particle. In such a case, the behavior of a group of many particles increasingly exhibits some particular characteristics that are different from a single particle as the collective particle mechanics becomes a governing factor. It is well known that the Boltzmann statistical averaging applied to a large number of molecules with an appropriate mean-free path can lead to field equations that closely resemble that of the continuum mechanics. It can also be applied to subatomic particles, such as neutrons, to obtain a transport theory for them. This can be done by first writing the balance equation for the particle density function, which is known as the Boltzmann transport equation. Then it is necessary to assume a form of the particle interaction term as well as stochastic characteristics of the particle density function. A simple model using two-molecular interaction was developed by Maxwell, thus the Boltzmann transport equation with the collision integral of Maxwell was called the Maxwell-Boltzmann equation. This equation became the foundation of the kinetic theory of gases. We recall that if the Maxwell-Boltzmann equation is multiplied by 1, particle velocity, or the kinetic energy $(1/2)\xi^2$ then averaged over the particle velocity field, it can be reduced to a form similar to the standard conservation equations of mass, momentum and energy in the continuum mechanics.

1.3 Various Averaging in Connection with Two-Phase Flow Analysis

In order to study two-phase flow systems, many of above averaging methods have been used by various researchers. The applications of averaging can be divided into two main categories

1. To define properties and then to correlate experimental data.
2. To obtain usable field and constitutive equations that can be used to predict macroscopic processes.

The most elementary use is to define mean properties and motions that include various kinds of concentrations, density, velocity and energy of each phase or of a mixture. These properly defined mean values then can be used for various experimental purposes and for developments of empirical correlations. The choice of averaging and instrumentations are closely coupled since, in general, measured quantities represent some kinds of mean values themselves.

Both the Eulerian time and spatial averaging are frequently in use, because experimenters incline to consider two-phase mixtures as quasi-continuum. Furthermore, they are usually the easiest mean values to measure in fluid flow systems. However, when a particular fluid particle can be distinguishable and traceable, as in the case of a bubbly or droplet flow, the Lagrangian mean values are also used. It is only natural that these mean values are obtained for stationary systems that can be considered to have steady-state characteristics in terms of mean values. Various correlations are then developed by further applying the statistical averaging among different data. This is the standard method of experimental physics to minimize errors.

Before we proceed to the second application of averaging, we discuss briefly two fundamentally different formulations of the macroscopic field equations; namely the two-fluid model and the drift-flux (mixture) model. The two-fluid model is formulated by considering each phase separately. Consequently, it is expressed by two sets of conservation equations of mass, momentum and energy. Each of these six field equations has invariably an interaction term coupling the two phases through jump conditions. The mixture model is formulated by considering the mixture as a whole. Thus, the model is expressed in terms of three-mixture conservation equations of mass, momentum, and energy with one additional diffusion (continuity) equation which takes account of the concentration changes. A mixture conservation equation can be obtained by adding two corresponding conservation equations for each phase with an appropriate jump condition. However, it should be noted that a proper mixture model should be formulated in terms of correctly defined mixture quantities. It can be said that the drift-flux model is an example of a mixture model that includes diffusion model, slip flow model and homogeneous flow model. However, for most practical applications, the drift-flux model is the best mixture model that is highly developed for normal gravity (Ishii, 1977) as well as micro-gravity conditions (Hibiki and Ishii, 2003b; Hibiki et al., 2004).

Now we proceed to a discussion of the second and more important application of these averaging. That is to obtain the macroscopic two-phase flow field equations and the constitutive equations in terms of mean values. Here, again, the Eulerian spatial and time averaging have been used extensively by various authors, though the Eulerian or Boltzmann statistical averaging have also been applied.

Using the *Eulerian volumetric averaging*, important contributions for an establishment of a three-dimensional model of highly dispersed flows has been made by Zuber (1964a), Zuber et al. (1964), Wundt (1967), Delhaye (1968) and Slattery (1972). These analyses were based on a volume element that included both phases at the same moment. Moreover, it was considered to be much smaller than the total system in interest, thus main applications were for highly dispersed flows.

It has long been realized that the *Eulerian area averaging* over a cross section of a duct is very useful for engineering applications, since field equations reduce to a one-dimensional model. By area averaging, the information on changes of variables in the direction normal to the main flow is basically lost. Therefore, the transfer of momentum and energy between the wall and the fluid should be expressed by empirical correlations or by simplified models which replace the exact interfacial conditions. We note that even in single-phase flow problems, the area-averaging method has been widely used because its simplicity is highly desirable in many practical engineering applications. For example, the use of the wall friction factor or the heat transfer coefficient is closely related to the concept of the area averaging. We also mention here its extensive use in compressible fluid flow analyses. A good review of single-phase flow area averaging as well as macroscopic equations that correspond to the open-system equations in thermodynamics can be found in Bird et al. (1960), Whitaker (1968) and Slattery (1972). The boundary layer integral method of von Karman is also an ingenious application of the area averaging. Furthermore, numerous examples of area averaging can be found in the literature on lubricating films, open channel flow and shell theories in mechanics.

However, in applications to two-phase flow systems, many authors used phenomenological approach rather than mathematically exact area averaging, thus the results of Martinelli and Nelson (1948), Kutateladze (1952), Brodkey (1967), Levy (1960) and Wallis (1969) were in disagreement with each other and none of them are complete (Kocamustafaogullari, 1971). The rational approach to obtain a one-dimensional model is to integrate single-phase differential field equations over the cross sectional area. Meyer (1960) was an early user of this method to obtain mixture equations, but his definitions of various mixture properties as well as the lack of a diffusion (continuity) equation were objectionable (Zuber, 1967).

A rigorous derivation of one-dimensional mixture field equations with an additional diffusion (continuity) equation, namely, the drift-flux model, was carried out by Zuber et al. (1964) and Zuber (1967). The result shows a significant similarity with the field equations for heterogeneous chemically reacting single-phase systems. The latter had been developed as the thermomechanical theory of diffusion based on the interacting continua occupying the same point at the same time but having two different velocities. Numerous authors have contributed in this theory, thus we only recall those of Fick (1855), Stefan (1871), von Karman (1950), Prigogine and Mazur (1951), Hirschfelder et al. (1954), Truesdell and Toupin (1960), and Truesdell (1969). A similar result obtained from an entirely different method of the kinetic theory of gas mixtures by Maxwell (1867) should also be noted here.

In contrast to the analysis of Zuber, the analysis of Delhaye (1968) and Vernier and Delhaye (1968) was directed to two-fluid model based on three field equations for each phase with three jump conditions that couple the two fields. A very systematic method was employed in deriving field equations from three different Eulerian spatial averaging as well as the statistical and the temporal averaging in Vernier and Delhaye (1968). This is apparently the first publication which shows important similarities as well as differences between the various averaging methods. The effect of surface tension, which is important for the analysis of interfacial stability and of flow regimes, has been included in the study of Kocamustafaogullari (1971). This study highlighted that the area averaged model is particularly suited for studying a separated flow regime and interfacial wave instabilities. However, it can be used in any type of flow regimes, provided the constitutive equations can be supplied (Bouré and Réocreux, 1972). Furthermore, in the former reference a clear separation of analytical methods between the drift-flux model and the two-fluid model has been given. Although this distinction had been well known for other kind of mixtures, for example in study of the super fluidity of helium II of Landau (1941), of the plasma dynamics of Pai (1962), and of the diffusion theory of Truesdell (1969), in two-phase flow analysis it had been vague. This shortcoming of traditional two-phase flow formulation was first pointed out by Zuber and Dougherty (1967). In subsequent analyses of Ishii (1971) using time averaging and of Kocamustafaogullari (1971) using area averaging a clear distinction between the two models has been made. This point was also discussed by Bouré and Réocreux (1972) in connection with the problem of the two-phase sound wave propagation and of choking phenomena.

The Eulerian time averaging, which has been widely applied in analyzing a single-phase turbulent flow, is also used for two-phase flow. In applying the time averaging method to a mixture, many authors coupled it with other space averaging procedures. Important contributions were made by Russian researchers (Teletov, 1945; Frankl, 1953; Teletov, 1957; Diunin, 1963) who have used the Eulerian time-volume mean values and obtained three-dimensional field equations.

An analysis based on the Eulerian time averaging alone was apparently initiated by Vernier and Delhaye (1968), however, a detailed study leading to a mathematical formulation was not given there. Furthermore, Panton (1968) obtained the mixture model by first integrating in a time interval then in a volume element. His analysis was more explicit in integration procedures than the works by the Russian researchers, but both results were quite similar. In Ishii (1971), a two-fluid model formulation including the surface source terms was obtained by using the time averaging alone, then the area averaging over a cross section of a duct was carried out. There, all the constitutive equations as well as boundary conditions, which should be specified in a standard one-dimensional two-phase flow model, were identified. We also note the extensive study by Drew (1971) who has used an Eulerian multiple mixed averaging procedure. In his analysis, two integrals over both space and time domains have been taken in order to smooth out higher order singularities. These multiple integral operations are equivalent to the continuum assumption, therefore, they are not necessary. The averaging should not be considered as a pure mathematical transformation, since the constitutive model can be only developed based on the continuum assumption. Here, readers should also refer to Delhaye (1969; 1970), where various models based on Eulerian space averaging as well as a comprehensive review on the subject could be found.

It can be said that the Eulerian time averaging is particularly useful for a turbulent two-phase flow or for a dispersed two-phase flow (Ishii, 1975; Ishii, 1977; Ishii and Mishima, 1984). In these flows, since the transport processes are highly dependent on the local fluctuations of variables about the mean, the constitutive equations are best obtainable for a time averaged model from experimental data. This is supported by the standard single-phase turbulent flow analysis.

An extensive study using *Eulerian statistical averaging* was carried out by Vernier and Delhaye (1968) in which they obtained an important conclusion. Under stationary flow condition, they concluded, the field equations from the true time averaging, namely, the temporal averaging with $\Delta t \rightarrow \infty$, and the ones from the statistical averaging are identical. Furthermore, the statistical averaging was combined with the spatial averaging, then supplemented with various constitutive assumptions to yield a practical two-dimensional model. The Boltzmann statistical averaging has also been used by several authors (Murray, 1954; Buevich, 1969; Buyevich, 1972; Culick, 1964; Kalinin, 1970; Pai, 1971) for a highly dispersed two-phase flow systems. In general, the particle density functions are considered, then the Boltzmann transport equation for the functions is written. Kalinin (1970) assumed that the particle density functions represent the expected number of particles of a particular mass and velocity, whereas Pai (1971) considered the radius, velocity and temperature as the arguments of the functions. Then a simplified version of Maxwell's equation of transfer for each phase has been obtained from the Maxwell-Boltzmann equation by integrating over the arguments of the particle density function except time and space variables. Since it involves assumptions on the distributions as well as on inter-particle and particle-gas interaction terms, the results are not general, but represent a special kind of continuum.

It is interesting, however, to note that three different methods and views of mechanics of mixtures in a local sense are represented: the Eulerian time or statistical averaging applied to two-phase mixtures; the thermomechanical theory of diffusion based on two continua; and the Boltzmann statistical averaging applied to gas mixtures or to highly dispersed flows. The first theory considers the mixture to be essentially a group of single-phase regions bounded by interfaces, whereas in the second theory the two components coexist at the same point and time. In contrast to the above two theories, which are established on the foundation of the continuum mechanics, the last theory is based on the statistical expectations and the probability. The importance, however, is that if each transfer terms of above models are correctly interpreted, the resulting field equations have very similar forms.

A preliminary study using *ensemble cell averaging* was carried out by Arnold et al. (1989) where they derived turbulent stress and interfacial pressure forces due to pressure variations over the surface of non-distorting bubbles for an idealized inviscid bubbly flow. They discussed deficiencies inherent in spatial averaging techniques, and recommended ensemble averaging for the formulation of two-fluid models of two-phase flows. Zhang and Prosperetti (1994a) derived averaged equations governing a mixture of equal spherical compressible bubbles in an inviscid liquid by the ensemble-averaging method. They concluded that the method was systematic and general because of no *ad hoc* closure relations required, and suggested the possibility that the method might be applied to a variety of thermo-fluid and solid mechanics situations. Zhang and Prosperetti (1994b) extended this method to the case of spheres with a variable radius. Zhang (1993) summarized the other applications to heat conduction and convection, Stokes flow, and thermocapillary process. Here, readers should also refer to Prosperetti (1999), in which some considerations on the modeling of disperse multiphase flows by averaged equations can be found. Kolev (2002) presented a two-phase flow formulation mostly for development of safety analysis codes based on multi-field approach.

Finally, we briefly discuss the application of the Lagrangian averaging to two-phase flow systems. This approach is useful for particulate flow, however, in general it encounters considerable difficulties and impracticabilities due to the diffusion and phase changes. For the particulate flow without phase changes, the Lagrangian equation of the mean particle motion can be obtained in detail for many practical cases. Thus, we note that the Lagrangian description of a single particle dynamics is frequently in use as a momentum equation for a particulate phase in a highly dispersed flow (Carrier, 1958; Zuber, 1964a). Many analyses on the bubble rise and terminal velocity use the Lagrangian time averaging implicitly, particularly in a case when the continuum phase is in the turbulent flow regime.

Chapter 4

BASIC RELATIONS IN TIME AVERAGING

The importance of the Eulerian time averaging in studying a single-phase turbulent flow is well known. Since the most useful information in analyzing standard fluid flow systems is the time mean values rather than the local instant responses of the fluid, its use both in experimental and analytical purposes is indispensable in turbulent flow studies. For example mean velocity, temperature and pressure or the heat transfer coefficient and the friction factor are the important mean values routinely required in standard problems. Furthermore, commonly used experimental methods and measurements are well suited for the application of the time average. Thus, a single-phase turbulent flow has been studied in great depth by using the time averaged field equations with the constitutive laws expressed by mean values. Although these models, which are based on time averaging, do not give answers to the fundamental origin, structure and transport mechanisms of turbulent flow, their applications to engineering systems are widely accepted as efficient means of solving problems.

In discussing the importance of the Eulerian time averaging applied to two-phase mixtures, we first recall that in two-phase flow the local instant fluctuations of variables are caused not only by turbulence but also by rapidly moving and deforming interfaces. Because of these complicated flow and fluctuations, the solutions from the local instant formulation are inaccessible, therefore in order to derive appropriate field and constitutive equations it is necessary to apply some averaging procedure to the original local instant formulation. In view of the above discussion on the importance and usefulness of the time average in a single-phase turbulent flow analysis, it is both natural and logical that we also apply the time averaging to two-phase flow.

It is expected that the averaged field equations distinctly exhibit macroscopic phenomena of the system from hopelessly complicated interfacial and turbulent fluctuations, since they enter the formulation only statistically. There are two notable consequences from the time averaging when it is applied to a two-phase mixture:

1. Smoothing out of turbulent fluctuations in same sense as in a single-phase flow;
2. Bringing two phases, which are alternately occupying a volume element, into continua simultaneously existing at same point with a properly defined probability for each phase.

Furthermore, it should be recognized that the constitutive laws appearing in the averaged field equations should be expressed through the time mean values. These constitutive laws can be developed from a simple modeling of two-phase transport phenomena together with various experimental data that are commonly expressed by the mean values.

In the following chapters, we develop a detailed theory of the thermo-fluid dynamics of two-phase flow using the time averaging. First, we assume that the occupant of any particular point is alternating randomly between the two phases and that the time-averaged functions are sufficiently smooth in the new coordinates. Namely, the time coordinate having a minimum scale of Δt based on the time interval of averaging below which a time differential operator has no physical meaning.

1.1 Time domain and definition of functions

First, we recall that the singular characteristic of two-phase or of two immiscible mixture is the presence of one or several interfaces between the phases or components. Furthermore, whereas single-phase flows can be classified according to the geometry of the flow in laminar, transitional and turbulent flow, the flow of two phases or of a mixture of immiscible liquids can be classified according to the geometry of the interface into three classes: separated flow; transitional or mixed flow; and dispersed flow. These classes of structured flow are shown in Table 1-1.

In any flow regime, various properties suffer discontinuous changes at phase interfaces, if these interfaces are considered as singular surfaces with their thickness being zero and the properties having jump discontinuities. This can be illustrated more dramatically by taking a fluid density ρ, as shown in Figs.4-1 and 4-2. Since in two-phase flow systems the mass of each phase is clearly separated by the interfaces and do not mix at the molecular level, the local instant fluid density shows stepwise discontinuities between ρ_1 and ρ_2. Figure 4-1 shows the instantaneous discontinuities of ρ in space, whereas Fig.4-2 exhibits the discontinuities in time at some fixed point $\boldsymbol{x}_0$.

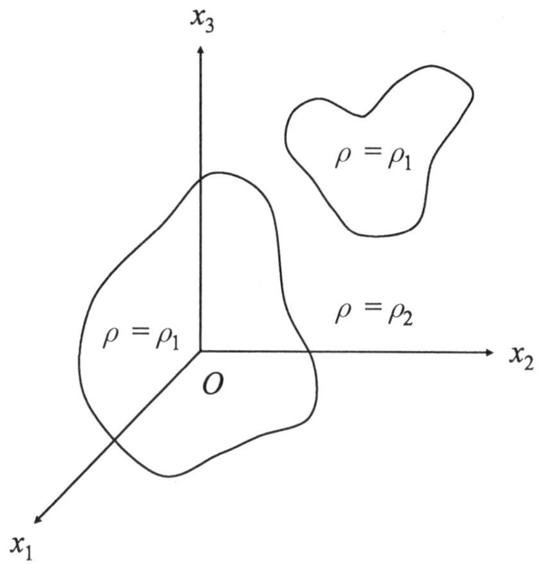

Figure 4-1. Fluid density in space at t=t_0 (Ishii, 1975)

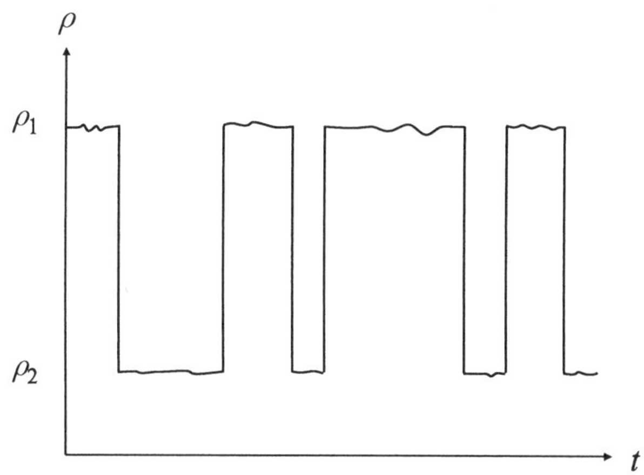

Figure 4-2. Fluid density in time at x=x_0 (Ishii, 1975)

For the purpose of time averaging, the observation from the time coordinate gives a more accurate picture of the problem. It can easily be seen that four distinct processes in terms of ρ may occur at any fixed point, which we can classify as follows:

1. $\rho = \rho_2(t)$ for all t ; always phase 2 at $\boldsymbol{x}_0$
2. $\rho = \rho_1(t)$ for all t ; always phase 1 at $\boldsymbol{x}_0$
3. ρ alternates between ρ_1 and ρ_2 ; phase 1 and 2 alternate at $\boldsymbol{x}_0$
4. ρ is neither ρ_1 and ρ_2 ; interface at $\boldsymbol{x}_0$ for some finite time

It is evident that, following a change of ρ from ρ_1 to ρ_2 or vice versa, all properties may change drastically because the phase occupying the point will be different. For the case of (1) and (2), since the time averaging at that point is trivial, we eliminate such cases. Furthermore, as case (4) is a rather singular configuration of case (3), it will be considered separately later. Hence, we examine the case in which the phase alternates stepwisely between 1 and 2.

Our purpose here is to average the fluid properties and field equations in order to treat two-phase flow as a mixture of continua. First, we take a fixed time interval Δt of the averaging and assume that it is large enough to smooth out the local variations of properties yet small compared to the macroscopic time constant of the unsteadiness of the bulk flow. This assumption is identical to that made in analyzing turbulent single-phase flow. After choosing any particular reference point and time $(\boldsymbol{x}_0, t_0)$, we have definite times, $t_1, t_2, \cdots, t_j, \cdots$ referring to the interfaces which pass the point $\boldsymbol{x}_0$ from time $\left(t_0 - \Delta t/2\right)$ to $\left(t_0 + \Delta t/2\right)$. By using the arbitrarily small interfacial thickness δ of the Section 1.2 of Chapter 2, the time intervals associated with each interface can be defined as

$$2\varepsilon = \frac{\delta}{v_{ni}} = \frac{\delta_1 + \delta_2}{v_{ni}} \tag{4-1}$$

which can vary among interfaces, thus we use ε_j for the jth- interface. Since we are going to treat the interfaces as a shell whose position is represented by a mathematical surface, we may take ε_j as a corresponding time interval for both δ_1 and δ_2. Then the assumption that an interface is a singular surface, or the interface thickness $\delta \to 0$, corresponds to

$$\lim_{\delta \to 0} \varepsilon_j = 0 \text{ for all } j \text{ if } \left|\boldsymbol{v}_{ni}\right| \neq 0. \tag{4-2}$$

In subsequent analyses, we frequently use this relation in order to derive macroscopic field equations. Now we define the set of time intervals, in which the characteristic of the interface dominates, as

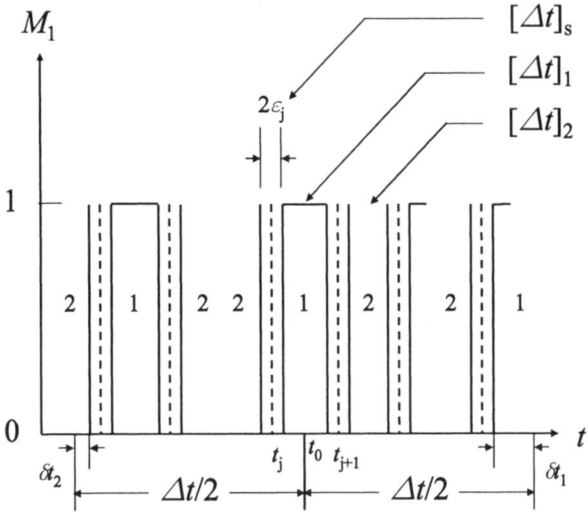

Figure 4-3. Various time intervals (Ishii, 1975)

$$[\Delta t]_S; \quad t \in \left[t_j - \varepsilon_j; t_j + \varepsilon_j\right] \quad \text{for } j = 1, ..., \text{n} \tag{4-3}$$

The remaining part of the time interval is given by $[\Delta t]_T$, which can be separated into intervals of phase 1 and 2. Thus

$$[\Delta t]_T = [\Delta t]_1 + [\Delta t]_2 . \tag{4-4}$$

By introducing

$$[\Delta t]; \quad t \in \left[t_0 - \frac{\Delta t}{2}; \ t_0 + \frac{\Delta t}{2}\right] \tag{4-5}$$

we have

$$[\Delta t] = [\Delta t]_S + [\Delta t]_T = [\Delta t]_S + \sum_{k=1}^{2} [\Delta t]_k . \tag{4-6}$$

These relations are shown in Fig.4-3.

Since in the course of analyses it becomes necessary to distinguish three states, namely, phase 1, phase 2 or interface, we assign state density functions M_1, M_2, and M_S which are defined as:

$$M_k(\boldsymbol{x},t)=1,\ M_S(\boldsymbol{x},t)=0\ \ (k=1 \text{ or } 2);$$
$$\text{A point occupied by } k^{\text{th}}\text{-phase}$$
$$M_S(\boldsymbol{x},t)=1,\ M_k(\boldsymbol{x},t)=0\ \ (k=1 \text{ and } 2);\tag{4-7}$$
$$\text{A point occupied by interface}$$

A general function F associated with two phases is considered to be continuously differentiable everywhere except in the interfacial regions of thickness δ. Then a general function of the kth-phase F_k at the point of averaging $\boldsymbol{x}_0$ is defined as

$$F_k\left(\boldsymbol{x}_0,t\right)=M_k\left(\boldsymbol{x}_0,t\right)F\quad\begin{cases}=F\left(\boldsymbol{x}_0,t\right) & \text{if } t\in[\Delta t]_k\\ =0 & \text{if } t\notin[\Delta t]_k.\end{cases}\tag{4-8}$$

This function F_k represents variables of each phase in a local instant formulation given in Chapter 2.

1.2 Local time fraction - Local void fraction

The time interval occupied by each phase is defined by taking the limit $\delta\rightarrow 0$ as (see Fig.4-3)

$$\begin{aligned}\Delta t_1&=\lim_{\delta\to 0}\left\{\sum_j\left[\left(t_{j+1}-\varepsilon_{j+1}\right)-\left(t_j+\varepsilon_j\right)\right]+\delta t_1\right\};\ j=2m-1\\ \Delta t_2&=\lim_{\delta\to 0}\left\{\sum_j\left[\left(t_{j+1}-\varepsilon_{j+1}\right)-\left(t_j+\varepsilon_j\right)\right]+\delta t_2\right\};\ j=2m.\end{aligned}\tag{4-9}$$

Hence, from the assumption (4-2) we have

$$\Delta t=\Delta t_1+\Delta t_2.\tag{4-10}$$

By recalling the previous assumption that the interfaces are not stationary and do not occupy a location $\boldsymbol{x}_0$ for finite time intervals, we can find the time averaged phase density function α_k as

$$\alpha_k\left(\boldsymbol{x}_0,t_0\right)\equiv\lim_{\delta\to 0}\frac{1}{\Delta t}\int_{[\Delta t]}M_k\left(\boldsymbol{x}_0,t\right)dt.\tag{4-11}$$

Hence, in view of Eq.(4-9), we get

$$\alpha_k = \frac{\Delta t_k}{\Delta t} \text{ for } k = 1 \text{ and } 2. \tag{4-12}$$

And from Eqs.(4-10) and (4-12) we obtain the following relation

$$\alpha_1 + \alpha_2 = 1 \tag{4-13}$$

which is the consequence that the averaged interface density function α_S is zero. We note here that α_S is defined in parallel with Eqs.(4-7) and (4-11) as

$$\alpha_S \equiv \lim_{\delta\to 0} \frac{1}{\Delta t} \int_{[\Delta t]} M_S\left(\boldsymbol{x}_0, t\right) dt. \tag{4-14}$$

The function α_k, which appears only after the integral operation, is a fundamental parameter in studying the time averaged field equations. Physically α_k represents a probability of finding the kth-phase, thus it expresses the geometrical (static) importance of that phase. Hereafter we call α_k as a *local time fraction* or a *local void fraction* of the kth-phase.

1.3 Time average and weighted mean values

In this section, we define the time average and weighted mean values of functions associated with two-phase flow fields.

Time Average

The Eulerian time average of the general function F is defined by

$$\overline{F}\left(\boldsymbol{x}_0, t_0\right) \equiv \lim_{\delta\to 0} \frac{1}{\Delta t} \int_{[\Delta t]_T} F\left(\boldsymbol{x}_0, t\right) dt. \tag{4-15}$$

Hereafter, the symbol $\overline{F}$ denotes the mathematical operation defined in the right-hand side of Eq.(4-15). Similarly, the mean value of the kth-phase general function F_k is given by

$$\overline{F_k}\left(\boldsymbol{x}_0, t_0\right) \equiv \lim_{\delta\to 0} \frac{1}{\Delta t} \int_{[\Delta t]_T} F_k\left(\boldsymbol{x}_0, t\right) dt. \tag{4-16}$$

Because of the special property associated with M_k it can be shown that

$$\alpha_k = \overline{M_k}. \tag{4-17}$$

In view of Eqs.(4-8), (4-15) and (4-16) we have

$$\overline{F} = \lim_{\delta \to 0} \left\{ \frac{1}{\Delta t} \int_{[\Delta t]_T} F_1\left(\boldsymbol{x}_0, t\right) dt + \frac{1}{\Delta t} \int_{[\Delta t]_T} F_2\left(\boldsymbol{x}_0, t\right) dt \right\}. \tag{4-18}$$

Hence we obtain an important relation

$$\overline{F} = \overline{F_1} + \overline{F_2}. \tag{4-19}$$

The functions of F_1 and F_2 are directly related to instant, local physical or flow variables of each phase; however, F_1 and F_2 are averaged over the total time interval Δt . Thus, they can be considered as superficially averaged values. From this point of view, we introduce various weighted mean values that preserve some of the important characteristics of original variables.

We start from a general case, then proceed to special cases. Hence by taking a non-zero scalar weight function *w*, we define the general weighted mean value of F as

$$\overline{\overline{F}}^{w} \equiv \frac{\overline{wF}}{\overline{w}} \tag{4-20}$$

where the function w also belongs to the group of a general function F defined in the previous section. Then it follows that the weighting function for each phase can also be defined through Eq.(4-8).

Consequently a general phase weighted mean value in parallel with Eq.(4-20) should be

$$\overline{\overline{F_k}}^{w_k} \equiv \frac{\overline{w_k F_k}}{\overline{w_k}}. \tag{4-21}$$

It follows that

$$\overline{F}^{w} = \frac{\sum_{k=1}^{2}\overline{w_k F_k}}{\overline{w}} = \frac{\sum_{k=1}^{2}\overline{w_k}\,\overline{F}^{w_k}}{\sum_{k=1}^{2}\overline{w_k}} \tag{4-22}$$

which relates the mixture and the phase mean values. Since these formulas are too general, we discuss in below some of the important special cases.

Phase Average $\overline{\overline{F_k}}$

The most natural mean value associated with each phase can be defined by taking the phase density function M_k as a weighting function in Eq.(4-21), hence we have

$$\overline{\overline{F_k}} \equiv \frac{\overline{M_k F_k}}{\overline{M_k}} = \frac{\overline{F_k}}{\alpha_k} = \frac{1}{\Delta t_k}\int_{[\Delta t]_k} F_k\, dt \tag{4-23}$$

where we used Eqs.(4-7), (4-8) and (4-17). As it is evident from the definition, the phase average denoted by F_k represents the simple average in the time interval $[\Delta t]_k$ of the phase. Hence, we have

$$\overline{F} = \sum_{k=1}^{2}\overline{F_k} = \sum_{k=1}^{2}\alpha_k \overline{\overline{F_k}}. \tag{4-24}$$

Mass Weighted Mean Value $\widehat{\psi}$ *and* $\widehat{\psi_k}$

In general, the volume, momentum, energy and entropy, etc. are considered to be extensive variables (Callen, 1960). If the function F is taken as a quantity per unit volume of the extensive characteristic, then they can also be expressed in terms of the variable per unit mass ψ as

$$F = \rho\psi \tag{4-25}$$

where ρ is the local instant fluid density. Hence, the properties for each phase F_k are given by

$$F_k = \rho_k \psi_k \quad k = 1 \text{ or } 2. \tag{4-26}$$

Here ρ_k and ψ_k denote k^{th}-phase local instant density and a quantity per unit mass, respectively. Then the appropriate mean values for ψ and ψ_k should be weighted by the densities as

$$\widehat{\psi} = \frac{\overline{\rho\psi}}{\overline{\rho}} \tag{4-27}$$

and

$$\widehat{\psi_k} = \frac{\overline{\rho_k \psi_k}}{\overline{\rho_k}} = \frac{\overline{\overline{\rho_k \psi_k}}}{\overline{\overline{\rho_k}}}. \tag{4-28}$$

The most important reason of weighting ψ by the density ρ is that the quantities represented by ψ are an additive set function of mass.

From the definition of the mass weighted mean values, we have

$$\overline{\rho\psi} = \sum_{k=1}^{2} \overline{\rho_k \psi_k}. \tag{4-29}$$

Hence, we obtain the most important relation between the mass weighted mixture property and that of two phase as

$$\widehat{\psi} = \frac{\sum_{k=1}^{2} \alpha_k \overline{\overline{\rho_k}} \widehat{\psi_k}}{\sum_{k=1}^{2} \alpha_k \overline{\overline{\rho_k}}} = \frac{\sum_{k=1}^{2} \overline{\rho_k} \widehat{\psi_k}}{\sum_{k=1}^{2} \overline{\rho_k}}. \tag{4-30}$$

We note here that the above result is analogous to the definitions used in the thermo-mechanical theory of diffusion (Truesdell, 1969) and in Maxwell's equation of transfer in the kinetic theory of gases (Maxwell, 1867). In particular, since the density is a property per unit volume, we have

$$\overline{\rho} = \sum_{k=1}^{2} \overline{\rho_k} = \sum_{k=1}^{2} \alpha_k \overline{\overline{\rho_k}}. \tag{4-31}$$

Fundamental Hypothesis on Smoothness of Mean Values

Our purpose of the averaging is to transform two phases, alternately occupying a point with discontinuities at interfaces, into two simultaneous continua. Consequently, the assumption on the continuity of derivatives is required. Thus, let us introduce here a fundamental hypothesis on the smoothness of the mean values $\overline{F}$ and $\overline{F_k}$.

By considering a macroscopic process in terms of the mean values, it is assumed that they are sufficiently smooth to have higher order derivatives as necessary except at some isolated singularities, if the time constant of the process is sufficiently larger than Δt. In other words any changes of mean values within the time interval Δt are considered to be infinitesimal. This can be visualized by considering the time differential operator in the average field as a finite difference operator with the time increment $\delta t \to \Delta t$ and not $\delta t \to 0$. If we apply averaging to a mean value we get

$$\overline{\left(\overline{F}\right)} = \frac{1}{\Delta t} \lim_{\delta \to 0} \int_{[\Delta t]_T} \overline{F}\left(\boldsymbol{x}_0, t\right) dt. \tag{4-32}$$

However, since $\overline{F}$ is continuous, we obtain from the integral mean value theorem

$$\overline{\left(\overline{F}\right)} = \overline{F}\left(\boldsymbol{x}_0, \tau_0\right) \tag{4-33}$$

where

$$t_0 - \frac{\Delta t}{2} \leqq \tau_0 \leqq t_0 + \frac{\Delta t}{2}.$$

Thus, in analogy with the fundamental hypothesis, it is assumed that in the macroscopic fields we have

$$\overline{\left(\overline{F}\right)} = \overline{F}\left(\boldsymbol{x}_0, t_0\right). \tag{4-34}$$

It states that the averaging does not alter the mean values. Then it is straightforward that we obtain

$$\begin{aligned} &\overline{\left(\overline{F}^{w}\right)}^{w} = \overline{F}^{w}; \quad \overline{\left(\overline{F}^{w}\right)} = \overline{F}^{w} \\ &\overline{\left(\overline{F_k}^{w_k}\right)}^{w_k} = \overline{F_k}^{w_k}; \quad \overline{\overline{\left(\overline{F_k}^{w_k}\right)}} = \overline{F_k}^{w_k}. \end{aligned} \tag{4-35}$$

And for a constant C we have

$$\overline{C} = \overline{\overline{C_k}} = \overline{C_k}^{w} = \overline{C_k}^{w_k} = \frac{\overline{C_k}}{\alpha_k} = C. \tag{4-36}$$

Fluctuating Component

As in the analyses of turbulent flows, it is a prime importance to introduce fluctuating components of variables in order to take into account these effects statistically. In general, they are defined as a difference between a local instant variable and its weighted mean value, thus we have

$$F_k' \equiv F_k - \overline{F_k}^{w_k}. \tag{4-37}$$

Since once a variable F_k is specified then the form of the weighted average will be given, the fluctuating component can be defined uniquely. From Eqs.(4-35) and (4-37) we immediately obtain

$$\overline{F_k'}^{w_k} = 0. \tag{4-38}$$

Furthermore, the mean value of the fluctuating component can be related to other parameters as

$$\overline{F_k} = \alpha_k \overline{\overline{F_k}} = \alpha_k \left(\overline{F_k}^{w_k} + \overline{\overline{F_k'}} \right) \tag{4-39}$$

and

$$\overline{F_k'} = \overline{F_k} - \overline{F_k}^{w_k} = \alpha_k \overline{\overline{F_k'}} - \left(1 - \alpha_k\right) \overline{F_k}^{w_k}. \tag{4-40}$$

These relations will be used in analyzing the two-phase turbulent fluxes in the averaged field equations.

1.4 Time average of derivatives

In this section, the relation between the average of the derivative and the derivative of the average is obtained. By the time derivative of the average, we mean

$$\frac{\partial \overline{F}\left(\boldsymbol{x}_0, t_0\right)}{\partial t_0} = \frac{\partial}{\partial t_0} \left\{ \frac{1}{\Delta t} \lim_{\delta \to 0} \int_{[\Delta t]_T} F\left(\boldsymbol{x}_0, t\right) dt \right\}. \tag{4-41}$$

Since the domain of the integration is discontinuous, we subdivide it and apply the Leibnitz rule to Eq.(4-41). Thus we have

$$\frac{\partial \overline{F}}{\partial t_0} = \frac{1}{\Delta t} \lim_{\delta \to 0} \left\{ \int_{[\Delta t]_T} \frac{\partial F(\boldsymbol{x}_0, t)}{\partial t} dt \right. \\ \left. + \sum_j \left[F(\boldsymbol{x}_0, t_j + \varepsilon_j) - F(\boldsymbol{x}_0, t_j - \varepsilon_j) \right] \right\}. \tag{4-42}$$

In view of Eq.(4-2) we define

$$\lim_{\varepsilon_j \to 0} t_j \pm \varepsilon_j \equiv t_j^{\pm} \tag{4-43}$$

and correspondingly

$$\lim_{\varepsilon_j \to 0} F(\boldsymbol{x}_0, t_j \pm \varepsilon_j) = F(\boldsymbol{x}_0, t_j^{\pm}) \equiv F^{\pm}(\boldsymbol{x}_0, t_j). \tag{4-44}$$

Hence, from Eq.(4-42) we get

$$\overline{\frac{\partial F}{\partial t}} = \frac{\partial \overline{F}(\boldsymbol{x}_0, t_0)}{\partial t_0} - \sum_j \frac{1}{\Delta t} \left\{ F^{+}(\boldsymbol{x}_0, t_j) - F^{-}(\boldsymbol{x}_0, t_j) \right\}. \tag{4-45}$$

The average of the space derivative at $\boldsymbol{x} = \boldsymbol{x}_0$ can be written as

$$\overline{\nabla F} = \lim_{\delta \to 0} \frac{1}{\Delta t} \int_{[\Delta t]_T} \nabla F(\boldsymbol{x}_0, t) dt. \tag{4-46}$$

Therefore, by applying the Leibnitz rule we obtain

$$\overline{\nabla F} = \lim_{\delta \to 0} \left[\nabla \left\{ \frac{1}{\Delta t} \int_{[\Delta t]_T} F(\boldsymbol{x}_0, t) dt \right\} \right. \\ + \frac{1}{\Delta t} \sum_j \left\{ -\nabla (t_j - \varepsilon_j) F(\boldsymbol{x}_0, t_j - \varepsilon_j) \right. \\ \left. \left. + \nabla (t_j + \varepsilon_j) F(\boldsymbol{x}_0, t_j + \varepsilon_j) \right\} \right]. \tag{4-47}$$

The physical significance of the last term of the right-hand side of the above equation is not clear and should be examined in more detail. For this purpose we introduce the equation of a surface given by

$$f(x,y,z,t)=0 \tag{4-48}$$

which passes the point $\boldsymbol{x}_0$ at $t=t_j$. Then

$$df=(\nabla f)\cdot d\boldsymbol{x}_0+\frac{\partial f}{\partial t}dt_j=0 \tag{4-49}$$

Thus, in view of Eq.(4-43), we have

$$\nabla t_j=-\frac{\nabla f}{\dfrac{\partial f}{\partial t}}. \tag{4-50}$$

However, the normal vector and the displacement velocity (Truesdell and Toupin, 1960) are given by

$$\boldsymbol{n}=\frac{\nabla f}{|\nabla f|} \tag{4-51}$$

and

$$\boldsymbol{v}_i\cdot\boldsymbol{n}=v_{ni}=-\frac{\dfrac{\partial f}{\partial t}}{|\nabla f|}. \tag{4-52}$$

Hence, by eliminating the surface function in Eq.(4-50) we obtain

$$\nabla t_j=\frac{\boldsymbol{n}}{\boldsymbol{v}_i\cdot\boldsymbol{n}}. \tag{4-53}$$

Then Eq.(4-47) becomes

$$\overline{\nabla F} = \nabla \overline{F}\left(\boldsymbol{x}_0, t_0\right)$$
$$+\sum_j \frac{1}{\Delta t}\frac{1}{v_{ni}}\left\{\boldsymbol{n}^{+}F\left(\boldsymbol{x}_0, t_j^{+}\right)+\boldsymbol{n}^{-}F\left(\boldsymbol{x}_0, t_j^{-}\right)\right\}. \tag{4-54}$$

The unit normal vector of the interface is defined such that

$$\boldsymbol{n}\cdot\boldsymbol{v}_i \equiv v_{ni} \geq 0. \tag{4-55}$$

Then

$$\boldsymbol{n}^{+}\cdot\boldsymbol{v}_i \geq 0; \quad \boldsymbol{n}^{-}\cdot\boldsymbol{v}_i \leq 0 \tag{4-56}$$

where $\boldsymbol{n}^{+}$ and $\boldsymbol{n}^{-}$ correspond to the limit outward normal vector of the fluid at each side of the interface.

Using the simplified notations, Eqs.(4-45) and (4-54) become

$$\overline{\frac{\partial F}{\partial t}} = \frac{\partial \overline{F}\left(\boldsymbol{x}_0, t_0\right)}{\partial t_0} - \frac{1}{\Delta t}\sum_j \frac{1}{v_{ni}}\left(F^{+}\boldsymbol{n}^{+}\cdot\boldsymbol{v}_i + F^{-}\boldsymbol{n}^{-}\cdot\boldsymbol{v}_i\right) \tag{4-57}$$

$$\overline{\nabla F} = \nabla \overline{F}\left(\boldsymbol{x}_0, t_0\right) + \frac{1}{\Delta t}\sum_j \frac{1}{v_{ni}}\left(\boldsymbol{n}^{+}F^{+} + \boldsymbol{n}^{-}F^{-}\right). \tag{4-58}$$

We note here that the function F can be a scalar, a vector or a tensor, and ∇ operator can be a divergence or a gradient operator with proper tensorial operation between $\boldsymbol{n}^{\pm}$ and $F^{\pm}$. These above two transformations and the definitions of various mean values are the basic tools to be used to obtain the macroscopic field equations in terms of mean values. In contrast to the case without discontinuities in the function F, the above transformations show the important contributions made by the moving interfaces in relating the average of the derivatives to the derivatives of the average.

As corollaries of Eqs.(4-57) and (4-58) we have

$$\overline{\frac{\partial F_k}{\partial t}} = \frac{\partial \overline{F_k}}{\partial t_0} - \frac{1}{\Delta t}\sum_j \frac{1}{v_{ni}}\left(F_k\boldsymbol{n}_k\cdot\boldsymbol{v}_i\right) \tag{4-59}$$

$$\overline{\nabla F_k} = \nabla \overline{F_k} + \frac{1}{\Delta t}\sum_j \frac{1}{v_{ni}}\left(\boldsymbol{n}_k F_k\right). \tag{4-60}$$

The special case of the above equations is for the time fraction α_k. In this case, directly from the original definition Eq.(4-11), we obtain

$$\frac{\partial \alpha_k}{\partial t} = \frac{1}{\Delta t}\sum_j \frac{1}{v_{ni}}\left(\boldsymbol{n}_k \cdot \boldsymbol{v}_i\right) \tag{4-61}$$

$$\nabla \alpha_k = -\frac{1}{\Delta t}\sum_j \frac{\boldsymbol{n}_k}{v_{ni}}. \tag{4-62}$$

These equations clearly demonstrate the existence of *microscopic singularities* explained in connection with the fundamental hypothesis of smoothness in the Section 1.3 of Chapter 4.

1.5 Concentrations and mixture properties

The local time fraction α_k has been defined in the Section 1.2 of Chapter 4. The parameter α_k signifies the physical events and the structures of the two-phase flow at any particular point. Therefore, it is anticipated that the local time fraction α_k appears in all field equations. Furthermore, as the two-phase constitutive laws should also depend on the physical structures of the flow, its importance in deriving these laws is expected.

Apart from the local time fraction α_k, another concentration based on mass can be defined. In analogy with the theory of diffusion, the mass fraction c_k is given by

$$c_k \equiv \frac{\overline{\rho_k}}{\overline{\rho}} = \frac{\alpha_k \overline{\overline{\rho_k}}}{\alpha_1 \overline{\overline{\rho_1}} + \alpha_2 \overline{\overline{\rho_2}}} = \frac{\alpha_k \overline{\overline{\rho_k}}}{\rho_m}. \tag{4-63}$$

It is the measure of the relative significance of the kth-phase mass with respect to the mixture mass. Since the momentum and energy are an additive set function of mass, it is expected that the mixture properties of these variables can be expressed by those of each phase with the mass fraction c_k as a weighting function. From Eq.(4-63) we have

$$\sum_{k=1}^{2} c_k = 1 \tag{4-64}$$

and

$$\frac{1}{\rho_m} = \sum_{k=1}^{2} \frac{c_k}{\overline{\overline{\rho_k}}}. \tag{4-65}$$

The above two parameters, namely, α_k and c_k, are static concentrations, because they represent the events, structures or masses in a two-phase flow. Furthermore, kinematic concentrations are defined through various mean velocity fields, thus they represent the relative importance of the amount of flows or fluxes. Because of this basic characteristic of the kinematic variables, they cannot generally be defined in a 3-dimensional formulation, since the flows and fluxes are vector and not scalar quantities. However, they can easily be defined for a one-dimensional model. For example, the quality x has been frequently used in the literature. In what follows, we define important mixture properties.

1) The mixture density

$$\rho_m = \sum_{k=1}^{2} \alpha_k \overline{\overline{\rho_k}} \tag{4-66}$$

where

$$\overline{\overline{\rho_k}} = \frac{\overline{\rho_k}}{\alpha_k}. \tag{4-67}$$

2) The mixture center of mass velocity

$$\boldsymbol{v}_m = \frac{\sum_{k=1}^{2} \alpha_k \overline{\overline{\rho_k}} \widehat{\boldsymbol{v}_k}}{\rho_m} = \sum_{k=1}^{2} c_k \widehat{\boldsymbol{v}_k} \tag{4-68}$$

where

$$\widehat{\boldsymbol{v}_k} = \frac{\overline{\overline{\rho_k \boldsymbol{v}_k}}}{\overline{\overline{\rho_k}}} = \frac{\overline{\rho_k \boldsymbol{v}_k}}{\overline{\rho_k}}. \tag{4-69}$$

3) The mixture energy

$$u_m = \frac{\sum_{k=1}^{2} \alpha_k \overline{\overline{\rho_k}} \widehat{u_k}}{\rho_m} = \sum_{k=1}^{2} c_k \widehat{u_k} \tag{4-70}$$

where

$$\widehat{u_k} = \frac{\overline{\overline{\rho_k u_k}}}{\overline{\overline{\rho_k}}} = \frac{\overline{\rho_k u_k}}{\overline{\rho_k}}. \tag{4-71}$$

4) The mixture pressure

$$p_m = \sum_{k=1}^{2} \alpha_k \overline{\overline{p_k}} \tag{4-72}$$

where

$$\overline{\overline{p_k}} = \frac{\overline{p_k}}{\alpha_k}. \tag{4-73}$$

5) The mixture enthalpy

$$i_m = \frac{\sum_{k=1}^{2} \alpha_k \overline{\overline{\rho_k}} \widehat{i_k}}{\rho_m} = \sum_{k=1}^{2} c_k \widehat{i_k} \tag{4-74}$$

where

$$\widehat{i_k} = \frac{\overline{\overline{\rho_k i_k}}}{\overline{\overline{\rho_k}}} = \frac{\overline{\rho_k i_k}}{\overline{\rho_k}}. \tag{4-75}$$

Then it can be shown that

$$i_m = u_m + \frac{p_m}{\rho_m} \tag{4-76}$$

and

$$\widehat{i_k} = \widehat{u_k} + \frac{\overline{\overline{p_k}}}{\overline{\overline{\rho_k}}}. \tag{4-77}$$

6) The mixture entropy

$$s_m = \frac{\sum_{k=1}^{2} \alpha_k \overline{\overline{\rho_k}}\, \widehat{s_k}}{\rho_m} = \sum_{k=1}^{2} c_k \widehat{s_k} \tag{4-78}$$

where

$$\widehat{s_k} = \frac{\overline{\overline{\rho_k s_k}}}{\overline{\overline{\rho_k}}} = \frac{\overline{\rho_k s_k}}{\overline{\rho_k}}. \tag{4-79}$$

7) The general mixture flux $\overline{\mathcal{J}}$

We recall here the general balance equation, Eq.(2-6), the generalized flux and the volume source defined in the Section 1.1 of Chapter 2. From the form of the balance equation, it is natural to define the mixture molecular diffusion flux $\overline{\mathcal{J}}$ as

$$\overline{\mathcal{J}} = \sum_{k=1}^{2} \alpha_k \overline{\overline{\mathcal{J}_k}} \tag{4-80}$$

where

$$\overline{\overline{\mathcal{J}_k}} = \frac{\overline{\mathcal{J}_k}}{\alpha_k}. \tag{4-81}$$

8) The mixture general source term ϕ_m

Since the source term ϕ is defined as the variable per unit mass, it should be weighed by the density. Hence, we have

$$\phi_m = \frac{\sum_{k=1}^{2} \alpha_k \overline{\overline{\rho_k}} \widehat{\phi_k}}{\rho_m} = \sum_{k=1}^{2} c_k \widehat{\phi_k} \tag{4-82}$$

where

$$\widehat{\phi_k} = \frac{\overline{\overline{\rho_k \phi_k}}}{\overline{\overline{\rho_k}}} = \frac{\overline{\rho_k \phi_k}}{\overline{\rho_k}}. \tag{4-83}$$

It can be seen that the variables based on unit mass are weighted by the mass concentrations, whereas the ones based on unit volume or surface are weighted by the time fractions.

1.6 Velocity field

In general, two-phase flow systems with transport of mass, momentum and energy are characterized by the existence of two different densities and velocities. Thus it is necessary to introduce two properly defined mean velocity fields in the formulation in order to take into account the effects of the relative motion between the phases, namely, the diffusion of mass, momentum, and energy. However, there are several velocity fields that are useful in analyzing various aspects of a two-phase flow problem. A selection of velocity fields for a particular problem depends upon characteristics and nature of the flow as well as on the forms of available constitutive laws. In what follows, we present these velocity fields that are important in studying various aspects of two-phase flow systems.

As it has been explained in the previous section, the definition of *the center of mass velocities* is based on the fundamental characteristic of linear momentum. First, we recall that it is an additive set function of mass. In other words, as it is well known as a fundamental theorem on the center of mass, the total momentum of a body is given by the momentum of the center of mass with the same mass as the body. It is the direct extension of the above idea into the averaging procedure that we obtain the mass-weighted mixture and phase velocities as proper mean velocities.

The concept of the center of mass in the time averaging is trivial and it has the form

$$\boldsymbol{x} = \frac{\lim\limits_{\delta\to 0}\int_{[\Delta t]_T} \rho\, \boldsymbol{x}_0 dt}{\lim\limits_{\delta\to 0}\int_{[\Delta t]_T} \rho\, dt}. \tag{4-84}$$

However $\boldsymbol{x}_0$ is kept constant during the integration, thus the center of mass is $\boldsymbol{x}_0$. Then the definition of the mixture density, Eq.(4-66), naturally follows. The fundamental theorem on the center of mass can be extended as

$$\lim_{\delta\to 0}\int_{[\Delta t]_T} \rho_m \boldsymbol{v}_m\, dt = \lim_{\delta\to 0}\int_{[\Delta t]_T} \rho\, \boldsymbol{v}\, dt. \tag{4-85}$$

Hence in view of the definitions of weighted mean values, it is straightforward to show that the center of mass velocities of the mixture and of each phase have been given correctly by Eqs.(4-68) and (4-69), respectively.

We define the relative velocity by

$$\boldsymbol{v}_r \equiv \widehat{\boldsymbol{v}_2} - \widehat{\boldsymbol{v}_1}. \tag{4-86}$$

And the *volumetric flux* of each phase are given by

$$\boldsymbol{j}_k \equiv \alpha_k \widehat{\boldsymbol{v}_k} \tag{4-87}$$

which can be considered as the velocity when one of the phases superficially occupies the entire interval Δt with the total amount of the flow fixed, and therefore it is also called the superficial velocity. Accordingly the mixture volumetric flux, namely, the velocity of the center of volume, is defined by

$$\boldsymbol{j} = \sum_{k=1}^{2} \boldsymbol{j}_k = \sum_{k=1}^{2} \alpha_k \widehat{\boldsymbol{v}_k}. \tag{4-88}$$

If the relative velocity between the phases exists, the velocities $\boldsymbol{v}_m$ and $\boldsymbol{j}$ are not equal because of the differences of the densities of the two phases. The *diffusion velocity* of each phase, namely, the relative velocity with respect to the mass center of the mixture, is defined by

$$\boldsymbol{V}_{km} = \widehat{\boldsymbol{v}_k} - \boldsymbol{v}_m \tag{4-89}$$

which are frequently used in the analyses of heterogeneous chemically reacting single-phase systems. The diffusion velocities can also be expressed by the relative velocity, though the symmetry between phases cannot be kept because of the definition, Eq.(4-86). Thus we have

$$\boldsymbol{V}_{1m} = -\frac{\alpha_2 \overline{\overline{\rho_2}}}{\rho_m} \boldsymbol{v}_r = -c_2 \boldsymbol{v}_r$$

and (4-90)

$$\boldsymbol{V}_{2m} = \frac{\alpha_1 \overline{\overline{\rho_1}}}{\rho_m} \boldsymbol{v}_r = c_1 \boldsymbol{v}_r.$$

In a two-phase flow system, the drift velocity of each phase, namely the relative velocity with respect to the center of volume, is important because the constitutive equations for these velocities in the mixture formulation is relatively simple and well developed (Zuber et al, 1964; Ishii, 1977). By definition, the *drift velocity* is given by

$$\boldsymbol{V}_{kj} = \widehat{\boldsymbol{v}_k} - \boldsymbol{j}. \tag{4-91}$$

In terms of the relative velocity, it becomes

$$\boldsymbol{V}_{1j} = -\alpha_2 \boldsymbol{v}_r$$

and (4-92)

$$\boldsymbol{V}_{2j} = \alpha_1 \boldsymbol{v}_r.$$

Several important relations between the above velocities can be obtained directly from the definitions. For example, from Eqs.(4-88), (4-89) and (4-90) we get

$$\boldsymbol{j} = \boldsymbol{v}_m + \alpha_1 \alpha_2 \frac{\left(\overline{\overline{\rho_1}} - \overline{\overline{\rho_2}}\right)}{\rho_m} \boldsymbol{v}_r$$

or (4-93)

$$= \boldsymbol{v}_m - \alpha_1 \frac{\left(\overline{\overline{\rho_1}} - \overline{\overline{\rho_2}}\right)}{\rho_m} \boldsymbol{V}_{1j}.$$

From Eqs.(4-90) and (4-92) we have

$$\sum_{k=1}^{2} c_k \boldsymbol{V}_{km} = 0 \tag{4-94}$$

and

$$\sum_{k=1}^{2} \alpha_k \boldsymbol{V}_{kj} = 0. \tag{4-95}$$

Finally, we note that if the relative velocity is zero, then

$$\boldsymbol{V}_{1m} = \boldsymbol{V}_{2m} = \boldsymbol{V}_{1j} = \boldsymbol{V}_{2j} = \boldsymbol{v}_r = 0 \tag{4-96}$$

and thus

$$\widehat{\boldsymbol{v}_1} = \widehat{\boldsymbol{v}_2} = \boldsymbol{v}_m = \boldsymbol{j} \tag{4-97}$$

which characterizes the homogenous velocity field.

Generally speaking, the velocities based on the center of mass are important for dynamic analyses because of the fundamental theorem of center of mass. However, the velocities based on the center of volume, namely, volumetric fluxes, are useful for kinematic analyses. This is particularly true if each phase has constant properties such as the constant densities, internal energies or enthalpies, etc.

1.7 Fundamental identity

In developing a drift-flux model based on the mixture properties, it is necessary to express an average convective flux by various mean values. In this section, we derive this relation directly from the definitions. From Eq.(4-29), the convective flux of the mixture becomes

$$\overline{\rho\psi\boldsymbol{v}} = \sum_{k=1}^{2} \overline{\rho_k \psi_k \boldsymbol{v}_k} = \sum_{k=1}^{2} \alpha_k \overline{\overline{\rho_k \psi_k \boldsymbol{v}_k}}. \tag{4-98}$$

Our purpose here is to split the right-hand side of the equation into the terms expressed by the mean values and the ones representing the statistical effects of the fluctuating components. Since the mean values of ψ and $\boldsymbol{v}$ are weighted by mass, the fluctuating components are given by Eq.(4-37) as

$$\rho_k = \overline{\overline{\rho_k}} + \rho_k', \quad \psi_k = \widehat{\psi_k} + \psi_k', \quad \boldsymbol{v}_k = \widehat{\boldsymbol{v}_k} + \boldsymbol{v}_k' \tag{4-99}$$

with

$$\overline{\overline{\rho_k'}} = 0, \quad \overline{\overline{\rho_k \psi_k'}} = 0, \quad \overline{\overline{\rho_k \boldsymbol{v}_k'}} = 0. \tag{4-100}$$

By substituting Eqs.(4-99) and (4-100) into Eq.(4-98) we obtain

$$\overline{\rho\psi\boldsymbol{v}} = \sum_{k=1}^{2} \alpha_k \overline{\overline{\rho_k}} \widehat{\psi_k} \widehat{\boldsymbol{v}_k} + \sum_{k=1}^{2} \alpha_k \overline{\overline{\rho_k \psi_k' \boldsymbol{v}_k'}}. \tag{4-101}$$

By using the definition of the mixture properties and that of the diffusion velocities, the above equation reduces to

$$\overline{\rho\psi\boldsymbol{v}} = \rho_m \psi_m \boldsymbol{v}_m + \sum_{k=1}^{2} \alpha_k \overline{\overline{\rho_k}} \widehat{\psi_k} \boldsymbol{V}_{km} + \sum_{k=1}^{2} \alpha_k \overline{\overline{\rho_k \psi_k' \boldsymbol{v}_k'}}. \tag{4-102}$$

It shows that the average convective flux can be split into three parts according to the different transport mechanisms: the mixture transport based on the mixture properties; the diffusion transport of $\widehat{\psi_k}$ due to the difference of the phase velocities; and the transport due to the two-phase and turbulent fluctuations. In order to distinguish these last two transport mechanisms we introduce special fluxes associated with them. Hence, we define the diffusion flux $\mathcal{J}^D$ as

$$\mathcal{J}^D \equiv \sum_{k=1}^{2} \alpha_k \overline{\overline{\rho_k}} \widehat{\psi_k} \boldsymbol{V}_{km} = \frac{\overline{\overline{\rho_1}}\,\overline{\overline{\rho_2}}}{\rho_m} \sum_{k=1}^{2} \alpha_k \widehat{\psi_k} \boldsymbol{V}_{kj} \tag{4-103}$$

whereas the covariance or the turbulent flux $\mathcal{J}_k^T$ is defined as

$$\mathcal{J}_k^T \equiv \overline{\overline{\rho_k \psi_k' \boldsymbol{v}_k'}}. \tag{4-104}$$

Thus, the mixture turbulent flux should be

$$\mathcal{J}^T = \sum_{k=1}^{2} \alpha_k \mathcal{J}_k^T = \sum_{k=1}^{2} \alpha_k \overline{\overline{\rho_k \psi_k' \boldsymbol{v}_k'}}. \tag{4-105}$$

By substituting Eqs.(4-103) and (4-105) into Eq.(4-101) we obtain a fundamental identity

$$\overline{\rho\psi\boldsymbol{v}} = \rho_m\psi_m\boldsymbol{v}_m + \mathcal{J}^D + \mathcal{J}^T = \sum_{k=1}^{2}\alpha_k\overline{\overline{\rho_k}}\widehat{\psi_k}\widehat{\boldsymbol{v}_k} + \sum_{k=1}^{2}\alpha_k\mathcal{J}_k^T. \quad \text{(4-106)}$$

Hence, from the definitions of the mean values we have

$$\frac{\partial\overline{\rho\psi}}{\partial t} + \nabla\cdot\left(\overline{\rho\psi\boldsymbol{v}}\right) = \frac{\partial\rho_m\psi_m}{\partial t} + \nabla\cdot\left(\rho_m\psi_m\boldsymbol{v}_m\right) + \nabla\cdot\left(\mathcal{J}^D + \mathcal{J}^T\right) \quad \text{(4-107)}$$

and for each phase we get

$$\frac{\partial\overline{\rho_k\psi_k}}{\partial t} + \nabla\cdot\overline{\left(\rho_k\psi_k\boldsymbol{v}_k\right)} = \frac{\partial\alpha_k\overline{\overline{\rho_k}}\widehat{\psi_k}}{\partial t} + \nabla\cdot\left(\alpha_k\overline{\overline{\rho_k}}\widehat{\psi_k}\widehat{\boldsymbol{v}_k}\right) + \nabla\cdot\left(\alpha_k\mathcal{J}_k^T\right). \quad \text{(4-108)}$$

We note that Eq.(4-108) shows a simple analogy with a single-phase turbulent flow averaging, therefore, the last term is called the Reynolds flux.

In view of Eq.(4-107), it can be seen that the left-hand side of the equation is not expressed by the averages of the derivatives, but by the derivatives of the averages. However, when we apply the Eulerian temporal averaging to the local formulation of two-phase flows, we first encounter with the averages of the derivatives. Now we recall that the important transformations between these two operations have been derived in the Section 1.4 of Chapter 4. Thus, by substituting Eqs.(4-57) and (4-58) into Eq.(4-107) we obtain

$$\overline{\left(\frac{\partial\rho\psi}{\partial t}\right)} + \overline{\nabla\cdot\left(\rho\psi\boldsymbol{v}\right)} = \frac{\partial\rho_m\psi_m}{\partial t} + \nabla\cdot\left(\rho_m\psi_m\boldsymbol{v}_m\right) + \nabla\cdot\left(\mathcal{J}^D + \mathcal{J}^T\right) + \frac{1}{\Delta t}\sum_j\left[\frac{1}{v_{ni}}\sum_{k=1}^{2}\left\{\boldsymbol{n}_k\cdot\rho_k\left(\boldsymbol{v}_k - \boldsymbol{v}_i\right)\psi_k\right\}\right]. \quad \text{(4-109)}$$

A similar relation for an individual phase follows, thus we have

$$\overline{\left(\frac{\partial \rho_k \psi_k}{\partial t}\right)} + \overline{\nabla \cdot \left(\rho_k \psi_k \boldsymbol{v}_k\right)} = \frac{\partial \alpha_k \overline{\overline{\rho_k}} \widehat{\psi_k}}{\partial t} + \nabla \cdot \left(\alpha_k \overline{\overline{\rho_k}} \widehat{\psi_k} \widehat{\boldsymbol{v}_k}\right)$$
$$+\nabla \cdot \left(\alpha_k \mathscr{J}_k^T\right) + \frac{1}{\Delta t} \sum_j \left\{\frac{1}{v_{ni}} \boldsymbol{n}_k \cdot \rho_k \left(\boldsymbol{v}_k - \boldsymbol{v}_i\right) \psi_k\right\}. \tag{4-110}$$

These above two equations show important contributions made by the interfacial transfer in addition to the statistical effects of fluctuations. Furthermore, in the mixture average, Eq.(4-107), the diffusion term $\mathscr{J}^D$ appears due to the differences in phase velocities.

Furthermore, from the definitions of mean values and of the diffusion flux we have

$$\frac{\partial \rho_m \psi_m}{\partial t} + \nabla \cdot \left(\rho_m \psi_m \boldsymbol{v}_m\right) + \nabla \cdot \mathscr{J}^D$$
$$= \sum_{k=1}^{2} \left\{\frac{\partial \alpha_k \overline{\overline{\rho_k}} \widehat{\psi_k}}{\partial t} + \nabla \cdot \left(\alpha_k \overline{\overline{\rho_k}} \widehat{\psi_k} \widehat{\boldsymbol{v}_k}\right)\right\}. \tag{4-111}$$

This relation enables us to transform the field equations for the two-fluid model to the ones for the drift-flux model.

Chapter 5

TIME AVERAGED BALANCE EQUATION

1.1 General balance equation

In the preceding chapter, the important definitions and basic relations between them have been given. We now apply them to the time averaging of the balance laws in the two-phase flow media. As it has been explained in the Section 1.1 of Chapter 4, it was necessary to introduce several sets of time intervals because of the discontinuous changes in the nature of fluid surrounding the point of average. Thus the domain of averaging has been divided into $[\Delta t]_S$ and $[\Delta t]_T$. During $[\Delta t]_T$, the standard balance equation (2-6) holds, since the fluid occupying the point $\boldsymbol{x}_0$ can be considered as a continuum. However, in $[\Delta t]_S$ the interfacial balance equation, namely, the jump condition of the Section 1.2 of Chapter 2, is valid because the characteristics of the interface dominates in this time interval.

Our purpose here is to average the balance laws in time by properly assigning appropriate balance equations of the bulk fluid and of an interface. Now let us first proceed with an analysis in $[\Delta t]_T$ when the point of averaging is occupied by one of the phases and not by an interface. For time $t \in [\Delta t]_T = [\Delta t]_1 + [\Delta t]_2$, we consider the balance of a quantity ψ in the following form

$$B_V = \frac{\partial \rho\psi}{\partial t} + \nabla\cdot(\rho\psi\boldsymbol{v}) + \nabla\cdot\mathcal{J} - \rho\phi = 0. \tag{5-1}$$

Here $\mathcal{J}$ and ϕ represent the generalized tensor efflux and the source of ψ, respectively. Since it is multiplied by the density ρ, the quantity ψ is expressed as the quantity per unit mass. Thus the above equation itself is a mathematical statement of the balance of the quantity *in a unit volume*. This is an important point to be remembered when we compare it to the surface balance equation in the course of the time averaging. In order to keep the volumetric origin of Eq.(5-1), the balance is denoted by B_V. Furthermore, we recall here that when Eq.(5-1) is applied for each phase, the subscripts that differentiate two fluids should appear with variables. For time $t \in [\Delta t]_S$, a different kind of balance equation should be used due to the special characteristics of an interface. Since the detailed derivation of the interfacial balance equation has been given in the Section 1.2 of Chapter 2, we simply recall those results. Thus, from Eq.(2-67), the balance of matter ψ at the interface becomes

$$\begin{aligned} B_S = \frac{1}{\delta}\Bigg\{ & \frac{d_s}{dt}\psi_a + \psi_a \nabla_s \cdot \boldsymbol{v}_i \\ & -\sum_{k=1}^{2}\left[\rho_k \boldsymbol{n}_k \cdot \left(\boldsymbol{v}_k - \boldsymbol{v}_i\right)\psi_k + \boldsymbol{n}_k \cdot \mathscr{J}_k\right] \\ & + g_{\ln} A^{\alpha\beta}\left(t_\alpha^n \mathscr{J}_a^{l\cdot}\right)_{,\beta} - \phi_a \Bigg\} = 0. \end{aligned} \tag{5-2}$$

In order to obtain Eq.(5-2), we have divided Eq.(2-67) by the interfacial thickness δ. Consequently, the above equation is the balance of ψ in an unit volume of the region.

The averaged balance per volume can be obtained by integrating the proper balance equations in the time domain. Now let us express the balance equation in general by

$$B = 0 \tag{5-3}$$

where

$$B = B_V = 0 \quad \text{for } t \in [\Delta t]_T \tag{5-4}$$

$$B = B_S = 0 \quad \text{for } t \in [\Delta t]_S. \tag{5-5}$$

By taking a time average of B we have

$$\frac{1}{\Delta t}\int_{[\Delta t]} B dt = 0. \tag{5-6}$$

Following the assumption previously made, we approximate the interfacial region with a singular surface by taking the limit $\delta \to 0$. Thus Eq.(5-6) becomes

$$\frac{1}{\Delta t}\lim_{\delta\to 0}\int_{[\Delta t]_T} B_V\,dt + \frac{1}{\Delta t}\lim_{\delta\to 0}\int_{[\Delta t]_S} B_S\,dt = 0. \tag{5-7}$$

The first part can be expressed in terms of the mean values defined in the Section 1.3 of Chapter 4, hence from Eq.(5-1) with Eqs.(4-25), (4-57) and (4-58) we obtain

$$\begin{aligned}&\frac{1}{\Delta t}\lim_{\delta\to 0}\int_{[\Delta t]_T} B_V\,dt = \frac{\partial\overline{\rho\psi}}{\partial t} + \nabla\cdot\overline{\rho\psi\boldsymbol{v}} + \nabla\cdot\overline{\mathcal{J}} - \overline{\rho\phi}\\ &+\frac{1}{\Delta t}\sum_j\left\{\frac{1}{v_{ni}}\sum_{k=1}^{2}\left[\boldsymbol{n}_k\cdot\rho_k\left(\boldsymbol{v}_k-\boldsymbol{v}_i\right)\psi_k+\boldsymbol{n}_k\cdot\mathcal{J}_k\right]\right\}=0\end{aligned} \tag{5-8}$$

or in terms of the mixture properties

$$\begin{aligned}&\frac{1}{\Delta t}\lim_{\delta\to 0}\int_{[\Delta t]_T} B_V\,dt = \frac{\partial\rho_m\psi_m}{\partial t} + \nabla\cdot\left(\rho_m\psi_m\boldsymbol{v}_m\right)\\ &+\nabla\cdot\left(\overline{\mathcal{J}}+\mathcal{J}^D+\mathcal{J}^T\right)-\rho_m\phi_m\\ &+\frac{1}{\Delta t}\sum_j\left\{\frac{1}{v_{ni}}\sum_{k=1}^{2}\left[\boldsymbol{n}_k\cdot\rho_k\left(\boldsymbol{v}_k-\boldsymbol{v}_i\right)\psi_k+\boldsymbol{n}_k\cdot\mathcal{J}_k\right]\right\}=0\end{aligned} \tag{5-9}$$

where the fundamental identity of the Section 1.7 of Chapter 4 has been used. From Eq.(4-2) with Eq.(4-1) the second part originating from interfaces becomes

$$\begin{aligned}&\frac{1}{\Delta t}\lim_{\delta\to 0}\int_{[\Delta t]_S} B_S\,dt = \frac{1}{\Delta t}\sum_j\frac{1}{v_{ni}}\left\{\frac{d_s}{dt}\left(\psi_a\right)+\psi_a\nabla_s\cdot\boldsymbol{v}_i-\phi_a\right.\\ &\left.+g_{\ln}A^{\alpha\beta}\left(t_\alpha^n\mathcal{J}_a^{l\cdot}\right)_{,\beta}-\sum_{k=1}^{2}\boldsymbol{n}_k\cdot\left[\rho_k\left(\boldsymbol{v}_k-\boldsymbol{v}_i\right)\psi_k+\mathcal{J}_k\right]\right\}=0.\end{aligned} \tag{5-10}$$

It is evident that the above equation is a time-averaged interfacial balance equation. In order to distinguish it from the local jump condition, we call it the *interfacial transfer condition* or the *macroscopic jump condition.*

In view of Eqs.(5-7), (5-9) and (5-10), we obtain a macroscopic balance equation for the mixture

$$\begin{aligned}&\frac{\partial \rho_m \psi_m}{\partial t}+\nabla\cdot\left(\rho_m\psi_m\boldsymbol{v}_m\right)+\nabla\cdot\left(\overline{\mathcal{J}}+\mathcal{J}^D+\mathcal{J}^T\right)-\rho_m\phi_m\\&+\frac{1}{\Delta t}\sum_j\frac{1}{v_{ni}}\left\{\frac{d_s}{dt}\left(\psi_a\right)+\psi_a\nabla_s\cdot\boldsymbol{v}_i-\phi_a\right\}\\&+\frac{1}{\Delta t}\sum_j\frac{1}{v_{ni}}g_{\ln}A^{\alpha\beta}\left(t_\alpha^n\mathcal{J}_a^{l\cdot}\right)_{,\beta}=0.\end{aligned}\tag{5-11}$$

The terms given by $\overline{\mathcal{J}}$, $\mathcal{J}^D$ and $\mathcal{J}^T$ represent the effluxes due to the average molecular diffusion, the macroscopic phase diffusions with respect to the mixture center of mass, and the statistical effects of the two-phase and turbulent fluctuations, whereas $\rho_m\phi_m$ is the mixture volumetric source. From the form of the balance equation, it is also possible to consider the interfacial terms as an additional source or sink.

It is generally accepted that the mass and momentum of the interface can be neglected. The surface energy, however, may not be insignificant because of the energy associated with the thermodynamic tension, namely, the surface tension. Thus, the first part of the interfacial term is important only in the energy balance equation. The surface line flux appears in the momentum and the energy balances, though the molecular diffusion transfers along an interface (namely the surface viscous stress and the surface heat flux) are neglected. This means that these line fluxes account for the effects of the surface tension only. When Eq.(5-11) is applied for the balance of mass, momentum and energy, appropriate forms corresponding to the simplified jump conditions of Eqs.(2-69), (2-72) and (2-73) should be used.

The averaged balance equations for each phase can be obtained by considering the function associated only with a particular phase, Eq.(4-8). Thus, in analogy with Eq.(5-9), we have

$$\begin{aligned}&\frac{\partial \alpha_k\overline{\overline{\rho_k}}\widehat{\psi_k}}{\partial t}+\nabla\cdot\left(\alpha_k\overline{\overline{\rho_k}}\widehat{\psi_k}\widehat{\boldsymbol{v}_k}\right)+\nabla\cdot\left[\alpha_k\left(\overline{\overline{\mathcal{J}_k}}+\mathcal{J}_k^T\right)\right]-\alpha_k\overline{\overline{\rho_k}}\widehat{\phi_k}\\&+\frac{1}{\Delta t}\sum_j\left\{\frac{1}{v_{ni}}\left[\boldsymbol{n}_k\cdot\rho_k\left(\boldsymbol{v}_k-\boldsymbol{v}_i\right)\psi_k+\boldsymbol{n}_k\cdot\mathcal{J}_k\right]\right\}=0\end{aligned}\tag{5-12}$$

where we have used the transformation (4-110) and the local instant general balance equation for the kth-phase, Eq.(2-6). For simplicity, let us define

$$I_k \equiv -\frac{1}{\Delta t}\sum_j \left\{ \frac{1}{v_{ni}} \boldsymbol{n}_k \cdot \left[\rho_k \left(\boldsymbol{v}_k - \boldsymbol{v}_i \right) \psi_k + \mathcal{J}_k \right] \right\} \tag{5-13}$$

$$I_m \equiv -\frac{1}{\Delta t}\sum_j \left\{ \frac{1}{v_{ni}} \left[\frac{d_s \psi_a}{dt} + \psi_a \nabla_s \cdot \boldsymbol{v}_i - \phi_a \right. \right. \\ \left. \left. + g_{\ln} A^{\alpha\beta} \left(t_\alpha^n \mathcal{J}_a^{l \cdot} \right)_{,\beta} \right] \right\}. \tag{5-14}$$

And the mixture total flux is given by

$$\mathcal{J}_m \equiv \overline{\mathcal{J}} + \mathcal{J}^D + \mathcal{J}^T. \tag{5-15}$$

Here I_k and I_m represent the interfacial source for the kth-phase and for the mixture, respectively. With these definitions the *mixture general balance equation* (5-11) reduces to

$$\frac{\partial \rho_m \psi_m}{\partial t} + \nabla \cdot \left(\rho_m \psi_m \boldsymbol{v}_m \right) = -\nabla \cdot \mathcal{J}_m + \rho_m \phi_m + I_m \tag{5-16}$$

whereas the *balance equation for the kth-phase* becomes

$$\frac{\partial \alpha_k \overline{\overline{\rho_k}} \widehat{\psi_k}}{\partial t} + \nabla \cdot \left(\alpha_k \overline{\overline{\rho_k}} \widehat{\psi_k} \widehat{\boldsymbol{v}_k} \right) = -\nabla \cdot \left[\alpha_k \left(\overline{\overline{\mathcal{J}_k}} + \mathcal{J}_k^T \right) \right] \\ + \alpha_k \overline{\overline{\rho_k}} \widehat{\phi_k} + I_k. \tag{5-17}$$

Furthermore, the interfacial transfer condition (5-10) can be rewritten as

$$\sum_{k=1}^{2} I_k - I_m = 0. \tag{5-18}$$

Each of these three macroscopic equations expresses the balance of matter ψ for the mixture, for the kth-phase, and at the interfaces, respectively. The mixture balance equation will be the foundation of the formulation of the drift-flux model. Furthermore, the phase balance equations and the interfacial transfer conditions are required for the two-fluid model formulation.

In view of Eqs.(5-16), (5-17) and (5-18), our fundamental purpose of averaging has been accomplished. Thus, the original two phases which are alternately occupying a point have been transformed into two co-existing continua. Moreover the hopelessly complicated two-phase and turbulent fluctuations have been smoothed out and their statistical macroscopic effects have been taken into account by the covariance (or turbulent flux) terms. In the next two sections we present balance equations of mass, momentum and energy for the *diffusion model* and for the *two-fluid model* separately.

1.2 Two-fluid model field equations

In this section, the macroscopic balance equation (5-17) and the interfacial transfer condition (5-18), which have been derived from the time averaging, are applied to the conservation laws of mass, momentum and energy. The choice of variables in these equations follows that of the local instant formulation of Chapter 2.

Mass Balance
In order to obtain mass balance equations, we set

$$\psi_k = 1, \quad \mathscr{J}_k = 0, \quad \phi_k = 0. \tag{2-7}$$

And in view of Eq.(2-69) we define

$$\Gamma_k \equiv I_k = -\frac{1}{\Delta t}\sum_j\left\{\frac{1}{v_{ni}}\boldsymbol{n}_k \cdot \rho_k\left(\boldsymbol{v}_k - \boldsymbol{v}_i\right)\right\} \tag{5-19}$$

$$I_m = 0. \tag{5-20}$$

Then by substituting Eq.(2-7) into Eqs.(5-17) and (5-18) we get

$$\frac{\partial \alpha_k \overline{\overline{\rho_k}}}{\partial t} + \nabla\cdot\left(\alpha_k \overline{\overline{\rho_k}}\widehat{\boldsymbol{v}_k}\right) = \Gamma_k \quad k = 1 \text{ and } 2 \tag{5-21}$$

and

$$\sum_{k=1}^{2}\Gamma_k = 0. \tag{5-22}$$

Equation (5-21) is the continuity equation for each phase with the interfacial mass source Γ_k appearing on the right-hand side due to phase changes, whereas the second equation, Eq.(5-22), expresses the conservation of mass at the interfaces.

Momentum Balance

The macroscopic momentum balance can be obtained from Eqs.(5-13), (5-14) and (5-17) by setting

$$\psi_k = \boldsymbol{v}_k, \quad \mathcal{J}_k = -\mathcal{T}_k = p_k\mathcal{I} - \mathcal{C}_k, \quad \phi_k = \boldsymbol{g}_k \tag{2-9}$$

and by defining the following terms in view of Eqs. (2-72) and (4-104)

$$\boldsymbol{M}_k \equiv \boldsymbol{I}_k = -\frac{1}{\Delta t}\sum_j \left\{\frac{1}{v_{ni}}\boldsymbol{n}_k \cdot \left[\rho_k\left(\boldsymbol{v}_k - \boldsymbol{v}_i\right)\boldsymbol{v}_k - \mathcal{T}_k\right]\right\} \tag{5-23}$$

$$\boldsymbol{M}_m \equiv \boldsymbol{I}_m = \frac{1}{\Delta t}\sum_j \left\{\frac{1}{v_{ni}}\left(\boldsymbol{t}_\alpha A^{\alpha\beta}\sigma\right)_{,\beta}\right\} \tag{5-24}$$

$$\mathcal{C}_k^T \equiv -\mathcal{J}_k^T = -\overline{\overline{\rho_k \boldsymbol{v}_k' \boldsymbol{v}_k'}}. \tag{5-25}$$

With these definitions we obtain from Eqs.(5-17) and (5-18)

$$\begin{aligned}&\frac{\partial \alpha_k \overline{\overline{\rho_k}}\widehat{\boldsymbol{v}_k}}{\partial t} + \nabla\cdot\left(\alpha_k \overline{\overline{\rho_k}}\widehat{\boldsymbol{v}_k}\widehat{\boldsymbol{v}_k}\right) = -\nabla\left(\alpha_k \overline{\overline{p_k}}\right)\\ &+\nabla\cdot\left[\alpha_k\left(\overline{\overline{\mathcal{C}_k}} + \mathcal{C}_k^T\right)\right] + \alpha_k\overline{\overline{\rho_k}}\widehat{\boldsymbol{g}_k} + \boldsymbol{M}_k\end{aligned} \tag{5-26}$$

and

$$\sum_{k=1}^{2}\boldsymbol{M}_k - \boldsymbol{M}_m = 0. \tag{5-27}$$

Here the terms $\mathcal{C}_k^T$ and $\boldsymbol{M}_k$ denote the turbulent flux and the kth-phase momentum source form the interfacial transfer, respectively, whereas the term $\boldsymbol{M}_m$ is the mixture momentum source due to the surface tension effect.

Energy Balance

The energy balances for the macroscopic fields can be obtained from Eq.(5-17) by first setting

$$\psi_k = u_k + \frac{v_k^2}{2}, \quad \mathscr{J}_k = \boldsymbol{q}_k - \mathscr{T}_k \cdot \boldsymbol{v}_k, \quad \phi_k = \boldsymbol{g}_k \cdot \boldsymbol{v}_k + \frac{\dot{q}_k}{\rho_k} \tag{2-12}$$

and by defining following terms in view of Eqs.(5-13), (5-14) and (2-73)

$$E_k \equiv I_k = -\frac{1}{\Delta t}\sum_j \left\{\frac{1}{v_{ni}} \boldsymbol{n}_k \cdot \left[\rho_k \left(\boldsymbol{v}_k - \boldsymbol{v}_i\right)\left(u_k + \frac{v_k^2}{2}\right) - \mathscr{T}_k \cdot \boldsymbol{v}_k + \boldsymbol{q}_k\right]\right\} \tag{5-28}$$

$$E_m \equiv I_m = \frac{1}{\Delta t}\sum_j \left\{\frac{1}{v_{ni}}\left[T_i \left\{\frac{d_s}{dt}\left(\frac{d\sigma}{dT}\right) + \left(\frac{d\sigma}{dT}\right)\nabla_s \cdot \boldsymbol{v}_i\right\} + \left(\boldsymbol{t}_\alpha A^{\alpha\beta}\sigma\right)_{,\beta} \cdot \boldsymbol{v}_i\right]\right\} \tag{5-29}$$

$$\boldsymbol{q}_k^T \equiv \mathscr{J}_k^T - \overline{\overline{\mathscr{T}_k \cdot \boldsymbol{v}_k'}} = \overline{\overline{\rho_k \left(u_k + \frac{v_k^2}{2}\right)' \boldsymbol{v}_k'}} - \overline{\overline{\mathscr{T}_k \cdot \boldsymbol{v}_k'}} + \overline{\overline{p_k \cdot \boldsymbol{v}_k'}} \tag{5-30}$$

$$\widehat{e_k} \equiv \widehat{u_k} + \frac{\widehat{\left(\boldsymbol{v}_k'\right)^2}}{2}. \tag{5-31}$$

Thus, we have from Eqs.(5-17) and (5-18)

$$\begin{aligned} &\frac{\partial}{\partial t}\left[\alpha_k \overline{\overline{\rho}}_k \left(\widehat{e_k} + \frac{\widehat{v_k}^2}{2}\right)\right] + \nabla \cdot \left[\alpha_k \overline{\overline{\rho}}_k \left(\widehat{e_k} + \frac{\widehat{v_k}^2}{2}\right)\widehat{\boldsymbol{v}_k}\right] \\ &= -\nabla \cdot \left[\alpha_k \left(\overline{\overline{\boldsymbol{q}_k}} + \boldsymbol{q}_k^T\right)\right] + \nabla \cdot \left(\alpha_k \overline{\overline{\mathscr{T}_k}} \cdot \widehat{\boldsymbol{v}_k}\right) \\ &+ \alpha_k \overline{\overline{\rho}}_k \widehat{\boldsymbol{g}_k} \cdot \widehat{\boldsymbol{v}_k} + E_k \end{aligned} \tag{5-32}$$

and

$$\sum_{k=1}^{2} E_k - E_m = 0 \tag{5-33}$$

where we have assumed

$$\boldsymbol{g}_k = \widehat{\boldsymbol{g}_k}. \tag{5-34}$$

We also note that the apparent internal energy $\widehat{e_k}$ consists of the standard thermal energy and the turbulent kinetic energy, see Eq.(5-31). The term E_k represents the interfacial supply of energy to the kth-phase, while E_m is the energy source for the mixture. This means that the energy can be stored at or released from interfaces. As it can be seen from the definition, the turbulent heat flux $\boldsymbol{q}_k^T$ takes account for the turbulent energy convection as well as for the turbulent work. For most two-phase flow problems, the internal heating $\dot{q}_k$ can be neglected.

The two-fluid model is based on the above *six field equations*, namely, two continuity, two momentum and two energy equations. The *interfacial transfer conditions* for mass, momentum and energy couple the transport processes of each phase. Since these nine equations basically express the conservation laws, they should be supplemented by various constitutive equations that specify molecular diffusions, turbulent transports, and interfacial transfer mechanisms as well as a relation between the thermodynamic state variables.

In solving problems, it is often useful to separate the mechanical and thermal effects in the total energy equation. Thus from the standard method of dotting the momentum equation by the velocity, we have the *mechanical energy equation*

$$\begin{aligned}
&\frac{\partial\left(\alpha_k\overline{\overline{\rho_k}}\dfrac{\widehat{v_k}^2}{2}\right)}{\partial t} + \nabla\cdot\left(\alpha_k\overline{\overline{\rho_k}}\frac{\widehat{v_k}^2}{2}\widehat{\boldsymbol{v}_k}\right) = -\widehat{\boldsymbol{v}_k}\cdot\nabla\left(\alpha_k\overline{\overline{p_k}}\right) \\
&+\widehat{\boldsymbol{v}_k}\cdot\nabla\cdot\left[\alpha_k\left(\overline{\overline{\mathcal{T}_k}} + \mathcal{T}_k^T\right)\right] + \alpha_k\overline{\overline{\rho_k}}\boldsymbol{g}_k\cdot\widehat{\boldsymbol{v}_k} + \boldsymbol{M}_k\cdot\widehat{\boldsymbol{v}_k} - \frac{\widehat{v_k}^2}{2}\Gamma_k.
\end{aligned} \tag{5-35}$$

Then by subtracting Eq.(5-35) from Eq.(5-32) the *internal energy equation* can be obtained, thus

$$\begin{aligned}&\frac{\partial \alpha_k \overline{\overline{\rho_k}}\widehat{e_k}}{\partial t}+\nabla\cdot\left(\alpha_k\overline{\overline{\rho_k}}\widehat{e_k}\widehat{\boldsymbol{v}_k}\right)=-\nabla\cdot\left(\alpha_k\overline{\overline{\boldsymbol{q}_k}}\right)\\&-\nabla\cdot\left\{\alpha_k\left(\boldsymbol{q}_k^T+\mathcal{T}_k^T\cdot\widehat{\boldsymbol{v}_k}\right)\right\}-\alpha_k\overline{\overline{p_k}}\nabla\cdot\widehat{\boldsymbol{v}_k}\\&+\alpha_k\left(\overline{\overline{\mathcal{T}_k}}+\mathcal{T}_k^T\right):\nabla\widehat{\boldsymbol{v}_k}\\&+\left(\frac{\widehat{\boldsymbol{v}_k}^2}{2}\Gamma_k-\boldsymbol{M}_k\cdot\widehat{\boldsymbol{v}_k}+E_k\right).\end{aligned}\tag{5-36}$$

Here we recall that the virtual internal energy $\widehat{e_k}$ includes the turbulent kinetic energy in addition to the standard internal energy.

For two-phase flow analyses, the enthalpy energy equation is important and it is frequently used to solve various engineering problems. Thus, in parallel with Eq.(5-31), we introduce a virtual enthalpy $\widehat{h_k}$ defined by

$$\widehat{h_k}\equiv\widehat{i_k}+\frac{\widehat{\left(\boldsymbol{v}_k'\right)^2}}{2}=\widehat{e_k}+\frac{\overline{\overline{p_k}}}{\overline{\overline{\rho_k}}}.\tag{5-37}$$

By substituting Eq.(5-37) into Eq. (5-36) we obtain

$$\begin{aligned}&\frac{\partial \alpha_k \overline{\overline{\rho_k}}\widehat{h_k}}{\partial t}+\nabla\cdot\left(\alpha_k\overline{\overline{\rho_k}}\widehat{h_k}\widehat{\boldsymbol{v}_k}\right)=-\nabla\cdot\left(\alpha_k\overline{\overline{\boldsymbol{q}_k}}\right)\\&-\nabla\cdot\left\{\alpha_k\left(\boldsymbol{q}_k^T+\mathcal{T}_k^T\cdot\widehat{\boldsymbol{v}_k}\right)\right\}+\frac{D_k}{Dt}\left(\alpha_k\overline{\overline{p_k}}\right)\\&+\alpha_k\left(\overline{\overline{\mathcal{T}_k}}+\mathcal{T}_k^T\right):\nabla\widehat{\boldsymbol{v}_k}+\left(\frac{\widehat{v_k}^2}{2}\Gamma_k-\boldsymbol{M}_k\cdot\widehat{\boldsymbol{v}_k}+E_k\right)\end{aligned}\tag{5-38}$$

where the substantial derivative D_k/Dt is taken by following the center of mass of kth-phase or moving with velocity $\widehat{\boldsymbol{v}_k}$, thus $D_k/Dt = \partial/\partial t+\widehat{\boldsymbol{v}_k}\cdot\nabla$. These thermal energy equations are extremely complicated due to the interactions between the mechanical terms from the turbulent fluctuations and the thermal terms. However, in many practical two-phase flow problems, the heat transfer and the phase change terms dominate the energy equations. In such a case, the above equations can be reduced to simple forms.

As it can be seen from Eq.(5-38), the interfacial transfer in the thermal energy equation has a special form which is expressed by a combination of the mass, momentum and energy transfer terms. Thus we define

$$\Lambda_k \equiv \frac{\widehat{v_k}^2 \Gamma_k}{2} - \boldsymbol{M}_k \cdot \widehat{\boldsymbol{v}_k} + E_k. \tag{5-39}$$

1.3 Diffusion (mixture) model field equations

The basic concept of the diffusion (mixture) model is to consider the mixture as a whole, therefore the field equations should be written for the balance of mixture mass, momentum and energy in terms of the mixture properties. These three macroscopic mixture conservation equations are then supplemented by a diffusion equation that takes account for the concentration changes.

Mixture Continuity and Diffusion Equations

From the mixture general balance equation, Eq.(5-16), with the definitions of ρ_m and $\boldsymbol{v}_m$, we obtain the mixture continuity equation

$$\frac{\partial \rho_m}{\partial t} + \nabla \cdot \left(\rho_m \boldsymbol{v}_m\right) = 0. \tag{5-40}$$

The above equation has exactly the same form as that for a continuum without internal discontinuities.

The diffusion equation, which expresses the changed in concentration α_1, can be derived from Eqs.(5-21) and (4-89)

$$\frac{\partial \alpha_1 \overline{\overline{\rho_1}}}{\partial t} + \nabla \cdot \left(\alpha_1 \overline{\overline{\rho_1}} \boldsymbol{v}_m\right) = \Gamma_1 - \nabla \cdot \left(\alpha_1 \overline{\overline{\rho_1}} \boldsymbol{V}_{1m}\right). \tag{5-41}$$

It has a mass source term Γ_1 that appears only after the continuity equation being averaged over the time interval because it accounts for the mass transfer at the interface. In addition, Eq.(5-41) has a diffusion term on the right-hand side, since the convective flux has been expressed by the mixture center of mass velocity $\boldsymbol{v}_m$.

Mixture Momentum Equation

By applying the general balance equation (5-16) to the conservation of momentum we obtain

$$\frac{\partial \rho_m \boldsymbol{v}_m}{\partial t} + \nabla \cdot \left(\rho_m \boldsymbol{v}_m \boldsymbol{v}_m\right) = -\nabla p_m + \nabla \cdot \left(\overline{\mathcal{T}} + \mathcal{T}^T + \mathcal{T}^D\right) + \rho_m \boldsymbol{g}_m + \boldsymbol{M}_m \tag{5-42}$$

where we have

$$
\begin{aligned}
p_m &= \sum_{k=1}^{2} \alpha_k \overline{\overline{p_k}} \\
\overline{\mathcal{T}} &= \sum_{k=1}^{2} \alpha_k \overline{\overline{\mathcal{T}_k}} \\
\mathcal{T}^T &= -\sum_{k=1}^{2} \alpha_k \overline{\overline{\rho_k \boldsymbol{v}_k' \boldsymbol{v}_k'}} \\
\mathcal{T}^D &= -\sum_{k=1}^{2} \alpha_k \overline{\overline{\rho_k}} \boldsymbol{V}_{km} \boldsymbol{V}_{km} \\
\boldsymbol{g}_m &= \frac{\sum_{k=1}^{2} \alpha_k \overline{\overline{\rho_k}} \boldsymbol{g}_k}{\rho_m}.
\end{aligned}
\tag{5-43}
$$

Furthermore the interfacial momentum source $\boldsymbol{M}_m$ is given by Eq.(5-24). Three tensor fluxes $\overline{\mathcal{T}}$, $\mathcal{T}^T$ and $\mathcal{T}^D$ represent the average viscous stress, the turbulent stress and the diffusion stress, respectively. It is evident that if the surface tension term is neglected, then there are not direct interfacial terms in the mixture momentum equation.

Mixture Total Energy Equation

The mixture energy equation can be obtained from Eq.(5-16) applied to the balance of the total energy, thus

$$
\begin{aligned}
&\frac{\partial}{\partial t}\left\{\rho_m\left[e_m+\left(\frac{v^2}{2}\right)_m\right]\right\}+\nabla\cdot\left\{\rho_m\left[e_m+\left(\frac{v^2}{2}\right)_m\right]\boldsymbol{v}_m\right\} \\
&= -\nabla\cdot\left(\overline{\boldsymbol{q}}+\boldsymbol{q}^T+\boldsymbol{q}^D\right)-\nabla\cdot\left(p_m\boldsymbol{v}_m\right)+\nabla\cdot\left(\overline{\mathcal{T}}\cdot\boldsymbol{v}_m\right) \\
&+\rho_m\boldsymbol{g}_m\cdot\boldsymbol{v}_m+\sum_{k=1}^{2}\alpha_k\overline{\overline{\rho_k}}\boldsymbol{g}_k\cdot\boldsymbol{V}_{km}+E_m
\end{aligned}
\tag{5-44}
$$

where we have

$$
\overline{\boldsymbol{q}} = \sum_{k=1}^{2} \alpha_k \overline{\overline{\boldsymbol{q}_k}}
\tag{5-45}
$$

$$\boldsymbol{q}^T = \sum_{k=1}^{2} \alpha_k \boldsymbol{q}_k^T = \sum_{k=1}^{2} \alpha_k \left\{ \overline{\overline{\rho_k \left(u_k + \frac{v_k^2}{2} \right)' \boldsymbol{v}_k'}} - \overline{\overline{\mathcal{T}_k' \cdot \boldsymbol{v}_k'}} \right\} \tag{5-46}$$

$$\begin{aligned} \boldsymbol{q}^D &= \mathcal{J}^D - \sum_{k=1}^{2} \alpha_k \overline{\overline{\mathcal{T}_k}} \cdot \boldsymbol{V}_{km} \\ &= \sum_{k=1}^{2} \alpha_k \left\{ \overline{\overline{\rho_k}} \left(\widehat{e_k} + \frac{\widehat{v_k}^2}{2} \right) \boldsymbol{V}_{km} - \overline{\overline{\mathcal{T}_k}} \cdot \boldsymbol{V}_{km} \right\}. \end{aligned} \tag{5-47}$$

And, by definition, we have following mixture properties

$$e_m = \frac{\sum_{k=1}^{2} \alpha_k \overline{\overline{\rho_k}} \widehat{e_k}}{\rho_m} = \frac{\sum_{k=1}^{2} \alpha_k \overline{\overline{\rho_k}} \left(\widehat{u_k} + \dfrac{\widehat{\boldsymbol{v}_k'^2}}{2} \right)}{\rho_m} \tag{5-48}$$

$$\left(\frac{v^2}{2} \right)_m = \frac{\sum_{k=1}^{2} \alpha_k \overline{\overline{\rho_k}} \dfrac{\widehat{v_k}^2}{2}}{\rho_m} = \frac{v_m^2}{2} + \frac{\sum_{k=1}^{2} \alpha_k \overline{\overline{\rho_k}} \dfrac{V_{km}^2}{2}}{\rho_m}. \tag{5-49}$$

The interfacial energy source E_m is given by Eq.(5-29). The important special case is when the body force field is constant

$$\boldsymbol{g}_k = \widehat{\boldsymbol{g}_k} = \boldsymbol{g}_m = \boldsymbol{g}. \tag{5-50}$$

Then the diffusion body work term becomes zero, thus

$$\sum_{k=1}^{2} \alpha_k \overline{\overline{\rho_k}} \boldsymbol{g}_k \cdot \boldsymbol{V}_{km} = 0 \tag{5-51}$$

where we used the identity, Eq.(4-94).

Hence, under the standard condition of the constant body force field, the total mixture energy equation reduces to

$$\frac{\partial}{\partial t}\left\{\rho_m\left[e_m+\left(\frac{v^2}{2}\right)_m\right]\right\}+\nabla\cdot\left\{\rho_m\left[e_m+\left(\frac{v^2}{2}\right)_m\right]\boldsymbol{v}_m\right\}$$
$$=-\nabla\cdot\left(\overline{\boldsymbol{q}}+\boldsymbol{q}^T+\boldsymbol{q}^D\right)-\nabla\cdot\left(p_m\boldsymbol{v}_m\right)+\nabla\cdot\left(\overline{\mathcal{T}}\cdot\boldsymbol{v}_m\right) \qquad (5\text{-}52)$$
$$+\rho_m\boldsymbol{g}_m\cdot\boldsymbol{v}_m+E_m.$$

It can be seen that the form of Eq.(5-52) is quite similar to the single-phase flow energy equation. The differences appear as additional heat fluxes, namely, the turbulent flux $\boldsymbol{q}^T$ and the diffusion flux $\boldsymbol{q}^D$, and the interfacial body source E_m. However the most interesting characteristic of the mixture can be found in the kinetic energy term, Eq.(5-49). We see from the equation that the total mixture kinetic energy consists of the kinetic energy of the center of mass plus the diffusion kinetic energies of both phases. It also should be remembered that the turbulent kinetic energy has been included in the virtual internal energy due to the great difficulties in separating it from the thermal effects. We also point out that if the surface tension effects are neglected, then the interfacial term does not appear in the mixture total energy equation (as in the case with the mixture momentum equation).

Mixture Thermal Energy Equation

In a single-phase flow, the separation of the mechanical and thermal energy can be carried out quite easily by subtracting the mechanical energy equation from the total energy balance. Exactly the same method could be used in the two-fluid model formulation, if we would include the turbulent kinetic energy in the virtual thermal energy as we have done in Eqs.(5-36) and (5-37). In the diffusion model formulation, however, it is further complicated by the existence of the *diffusion kinetic energy* transport. Consequently, there is no clear-cut method to obtain a corresponding thermal energy equation for the mixtures. In the following, we demonstrate two distinct methods which give quite different results.

The first method is to subtract the sum of the kinetic energy equations of both phases from the total energy equation (5-44). In this way, the diffusion kinetic energy can be eliminated. Thus, from Eqs.(5-35) and (5-44), we obtain

$$\frac{\partial\rho_m h_m}{\partial t}+\nabla\cdot\left(\rho_m h_m\boldsymbol{v}_m\right)=-\nabla\cdot\left(\overline{\boldsymbol{q}}+\boldsymbol{q}^T\right)$$
$$-\nabla\cdot\left(\sum_{k=1}^{2}\alpha_k\overline{\overline{\rho_k}}\widehat{h_k}\boldsymbol{V}_{km}\right)+\frac{D}{Dt}p_m+\sum_{k=1}^{2}\alpha_k\overline{\overline{\mathcal{T}_k}}:\nabla\widehat{\boldsymbol{v}_k} \qquad (5\text{-}53)$$

$$+\sum_{k=1}^{2}\Lambda_k+\sum_{k=1}^{2}\left\{\boldsymbol{V}_{km}\cdot\nabla\left(\alpha_k\overline{\overline{p_k}}\right)-\widehat{\boldsymbol{v}_k}\cdot\nabla\cdot\left(\alpha_k\overline{\overline{\mathcal{T}_k}}^T\right)\right\}.$$

Here from the definition, the mixture enthalpy h_m is given by

$$h_m\equiv\frac{\sum_{k=1}^{2}\alpha_k\overline{\overline{\rho_k}}\widehat{h_k}}{\rho_m}=\frac{\sum_{k=1}^{2}\alpha_k\overline{\overline{\rho_k}}\left(\widehat{i_k}+\frac{\widehat{\boldsymbol{v}_k'^2}}{2}\right)}{\rho_m}. \tag{5-54}$$

The same equation (5-53) can also be obtained by adding the enthalpy equation of each phase, Eq.(5-38). The form of the equation is reasonably simple except the last term, but we should realize that the interfacial term $\sum_{k=1}^{2}\Lambda_k$ involves complicated exchanges between the total and the mechanical energies.

It is noted that by using the mixture kinetic energy equation in terms of the center of mass velocity $\boldsymbol{v}_m$, this difficulty in the interfacial term can be avoided. The resulting thermal energy equation, however, has additional terms from the diffusion kinetic energy. By subtracting the mixture mechanical energy equation, namely, the momentum equation (5-42) dotted by $\boldsymbol{v}_m$, from Eq.(5-52), we obtain

$$\begin{aligned}&\frac{\partial\rho_m h_m}{\partial t}+\nabla\cdot\left(\rho_m h_m\boldsymbol{v}_m\right)=-\nabla\cdot\left(\overline{\boldsymbol{q}}+\boldsymbol{q}^T\right)\\&-\nabla\cdot\left(\sum_{k=1}^{2}\alpha_k\overline{\overline{\rho_k}}\widehat{h_k}\boldsymbol{V}_{km}\right)+\frac{Dp_m}{Dt}\\&-\left\{\rho_m\frac{D}{Dt}\left(\sum_{k=1}^{2}\frac{\alpha_k\overline{\overline{\rho_k}}}{\rho_m}\frac{V_{km}^2}{2}\right)+\nabla\cdot\sum_{k=1}^{2}\alpha_k\overline{\overline{\rho_k}}\frac{V_{km}^2}{2}\boldsymbol{V}_{km}\right\}\\&+\left(\overline{\mathcal{T}}+\mathcal{T}^D\right):\nabla\boldsymbol{v}_m+\left(E_m-\boldsymbol{M}_m\cdot\boldsymbol{v}_m\right)-\boldsymbol{v}_m\cdot\left(\nabla\cdot\mathcal{T}^T\right)\\&+\nabla\cdot\left(\sum_{k=1}^{2}\alpha_k\overline{\overline{\mathcal{T}_k}}\cdot\boldsymbol{V}_{km}\right).\end{aligned} \tag{5-55}$$

In view of these two thermal energy equations, namely, Eqs.(5-53) and (5-55), it can be concluded that the mixture energy transfer is highly complicated due to the diffusion of each phase with respect to the mass center. The form of the right-hand side of each equation suggests that if the effects of the mechanical terms originated from the diffusion are important, then the constitutive laws for the diffusion (or mixture) model cannot be simple. Thus, in such a case, the two-fluid model may be more suitable. However, in most two-phase problems with large heat additions, these mechanical effects from the diffusions are insignificant. The only important effect to be taken into account is the diffusion transport of thermal energy because of the large difference on the phase enthalpies, namely, the latent heat.

1.4 Singular case of v_{ni}=0 (quasi-stationary interface)

In the preceding analyses it has been assumed that the interfacial displacement velocity v_{ni} is non-zero, however in reality it can be zero at isolated singularities. For example, it happens when an interface is stationary or the motion of the interface is purely tangential to it. Since $v_{ni} = 0$ is an important singularity associated with all the interfacial terms in the balance equations such as I_k and I_m, we study it in some detail.

In connection with this singularity, we first introduce a surface area concentration per volume. By considering only one interface it can be given as

$$a_{ij} = \frac{1}{L_j} = \frac{1}{\Delta t}\lim_{\delta\to 0}\frac{2\varepsilon_j}{\delta} = \frac{1}{\Delta t}\left(\frac{1}{v_{ni}}\right)_j \tag{5-56}$$

whereas the total area concentration is given by

$$\begin{aligned} a_i &= \frac{1}{L_S} = \frac{1}{\Delta t}\lim_{\delta\to 0}\int_{[\Delta t]_s}\frac{1}{\delta}dt = \frac{1}{\Delta t}\sum_j\left(\frac{1}{v_{ni}}\right)_j \\ &= \sum_j\frac{1}{L_j} = \sum_j\frac{1}{a_{ij}} \end{aligned} \tag{5-57}$$

The reciprocals of the area concentrations have the dimension of length and they are denoted by L_j and L_S for a single interface and for combined interfaces, respectively. a_{ij} is the surface area concentration for the jth-interface. We note here that the total interfacial length scale L_S has an important physical significance comparative to those of the molecular mean free path and the mixing length. Consequently all the interfacial terms appearing in the field equations, I_k and I_m, are expressed as an addition of contributions from each interface with a_{ij} as a weighting factor. This fact is clearly demonstrated in Eqs.(5-13) and (5-14). Furthermore, the derivatives of the time fraction α_k are closely related to the area concentrations, as it can be seen from Eqs.(4-61) and (4-62). The importance of the interfacial area concentration a_i or the length scale L_S should be noted. a_i has the physical significance of the interfacial area per unit volume and it is the most important geometrical factor affecting the interfacial transfer. The inverse of a_i given by L_S is the internal length scale of the two-phase flow. This variable is discussed in detail in Chapter 10.

Now we return to the singular case of $v_{ni} = 0$. In view of Eqs.(5-56) and (5-57), it can be said that the transport length L_S becomes also singular in such a case and, thus it loses its physical significance. As a consequence all the interfacial terms in the balance equations, I_k and I_m, are also singular and the time fraction α_S or its derivative may suffer discontinuities. In order to cope with this difficulty, we first recall that α_S is the time fraction of the interfaces. We note here that from Eq.(4-14), the case of $v_{ni} \neq 0$ corresponds to $\alpha_S = 0$. Furthermore, if the normal velocity of an interface v_{ni} is zero, then it may stay at $\boldsymbol{x}_0$ for some finite time. Thus we have

$$\alpha_S = \frac{\Delta t_S}{\Delta t} \geq 0 \tag{5-58}$$

where Δt_S is the total time occupied by interfaces in the interval of Δt. And therefore

$$\Delta t = \Delta t_S + \sum_{k=1}^{2} \Delta t_k \tag{5-59}$$

or in terms of the time fractions

$$1 = \alpha_S + \alpha_1 + \alpha_2. \tag{5-60}$$

Now it is clear that we have two distinct cases of singularities, namely

1. $v_{ni} = 0$ and $\alpha_S = 0$;
2. $v_{ni} = 0$ and $\alpha_S > 0$.

As the definition of α_k shows, Case 1 does not bring discontinuities in α_k, but only in its derivatives in the microscopic sense. However, it has been explained in the Section 1.3 of Chapter 4 in connection to the *fundamental hypothesis* on smoothness of mean values, these microscopic singularities should be neglected when macroscopic problems are concerned. Thus, unless an interface stays at a point for a finite time interval, the field is considered to be occupied by two continua with continuous interfacial transfer and source term I_k and I_m. Hence the singular surface of the interfacial origin appears in the macroscopic formulation only when $\alpha_s > 0$.

The above discussion also clearly indicates that the interfacial terms I_k and I_m given by Eq.(5-13) and Eq.(5-14) are not the constitutive relations to be used in the macroscopic formulation. These equations still retain all the details of the local instant variables which should not appear in the averaged formulation. Consequently, it is necessary to transform these equations in terms of the macroscopic variables.

We note that the macroscopic interface represented by Case 2 is most easily exemplified by the stationary interface such as a solid wall. Since for $\alpha_s > 0$ it is not possible to consider the average of the volumetric balance equation, the correct form of the macroscopic balance is simply the time average of the jump condition in the time interval of Δt. Thus we have

$$\frac{1}{\Delta t}\lim_{\delta\to 0}\int_{[\Delta t]_s}(B_s\delta)dt = 0. \tag{5-61}$$

Case 2 in normal two-phase flow corresponds to the discontinuity in α_k, and therefore, it can be treated as a concentration shock, see Eqs.(4-61) and (4-62). For most of the flow field it is assumed to be continuous. As the time-averaged macroscopic formulation is intended to be applied for such two-phase flows, these interfacial singularities can be neglected in most applications.

1.5 Macroscopic jump conditions

We have discussed the singularity related to a quasi-stationary interface in the preceding section. Now we study important singularities related to *macroscopic shock discontinuities* in the time averaged field. The essential part of the analysis can be developed in parallel with the Section 1.2 of Chapter 2 where the standard jump conditions at an interface have been derived.

In single-phase flows as well as in two-phase flows, the existence of regions where various properties suffer extremely large changes is well known. It can be exemplified by shock waves due to compressibility effects or by concentration shocks in mixtures. The unusually high gradients in these regions require special considerations on the constitutive laws in order to treat them as a part of continuum mechanics. However, as it has been

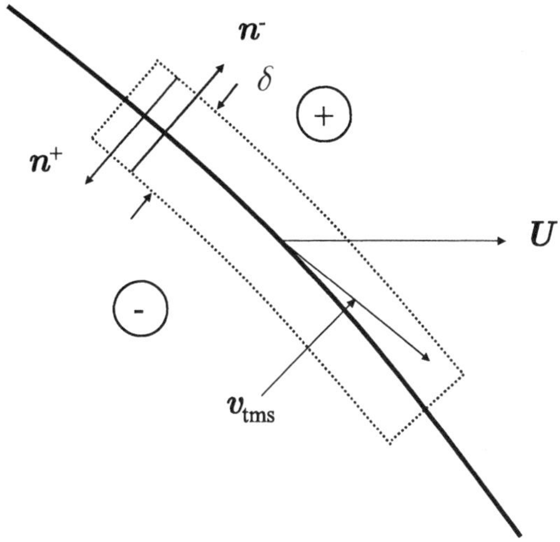

Figure 5-1. Macroscopic discontinuity (Ishii, 1975)

mentioned in the Section 1.2 of Chapter 2, for most practical flow problems replacements of these regions by surfaces of discontinuity with the jump conditions yield sufficiently accurate models. Hence in this section we derive macroscopic jump conditions which stand as balance equations at a surface of discontinuities in the averaged field. By considering the time-averaged macroscopic field as a continuum, the analysis of the Section of Chapter 2 can be directly applied here. The velocity of the surface $\boldsymbol{U}$ is defined in analogy with $\boldsymbol{v}_i$ of Eqs.(2-49) and (2-50), thus its normal component is the surface displacement velocity, whereas its tangential component is the mean mixture tangential velocity in the region with the thickness δ. By denoting each side of the region by + and -, we have from Eq.(2-64)

$$\begin{aligned}&\frac{d_s}{dt}\left(\rho_{ma}\psi_{ms}\right)+\rho_{ma}\psi_{ms}\nabla_s\cdot\boldsymbol{U}\\&=\sum_{+,-}\left\{\rho_m\psi_m\boldsymbol{n}\cdot\left(\boldsymbol{v}_m-\boldsymbol{U}\right)+\boldsymbol{n}\cdot\mathcal{J}_m\right\}\\&-g_{\ln}A^{\alpha\beta}\left(t_\alpha^n\mathcal{J}_{ma}^{l\bullet}\right)_{,\beta}+\rho_{ma}\phi_{ms}+I_{ma}\end{aligned} \tag{5-62}$$

where the summation stands for both sides of the region, the subscript a and s represent the mean values in the region as defined by Eqs.(2-45) to (2-54) and the surface derivatives have the standard significances. Equation (5-62) is a balance of ψ_m at the region of a shock. The left-hand side of the equation accounts for the change of the matter ψ_m in the region, and each term on the right-hand side represents: the flux from the bulk fluid; the flux from the periphery with the thickness δ; the body source; and the interfacial source term.

It is evident that the balance equations similar to Eq.(5-62) can be obtained also for each of the two phases, thus

$$\begin{aligned}&\frac{d_s}{dt}\left\{\left(\alpha_k\overline{\overline{\rho_k}}\right)_a\widehat{\psi_{ks}}\right\}+\left(\alpha_k\overline{\overline{\rho_k}}\right)_a\widehat{\psi_{ks}}\nabla_s\cdot U\\&=\sum_{+,-}\left\{\alpha_k\overline{\overline{\rho_k}}\widehat{\psi_k}\boldsymbol{n}\cdot\left(\widehat{\boldsymbol{v}_k}-\boldsymbol{U}\right)+\boldsymbol{n}\cdot\left[\alpha_k\left(\overline{\overline{\mathcal{J}_k}}+\mathcal{J}_k^T\right)\right]\right\}\\&-g_{\ln}A^{\alpha\beta}\left(t_\alpha^n\mathcal{J}_{ka}^{l\cdot}\right)_{,\beta}+\left(\alpha_k\overline{\overline{\rho_k}}\right)_a\widehat{\phi_{ks}}+I_{ka}.\end{aligned}\tag{5-63}$$

From the definition of the mean values at the region it is easy to show that if each term in Eq.(5-63) is summed up for both phases, the resulting term becomes that of the mixture appearing in Eq.(5-62).

Since Eqs.(5-62) and (5-63) introduce new variables associated with the discontinuities, namely, the surface properties and the line fluxes, it is necessary to make some specific assumptions on these terms or to give sufficient constitutive laws. A simple result of practical importance can be obtained by considering the limit of $\delta\rightarrow 0$ and, furthermore, by neglecting the surface energy of shocks and the associated thermodynamic tension. Under these conditions, we have for the mixture

$$\sum_{+,-}\left\{\rho_m\psi_m\boldsymbol{n}\cdot\left(\boldsymbol{v}_m-\boldsymbol{U}\right)+\boldsymbol{n}\cdot\mathcal{J}_m\right\}+I_{ma}=0.\tag{5-64}$$

And for each phase with $k=1$ and 2

$$\sum_{+,-}\left\{\alpha_k\overline{\overline{\rho_k}}\widehat{\psi_k}\boldsymbol{n}\cdot\left(\widehat{\boldsymbol{v}_k}-\boldsymbol{U}\right)+\boldsymbol{n}\cdot\alpha_k\left(\overline{\overline{\mathcal{J}_k}}+\mathcal{J}_k^T\right)\right\}+I_{ka}=0.\tag{5-65}$$

Here, we note the importance of the terms I_{ma} and I_{ka} that permit the exchange of mass, momentum and energy within the shock layers. It is incorrect to neglect these terms simply because they appear as volumetric sources. Depending on the constitutive laws expressing them, it is possible that they take a form of fluxes. Furthermore, from the physical points of view it is natural to have these interfacial transfer terms I_{ka} in the macroscopic jump conditions because of the highly non-equilibrium state of two-phase shock layers.

1.6 Summary of macroscopic field equations and jump conditions

The field equations for the mixture and for each phase were given by Eqs. (5-16) and (5-17)

$$\frac{\partial \rho_m \psi_m}{\partial t} + \nabla \cdot (\rho_m \psi_m \boldsymbol{v}_m) = -\nabla \cdot \mathcal{J}_m + \rho_m \phi_m + I_m \tag{5-16}$$

and

$$\frac{\partial \alpha_k \overline{\overline{\rho_k}} \widehat{\psi_k}}{\partial t} + \nabla \cdot \left(\alpha_k \overline{\overline{\rho_k}} \widehat{\psi_k} \widehat{\boldsymbol{v}_k}\right) = -\nabla \cdot \left[\alpha_k \left(\overline{\overline{\mathcal{J}_k}} + \mathcal{J}_k^T\right)\right]$$
$$+\alpha_k \overline{\overline{\rho_k}} \widehat{\phi_k} + I_k. \tag{5-17}$$

The mixture total flux has been defined by

$$\mathcal{J}_m \equiv \overline{\mathcal{J}} + \mathcal{J}^D + \mathcal{J}^T \tag{5-15}$$

with

$$\overline{\mathcal{J}} = \sum_{k=1}^{2} \alpha_k \overline{\overline{\mathcal{J}_k}} \tag{4-24}$$

$$\mathcal{J}^D = \sum_{k=1}^{2} \alpha_k \overline{\overline{\rho_k}} \widehat{\psi_k} \boldsymbol{V}_{km} \tag{4-103}$$

$$\mathcal{J}^T = \sum_{k=1}^{2} \alpha_k \mathcal{J}_k^T = \sum_{k=1}^{2} \alpha_k \overline{\overline{\rho_k \psi_k' \boldsymbol{v}_k'}}. \tag{4-105}$$

And the interfacial transfer condition was

$$I_m = \sum_{k=1}^{2} I_k. \tag{5-18}$$

The macroscopic jump conditions at shock waves in their simplified forms were

$$\sum_{+,-}\left\{\rho_m\psi_m\boldsymbol{n}\cdot\left(\boldsymbol{v}_m-\boldsymbol{U}\right)+\boldsymbol{n}\cdot\mathcal{J}_m\right\}+I_{ma}=0 \tag{5-64}$$

and

$$\sum_{+,-}\left\{\alpha_k\overline{\overline{\rho_k}}\widehat{\psi_k}\boldsymbol{n}\cdot\left(\widehat{\boldsymbol{v}_k}-\boldsymbol{U}\right)+\boldsymbol{n}\cdot\alpha_k\left(\overline{\overline{\mathcal{J}_k}}+\mathcal{J}_k^T\right)\right\}+I_{ka}=0 \tag{5-65}$$

where $\boldsymbol{n}$ (with + and -) is the outward unit normal vector at each side of the surface, and $\boldsymbol{U}$ denotes shock surface velocity.

1.7 Alternative form of turbulent heat flux

The energy equations in the Sections 1.2 and 1.3 of Chapter 5 are derived based on the definition of turbulent heat flux of Eq.(5-30). Here, an alternative definition of the turbulent heat flux of Eq.(5-30)' may be possible as

$$\boldsymbol{q}_k^T\equiv\overline{\overline{\rho_k\left\{u_k'+\left(\frac{v_k^2}{2}\right)'\right\}\boldsymbol{v}_k'}}-\overline{\overline{\mathcal{T}_k\cdot\boldsymbol{v}_k'}}. \tag{5-30)'$$

Then, we can recast the stagnation internal energy equation, Eq.(5-32), the internal energy equation, Eq.(5-36), and the stagnation enthalpy equation, Eq.(5-38) as given by Eqs.(5-32)', (5-36)', and (5-38)', respectively.

$$\begin{aligned}&\frac{\partial}{\partial t}\left[\alpha_k\overline{\overline{\rho_k}}\left(\widehat{e_k}+\frac{\widehat{v_k}^2}{2}\right)\right]+\nabla\cdot\left[\alpha_k\overline{\overline{\rho_k}}\left(\widehat{e_k}+\frac{\widehat{v_k}^2}{2}\right)\widehat{\boldsymbol{v}_k}\right]\\&=-\nabla\cdot\left[\alpha_k\left(\overline{\overline{\boldsymbol{q}_k}}+\boldsymbol{q}_k^T\right)\right]-\nabla\cdot\left(\alpha_k\overline{\overline{p_k}}\widehat{\boldsymbol{v}_k}\right)\\&+\nabla\cdot\left[\alpha_k\left(\overline{\overline{\mathcal{T}_k}}+\mathcal{T}_k^T\right)\cdot\widehat{\boldsymbol{v}_k}\right]+\alpha_k\overline{\overline{\rho_k}}\widehat{\boldsymbol{g}_k}\cdot\widehat{\boldsymbol{v}_k}+E_k\end{aligned} \tag{5-32)'$$

$$\frac{\partial \alpha_k \overline{\overline{\rho_k}} \widehat{e_k}}{\partial t} + \nabla \cdot \left(\alpha_k \overline{\overline{\rho_k}} \widehat{e_k} \widehat{\boldsymbol{v}_k} \right) = -\nabla \cdot \left(\alpha_k \overline{\overline{\boldsymbol{q}_k}} \right) - \nabla \cdot \left(\alpha_k \boldsymbol{q}_k^T \right)$$
$$- \alpha_k \overline{\overline{p_k}} \nabla \cdot \widehat{\boldsymbol{v}_k} + \alpha_k \left(\overline{\overline{\mathcal{T}_k}} + \mathcal{T}_k^T \right) : \nabla \widehat{\boldsymbol{v}_k}$$
$$+ \left[\Gamma_k \widehat{h_{ki}} + \frac{\overline{\overline{q''_{ki}}}}{L_s} + W_{ki}^T + \left(\boldsymbol{M}_k - \Gamma_k \widehat{\boldsymbol{v}_{ki}} \right) \cdot \left(\widehat{\boldsymbol{v}_{ki}} - \widehat{\boldsymbol{v}_k} \right) \right] \tag{5-36)'$$
$$- p_{ki} \left[\frac{\partial \alpha_k}{\partial t} + \widehat{\boldsymbol{v}_{ki}} \cdot \nabla \alpha_k \right]$$

$$\frac{\partial \alpha_k \overline{\overline{\rho_k}} \widehat{h_k}}{\partial t} + \nabla \cdot \left(\alpha_k \overline{\overline{\rho_k}} \widehat{h_k} \widehat{\boldsymbol{v}_k} \right) = -\nabla \cdot \left(\alpha_k \overline{\overline{\boldsymbol{q}_k}} \right) - \nabla \cdot \left(\alpha_k \boldsymbol{q}_k^T \right)$$
$$+ \frac{D_k}{Dt} \left(\alpha_k \overline{\overline{p_k}} \right) + \alpha_k \left(\overline{\overline{\mathcal{T}_k}} + \mathcal{T}_k^T \right) : \nabla \widehat{\boldsymbol{v}_k} \tag{5-38)'$$
$$+ \left(\frac{\widehat{v_k}^2}{2} \Gamma_k - \boldsymbol{M}_k \cdot \widehat{\boldsymbol{v}_k} + E_k \right).$$

We can also recast the total mixture energy equation, Eq.(5-44), and the mixture stagnation enthalpy equation, Eq.(5-53) as given by Eqs.(5-44)' and (5-53)', respectively.

$$\frac{\partial}{\partial t} \left\{ \rho_m \left[e_m + \left(\frac{v^2}{2} \right)_m \right] \right\} + \nabla \cdot \left\{ \rho_m \left[e_m + \left(\frac{v^2}{2} \right)_m \boldsymbol{v}_m \right] \right\}$$
$$= -\nabla \cdot \left(\overline{\boldsymbol{q}} + \boldsymbol{q}^T + \boldsymbol{q}^D \right) - \nabla \cdot \left(p_m \boldsymbol{v}_m \right)$$
$$+ \nabla \cdot \left\{ \left(\overline{\mathcal{T}} + \mathcal{T}^T \right) \cdot \boldsymbol{v}_m \right\} + \rho_m \boldsymbol{g}_m \cdot \boldsymbol{v}_m \tag{5-44)'$$
$$+ \sum_{k=1}^{2} \alpha_k \overline{\overline{\rho_k}} \boldsymbol{g}_k \cdot \boldsymbol{V}_{km} + E_m$$

where we have

$$\boldsymbol{q}^T = \sum_{k=1}^{2} \alpha_k \boldsymbol{q}_k^T = \sum_{k=1}^{2} \alpha_k \left[\overline{\overline{\rho_k \left\{ u_k' + \left(\frac{v_k^2}{2} \right)' \right\} \boldsymbol{v}_k'}} - \overline{\overline{\mathcal{T}_k \cdot \boldsymbol{v}_k'}} \right] \tag{5-46'}$$

$$\begin{aligned} \boldsymbol{q}^D &= \mathcal{J}^D - \sum_{k=1}^{2} \alpha_k \overline{\overline{\mathcal{T}_k}} \cdot \boldsymbol{V}_{km} \\ &= \sum_{k=1}^{2} \alpha_k \left\{ \overline{\overline{\rho_k}} \left(\widehat{e_k} + \frac{\widehat{v_k}^2}{2} \right) \boldsymbol{V}_{km} - \left(\overline{\overline{\mathcal{T}_k}} + \mathcal{T}_k^T \right) \cdot \boldsymbol{V}_{km} \right\} \end{aligned} \tag{5-47'}$$

$$\begin{aligned} &\frac{\partial \rho_m h_m}{\partial t} + \nabla \cdot \left(\rho_m h_m \boldsymbol{v}_m \right) = -\nabla \cdot \left(\overline{\boldsymbol{q}} + \boldsymbol{q}^T \right) \\ &-\nabla \cdot \left(\sum_{k=1}^{2} \alpha_k \overline{\overline{\rho_k}} \widehat{h_k} \boldsymbol{V}_{km} \right) + \frac{D}{Dt} p_m + \sum_{k=1}^{2} \alpha_k \left(\overline{\overline{\mathcal{T}_k}} + \mathcal{T}_k^T \right) : \nabla \widehat{\boldsymbol{v}_k} \\ &+ \sum_{k=1}^{2} \Lambda_k + \sum_{k=1}^{2} \left\{ \boldsymbol{V}_{km} \cdot \nabla \left(\alpha_k \overline{\overline{p_k}} \right) \right\}. \end{aligned} \tag{5-53'}$$

Correspondingly, Eqs.(5-52) and (5-55) can be recast as given by Eqs.(5-52)' and (5-55)', respectively.

$$\begin{aligned} &\frac{\partial}{\partial t} \left\{ \rho_m \left[e_m + \left(\frac{v^2}{2} \right)_m \right] \right\} + \nabla \cdot \left\{ \rho_m \left[e_m + \left(\frac{v^2}{2} \right)_m \right] \boldsymbol{v}_m \right\} \\ &= -\nabla \cdot \left(\overline{\boldsymbol{q}} + \boldsymbol{q}^T + \boldsymbol{q}^D \right) - \nabla \cdot \left(p_m \boldsymbol{v}_m \right) \\ &+ \nabla \cdot \left\{ \left(\overline{\mathcal{T}} + \mathcal{T}^T \right) \cdot \boldsymbol{v}_m \right\} + \rho_m \boldsymbol{g}_m \cdot \boldsymbol{v}_m + E_m \end{aligned} \tag{5-52'}$$

$$\begin{aligned} &\frac{\partial \rho_m h_m}{\partial t} + \nabla \cdot \left(\rho_m h_m \boldsymbol{v}_m \right) = -\nabla \cdot \left(\overline{\boldsymbol{q}} + \boldsymbol{q}^T \right) \\ &-\nabla \cdot \left(\sum_{k=1}^{2} \alpha_k \overline{\overline{\rho_k}} \widehat{h_k} \boldsymbol{V}_{km} \right) + \frac{D p_m}{Dt} \end{aligned} \tag{5-55'}$$

$$-\left\{\rho_m \frac{D}{Dt}\left(\sum_{k=1}^{2}\frac{\alpha_k\overline{\overline{\rho_k}}}{\rho_m}\frac{V_{km}^2}{2}\right)+\nabla\cdot\sum_{k=1}^{2}\alpha_k\overline{\overline{\rho_k}}\frac{V_{km}^2}{2}\boldsymbol{V}_{km}\right\}$$

$$+\left(\overline{\mathscr{T}}+\mathscr{T}^T+\mathscr{T}^D\right):\nabla\boldsymbol{v}_m$$

$$+\left(E_m-\boldsymbol{M}_m\cdot\boldsymbol{v}_m\right)+\nabla\cdot\left\{\sum_{k=1}^{2}\alpha_k\left(\overline{\overline{\mathscr{T}_k}}+\mathscr{T}^T\right)\cdot\boldsymbol{V}_{km}\right\}.$$

With the alternative definition of the turbulent heat flux such as Eq.(5-30)', the energy equations become symmetrical about the stress tensor term. The energy equations with the definition of Eq.(5-30)' are identical with those with the definition of Eq.(5-30). However, the energy equations with the definition of Eq.(5-30)' may be controversial and misleading on the following grounds. For example, the apparent energy dissipation term due to the turbulent stress such as $\alpha_k\mathscr{T}_k^T:\nabla\widehat{\boldsymbol{v}_k}$ appears in the internal energy equation of Eq.(5-36)'. Since this term is due to convection or turbulence, it may take a negative or positive value. Thus, the term due to turbulent work may not be regarded as the energy dissipation, which should invariably be positive. In order for the term due to turbulent work to be invariably positive, some additional constraints may be applied to the entropy inequality. Thus, the energy equations with the definition of Eq.(5-30)' are questionable to explain irreversible thermodynamics.

Chapter 6

CONNECTION TO OTHER STATISTICAL AVERAGES

1.1 Eulerian statistical average (ensemble average)

The basic concept of the Eulerian statistical averaging has been explained in Chapter 3. By considering a set of N similar samples or systems, a statistical mean value is defined by a simple arithmetic mean among them, Eq.(3-7). Thus, the mathematical operation of integration for the time averaging should be replaced by that of summation in the statistical averaging. It is evident that the entire derivation of the field theory based on the statistical averaging can be carried out in parallel with Chapters 4 and 5 by simply substituting the finite statistical mean operator, Eq.(3-7), in the place of the time averaging, Eq.(3-2).

The most important parameters are N_k and N_S which represent the number of occurrences of the kth-phase and the interfaces if the interfacial thickness is δ. Then the void fraction can be defined as a ratio of N_k to N and taking the limit $\delta \rightarrow 0$. The general function $(F)_n$ and $(F_k)_n$ can be defined in space-time domain in analogy with Eq.(4-8).

Since the derivation of the statistically averaged equations follows exactly the same steps as in the case of the time averaging, we only list the most important and characteristic relations between them. In Table 6-1 we see four basic parameters of averaging, namely: the void fraction α_k; the average of the two-phase general function F; the average of the kth-phase general function F_k; and the interfacial area concentration in a unit volume.

It is evident that the general balance equations of the form of Eqs.(5-16) and (5-17) can be obtained from the statistical averaging applied to a set of N similar samples. We note here that important differences between the time and statistical averaging exist not in the resulting form of the balance

Table 6-1. Relations between time and statistical averages (Ishii, 1975)

Variable	Time Average	Statistical Average
α_k	$\dfrac{\Delta t_k}{\Delta t}$	$\lim\limits_{\delta\to 0}\dfrac{N_k}{N}$
$\overline{F}$	$\lim\limits_{\delta\to 0}\dfrac{1}{\Delta t}\int_{[\Delta t]_T} F\,dt$	$\lim\limits_{\delta\to 0}\dfrac{1}{N}\sum\limits_n (F)_n$
$\overline{F_k}$	$\lim\limits_{\delta\to 0}\dfrac{1}{\Delta t}\int_{[\Delta t]_T} F_k\,dt$	$\lim\limits_{\delta\to 0}\dfrac{1}{N}\sum\limits_n (F_k)_n$
a_i	$\lim\limits_{\delta\to 0}\dfrac{1}{\Delta t}\int_{[\Delta t]_S}\dfrac{1}{\delta}\,dt$	$\lim\limits_{\delta\to 0}\dfrac{1}{N}\sum^{N_S}\left(\dfrac{1}{\delta}\right)$

equations but in the interpretation of the variables with respect to an actual flow, as it has been discussed in detail in the Section 1.3 of Chapter 3.

1.2 Boltzmann statistical average

Because of its unique characteristic among various averaging procedures, we discuss in detail the Boltzmann statistical averaging applied to two-phase flow systems. First we recall that $f(\boldsymbol{x},t,\boldsymbol{\xi})$ is the particle density function where $\boldsymbol{x}$, t and $\boldsymbol{\xi}$ represents the position, time and velocity of a particle, respectively. In the standard analysis of the kinetic theory of gases, the particle mass of each component is considered to be constant because it represents the molecular mass. However, in applications to two-phase highly dispersed flows, it may be necessary to assume that the particle mass varies. Thus, by taking into account the existence of variable particle mass, we define

$$f_{kn} = f_{kn}(\boldsymbol{x},t,\boldsymbol{\xi}) \tag{6-1}$$

where f_{kn} is the particle density function of kth-phase particles having m_{kn} mass. It can be said that m_{kn} is a multiple of the single molecular mass. The total number of kth-phase particles in the phase space element $d\boldsymbol{x}d\boldsymbol{\xi}$ at $\boldsymbol{x}$ and $\boldsymbol{\xi}$ is given by

$$\sum_n f_{kn}\left(\boldsymbol{x},t,\boldsymbol{\xi}\right)d\boldsymbol{x}d\boldsymbol{\xi}. \tag{6-2}$$

The Boltzmann equation for the m_{kn} particles can be obtained by introducing $\boldsymbol{g}_k$, the external force field, and taking the balance on the number of particles, thus

$$\frac{\partial f_{kn}}{\partial t} + \frac{\partial}{\partial \boldsymbol{x}} \cdot \left(\boldsymbol{\xi} f_{kn} \right) + \frac{\partial}{\partial \boldsymbol{\xi}} \cdot \left(\boldsymbol{g}_{kn} f_{kn} \right) = C_{kn}^{\ +} - C_{kn}^{\ -} \tag{6-3}$$

where C_{kn}^{+} and C_{kn}^{-} represent the source and sink terms, namely, the gain and loss of the m_{kn} particles caused by the changes in the particle mass and by the collisions that throw the particles in and out of the phase element $d\boldsymbol{\xi}$. If $\boldsymbol{g}_{kn}$ is independent of the velocity, then we have

$$\frac{\partial f_{kn}}{\partial t} + \boldsymbol{\xi} \cdot \frac{\partial f_{kn}}{\partial \boldsymbol{x}} + \boldsymbol{g}_{kn} \cdot \frac{\partial f_{kn}}{\partial \boldsymbol{\xi}} = C_{kn}^{\ +} - C_{kn}^{\ -}. \tag{6-4}$$

The above Boltzmann transport equation with the collision terms expressed by the simple model of Maxwell's binary collision integral is called the *Maxwell-Boltzmann equation.* It is the foundation of the kinetic theory of gases. A similar approach can be used for neutron transport.

The partial density of the m_{kn} particles is given by

$$\rho_{kn} = \int m_{kn} f_{kn} d\boldsymbol{\xi}. \tag{6-5}$$

Thus, the expectations based upon the probability f_{kn} can be defined as

$$\widehat{\psi_{kn}}\left(\boldsymbol{x}, t\right) = \frac{\int m_{kn} f_{kn} \psi_{kn} d\boldsymbol{\xi}}{\int m_{kn} f_{kn} d\boldsymbol{\xi}}. \tag{6-6}$$

Since the total mass is the sum of the mass of each particle, the partial density of the kth-phase is given by

$$\overline{\rho_k} = \sum_n \rho_{kn} = \sum_n \int m_{kn} f_{kn} d\boldsymbol{\xi}. \tag{6-7}$$

It follows that the mass weighted mean value is defined by

$$\widehat{\psi_k}\left(\boldsymbol{x},t\right)=\frac{\sum_n\int m_{kn}f_{kn}\psi_{kn}\,d\boldsymbol{\xi}}{\sum_n\int m_{kn}f_{kn}\,d\boldsymbol{\xi}}=\frac{\sum_n\rho_{kn}\widehat{\psi_{kn}}}{\overline{\rho_k}}. \tag{6-8}$$

Denoting the mean velocity by $\widehat{\boldsymbol{v}_k}$, the peculiar velocity of each particle is given by

$$\boldsymbol{V}_{kn}=\boldsymbol{\xi}_{kn}-\widehat{\boldsymbol{v}_k}. \tag{6-9}$$

If we multiply the Maxwell-Boltzmann transport equation, i.e. Eq.(6-4), by $m_{kn}\psi_{kn}$, integrate it over the phase velocity $\boldsymbol{\xi}$ and then sum it up for all kinds of particles of the kth-phase, we obtain *Maxwell's equation of transfer* in terms of mean values. Hence,

$$\begin{aligned}\frac{\partial}{\partial t}\left(\overline{\rho_k}\widehat{\psi_k}\right)+\nabla\cdot\left(\overline{\rho_k}\widehat{\psi_k}\widehat{\boldsymbol{v}_k}\right)&=-\nabla\cdot\sum_n\rho_{kn}\widehat{\psi_{kn}\boldsymbol{V}_{kn}}\\&-\sum_n\int m_{kn}\psi_{kn}\boldsymbol{g}_{kn}\cdot\frac{\partial f_{kn}}{\partial\boldsymbol{\xi}}\,d\boldsymbol{\xi}\\&+\sum_n\int m_{kn}f_{kn}\left[\frac{\partial\psi_{kn}}{\partial t}+\boldsymbol{\xi}\cdot\nabla\psi_{kn}\right]d\boldsymbol{\xi}\\&+\sum_n\int\left[C_{kn}^{+}-C_{kn}^{-}\right]m_{kn}\psi_{kn}\,d\boldsymbol{\xi}.\end{aligned} \tag{6-10}$$

We note here that each term in the right-hand side of Eq.(6-10) represents the transfer due to inter-particle diffusions, the source due to the body force field, the multi-molecular particle effect which arises when the transport property ψ_{kn} is a function of the phase velocity $\boldsymbol{\xi}$, and the source due to phase changes and/or collisions.

Conservation of Mass

The mass field equation can be obtained by setting $\psi_{kn}=1$ in Eq.(6-10), thus we have

$$\frac{\partial\overline{\rho_k}}{\partial t}+\nabla\cdot\left(\overline{\rho_k}\widehat{\boldsymbol{v}_k}\right)=\Gamma_k \tag{6-11}$$

where

$$\Gamma_k = \sum_n \int \left[C_{kn}^{\ +} - C_{kn}^{\ -} \right] m_{kn} d\boldsymbol{\xi}. \tag{6-12}$$

The assumptions of over all conservation of mass gives

$$\sum_{k=1}^{2} \Gamma_k = 0. \tag{6-13}$$

From Eqs.(6-11) and (6-13) we have the mixture continuity equation

$$\frac{\partial \rho_m}{\partial t} + \nabla \cdot \left(\rho_m \boldsymbol{v}_m \right) = 0 \tag{6-14}$$

where

$$\rho_m = \sum_{k=1}^{2} \overline{\rho_k} \quad \text{and} \quad \boldsymbol{v}_m = \sum_{k=1}^{2} \frac{\overline{\rho_k} \widehat{\boldsymbol{v}_k}}{\rho_m}. \tag{6-15}$$

Recalling that ρ_k is a partial density of the kth-phase, we find complete similarity between the results of the time averaging, Eqs.(5-21), (5-22) and (5-40), and those of the Boltzmann statistical averaging, Eqs.(6-11), (6-13) and (6-14).

Conservation of Momentum

The linear momentum equation can be obtained from Eq.(6-10) by setting $\psi_{kn} = \boldsymbol{\xi}$, thus

$$\frac{\partial \overline{\rho_k} \widehat{\boldsymbol{v}_k}}{\partial t} + \nabla \cdot \left(\overline{\rho_k} \widehat{\boldsymbol{v}_k} \widehat{\boldsymbol{v}_k} \right) = -\nabla \cdot \overline{\mathcal{P}_k} + \overline{\rho_k} \widehat{\boldsymbol{g}_k} + \boldsymbol{M}_k \tag{6-16}$$

where the partial pressure tensor $\overline{\mathcal{P}_k}$ is defined by

$$\overline{\mathcal{P}_k} \equiv \sum_n \rho_{kn} \widehat{\boldsymbol{V}_{kn} \boldsymbol{V}_{kn}} \tag{6-17}$$

and the momentum supply $\boldsymbol{M}_k$ by

$$\boldsymbol{M}_k \equiv \sum_n \int \left[C_{kn}^{\ +} - C_{kn}^{\ -} \right] m_{kn} \boldsymbol{\xi}\, d\boldsymbol{\xi}. \tag{6-18}$$

We note here that if the negative of the partial pressure tensor, $-\overline{\mathcal{P}_k}$, is interpreted as the combined stresses acting on the kth-phase, the momentum equation (6-16) has exactly the same form as that obtained from the time averaging, namely, Eq.(5-26). However, the physical meaning of the flux term and the momentum source $\boldsymbol{M}_k$ in this equation can be significantly different from those of the time-averaged equation.

As an example, let us consider a dispersed two-phase flow system with phase 1 as a dispersed phase. If the particle sizes are considerably small and the continuous phase is a dilute gas, then the formulation essentially reduces to that of reacting gas mixtures. In this case, the effects of the collisions of particles with the molecules and/or particles that lie outside of a particular volume element in real space can be neglected. Thus, the total collision term $\sum_{k=1}^{2}\boldsymbol{M}_k$ can be taken as zero. Hence, as we can see from Eq.(6-18), the collision term of each phase consists of the momentum source due to mass transfer and the drag forces resulting from the momentum exchange during collisions.

Furthermore, if the particles have definite volumes in contrast to the previous point mass assumption, the multi-collisions of particles with molecules and with other particles whose centers lie outside of the volume element become important. In this case, the collision term for the dispersed phase can be split to three different parts: the internal momentum transfer due to collisions; the effects of the changes of phases; and the external collision effect. Consequently, the introduction of the void fraction α_k in the formulation is necessary, where α_1 is the ratio of the volume occupied by the particles to the total volume element and thus $\alpha_2 = 1 - \alpha_1$. Nevertheless, in order to integrate the momentum collision term, it is necessary to introduce models of molecule-particle and particle-particle collision processes. Because of the multi-collisions between a particle and a cloud of molecules and of the effects of the phase changes, these models will be extremely complicated. Thus we do not go into the detail of the collision integrals here. If the standard fluid mechanic viewpoint is introduced, it is possible to interpret the momentum source $\boldsymbol{M}_k$ into relevant physical terms. From the above discussion, we expand the total collision term of each phase in the following form

$$\boldsymbol{M}_k = \boldsymbol{M}_k^i + \boldsymbol{M}_k^\Gamma + \boldsymbol{M}_k^e. \tag{6-19}$$

Each term on the right-hand side represents the internal collision force interaction, the momentum source due to phase changes and the external collision force interaction because of the finite particle size, respectively. The first term can be considered as the standard drag force and the pressure effect due to the void fraction gradient, whereas the last term gives extra flux from the particles having their centers outside of a volume element.

For the mixture, the total collision term $\sum_{k=1}^{2}\boldsymbol{M}_k$ is not zero due to the external effects $\boldsymbol{M}_k^e$. Thus, we can write approximately

$$\sum_{k=1}^{2}\boldsymbol{M}_k \doteq \boldsymbol{M}_m - \left(\nabla\cdot\frac{\alpha_1}{\alpha_2}\overline{\mathcal{P}_2}\right) \tag{6-20}$$

where $\boldsymbol{M}_m$ can be considered as the inter-particle collision effect. The second term is the effect of the continuous phase on particles which lie on the boundary of a volume element. The real pressure tensor for the continuous phase 2 should be

$$\overline{\overline{\mathcal{P}_2}} = \frac{\overline{\mathcal{P}_2}}{\alpha_2}. \tag{6-21}$$

Then the mixture momentum equation becomes

$$\begin{aligned}&\frac{\partial\rho_m\boldsymbol{v}_m}{\partial t} + \nabla\cdot\left(\rho_m\boldsymbol{v}_m\boldsymbol{v}_m\right)\\ &= -\nabla\cdot\left(\overline{\mathcal{P}_1} + \overline{\overline{\mathcal{P}_2}} + \sum_{k=1}^{2}\overline{\rho_k}\widehat{\boldsymbol{V}_{km}\boldsymbol{V}_{km}}\right) + \rho_m\boldsymbol{g}_m + \boldsymbol{M}_m.\end{aligned} \tag{6-22}$$

Thus, the mixture total stress consists of the partial stress of the dispersed phase, the real stress of the continuous phase and the diffusion stress due to the relative motion between two phases. Here, the definition of the average diffusion velocity $\boldsymbol{V}_{km}$ takes the form of Eq.(4-89). It is evident that significant differences exist between the physical meaning of the stress tensor term of the present mixture momentum equation based on the Boltzmann transport equation and that of the one obtained from the time averaging, namely, Eq.(5-42). This basic difference arises because, in the former approach, the stresses are defined from the motions of the particles. Thus, the stress inside the particles has no place in the analysis. In the latter approach, however, the stresses are defined everywhere in the system. We may add here that the Boltzmann statistical average can be easily extended to include the turbulent fluctuations in the continuous phase by considering the eddy transports as the particle transports.

Conservation of Energy

In contrast to the simple kinetic theory, the energy transfer by multi-molecular particles is considerably complicated due to the internal freedom of the energy state of the particles. It is obvious that the kinetic theory translational temperature based on the diffusion kinetic energy is not useful if each particle consists of a large number of molecules.

Now let us suppose that ψ_{kn} is the total energy carried by the kth-phase particles having mass m_{kn}, thus

$$\psi_{kn} = \frac{1}{2}\xi_{kn}^2 + u_{kn} \tag{6-23}$$

where the first term is the translational kinetic energy and u_{kn} the internal energy contained by molecules in the particles. Since the term u_{kn} is not a function of $\boldsymbol{\xi}_{kn}$, the transport equation (6-10) is not a velocity moment of the Boltzmann transport equation.

We introduce the average energy and the flux of the kth-phase as

$$\widehat{e_k} = \frac{\sum_n \rho_{kn}\left(\frac{1}{2}\widehat{V_{kn}^2} + \widehat{u_{kn}}\right)}{\overline{\rho_k}} \tag{6-24}$$

and

$$\overline{\boldsymbol{q}_k} = \sum_n \rho_{kn}\overline{\left(\frac{1}{2}V_{kn}^2 + u_{kn}\right)\boldsymbol{V}_{kn}} \tag{6-25}$$

where the definitions of the right-hand side averages follow Eq.(6-8). Substituting Eqs.(6-23), (6-24) and (6-25) into Eq.(6-10), we obtain the total energy equation for the kth-phase

$$\begin{aligned} &\frac{\partial}{\partial t}\left\{\overline{\rho_k}\left(\widehat{e_k} + \frac{1}{2}\widehat{v_k}^2\right)\right\} + \nabla\cdot\left\{\overline{\rho_k}\left(\widehat{e_k} + \frac{1}{2}\widehat{v_k}^2\right)\widehat{\boldsymbol{v}_k}\right\} \\ &= -\nabla\cdot\overline{\boldsymbol{q}_k} - \nabla\cdot\left(\overline{\mathcal{P}_k}\cdot\widehat{\boldsymbol{v}_k}\right) + \overline{\rho_k}\widehat{\boldsymbol{g}_k}\cdot\widehat{\boldsymbol{v}_k} \\ &+\sum_n \int m_{kn} f_{kn}\left(\frac{\partial u_{kn}}{\partial t} + \boldsymbol{\xi}\cdot\nabla u_{kn}\right)d\boldsymbol{\xi} + E_k. \end{aligned} \tag{6-26}$$

The average internal energy $\widehat{e_k}$ is the sum of the random thermal translational energy and the true internal energy of the particles. This definition is in complete analogy with Eq.(5-31) of the time averaging. The most important characteristic appears in the last two terms of Eq.(6-26). It is evident that the change of the individual particle internal energy given by the second term from the last is coupled with the collision term E_k since, in the absence of the long range energy exchanges, the particle internal energy changes only by the interactions with the surrounding molecules and particles. Thus, in analogy with the momentum exchange given by Eq.(6-20), the total energy interaction can be given as

$$\sum_{k=1}^{2}\left\{E_k + \sum_n \int m_{kn} f_{kn}\left(\frac{\partial u_{kn}}{\partial t} + \boldsymbol{\xi}\cdot\nabla u_{kn}\right)d\xi\right\} \sim -\nabla\cdot\left\{\frac{\alpha_1}{\alpha_2}\left(\overline{\boldsymbol{q}_2} + \overline{\mathcal{P}_2}\cdot\widehat{\boldsymbol{v}_2}\right)\right\} + E_m. \tag{6-27}$$

Here the first term of the right-hand side of Eq.(6-27) takes into account the particle-molecular collisions for finite volume particles. The term denoted by E_m represents such effects as the inter-particle collision transport of energy. By adding two energy equations for each phase, the mixture equation can be obtained

$$\begin{aligned}
&\frac{\partial}{\partial t}\left\{\rho_m\left[e_m + \left(\frac{v^2}{2}\right)_m\right]\right\} + \nabla\cdot\left\{\rho_m\left[e_m + \left(\frac{v^2}{2}\right)_m\right]\boldsymbol{v}_m\right\} \\
&= -\nabla\cdot\left[\overline{\boldsymbol{q}_1} + \overline{\overline{\boldsymbol{q}_2}} + \sum_{k=1}^{2}\overline{\rho_k}\left(\widehat{e_k} + \frac{\widehat{v_k}^2}{2}\right)\boldsymbol{V}_{km}\right] + \rho_m\boldsymbol{g}_m\cdot\boldsymbol{v}_m \\
&\quad -\nabla\cdot\left[\overline{\mathcal{P}_1}\cdot\widehat{\boldsymbol{v}_1} + \overline{\overline{\mathcal{P}_2}}\cdot\widehat{\boldsymbol{v}_2}\right] + \sum_k \overline{\rho_k}\boldsymbol{g}_k\cdot\boldsymbol{V}_{km} + E_m.
\end{aligned} \tag{6-28}$$

Here, we again note that phase 1 is the dispersed phase. The physical significance of each term of the total energy equation parallels that of the momentum equation. The partial fluxes $\overline{\boldsymbol{q}_1}$ and $\overline{\mathcal{P}_1}$ for the dispersed phase, and the real fluxes $\overline{\overline{\boldsymbol{q}_2}}$ and $\overline{\overline{\mathcal{P}_2}}$ for the continuous phase appear because of the finite volume occupied by the particles of phase 1.

As in the case of the mixture momentum equation, significant differences between the Boltzmann statistical model and the time-averaged model appear in the total flux terms of heat and work. Since the heat transfer and work inside the particles are not considered in the former model, the partial heat flux and the partial pressure tensor for the dispersed phase represent the transfer due to inter-particle diffusions. The molecules of the continuous phase 2 are treated as the point mass, thus *the fluxes of each phase do not appear symmetrically.*

The interest of the Boltzmann statistical average applied to two-phase flow systems lies mainly on the study of constitutive equations from a simple model for the collision terms together with stochastic assumptions. It is highly improbable that this can be a general model for a dispersed two-phase flow system, since the inclusion of the effects of particle shapes and deformations in the Boltzmann transport equation brings considerable difficulties in the analysis. Even without these effects, it is anticipated that the collision terms for each phase are very complicated due to three effects: 1) the inter-particle collisions, coalescences and disintegrations; 2) the multi-collisions between a particle and a large number of molecules; 3) the existence of phase changes. Furthermore as it can be seen from Eq.(6-27), the dispersed phase energy transfer term requires a special constitutive law for the heat transfer between particles and fluid.

In summarizing the section, it can be said that the Boltzmann statistical averaging is useful for a highly dispersed flow where each particle is considered as a lumped entity rather than as a distributed system itself. For example, the mixture stress tensor for a particle flow has a more natural form in the present model than the one from the time averaging. This is because in such two-phase flow it is not practical to introduce stresses inside the particles. However, the number of particles in a volume element should be significantly large for a statistical treatment of the number density to be realistic. Furthermore, if the deformations of the interfaces and the changes of properties within the particles are important, the Boltzmann statistical method cannot be used. In contrast to the case of dilute gases, the collision integral terms for a dispersed two-phase system with finite particle sizes are extremely complicated, thus to obtain constitutive laws from the statistical mechanics is very difficult. In many cases these collision terms should be supplied from the continuum mechanics considerations. It is, therefore, also possible to construct a model on a combination of the continuum theory and the Boltzmann statistical method. For example, we take the statistical average only for the dispersed phase with the drag forces included in the body force. Then we use the standard volume averaged field equations for the other phase with the corresponding interaction terms. Such a formulation is useful to dispersed two-phase systems as well as to fluidized beds. Furthermore, the Boltzmann statistical method can be very useful for obtaining some constitutive laws as demonstrated in Chapter 10.

Chapter 7

KINEMATICS OF AVERAGED FIELDS

1.1 Convective coordinates and convective derivatives

The time-mean values are consistently expressed by the spatial description as shown by the definitions (4-15) and (4-16), and the idea of the particle coordinates for the averaged two-phase flow fields is not clear nor trivial due to the phase changes and the diffusions. The phase change corresponds to the production or disappearance of fluid particles for each phase throughout the field. The difficulty arises because each phase itself does not apparently obey the corollary of the axiom of continuity, namely, the permanence of matter. However, the diffusion of each phase permits the penetration of mixture particles by other fluid particles. It is clear that the material coordinates, which is the base of the standard continuum mechanics, is not inherent to a general two-phase flow field obtained from the time averaging. However, it is possible to introduce mathematically special convective coordinates which are useful in studying the kinematics of each phase and of the mixture.

The path line of each phase is defined by the integral curve of the system

$$d\boldsymbol{x} = \widehat{\boldsymbol{v}_k}(\boldsymbol{x}, t)dt \tag{7-1}$$

with the initial condition

$$\boldsymbol{x} = \boldsymbol{X}_k \text{ at } t = t_0 \tag{7-2}$$

where we define $\boldsymbol{X}_k$ as the convective coordinates of the kth-phase. Hence, upon integration of Eq.(7-1), we obtain

$$\boldsymbol{x} = \boldsymbol{x}(\boldsymbol{X}_k, t) \tag{7-3}$$

This equation gives the path line of the fixed point on the convective coordinates $\boldsymbol{X}_k$, which are moving with the particle velocity $\widehat{\boldsymbol{v}_k}$.

With the standard assumption of smoothness, or the existence of the Jacobian, we can transform Eq.(7-3) to

$$\boldsymbol{X}_k = \boldsymbol{X}_k(\boldsymbol{x}, t) \tag{7-4}$$

This equation expresses the position of the imaginary particle that moves with the local mean velocity of the kth-phase $\widehat{\boldsymbol{v}_k}$. The formulation of problems in which $\boldsymbol{x}$ and t are taken as independent variables is called the spatial description, whereas if $\boldsymbol{X}_k$ and t are taken as the independent variables, it is called the convective description. In general, the contents of the particles occupying the neighborhood of $\boldsymbol{x} = \boldsymbol{x}(\boldsymbol{X}_k, t)$ can be different from the initial particles due to phase changes. Thus, it is not possible to consider the change with fixed particles. However, it is simple to observe a process with fixed $\boldsymbol{X}_k$. The velocity of the kth-phase, for example, can be given in analogy with a single-phase flow as

$$\widehat{\boldsymbol{v}_k} = \left.\frac{\partial \boldsymbol{x}}{\partial t}\right)_{\boldsymbol{X}_k} \quad k = 1 \text{ and } 2. \tag{7-5}$$

The above analysis can also be applied to the mixture center of mass, thus we define the mixture path line by

$$d\boldsymbol{x} = \boldsymbol{v}_m(\boldsymbol{x}, t)\, dt \tag{7-6}$$

with $\boldsymbol{x} = \boldsymbol{X}_m$ at $t = t_0$. By integrating Eq.(7-6) we obtain the path line

$$\boldsymbol{x} = \boldsymbol{x}(\boldsymbol{X}_m, t) \tag{7-7}$$

and with the inverse transformation we get

$$\boldsymbol{X}_m = \boldsymbol{X}_m(\boldsymbol{x}, t). \tag{7-8}$$

Hence, if the mixture convective coordinates are fixed in Eq.(7-7), the observer moves with the local mixture velocity $\boldsymbol{v}_m$. However, due to the diffusion of each phase with respect to the mass center, the particles at fixed $\boldsymbol{X}_m$ are continually changing along the path line.

Furthermore, it is interesting to note that if the flow field is homogeneous, or $\widehat{\boldsymbol{v}_1} = \widehat{\boldsymbol{v}_2} = \boldsymbol{v}_m$, then the mixture convective coordinates become the material coordinates regardless of the phase changes. From Eqs.(7-6) and (7-7) the mixture velocity can be given symbolically as

$$\boldsymbol{v}_m = \left.\frac{\partial \boldsymbol{x}}{\partial t}\right)_{\boldsymbol{X}_m} . \tag{7-9}$$

It is easily seen from Eqs.(7-1) and (7-6) that the path line for each phase and for the mixture can intersect each other.

Since the Eulerian time mean values are in spatial description, the time rate of change at fixed point is denoted by

$$\frac{\partial}{\partial t} \equiv \left(\frac{\partial}{\partial t}\right)_{\boldsymbol{x}} . \tag{7-10}$$

However, the rate of change seen from the observer moving with the fluid velocity is called the convective or substantial derivative. It is given by

$$\frac{D_k}{Dt} = \left.\frac{\partial}{\partial t}\right)_{\boldsymbol{X}_k} = \frac{\partial}{\partial t} + \widehat{\boldsymbol{v}_k} \cdot \nabla \tag{7-11}$$

and

$$\frac{D}{Dt} \equiv \left.\frac{\partial}{\partial t}\right)_{\boldsymbol{X}_m} = \frac{\partial}{\partial t} + \boldsymbol{v}_m \cdot \nabla . \tag{7-12}$$

The convective derivatives of Eq.(7-11) and Eq.(7-12) are taken by following the center of mass of the kth-phase and that of the mixture, respectively.

If the phase convective derivative is applied to express the left-hand side of the field equation (5-17), we obtain

$$\frac{\partial \alpha_k \overline{\overline{\rho_k}} \widehat{\psi_k}}{\partial t} + \nabla \cdot \left(\alpha_k \overline{\overline{\rho_k}} \widehat{\psi_k} \widehat{\boldsymbol{v}_k}\right) = \alpha_k \overline{\overline{\rho_k}} \frac{D_k \widehat{\psi_k}}{Dt} + \Gamma_k \widehat{\psi_k} \tag{7-13}$$

where we have used the continuity equation (5-21). Similarly for the mixture, we get from Eqs.(7-12) and (5-40) the following result

$$\frac{\partial \rho_m \psi_m}{\partial t} + \nabla \cdot \left(\rho_m \psi_m \boldsymbol{v}_m \right) = \rho_m \frac{D\psi_m}{Dt}. \tag{7-14}$$

We note here that the contribution of the mass source term appears in Eq.(7-13), since the amount of mass within a volume having the surface velocity of $\widehat{\boldsymbol{v}_k}$ is not constant. By combining the corollary of the fundamental identity, Eq.(4-111), and above two relations, we have an important transformation between the mixture and phase convective derivatives, thus

$$\rho_m \frac{D\psi_m}{Dt} + \nabla \cdot \mathcal{J}^D = \sum_{k=1}^{2} \left(\alpha_k \overline{\overline{\rho_k}} \frac{D_k \widehat{\psi_k}}{Dt} + \Gamma_k \widehat{\psi_k} \right). \tag{7-15}$$

1.2 Streamline

The stagnation point is defined as a point where all velocities vanish, thus

$$\widehat{\boldsymbol{v}_1} = \widehat{\boldsymbol{v}_2} = \boldsymbol{v}_m = 0. \tag{7-16}$$

And the point where $\widehat{\boldsymbol{v}_k}$ is zero for one of the phases is called the kth-phase stagnation point. If the mixture velocity $\boldsymbol{v}_m$ is zero at a point, we call it as a pseudo-stagnation point. At such a point the motions of two phases are pure diffusions. The flow is completely steady if each of the phase velocities is independent of time as

$$\widehat{\boldsymbol{v}_k} = \widehat{\boldsymbol{v}_k}(\boldsymbol{x}) \text{ for both } k = 1 \text{ and } 2. \tag{7-17}$$

The mixture motion is steady if $\boldsymbol{v}_m = \boldsymbol{v}_m(\boldsymbol{x})$, however it does not correspond to the complete steady motion because the diffusion velocities can be a function of time.

The vector line of a vector field is a curve that is everywhere tangent to that vector. In particular, the vector line of the velocity field $\widehat{\boldsymbol{v}_k}$ is called the streamline of the kth-phase. Thus it can be given by an integral curve of the simultaneous equations

$$d\boldsymbol{x} = \widehat{\boldsymbol{v}_k} dl \text{ at } t = t_0 \tag{7-18}$$

where l is a parameter along the streamline. In general, the streamline is a

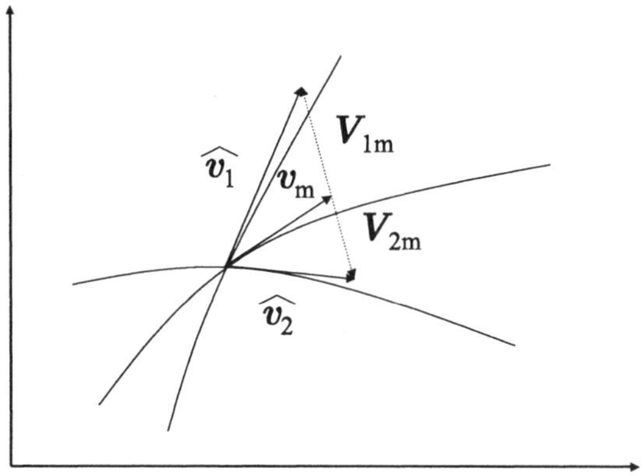

Figure 7-1. Streamlines (Ishii, 1975)

function of time and does not coincide with the path line. The two streamlines for each phase are also different because two velocity fields are not parallel. The streamline of the mixture can be defined similarly as

$$d\boldsymbol{x} = \widehat{\boldsymbol{v}_m} dl \quad \text{at } t = t_0. \tag{7-19}$$

The relations between various streamlines are illustrated in Figure 7-1. We also note here that the coincidence of the streamlines of each phase does not signify the homogenous flow field.

1.3 Conservation of mass

Formulation Based on Center of Mass Velocities

The continuity equations for each phase have been derived in the Section 1.2 of Chapter 5, thus we have

$$\frac{\partial \alpha_k \overline{\overline{\rho_k}}}{\partial t} + \nabla \cdot \left(\alpha_k \overline{\overline{\rho_k}} \widehat{\boldsymbol{v}_k} \right) = \Gamma_k \quad k = 1 \text{ and } 2 \tag{7-20}$$

and the interfacial mass transfer condition is given by

$$\sum_{k=1}^{2} \Gamma_k = 0. \tag{7-21}$$

Equation (7-20) simply states that the local time rate of change of the partial density $\alpha_k \overline{\overline{\rho_k}}$ per unit volume equals the net mass influx of the kth-phase

$-\nabla\cdot\left(\alpha_k\overline{\overline{\rho_k}}\widehat{\boldsymbol{v}_k}\right)$ plus the mass source due to the phase changes. As it has been shown in the Section 1.2 of Chapter 5, by adding these three equations we obtain

$$\frac{\partial\rho_m}{\partial t}+\nabla\cdot\left(\rho_m\boldsymbol{v}_m\right)=0 \tag{7-22}$$

which is the equation of continuity for the mixture.

In order to specify the conservation of mass in two-phase mixtures, it is necessary to employ two continuity equations. We have expressed these relations through the *center of mass velocity of each phase* in Eq.(7-20), however, it is interesting now to consider alternative forms of the continuity equations by introducing different views of observations. If the observer moves with the mixture center of mass, the diffusion terms explicitly appear in the phase continuity equations, thus we have

$$\frac{\partial\alpha_k\overline{\overline{\rho_k}}}{\partial t}+\nabla\cdot\left(\alpha_k\overline{\overline{\rho_k}}\boldsymbol{v}_m\right)=\Gamma_k-\nabla\cdot\left(\alpha_k\overline{\overline{\rho_k}}\boldsymbol{V}_{km}\right). \tag{7-23}$$

Formulation Based on Mass Fractions

Instead of using the time fraction α_k, we may also express the above equation in terms of the mass fraction c_k defined by Eq.(4-63) as

$$\frac{\partial c_k}{\partial t}+\boldsymbol{v}_m\cdot\nabla c_k=\frac{\Gamma_k}{\rho_m}-\frac{1}{\rho_m}\nabla\cdot\left(c_k\rho_m\boldsymbol{V}_{km}\right) \tag{7-24}$$

in which we have used the mixture continuity equation. Furthermore, the diffusion coefficient D_k may be used to express the diffusion flux in analogy with a heterogeneous single-phase mixture, hence we set

$$c_k\rho_m\boldsymbol{V}_{km}=-\rho_m D_k\nabla c_k. \tag{7-25}$$

We note here that Eq.(7-25) is correct only when the diffusion is due to the concentration gradient and it can be expressed by a linear constitutive law. However, it is expected that Fick's Law of Diffusion may not hold for a general two-phase flow system because the interfacial geometry, the body force field and the interfacial momentum transfer term are the significant factors affecting the diffusion of phases. The linear constitutive law given by Eq.(7-25) is in complete analogy with Newton's Law of Viscosity and with Fourier's Law of Heat Conduction. These linear constitutive laws are applicable for the molecular transport phenomena. However, it has to be remembered that the latter two constitutive equations express the *microscopic* molecular diffusions of momentum and energy, whereas the diffusion of phases in two-phase flow is *macroscopic*.

By considering a very simplified form of Eq.(7-24), it is possible to show that the diffusion equation with the linear constitutive law of Eq.(7-25) exhibits the *diffusive characteristic* of the concentration c_k due to the second order derivative of c_k in the equation. This is in direct contrast with the formulation based on the kinematic wave velocity in the later part of this section, which exhibits the *characteristic of propagations*.

Formulation Based on Volumetric Flux

The continuity relations also can be expressed through the volumetric flux $\boldsymbol{j}$ and the drift velocities $\boldsymbol{V}_{kj}$. Thus, from Eqs.(7-20) and (4-91) we have

$$\frac{\partial \alpha_k \overline{\overline{\rho_k}}}{\partial t} + \nabla \cdot \left(\alpha_k \overline{\overline{\rho_k}} \boldsymbol{j}\right) = \Gamma_k - \nabla \cdot \left(\alpha_k \overline{\overline{\rho_k}} \boldsymbol{V}_{kj}\right). \tag{7-26}$$

The last term on the right-hand side of the above equation represents the drift of kth-phase mass with respect to the mixture volume center. By differentiating by part of the left-hand side of Eq.(7-20), we get

$$\frac{\partial \alpha_k}{\partial t} + \nabla \cdot \left(\alpha_k \widehat{\boldsymbol{v}_k}\right) = \frac{\Gamma_k}{\overline{\overline{\rho_k}}} - \frac{\alpha_k}{\overline{\overline{\rho_k}}} \frac{D_k \overline{\overline{\rho_k}}}{Dt} \tag{7-27}$$

where the substantial derivative is defined by Eq.(7-11). The above equation can be considered as the continuity equation in terms of the time fraction or the void fraction, therefore it represents the volumetric transport. Thus, from the point of view of α_k, the continuity equation has a source term due to the mass transfer and a sink term due to the true compressibility of the phase. Furthermore, if we use Eq.(4-87), we obtain

$$\frac{\partial \alpha_k}{\partial t} + \nabla \cdot \boldsymbol{j}_k = \frac{\Gamma_k}{\overline{\overline{\rho_k}}} - \frac{\alpha_k}{\overline{\overline{\rho_k}}} \frac{D_k \overline{\overline{\rho_k}}}{Dt} \tag{7-28}$$

By adding these two equations for each phase, we get

$$\nabla \cdot \boldsymbol{j} = \sum_{k=1}^{2} \left\{ \frac{\Gamma_k}{\overline{\overline{\rho}}_k} - \frac{\alpha_k}{\overline{\overline{\rho}}_k} \frac{D_k \overline{\overline{\rho}}_k}{Dt} \right\} \tag{7-29}$$

which describes the divergence of the center of volume velocity. The first term of the right-hand side is the volume source due to the phase changes and the second term is the volume sink due to the compressibility.

The formulation based on the volumetric fluxes is important if each phase undergoes the incompressible or isochoric process defined by

$$\frac{D_k \overline{\overline{\rho}}_k}{Dt} = 0 \quad k = 1 \text{ and } 2. \tag{7-30}$$

In this case Eq.(7-29) reduces to

$$\nabla \cdot \boldsymbol{j} = \sum_{k=1}^{2} \frac{\Gamma_k}{\overline{\overline{\rho}}_k} \text{ (isochoric)} \tag{7-31}$$

which simply states that the divergence of the volumetric flux is proportional to the amount of phase changes and to the difference between the specific volumes of each phase. We recall that for an incompressible single-phase flow, the divergence of the velocity is zero. Thus, the two-phase flow equivalence is expressed by the velocity of a center of volume instead of that of mass and, furthermore, it has a source due to phase changes. It is seen that in the absence of the mass transfer, the above equation reduces to $\nabla \cdot \boldsymbol{j} = 0$. Since in many practical two-phase flow problems the incompressible fluid assumption is valid and the rate of phase change Γ_k can be expressed as a known function of position and time, Eq.(7-31) can play an important role in solving these problems.

Kinematic Wave Velocity and Void Propagation Equation

As in the thermomechanical theory of diffusion for single-phase mixtures, it is one of the basic assumptions of drift-flux (or mixture) model formulation that the relative motions between two phases can be expressed by *a constitutive law* rather than by two momentum equations. In this connection we already discussed Fick's Law of Diffusion which effectively eliminates one of the two momentum equations. It is quite clear from Eq.(7-25) that the constitutive law gives special kinematic relation between $\widehat{\boldsymbol{v}}_k$ and $\boldsymbol{v}_m$, thus the kth-phase momentum equation becomes redundant in the presence of the mixture momentum equation.

However, we have noted there that in general, the use of Fick's Law for two-phase mixtures is *not correct* and thus a different type of constitutive laws for diffusion should be used as in the drift-flux (or mixture) model formulation. One of the more useful constitutive laws for the relative motions between phases is to express it in terms of the drift velocity $\boldsymbol{V}_{kj}$ (Zuber et al., 1964, Ishii, 1977, Kataoka and Ishii, 1987, Hibiki and Ishii, 2003a, 2003b; Hibiki et al., 2003, Goda et al., 2003; Hibiki et al., 2004).

In particular if the drift velocity is a function only of the time concentration α_k, one of the very important theories in fluid mechanics, namely, the theory of kinematic waves (Kynch, 1952; Lighthill and Whitham, 1955; Hayes, 1970), can be applied to the two-phase flow systems. It was shown by Zuber (1964b) that such was the case for many flow regimes of practical interest and it was particularly useful for a dispersed flow regime.

Under the incompressible fluid assumption, Eq.(7-27) can be expressed by $\boldsymbol{j}$ and $\boldsymbol{V}_{kj}$ in the following form

$$\frac{\partial \alpha_k}{\partial t} + \boldsymbol{j} \cdot \nabla \alpha_k + \nabla \cdot \left(\alpha_k \boldsymbol{V}_{kj} \right) = \frac{\rho_m \Gamma_k}{\rho_1 \rho_2}. \tag{7-32}$$

And if the drift velocity can be approximated as a function of α_k only, then

$$\boldsymbol{V}_{kj} \approx \boldsymbol{V}_{kj} \left(\alpha_k \right). \tag{7-33}$$

Substituting Eq.(7-33) into Eq.(7-32) we obtain the *void propagation equation*

$$\frac{\partial \alpha_k}{\partial t} + \boldsymbol{C}_K \cdot \nabla \alpha_k = \frac{\rho_m \Gamma_k}{\rho_1 \rho_2} \tag{7-34}$$

where the kinematic wave velocity $\boldsymbol{C}_K$ is defined by

$$\boldsymbol{C}_K \equiv \boldsymbol{j} + \frac{\partial}{\partial \alpha_k} \left(\alpha_k \boldsymbol{V}_{kj} \right). \tag{7-35}$$

Hence, denoting the special convective derivative following the kinematic wave by

$$\frac{D_c}{Dt} \equiv \frac{\partial}{\partial t} + \boldsymbol{C}_K \cdot \nabla \tag{7-36}$$

the void propagation equation reduces to

$$\frac{D_c}{Dt}(\alpha_k) = \frac{\rho_m \Gamma_k}{\overline{\overline{\rho_1}}\,\overline{\overline{\rho_2}}}. \tag{7-37}$$

Thus if we observe the time rate of change of α_k by moving with the kinematic wave velocity, it is proportional to the source term due to phase changes. In the absence of mass transfer between the phases, the disturbance of α_k propagates with the kinematic wave velocity. Furthermore, under the condition of *constant* $\overline{\overline{\rho_1}}$ *and* $\overline{\overline{\rho_2}}$, we can express Eq.(7-34) in terms of the mixture density as follows

$$\frac{\partial \rho_m}{\partial t} + \boldsymbol{C}_K \cdot \nabla \rho_m = \frac{\rho_m}{\overline{\overline{\rho_1}}\,\overline{\overline{\rho_2}}} \sum_{k=1}^{2} \Gamma_k \overline{\overline{\rho_k}}$$

or (7-38)

$$\frac{\partial (\ln \rho_m)}{\partial t} + \boldsymbol{C}_K \cdot \nabla (\ln \rho_m) = \frac{1}{\overline{\overline{\rho_1}}\,\overline{\overline{\rho_2}}} \sum_{k=1}^{2} \Gamma_k \overline{\overline{\rho_k}}$$

which is called the *density propagation equation.*

Kinematic Shock Wave

It has been shown that if the drift velocity is a function only of α_k, the void fraction equation can be transformed into the void propagation equation. In contrast, if the diffusion of phases can be expressed by the constitutive equation having the form of Fick's Law of Diffusion, the field exhibits the characteristic of diffusive media and the clear-cut void propagation cannot be observed due to the second-order derivative in space.

The former phenomenon of the void propagation is known for several types of mixtures. For example, they are important in open channel, bubbly two-phase and highway traffic flows. In such systems, it is observed that under certain conditions these kinematic wave propagations lead to a formation of a concentration shock. Because of its origin and a necessity to differentiate it from a shock wave due to compressibility effects, we refer it as a *kinematic shock wave*. As it has been shown by Lighthill and Whitham (1955) and Kynch (1952), this phenomenon can be analyzed by a kinematic consideration with a simple constitutive law for the flux of matter which depends on the concentration. However, without going into a detailed discussion, it is possible to write conditions that should be satisfied at the kinematic shock wave. This can be done by utilizing the macroscopic jump conditions of the Section 1.5 of Chapter 5. Thus we apply Eqs.(5-64) and (5-65) to the balance of mass at the shock front, then

$$\sum_{+,-} \rho_m \boldsymbol{n} \cdot \left(\boldsymbol{v}_m - \boldsymbol{U} \right) = 0 \tag{7-39}$$

and

$$\sum_{+,-} \alpha_k \overline{\overline{\rho_k}} \boldsymbol{n} \cdot \left(\widehat{\boldsymbol{v}_k} - \boldsymbol{U} \right) + \Gamma_{ka} = 0. \tag{7-40}$$

It is evident that Eq.(7-39) expresses the conservation of total mass, whereas Eq.(7-40) gives the balance of kth-phase mass across the shock. Here Γ_{ka} denotes the amount of phase changes within the shock layer. Thus Eq.(7-40) states that the kth-phase mass fluxes from each side of the shock wave balance with the mass production due to phase changes in the shock.

By solving Eq.(7-40) for the displacement velocity of the shock, we obtain

$$\boldsymbol{n}^+ \cdot \boldsymbol{U} = \frac{\left(\alpha_k^+ \overline{\overline{\rho_k}}^+ \widehat{\boldsymbol{v}_k}^+ - \alpha_k^- \overline{\overline{\rho_k}}^- \widehat{\boldsymbol{v}_k}^- \right) \cdot \boldsymbol{n}^+ + \Gamma_{ka}}{\left(\alpha_k^+ \overline{\overline{\rho_k}}^+ - \alpha_k^- \overline{\overline{\rho_k}}^- \right)} \tag{7-41}$$

where + and - denote each side of the shock layer. It should be remembered that the condition given by Eq.(7-41) is applicable not only to a kinematic shock wave but also to a dynamic shock wave due to compressibility effects.

By limiting our case to a strictly kinematic phenomenon, we assume here that the phase densities are continuous across the shock and there is no change of phase in the layer. Thus we have

$$\overline{\overline{\rho_k}}^+ = \overline{\overline{\rho_k}}^- \tag{7-42}$$

and

$$\Gamma_{ka} = 0. \tag{7-43}$$

Hence Eq.(7-40) reduces to

$$\sum_{+,-} \alpha_k \boldsymbol{n} \cdot \left(\widehat{\boldsymbol{v}_k} - \boldsymbol{U} \right) = 0. \tag{7-44}$$

It follows that the kinematic shock wave velocity $\boldsymbol{U}$ should satisfy

$$\boldsymbol{n}^+ \cdot \boldsymbol{U} = \frac{\boldsymbol{n}^+ \cdot \left[\alpha_k^+ \widehat{\boldsymbol{v}_k}^+ - \alpha_k^- \widehat{\boldsymbol{v}_k}^- \right]}{\left(\alpha_k^+ - \alpha_k^- \right)}. \tag{7-45}$$

If we use the definition of the volumetric flux $\boldsymbol{j}_k$ defined by Eq.(4-87), we have

$$\boldsymbol{n}^+ \cdot \boldsymbol{U} = \frac{\boldsymbol{n}^+ \cdot \left(\boldsymbol{j}_k^+ - \boldsymbol{j}_k^- \right)}{\left(\alpha_k^+ - \alpha_k^- \right)}. \tag{7-46}$$

From Eq.(7-46) it can be shown

$$\boldsymbol{n}^+ \cdot \boldsymbol{j}^+ + \boldsymbol{n}^- \cdot \boldsymbol{j}^- = 0. \tag{7-47}$$

This means that across the simple kinematic shock, namely, the phase densities being continuous and no phase changes in the shock, the total volumetric flux $\boldsymbol{j}$ is conserved. In view of Eq.(7-47), we can transform Eq.(7-46) into the following form

$$\boldsymbol{n}^+ \cdot \boldsymbol{U} = \boldsymbol{n}^+ \cdot \boldsymbol{j} + \frac{\boldsymbol{n}^+ \cdot \left(\alpha_k^+ \boldsymbol{V}_{kj}^+ - \alpha_k^- \boldsymbol{V}_{kj}^- \right)}{\left(\alpha_k^+ - \alpha_k^- \right)}. \tag{7-48}$$

Here we see the close connection between the kinematic wave velocity given by Eq.(7-35) and the displacement velocity of the shock given by Eq.(7-48).

1.4 Dilatation

The Jacobian of the convective and spatial coordinates of the Section 1.1 of Chapter 7 is given by

$$J_k = \frac{\partial (\boldsymbol{x})}{\partial (\boldsymbol{X}_k)} = \frac{\partial \left(x_1, x_2, x_3 \right)}{\partial \left(X_{1k}, X_{2k}, X_{3k} \right)}. \tag{7-49}$$

Since $\boldsymbol{X}_k$ denotes the initial position, the Jacobian J_k gives the relation between the initial and the present volumes if the surface of the volume element moves with the center of mass velocity $\widehat{\boldsymbol{v}_k}$. Hence we have

$$dV_k = J_k dV_{k0} \tag{7-50}$$

where V_k and V_{k0} denotes the present volume and its initial volume, respectively. In view of Eq.(7-5) we have

$$\frac{1}{J_k}\frac{D_k J_k}{Dt} = \nabla \cdot \widehat{\boldsymbol{v}_k} \tag{7-51}$$

which gives an important physical interpretation of the divergence of the phase velocity. Recalling the continuity equation for the kth-phase, it also can be expressed as

$$\nabla \cdot \widehat{\boldsymbol{v}_k} = \frac{1}{\alpha_k \overline{\overline{\rho_k}}}\left(\Gamma_k - \frac{D_k \alpha_k \overline{\overline{\rho_k}}}{Dt}\right). \tag{7-52}$$

These two equations show that the divergence of $\widehat{\boldsymbol{v}_k}$ is directly related to the dilatation of a volume element rather than the density changes within the volume. Furthermore, we see that the dilatation is cause by three effects, namely, the phase change Γ_k, the phase redistribution $D_k \alpha_k / Dt$ and the real compressibility of the fluid $D_k \overline{\overline{\rho_k}} / Dt$.

For a mixture as a whole, the characteristic of the dilatation and of $\nabla \cdot \boldsymbol{v}_m$ basically reduces to that of a single-phase flow. Thus we have

$$\nabla \cdot \boldsymbol{v}_m = \frac{1}{J_m}\frac{DJ_m}{Dt} = -\frac{1}{\rho_m}\frac{D\rho_m}{Dt} \tag{7-53}$$

where J_m is the Jacobian between $\boldsymbol{x}$ and $\boldsymbol{X}_m$ given by

$$J_m = \frac{\partial(\boldsymbol{x})}{\partial(\boldsymbol{X}_m)}. \tag{7-54}$$

Chapter 8

INTERFACIAL TRANSPORT

The exact forms of the interfacial transport terms I_k and I_m for mass, momentum and energy interchanges have been given in the Section 1.2 of Chapter 5. However, they are expressed by the local instant variables, thus it is not possible to use them as the constitutive laws in the averaged field equations. It is evident that we need to understand the physical meaning of these terms in detail before constructing any particular constitutive equations for two-phase flow systems. With this in mind we clarify different physical mechanisms controlling these terms as well as to identify important parameters on which they depend. Furthermore, it is important to accept that not all the characteristics inherent to the local instant two-phase flow can be brought into the time-averaged model. We consider that the averaged field equations express general physical principles governing the macroscopic two-phase flows while the constitutive equations approximate the material responses of a particular group of systems with simple mathematical models. In this connection, we make a number of assumptions in the interfacial transfer terms in order to both distinguish the dominant transfer mechanisms and also eliminate some of the complicated terms that have insignificant effects in the macroscopic field.

1.1 Interfacial mass transfer

The interfacial mass transfer term Γ_k has been given by Eq.(5-19), thus in view of Eqs.(2-70) and (5-57) we have

$$\Gamma_k = -\sum_j \frac{1}{\Delta t}\frac{1}{v_{ni}}\left\{\boldsymbol{n}_k \cdot \rho_k\left(\boldsymbol{v}_k - \boldsymbol{v}_i\right)\right\} = -\sum_j a_{ij}\dot{m}_k \tag{8-1}$$

where $\dot{m}_k$ is the rate of mass loss per interfacial area in unit time from k^{th}-

phase, and $a_{ij}\left(\equiv 1/L_j\right)$ is the surface area concentration for the jth-interface.

We define the surface mean value as

$$\overline{\overline{F}}_{(i)} \equiv \left(\sum_j \frac{F}{L_j}\right) L_S = \frac{\sum_j a_{ij} F}{a_i}. \tag{8-2}$$

Hereafter, we use the subscript (i) only for the variables that may be confused with the bulk fluid properties, and we omit it for the variables that appear only at the interface, for example σ and H_{21}, etc. The mean value defined by Eq.(8-2) corresponds to the phase average of Eq.(4-23), thus we use the same symbol with the subscript i.

Similarly it is also possible to define a surface mean value in analogy with the phase mass weighted mean value of Eq.(4-28). However, at the interfaces these variables of the extensive characteristic appear always in the flux terms, thus it is more convenient to define a mean value weighted by the mass transfer rate $\dot{m}_k$ rather than by the density. Hence we have

$$\widehat{\psi_{ki}} \equiv \frac{\sum_j \frac{1}{\Delta t}\frac{1}{v_{ni}}\dot{m}_k\psi_k}{\sum_j \frac{1}{\Delta t}\frac{1}{v_{ni}}\dot{m}_k} = \frac{\sum a_{ij}\dot{m}_k\psi_k}{\sum a_{ij}\dot{m}_k}. \tag{8-3}$$

By using the definition of Eq.(8-2), the mean mass transfer per unit surface area becomes

$$\overline{\overline{\dot{m}_k}} \equiv \frac{\sum_j a_{ij}\dot{m}_k}{a_i}. \tag{8-4}$$

From Eqs.(8-1) and (8-4), the interfacial mass transfer condition can be rewritten as

$$\sum_{k=1}^{2} \Gamma_k = 0 \quad \text{with} \quad \Gamma_k = -a_i\overline{\overline{\dot{m}_k}} \tag{8-5}$$

1.2 Interfacial momentum transfer

We recall that the macroscopic interfacial momentum transfer term $\boldsymbol{M}_k$ has been obtained in the Section 1.2 of Chapter 5. Thus, in view of Eqs.(2-9), (2-70), (5-23) and (5-57), we have

$$\boldsymbol{M}_k = -\sum_j a_{ij}\left(\dot{m}_k \boldsymbol{v}_k + p_k \boldsymbol{n}_k - \boldsymbol{n}_k \cdot \mathcal{T}_k\right) \tag{8-6}$$

where the term inside the bracket is the rate of the interfacial momentum loss per area from the kth-phase. Since $\boldsymbol{M}_k$ represents the net interfacial momentum gain, it is weighted by the surface area concentration a_{ij}. Similarly, the mixture momentum source from the interfaces is given by

$$\boldsymbol{M}_m = \sum_j a_{ij}\left\{A^{\alpha\beta}\left(\boldsymbol{t}_\alpha\right),_\beta \sigma + A^{\alpha\beta}\boldsymbol{t}_\alpha (\sigma),_\beta\right\} \tag{8-7}$$

in which the two terms on the right-hand side of the above equation represent the effects of the mean curvature and of the surface-tension gradient.

Before we study the vectorial form of the interfacial momentum transfer equation, let us examine the normal component of the momentum jump condition, Eq.(2-91). This is because the original jump condition, Eq.(2-72), contains two distinct pieces of information; one in normal direction and the other in the tangential direction. We should pay special attention in order to preserve this characteristic in the interfacial transfer equation. By dotting the normal jump condition, Eq.(2-91), by a unit normal vector $\boldsymbol{n}_1$ and taking the time average we obtain

$$\begin{aligned}\sum_j a_{ij}&\left\{\left(\frac{\dot{m}_1^2}{\rho_1} - \frac{\dot{m}_2^2}{\rho_2}\right) + \left(p_1 - p_2\right)\right.\\ &\left.-\left(\tau_{nn1} - \tau_{nn2}\right) + 2H_{21}\sigma\right\} = 0.\end{aligned} \tag{8-8}$$

Now we express this equation with the surface mean values defined by Eq.(8-2). In order to simplify the result, we assume

$$\dot{m}_k \approx \overline{\overline{\dot{m}_k}}; \quad \sigma \approx \overline{\overline{\sigma}}; \quad \rho_k \approx \overline{\overline{\rho_{ki}}} \quad \text{at } t \in [\Delta t]_S. \tag{8-9}$$

Thus the mass transfer rate $\dot{m}_k$, the surface tension σ and the density at the interfaces remain approximately constant during the time interval of averaging Δt. Then by neglecting the normal stress terms, we obtain

$$\frac{\Gamma_1^2}{a_i^2}\left(\frac{1}{\overline{\overline{\rho_{1i}}}}-\frac{1}{\overline{\overline{\rho_{2i}}}}\right)+\left(\overline{\overline{p_{1i}}}-\overline{\overline{p_{2i}}}\right)+2\overline{\overline{H_{21}}}\,\overline{\overline{\sigma}}\doteq 0. \tag{8-10}$$

We note here that under similar assumptions we should be able to recover Eq.(8-10) from the vectorial interfacial transfer equation. By using the surface mean values, the kth-phase interfacial momentum gain $\boldsymbol{M}_k$ becomes

$$\boldsymbol{M}_k=\boldsymbol{M}_k^\Gamma+\boldsymbol{M}_k^n+\overline{\overline{p_{ki}}}\nabla\alpha_k+\boldsymbol{M}_k^t-\nabla\alpha_k\cdot\overline{\overline{\mathcal{T}_{ki}}} \tag{8-11}$$

where

$$\begin{aligned}
\boldsymbol{M}_k^\Gamma &= \Gamma_k\widehat{\boldsymbol{v}_{ki}}\\
\boldsymbol{M}_k^n &\doteq \sum_j a_{ij}\left(\overline{\overline{p_{ki}}}-p_k\right)\boldsymbol{n}_k\\
\boldsymbol{M}_k^t &\doteq \sum_j a_{ij}\boldsymbol{n}_k\cdot\left(\mathcal{T}_k-\overline{\overline{\mathcal{T}_{ki}}}\right).
\end{aligned} \tag{8-12}$$

It is noted here that the shear at the interface can be decomposed into the normal and tangential components, thus $\boldsymbol{n}_k\cdot\mathcal{T}_k=\boldsymbol{\tau}_{nk}+\boldsymbol{\tau}_{tk}$. However, the normal stress is negligibly small, therefore it can be assumed that $\boldsymbol{n}_k\cdot\mathcal{T}_k=\boldsymbol{\tau}_{tk}$. The first three terms on the right-hand side of Eq.(8-11) are originally the normal components, whereas the last two terms are essentially tangential components. Here the concentration gradient appears because of Eq.(4-62).

The mixture momentum source $\boldsymbol{M}_m$ becomes

$$\begin{aligned}
\boldsymbol{M}_m &= 2\overline{\overline{H_{21}}}\,\overline{\overline{\sigma}}\nabla\alpha_2+\sum_j 2a_{ij}\left(H_{21}-\overline{\overline{H_{21}}}\right)\overline{\overline{\sigma}}\boldsymbol{n}_1\\
&+\sum_j a_{ij}A^{\alpha\beta}\boldsymbol{t}_\alpha(\sigma)_{,\beta}.
\end{aligned} \tag{8-13}$$

The second term takes into account the effect of the changes of the mean curvature. However, the gradient of σ in the microscopic scale is considered to be small and its vectorial direction is quite random, thus the last term may be neglected. Hence we approximate

$$\boldsymbol{M}_m = 2\overline{\overline{H_{21}}}\,\overline{\overline{\sigma}}\nabla\alpha_2 + \boldsymbol{M}_m^H \tag{8-14}$$

where $\boldsymbol{M}_m^H$ is the effect of the change of the mean curvature on the mixture momentum source.

It is easy to show in view of Eq.(8-11) and Eq.(8-14) that the normal component of the interfacial momentum transfer condition, Eq.(5-27), can be given by

$$\sum_{k=1}^{2}\left\{\left(\frac{\Gamma_k^2}{\overline{\overline{\rho_{ki}}}a_i^2} + \overline{\overline{p_{ki}}}\right)\nabla\alpha_k + \boldsymbol{M}_k^n\right\} - 2\overline{\overline{H_{21}}}\,\overline{\overline{\sigma}}\nabla\alpha_2 - \boldsymbol{M}_m^H = 0. \tag{8-15}$$

Hence, by comparing the scalar form of the normal component Eq.(8-10) to the vectorial form Eq.(8-15), we obtain

$$\sum_{k=1}^{2}\boldsymbol{M}_k^n = \boldsymbol{M}_m^H. \tag{8-16}$$

Here $\boldsymbol{M}_k^n$ represents the form drag and lift force arising from the pressure imbalance at the interface. $\boldsymbol{M}_k^t$ represents the skin drag due to the imbalance of shear forces. The shear components, thus, should satisfy

$$\sum_{k=1}^{2}\boldsymbol{M}_k^t = 0 \text{ with } \overline{\overline{\mathcal{T}_{1i}}} \doteq \overline{\overline{\mathcal{T}_{2i}}} = \overline{\overline{\mathcal{T}_i}}. \tag{8-17}$$

Equation (8-17) shows that there exists an action-reaction relation between the skin drag forces of each phase as well as between the interfacial shear forces.

For simplicity, we combine these two drag forces and define the total generalized drag forces $\boldsymbol{M}_{ik}$ by

$$\boldsymbol{M}_{ik} = \boldsymbol{M}_k^n + \boldsymbol{M}_k^t \tag{8-18}$$

where

$$\boldsymbol{M}_k^n = \sum_j a_{ij}\left(\overline{\overline{p_{ki}}} - p_k\right)\boldsymbol{n}_k$$

$$\boldsymbol{M}_k^t = \sum_j a_{ij}\boldsymbol{n}_k\cdot\left(\mathcal{T}_k - \overline{\overline{\mathcal{T}_{ki}}}\right).$$

Hence,

$$\sum_{k=1}^{2} \boldsymbol{M}_{ik} = \boldsymbol{M}_m^H. \tag{8-19}$$

Furthermore, from the straightforward analysis on the mass transfer rate $\overline{\overline{\dot{m}_k}}$ with the relations given by Eqs.(4-61) and (4-62), we can show

$$\widehat{\boldsymbol{v}_{ki}} = \widehat{\boldsymbol{v}_i} + \frac{\Gamma_k}{\overline{\overline{\rho_{ki}}} a_i^2} \nabla \alpha_k \tag{8-20}$$

which enables us to replace $\widehat{\boldsymbol{v}_{1i}}$ and $\widehat{\boldsymbol{v}_{2i}}$ by a single parameter $\widehat{\boldsymbol{v}_i}$.

As a summary of the interfacial momentum transfer condition, we have the following relations

$$\boldsymbol{M}_k = \boldsymbol{M}_k^\Gamma + \overline{\overline{p_{ki}}} \nabla \alpha_k + \boldsymbol{M}_{ik} - \nabla \alpha_k \cdot \overline{\overline{\mathcal{T}_{ki}}} \tag{8-21}$$

where $\boldsymbol{M}_{ik}$ includes the effects of form drag, lift force and skin drag.

$$\boldsymbol{M}_m = 2\overline{\overline{H_{21}}}\, \overline{\overline{\sigma}} \nabla \alpha_2 + \boldsymbol{M}_m^H \tag{8-22}$$

$$\sum_{k=1}^{2} \boldsymbol{M}_k = \boldsymbol{M}_m \tag{8-23}$$

with

$$\boldsymbol{M}_k^\Gamma = \widehat{\boldsymbol{v}_{ki}} \Gamma_k = \left(\widehat{\boldsymbol{v}_i} + \frac{\Gamma_k}{\overline{\overline{\rho_{ki}}} a_i^2} \nabla \alpha_k \right) \Gamma_k \tag{8-24}$$

$$\sum_{k=1}^{2} \boldsymbol{M}_{ik} = \boldsymbol{M}_m^H. \tag{8-25}$$

If we assume that $\overline{\overline{p_{ki}}}$, $\overline{\overline{\sigma}}$, $\boldsymbol{M}_m^H$, a_i, Γ_k, and $\overline{\overline{\rho_{ki}}}$ are known, then three constitutive laws should be specified for $\overline{\overline{H_{21}}}$, $\widehat{\boldsymbol{v}_i}$ and $\boldsymbol{M}_{i1}$ identifying the interfacial geometry, motion and generalized drag forces. Furthermore, we note that the total generalized drag force consists of the form drag, the skin drag as well as the lift force.

1.3 Interfacial energy transfer

The macroscopic interfacial total energy transfer for the kth-phase is denoted by E_k which appears only after the phase energy equation has been averaged, whereas the mixture interfacial energy source term is E_m. These three terms should satisfy the interfacial energy transfer condition that is a balance equation at the interfaces. Since the relations for E_k and E_m given by Eqs.(5-28) and (5-29) are expressed by the local instant variables, they cannot be used in the macroscopic formulation in their original forms. Now we transform these relations in terms of the macroscopic variables as a first step to establish the constitutive laws at the interfaces.

Because of its practical importance, we start from the analysis on the interfacial thermal energy transfer term Λ_k, then we proceed to the study of E_k. From the definition of Eq.(5-39) and Eqs.(5-19), (5-23) and (5-28), we have

$$\begin{aligned}\Lambda_k &= \frac{\widehat{v_k}^2}{2}\Gamma_k - \boldsymbol{M}_k\cdot\widehat{\boldsymbol{v}_k} + E_k \\ &= \sum_j \frac{1}{\Delta t}\frac{1}{v_{ni}}\left\{-\dot{m}_k\left(u_k + \frac{v_k^2}{2} - \boldsymbol{v}_k\cdot\widehat{\boldsymbol{v}_k} + \frac{\widehat{v_k}^2}{2}\right)\right. \\ &\left. +\boldsymbol{n}_k\cdot\mathcal{T}_k\cdot\left(\boldsymbol{v}_k - \widehat{\boldsymbol{v}_k}\right) - \boldsymbol{n}_k\cdot\boldsymbol{q}_k\right\}.\end{aligned} \tag{8-26}$$

We define the virtual internal energy at the interfaces in analogy with Eq.(5-31), thus

$$\widehat{e_{ki}} \equiv \frac{\sum_j\left\{a_{ij}\dot{m}_k\left(u_k + \dfrac{\left|\boldsymbol{v}_k - \widehat{\boldsymbol{v}_k}\right|^2}{2}\right)\right\}}{\sum_j a_{ij}\dot{m}_k}. \tag{8-27}$$

And the heat input per unit interfacial area is defined by

$$\overline{\overline{q''_{ki}}} = -\left(\sum_j a_{ij}\boldsymbol{n}_k\cdot\boldsymbol{q}_k\right)\frac{1}{a_i}. \tag{8-28}$$

Then Eq.(8-26) can be rewritten as

$$\Lambda_k = \left(\Gamma_k \widehat{e_{ki}} + a_i \overline{\overline{q''_{ki}}}\right) + \sum_j \left\{a_{ij} \boldsymbol{n}_k \cdot \mathcal{T}_k \cdot \left(\boldsymbol{v}_k - \widehat{\boldsymbol{v}_k}\right)\right\}. \tag{8-29}$$

In order to examine the second group on the right-hand side of the above equation, we introduce fluctuating components defined by

$$p'_{ki} = p_k - \overline{\overline{p_{ki}}};\ \boldsymbol{v}'_{ki} = \boldsymbol{v}_k - \widehat{\boldsymbol{v}_{ki}}. \tag{8-30}$$

Then we have

$$\begin{aligned}&\sum_j a_{ij} \boldsymbol{n}_k \cdot \mathcal{T}_k \cdot \left(\boldsymbol{v}_k - \widehat{\boldsymbol{v}_k}\right) \\ &= \sum_j a_{ij} \left\{-\overline{\overline{p_{ki}}} \boldsymbol{n}_k \cdot \left(\boldsymbol{v}_k - \widehat{\boldsymbol{v}_k}\right)\right\} + \sum_j a_{ij} \left\{\boldsymbol{n}_k \cdot \overline{\overline{\mathcal{T}_{ki}}}\right. \\ &\left. + \boldsymbol{n}_k \cdot \left(\mathcal{T}_k - \overline{\overline{\mathcal{T}_{ki}}}\right) + \left(\overline{\overline{p_{ki}}} - p_k\right) \boldsymbol{n}_k\right\} \cdot \left(\boldsymbol{v}_k - \widehat{\boldsymbol{v}_k}\right).\end{aligned} \tag{8-31}$$

Since we have

$$\sum_j a_{ij} \left(\boldsymbol{v}_k - \widehat{\boldsymbol{v}_k}\right) \cdot \boldsymbol{n}_k = \frac{D_k \alpha_k}{Dt} - \frac{\Gamma_k}{\overline{\overline{\rho_{ki}}}} \tag{8-32}$$

the first term on the right-hand side of Eq.(8-31) becomes

$$\sum_j a_{ij} \left\{-\overline{\overline{p_{ki}}} \left(\boldsymbol{v}_k - \widehat{\boldsymbol{v}_k}\right) \cdot \boldsymbol{n}_k\right\} = \overline{\overline{p_{ki}}} \left(\frac{\Gamma_k}{\overline{\overline{\rho_{ki}}}} - \frac{D_k \alpha_k}{Dt}\right). \tag{8-33}$$

The second term can be rearranged to the following form

$$\begin{aligned}&\sum_j a_{ij} \left\{\boldsymbol{n}_k \cdot \overline{\overline{\mathcal{T}_{ki}}} + \boldsymbol{n}_k \cdot \left(\mathcal{T}_k - \overline{\overline{\mathcal{T}_{ki}}}\right) + \left(\overline{\overline{p_{ki}}} - p_k\right) \boldsymbol{n}_k\right\} \cdot \left(\boldsymbol{v}_k - \widehat{\boldsymbol{v}_k}\right) \\ &= \boldsymbol{M}_{ik} \cdot \left(\widehat{\boldsymbol{v}_{ki}} - \widehat{\boldsymbol{v}_k}\right) - \nabla \alpha_k \cdot \overline{\overline{\mathcal{T}_{ki}}} \cdot \left(\widehat{\boldsymbol{v}_{ki}} - \widehat{\boldsymbol{v}_k}\right) \\ &+ \sum_j a_{ij} \left\{\left(\boldsymbol{M}_{ik}\right)' + \left(\boldsymbol{\tau}_i\right)'\right\} \cdot \widehat{\boldsymbol{v}'_{ki}}\end{aligned} \tag{8-34}$$

where $(\boldsymbol{M}_{ik})'$ and $(\boldsymbol{\tau}_i)'$ are defined by

$$
\begin{aligned}
&(\boldsymbol{M}_{ik})' = (\boldsymbol{M}_k^n)' + (\boldsymbol{M}_k^t)' \\
&(\boldsymbol{M}_k^n)' \equiv -\left(p_k - \overline{\overline{p_{ki}}}\right)\boldsymbol{n}_k - \frac{\boldsymbol{M}_k^n}{a_i} \\
&(\boldsymbol{M}_k^t)' \equiv \boldsymbol{n}_k \cdot \left(\mathcal{T}_k - \overline{\overline{\mathcal{T}_{ki}}}\right) - \frac{\boldsymbol{M}_k^t}{a_i} \\
&(\boldsymbol{\tau}_i)' = \boldsymbol{n}_k \cdot \overline{\overline{\mathcal{T}_{ki}}} - \frac{\left(-\nabla\alpha_k \cdot \overline{\overline{\mathcal{T}_{ki}}}\right)}{a_i}.
\end{aligned}
\tag{8-35}
$$

Thus, it represents the fluctuating component of the total drag force. Consequently, we define the turbulent flux of work due to drag force W_{ki}^T as

$$
W_{ki}^T \equiv \sum_j a_{ij}\left\{(\boldsymbol{M}_{ik})' + (\boldsymbol{\tau}_i)'\right\} \cdot \widehat{\boldsymbol{v}_{ki}'}. \tag{8-36}
$$

Substituting Eqs.(8-33) and (8-34) with Eq.(8-36) into Eq.(8-31), we obtain

$$
\begin{aligned}
&\sum_j a_{ij}\boldsymbol{n}_k \cdot \mathcal{T}_k \cdot \left(\boldsymbol{v}_k - \widehat{\boldsymbol{v}_k}\right) = \overline{\overline{p_{ki}}}\left(\frac{\Gamma_k}{\overline{\overline{\rho_{ki}}}} - \frac{D_k\alpha_k}{Dt}\right) \\
&+\boldsymbol{M}_{ik} \cdot \left(\widehat{\boldsymbol{v}_{ki}} - \widehat{\boldsymbol{v}_k}\right) - \nabla\alpha_k \cdot \overline{\overline{\mathcal{T}_{ki}}} \cdot \left(\widehat{\boldsymbol{v}_{ki}} - \widehat{\boldsymbol{v}_k}\right) + W_{ki}^T.
\end{aligned}
\tag{8-37}
$$

In view of Eqs.(8-29) and (8-37), the macroscopic interfacial thermal energy transfer Λ_k becomes

$$
\begin{aligned}
&\Lambda_k = \left(\Gamma_k\widehat{e_{ki}} + a_i\overline{\overline{q_{ki}''}}\right) + \overline{\overline{p_{ki}}}\left(\frac{\Gamma_k}{\overline{\overline{\rho_{ki}}}} - \frac{D_k\alpha_k}{Dt}\right) \\
&+\boldsymbol{M}_{ik} \cdot \left(\widehat{\boldsymbol{v}_{ki}} - \widehat{\boldsymbol{v}_k}\right) - \nabla\alpha_k \cdot \overline{\overline{\mathcal{T}_{ki}}} \cdot \left(\widehat{\boldsymbol{v}_{ki}} - \widehat{\boldsymbol{v}_k}\right) + W_{ki}^T.
\end{aligned}
\tag{8-38}
$$

Now we introduce the virtual enthalpy of the kth-phase at interfaces in analogy with Eq.(8-27) and Eq.(5-37), thus

$$
\widehat{h_{ki}} = \widehat{e_{ki}} + \frac{\overline{\overline{p_{ki}}}}{\overline{\overline{\rho_{ki}}}}. \tag{8-39}
$$

Then we have

$$
\begin{aligned}
\Lambda_k &= \left(\Gamma_k \widehat{h_{ki}} + a_i \overline{\overline{q_{ki}''}}\right) - \overline{\overline{p_{ki}}} \frac{D_k \alpha_k}{Dt} + \boldsymbol{M}_{ik} \cdot \left(\widehat{\boldsymbol{v}_{ki}} - \widehat{\boldsymbol{v}_k}\right) \\
&- \nabla \alpha_k \cdot \overline{\overline{\mathcal{T}_{ki}}} \cdot \left(\widehat{\boldsymbol{v}_{ki}} - \widehat{\boldsymbol{v}_k}\right) + W_{ki}^T .
\end{aligned}
\tag{8-40}
$$

It is straightforward to obtain E_k from the relations for Λ_k, $\boldsymbol{M}_k$ and Γ_k, therefore we have from Eqs.(8-21) and (8-40) the following result

$$
\begin{aligned}
E_k &= \Gamma_k \left(\widehat{h_{ki}} + \widehat{\boldsymbol{v}_{ki}} \cdot \widehat{\boldsymbol{v}_k} - \frac{\widehat{v_k}^2}{2}\right) + a_i \overline{\overline{q_{ki}''}} - \overline{\overline{p_{ki}}} \frac{\partial \alpha_k}{\partial t} \\
&+ \boldsymbol{M}_{ik} \cdot \widehat{\boldsymbol{v}_{ki}} - \nabla \alpha_k \cdot \overline{\overline{\mathcal{T}_{ki}}} \cdot \widehat{\boldsymbol{v}_{ki}} + W_{ki}^T .
\end{aligned}
\tag{8-41}
$$

The expressions for Λ_k and E_k give the kth-phase interfacial fluxes of thermal energy and the total energy in terms of the mean values at the interfaces.

Now we proceed to the analysis of the mixture energy source term E_m. By assuming that

$$
\frac{d\sigma}{dT} \approx \text{constant} \tag{8-42}
$$

Eq.(5-29) can be approximated by

$$
E_m = \sum_j a_{ij} \left\{ T_i \left(\frac{d\sigma}{dT}\right) \nabla_s \cdot v_i + \left(\boldsymbol{t}_\alpha A^{\alpha\beta} \sigma\right)_{,\beta} \cdot \boldsymbol{v}_i \right\}. \tag{8-43}
$$

We recall here that the surface divergence of the interfacial velocity is the surface area dilatation (Aris, 1962). Therefore, we have

$$
\frac{1}{(dA)} \frac{d_s}{dt} (dA) = \nabla_s \cdot \boldsymbol{v}_i . \tag{8-44}
$$

Hence, together with the assumption that the surface tension gradient is small, we may approximate Eq.(8-43) by

$$E_m \doteq \overline{\overline{T}}_i \left(\frac{d\sigma}{dT} \right) \frac{D_i}{Dt}(a_i) + 2\overline{\overline{H_{21}}}\, \overline{\overline{\sigma}} \frac{\partial \alpha_1}{\partial t} + E_m^H \tag{8-45}$$

where the convective derivative D_i/Dt is defined by

$$\frac{D_i}{Dt} = \frac{\partial}{\partial t} + \widehat{\boldsymbol{v}}_i \cdot \nabla. \tag{8-46}$$

We note that the first term on the right-hand side of Eq.(8-45) takes into account the effects of the surface energy change associated with the changes in area, whereas the second and the last terms stand for the average work done by the surface tension. The last term represents the effect of the changes of the mean curvature on the mixture energy source. By combining Eqs.(8-21) and (8-26) we have

$$\begin{aligned} \sum_{k=1}^{2} \Lambda_k &= \sum_{k=1}^{2} \Gamma_k \left(\frac{\widehat{v_k}^2}{2} - \widehat{\boldsymbol{v}_{ki}} \cdot \widehat{\boldsymbol{v}_k} \right) - \sum_{k=1}^{2} \boldsymbol{M}_{ik} \cdot \widehat{\boldsymbol{v}_k} \\ &+ \sum_{k=1}^{2} \left(\overline{\overline{\boldsymbol{\mathcal{T}}_{ki}}} \cdot \widehat{\boldsymbol{v}_k} - \overline{\overline{p_{ki}}}\widehat{\boldsymbol{v}_k} \right) \cdot \nabla \alpha_k + E_m. \end{aligned} \tag{8-47}$$

As a summary on interfacial energy transfer we have the following relations.

Total energy transfer condition

$$\begin{aligned} E_k &= \Gamma_k \left(\widehat{h_{ki}} + \widehat{\boldsymbol{v}_{ki}} \cdot \widehat{\boldsymbol{v}_k} - \frac{\widehat{v_k}^2}{2} \right) + a_i \overline{\overline{q''_{ki}}} - \overline{\overline{p_{ki}}} \frac{\partial \alpha_k}{\partial t} \\ &+ \boldsymbol{M}_{ik} \cdot \widehat{\boldsymbol{v}_{ki}} - \nabla \alpha_k \cdot \overline{\overline{\boldsymbol{\mathcal{T}}_{ki}}} \cdot \widehat{\boldsymbol{v}_{ki}} + W_{ki}^T \quad \text{with} \\ E_m &= \sum_{k=1}^{2} E_k = \overline{\overline{T}}_i \left(\frac{d\sigma}{dT} \right) \frac{D_i}{Dt}(a_i) + 2\overline{\overline{H_{21}}}\, \overline{\overline{\sigma}} \frac{\partial \alpha_1}{\partial t} + E_m^H \end{aligned} \tag{8-48}$$

Thermal energy transfer condition

$$\begin{aligned} \Lambda_k &= \left(\Gamma_k \widehat{h_{ki}} + a_i \overline{\overline{q''_{ki}}} \right) - \overline{\overline{p_{ki}}} \frac{D_k \alpha_k}{Dt} + \boldsymbol{M}_{ik} \cdot \left(\widehat{\boldsymbol{v}_{ki}} - \widehat{\boldsymbol{v}_k} \right) \\ &- \nabla \alpha_k \cdot \overline{\overline{\boldsymbol{\mathcal{T}}_{ki}}} \cdot \left(\widehat{\boldsymbol{v}_{ki}} - \widehat{\boldsymbol{v}_k} \right) + W_{ki}^T \quad \text{with} \end{aligned} \tag{8-49}$$

$$\sum_{k=1}^{2} \Lambda_k = \overline{\overline{T_i}} \left(\frac{d\sigma}{dT} \right) \frac{D_i}{Dt} (a_i) + 2\overline{\overline{H_{21}}}\, \overline{\overline{\sigma}} \frac{\partial \alpha_1}{\partial t}$$

$$+E_m^H + \sum_{k=1}^{2} \left(\overline{\overline{\mathcal{T}_{ki}}} \cdot \widehat{\boldsymbol{v}_k} - \overline{\overline{p_{ki}}} \widehat{\boldsymbol{v}_{ki}} \right) \cdot \nabla \alpha_k$$

$$+\sum_{k=1}^{2} \left\{ \Gamma_k \left(\frac{\widehat{\boldsymbol{v}_k}^2}{2} - \widehat{\boldsymbol{v}_{ki}} \cdot \widehat{\boldsymbol{v}_k} \right) - \boldsymbol{M}_{ik} \cdot \widehat{\boldsymbol{v}_k} \right\}$$

Since these relations are now expressed by the mean values of the bulk fluid and of the interfaces, they can be considered as having the macroscopic forms. The constitutive equations can be obtained by relating the interfacial variables to the bulk fluid mean values and other characteristics parameters such as a_i.

In view of Eqs.(8-5), (8-21) and (8-48) we recognize considerable differences between the necessary interfacial constitutive laws for the two-fluid model and those for the drift-flux (mixture) model. For the former model it is necessary to specify Γ_1, $\boldsymbol{M}_1$, E_1, $\boldsymbol{M}_m$ and E_m by constitutive equations, whereas for the latter model it is sufficient to supply only Γ_1, $\boldsymbol{M}_m$ and E_m (or $\sum_{k=1}^{2} \Lambda_k$). Indeed this makes the drift-flux model quite simpler than the two-fluid model. In the diffusion or drift-flux model we supply the relation between the velocities of each phase, thus only one momentum equation is required. However, in the two-fluid model we specify the momentum exchange term $\boldsymbol{M}_k$ and then solve two momentum equations simultaneously. We also note that the sum of Λ_k for two phases does not reduce to a simple form as E_m without making assumptions, thus it is expected that special attention should be paid in using the thermal energy equation in the drift-flux model.

Chapter 9

TWO-FLUID MODEL

The two-fluid model (Ishii, 1975, Ishii and Mishima, 1984) is formulated by considering each phase separately. Thus, the model is expressed in terms of two sets of conservation equations governing the balance of mass, momentum and energy in each phase. However, since the averaged fields of one phase are not independent of the other phase, we have interaction terms appearing in these balance equations. The terms denoted by Γ_k, $\boldsymbol{M}_k$ and E_k are the mass, momentum and energy transfers to the kth-phase from the interfaces. As these quantities also should obey the balance laws at the interfaces, we have derived the interfacial transfer conditions from the local jump conditions. Consequently six differential field equations with three interfacial transfer conditions govern the macroscopic two-phase flow systems.

In the two-fluid model formulation, the transfer processes of each phase are expressed by their own balance equations, thus it is anticipated that the model can predict more detailed changes and phase interactions than the drift-flux (or mixture) model. However, this means that the two-fluid model is far more complicated not only in terms of the number of field equations involved but also in terms of the necessary constitutive equations. It is evident that these constitutive equations should be accurate to display the usefulness of the model. This is particularly true with the interaction terms Γ_k, $\boldsymbol{M}_k$ and E_k since, without these interfacial exchanges in the field equations, the two phases are essentially independent. These interaction terms decide the degree of coupling between the phases, thus the transfer processes in each phase are greatly influenced by these terms.

The real importance of the two-fluid model is that it can take into account the dynamic and non-equilibrium interactions between phases. This is accomplished by using the momentum equations for each phase and two independent velocity fields as well as the two energy equations in the formulation. Thus, it is expected that two-fluid model can be useful to the analyses of transient phenomena, wave propagations and of the flow regime changes. Particularly if the two phases are weakly coupled such that the inertia of each phase changes rapidly, the two-fluid model should be used to study these phenomena.

However, if the two phases are coupled strongly (in which the responses of phases are simultaneous such that two phases are close to mechanical and thermal equilibrium or the wave propagations are firmly interlocked), the two-fluid model brings into the system unnecessary complications for practical applications. Furthermore, it can be said that the two-fluid model is well suited to the studies of the local wave propagations and related stability problems. However, if one is concerned with the total response of the two-phase mixture in a system rather than the local behaviors of each phase, the drift-flux model is simpler and in most cases effective for solving problems. For general three-dimensional flow, the two-fluid model is better than the mixture model because the relative velocity correlation is extremely difficult to develop in a general three-dimensional form.

In what follows, we study a general formulation of the two-fluid model as well as various constitutive equations that are necessary to close the system of equations. It should be noted, however, that closing the system of differential equations by making the number of unknowns and equations the same does not imply the existence of a solution nor guarantee its uniqueness. However, it is a necessary condition for a properly set mathematical model that represents the physical systems to be analyzed.

It should also be remembered that mathematical models of two-phase flow systems are in no sense firmly established and some additional research is required to complete the three-dimensional model for a general two-phase flow. In order to appreciate the difficulties confronting us, we recall that even in a single-phase turbulent flow, the general constitutive equations for turbulent fluxes are not developed yet. In view of the present state of the art on the three dimensional two-fluid model formulation, the necessary form of the model is discussed in general terms first. This chapter, therefore, should be considered as a framework and guidance to formulate the constitutive equations from them. The following diagram summarizes the establishment of the two-fluid model formulation.

1.1 Two-fluid model field equations

Two-Fluid Model Continuity Equations

The two-fluid model is characterized by two independent velocity fields which specify the motions of each phase. The most natural choice of velocity fields is obviously the mass-weighted mean phase velocities $\widehat{v_k}$.

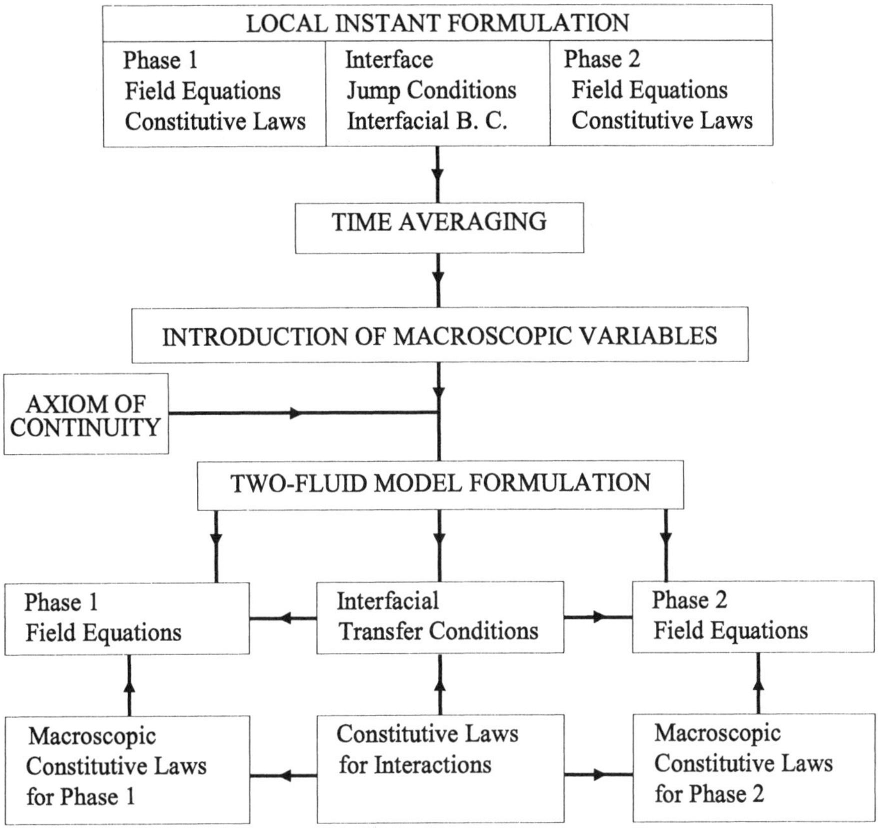

Thus the suitable form of the continuity equations to be used in the model should be Eq.(5-21)

$$\frac{\partial \alpha_k \overline{\overline{\rho_k}}}{\partial t} + \nabla \cdot \left(\alpha_k \overline{\overline{\rho_k}} \widehat{\boldsymbol{v}_k} \right) = \Gamma_k \tag{9-1}$$

with the interfacial mass transfer condition from Eq.(5-22)

$$\sum_{k=1}^{2} \Gamma_k = 0. \tag{9-2}$$

Thus, the term denoted by Γ_k represents the rate of production of kth-phase mass from the phase changes at the interfaces per unit volume. It appears because the local continuity equation has been integrated in time to obtain a macroscopic field equation. Furthermore, it can be said that Eqs.(9-1) and (9-2) are the general statements of the conservation of mass in the macroscopic field, as a result they hold regardless of the mechanism of phase changes.

In terms of the convective derivative of Eq.(7-11) the continuity equation becomes

$$\frac{D_k}{Dt}\left(\alpha_k\overline{\overline{\rho_k}}\right)+\alpha_k\overline{\overline{\rho_k}}\nabla\cdot\widehat{\boldsymbol{v}_k}=\Gamma_k. \tag{9-3}$$

And thus we have

$$\alpha_k\frac{D_k\overline{\overline{\rho_k}}}{Dt}+\overline{\overline{\rho_k}}\frac{D_k\alpha_k}{Dt}+\alpha_k\overline{\overline{\rho_k}}\nabla\cdot\widehat{\boldsymbol{v}_k}=\Gamma_k. \tag{9-4}$$

For a steady state flow, the time derivative of Eq.(9-1) drops, hence we have

$$\nabla\cdot\left(\alpha_k\overline{\overline{\rho_k}}\widehat{\boldsymbol{v}_k}\right)=\Gamma_k. \tag{9-5}$$

If each phase is originally incompressible, then the mean density $\overline{\overline{\rho_k}}$ is constant. Thus we have

$$\frac{\partial\alpha_k}{\partial t}+\nabla\cdot\left(\alpha_k\widehat{\boldsymbol{v}_k}\right)=\frac{\Gamma_k}{\overline{\overline{\rho_k}}}. \tag{9-6}$$

And furthermore, if there is no change of phases, the continuity equation reduces to

$$\frac{\partial\alpha_k}{\partial t}+\nabla\cdot\left(\alpha_k\widehat{\boldsymbol{v}_k}\right)=0 \tag{9-7}$$

which can be used in a low speed two-phase flow without phase changes. Under these conditions the kinematics of the two-phase system is completely governed by the phase redistribution, namely, by the convection and diffusion. The form of the above equation is analogous to that of a single-phase compressible flow.

The general form of the phase continuity equation has been given in the vector notation by Eq.(9-1). In view of a practical importance, we express the equation in rectangular and cylindrical coordinate systems. Thus, in *rectangular coordinates* (x,y,z) we have

$$\frac{\partial}{\partial t}\left(\alpha_k \overline{\overline{\rho_k}}\right) + \frac{\partial}{\partial x}\left(\alpha_k \overline{\overline{\rho_k}} \widehat{v_{xk}}\right) + \frac{\partial}{\partial y}\left(\alpha_k \overline{\overline{\rho_k}} \widehat{v_{yk}}\right) + \frac{\partial}{\partial z}\left(\alpha_k \overline{\overline{\rho_k}} \widehat{v_{zk}}\right) = \Gamma_k. \tag{9-8}$$

If the flow is restricted to two dimensions, then it represents a plane flow. In this case, the partial derivative with respect to x can be dropped from Eq.(9-8). We also note here that, for a steady plane flow with no phase changes, it is possible to introduce a stream function.

The continuity equation in *cylindrical coordinates* becomes

$$\frac{\partial}{\partial t}\left(\alpha_k \overline{\overline{\rho_k}}\right) + \frac{1}{r}\frac{\partial}{\partial r}\left(r\alpha_k \overline{\overline{\rho_k}} \widehat{v_{rk}}\right) + \frac{1}{r}\frac{\partial}{\partial \theta}\left(\alpha_k \overline{\overline{\rho_k}} \widehat{v_{\theta k}}\right) + \frac{\partial}{\partial z}\left(\alpha_k \overline{\overline{\rho_k}} \widehat{v_{zk}}\right) = \Gamma_k. \tag{9-9}$$

Flow is said to be axisymmetric, if there is no dependences on θ -direction, thus for such flows we have

$$\frac{\partial}{\partial t}\left(r\alpha_k \overline{\overline{\rho_k}}\right) + \frac{\partial}{\partial r}\left(r\alpha_k \overline{\overline{\rho_k}} \widehat{v_{rk}}\right) + \frac{\partial}{\partial z}\left(r\alpha_k \overline{\overline{\rho_k}} \widehat{v_{zk}}\right) = r\Gamma_k. \tag{9-10}$$

A stream function can also be introduced for a steady axisymmetric flow with no phase changes, making it is possible to eliminate the continuity equation from the formulation.

Two-Fluid Model Momentum Equations

In the two-fluid model formulation, the conservation of momentum is expressed by two momentum equations with the interfacial momentum transfer condition. As it was mentioned before, the appropriate field equations should be expressed by the center of mass or the barycentric velocity of each phase $\widehat{\boldsymbol{v}_k}$, thus from Eq.(5-26) we have two momentum equations given by

$$\frac{\partial \alpha_k \overline{\overline{\rho_k}} \widehat{\boldsymbol{v}_k}}{\partial t} + \nabla\cdot\left(\alpha_k \overline{\overline{\rho_k}} \widehat{\boldsymbol{v}_k}\widehat{\boldsymbol{v}_k}\right) = -\nabla\left(\alpha_k \overline{\overline{p_k}}\right) + \nabla\cdot\left[\alpha_k\left(\overline{\overline{\mathcal{T}}}_k + \mathcal{T}_k^T\right)\right] + \alpha_k \overline{\overline{\rho_k}} \widehat{\boldsymbol{g}_k} + \boldsymbol{M}_k. \tag{9-11}$$

The interfacial transfer condition (8-23) has the form

$$\sum_{k=1}^{2} \boldsymbol{M}_k = \boldsymbol{M}_m \tag{9-12}$$

with

$$\boldsymbol{M}_m = 2\overline{\overline{H_{21}}}\,\overline{\overline{\sigma}}\nabla\alpha_2 + \boldsymbol{M}_m^H. \tag{9-13}$$

We note here that the momentum equation for each phase has an interfacial source term $\boldsymbol{M}_k$ that couples the motions of two phases through Eq.(9-12). Here, $\overline{\overline{H_{21}}}$ and $\overline{\overline{\sigma}}$ are the average mean curvature of interfaces and the surface tension, whereas the term given by $\boldsymbol{M}_m$ takes account for the effect of the changes in the mean curvature.

In view of the Section 1.2 of Chapter 8, Eq.(9-11) can be rewritten as

$$\begin{aligned}
&\frac{\partial}{\partial t}\left(\alpha_k\overline{\overline{\rho_k}}\widehat{\boldsymbol{v}_k}\right) + \nabla\cdot\left(\alpha_k\overline{\overline{\rho_k}}\widehat{\boldsymbol{v}_k}\widehat{\boldsymbol{v}_k}\right) = -\nabla\left(\alpha_k\overline{\overline{p_k}}\right)\\
&+\nabla\cdot\left[\alpha_k\left(\overline{\overline{\mathcal{T}_k}} + \mathcal{T}_k^T\right)\right] + \alpha_k\overline{\overline{\rho_k}}\widehat{\boldsymbol{g}_k}\\
&+\left(\widehat{\boldsymbol{v}_{ki}}\Gamma_k + \overline{\overline{p_{ki}}}\nabla\alpha_k + \boldsymbol{M}_{ik} - \nabla\alpha_k\cdot\overline{\overline{\mathcal{T}_{ki}}}\right).
\end{aligned} \tag{9-14}$$

Hence, by using the convective derivative of Eq.(7-11) the kth-phase *equation of motion* becomes

$$\begin{aligned}
&\alpha_k\overline{\overline{\rho_k}}\frac{D_k\widehat{\boldsymbol{v}_k}}{Dt} = -\alpha_k\nabla\overline{\overline{p_k}} + \nabla\cdot\left[\alpha_k\left(\overline{\overline{\mathcal{T}_k}} + \mathcal{T}_k^T\right)\right] + \alpha_k\overline{\overline{\rho_k}}\widehat{\boldsymbol{g}_k}\\
&+\left(\overline{\overline{p_{ki}}} - \overline{\overline{p_k}}\right)\nabla\alpha_k + \left(\widehat{\boldsymbol{v}_{ki}} - \widehat{\boldsymbol{v}_k}\right)\Gamma_k + \boldsymbol{M}_{ik} - \nabla\alpha_k\cdot\overline{\overline{\mathcal{T}_{ki}}}
\end{aligned} \tag{9-15}$$

or by substituting Eq.(8-20) into Eq.(9-15) we have

$$\begin{aligned}
&\alpha_k\overline{\overline{\rho_k}}\frac{D_k\widehat{\boldsymbol{v}_k}}{Dt} = -\alpha_k\nabla\overline{\overline{p_k}} + \nabla\cdot\left[\alpha_k\left(\overline{\overline{\mathcal{T}_k}} + \mathcal{T}_k^T\right)\right] + \alpha_k\overline{\overline{\rho_k}}\widehat{\boldsymbol{g}_k}\\
&+\left(\overline{\overline{p_{ki}}} - \overline{\overline{p_k}}\right)\nabla\alpha_k + \left(\widehat{\boldsymbol{v}_i} - \widehat{\boldsymbol{v}_k} + \frac{\Gamma_k}{\overline{\overline{\rho_{ki}}}a_i^2}\nabla\alpha_k\right)\Gamma_k + \boldsymbol{M}_{ik}\\
&-\nabla\alpha_k\cdot\overline{\overline{\mathcal{T}_{ki}}}.
\end{aligned} \tag{9-16}$$

In general the equation of motion is a vectorial equation, thus we have three components or three corresponding scalar equations. In what follows, we express them in two coordinate systems of practical importances.

The equation of motion in *rectangular coordinates* (x, y, z) can be given as follows

x -component

$$\begin{aligned}
&\alpha_k \overline{\overline{\rho_k}} \left(\frac{\partial \widehat{v_{xk}}}{\partial t} + \widehat{v_{xk}} \frac{\partial \widehat{v_{xk}}}{\partial x} + \widehat{v_{yk}} \frac{\partial \widehat{v_{xk}}}{\partial y} + \widehat{v_{zk}} \frac{\partial \widehat{v_{xk}}}{\partial z} \right) = -\alpha_k \frac{\partial \overline{\overline{p_k}}}{\partial x} \\
&+\alpha_k \overline{\overline{\rho_k}} \widehat{g_{xk}} + \left\{ \frac{\partial}{\partial x} \alpha_k \left(\overline{\overline{\tau_{xxk}}} + \tau_{xxk}^T \right) + \frac{\partial}{\partial y} \alpha_k \left(\overline{\overline{\tau_{yxk}}} + \tau_{yxk}^T \right) \right. \\
&\left. + \frac{\partial}{\partial z} \alpha_k \left(\overline{\overline{\tau_{zxk}}} + \tau_{zxk}^T \right) \right\} + \left(\overline{\overline{p_{ki}}} - \overline{\overline{p_k}} \right) \frac{\partial \alpha_k}{\partial x} + \left(\widehat{v_{xki}} - \widehat{v_{xk}} \right) \Gamma_k \\
&+M_{ixk} - \left(\frac{\partial \alpha_k}{\partial x} \overline{\overline{\tau_{xxki}}} + \frac{\partial \alpha_k}{\partial y} \overline{\overline{\tau_{yxki}}} + \frac{\partial \alpha_k}{\partial z} \overline{\overline{\tau_{zxki}}} \right)
\end{aligned} \tag{9-17}$$

y -component

$$\begin{aligned}
&\alpha_k \overline{\overline{\rho_k}} \left(\frac{\partial \widehat{v_{yk}}}{\partial t} + \widehat{v_{xk}} \frac{\partial \widehat{v_{yk}}}{\partial x} + \widehat{v_{yk}} \frac{\partial \widehat{v_{yk}}}{\partial y} + \widehat{v_{zk}} \frac{\partial \widehat{v_{yk}}}{\partial z} \right) = -\alpha_k \frac{\partial \overline{\overline{p_k}}}{\partial y} \\
&+\alpha_k \overline{\overline{\rho_k}} \widehat{g_{yk}} + \left\{ \frac{\partial}{\partial x} \alpha_k \left(\overline{\overline{\tau_{xyk}}} + \tau_{xyk}^T \right) + \frac{\partial}{\partial y} \alpha_k \left(\overline{\overline{\tau_{yyk}}} + \tau_{yyk}^T \right) \right. \\
&\left. + \frac{\partial}{\partial z} \alpha_k \left(\overline{\overline{\tau_{zyk}}} + \tau_{zyk}^T \right) \right\} + \left(\overline{\overline{p_{ki}}} - \overline{\overline{p_k}} \right) \frac{\partial \alpha_k}{\partial y} + \left(\widehat{v_{yki}} - \widehat{v_{yk}} \right) \Gamma_k \\
&+M_{iyk} - \left(\frac{\partial \alpha_k}{\partial x} \overline{\overline{\tau_{xyki}}} + \frac{\partial \alpha_k}{\partial y} \overline{\overline{\tau_{yyki}}} + \frac{\partial \alpha_k}{\partial z} \overline{\overline{\tau_{zyki}}} \right)
\end{aligned} \tag{9-18}$$

z -component

$$\alpha_k \overline{\overline{\rho_k}} \left(\frac{\partial \widehat{v_{zk}}}{\partial t} + \widehat{v_{xk}} \frac{\partial \widehat{v_{zk}}}{\partial x} + \widehat{v_{yk}} \frac{\partial \widehat{v_{zk}}}{\partial y} + \widehat{v_{zk}} \frac{\partial \widehat{v_{zk}}}{\partial z} \right) = -\alpha_k \frac{\partial \overline{\overline{p_k}}}{\partial z}$$

$$
\begin{aligned}
&+\alpha_k \overline{\overline{\rho_k}} \widehat{g_{zk}} + \left\{ \frac{\partial}{\partial x} \alpha_k \left(\overline{\overline{\tau_{xzk}}} + \tau_{xzk}^T \right) + \frac{\partial}{\partial y} \alpha_k \left(\overline{\overline{\tau_{yzk}}} + \tau_{yzk}^T \right) \right. \\
&\left. + \frac{\partial}{\partial z} \alpha_k \left(\overline{\overline{\tau_{zzk}}} + \tau_{zzk}^T \right) \right\} + \left(\overline{\overline{p_{ki}}} - \overline{\overline{p_k}} \right) \frac{\partial \alpha_k}{\partial z} + \left(\widehat{v_{zki}} - \widehat{v_{zk}} \right) \Gamma_k \qquad \text{(9-19)} \\
&+ M_{izk} - \left(\frac{\partial \alpha_k}{\partial x} \overline{\overline{\tau_{xzki}}} + \frac{\partial \alpha_k}{\partial y} \overline{\overline{\tau_{yzki}}} + \frac{\partial \alpha_k}{\partial z} \overline{\overline{\tau_{zzki}}} \right).
\end{aligned}
$$

We note here that for a plane flow, the x-component of the equation of motion drops. Furthermore, all the partial derivatives with respect to x should be eliminated from the y and z-components of the equation of motion, namely, Eqs.(9-18) and (9-19).

The equation of motion in *cylindrical coordinates* (r, θ, z) becomes

r-component

$$
\begin{aligned}
&\alpha_k \overline{\overline{\rho_k}} \left(\frac{\partial \widehat{v_{rk}}}{\partial t} + \widehat{v_{rk}} \frac{\partial \widehat{v_{rk}}}{\partial r} + \frac{\widehat{v_{\theta k}}}{r} \frac{\partial \widehat{v_{rk}}}{\partial \theta} - \frac{\widehat{v_{\theta k}}^2}{r} + \widehat{v_{zk}} \frac{\partial \widehat{v_{rk}}}{\partial z} \right) \\
&= -\alpha_k \frac{\partial \overline{\overline{p_k}}}{\partial r} + \alpha_k \overline{\overline{\rho_k}} \widehat{g_{rk}} + \left\{ \frac{1}{r} \frac{\partial}{\partial r} r \alpha_k \left(\overline{\overline{\tau_{rrk}}} + \tau_{rrk}^T \right) \right. \\
&+ \frac{1}{r} \frac{\partial}{\partial \theta} \alpha_k \left(\overline{\overline{\tau_{r\theta k}}} + \tau_{r\theta k}^T \right) - \frac{1}{r} \alpha_k \left(\overline{\overline{\tau_{\theta\theta k}}} + \tau_{\theta\theta k}^T \right) \qquad \text{(9-20)} \\
&\left. + \frac{\partial}{\partial z} \alpha_k \left(\overline{\overline{\tau_{rzk}}} + \tau_{rzk}^T \right) \right\} + \left(\overline{\overline{p_{ki}}} - \overline{\overline{p_k}} \right) \frac{\partial \alpha_k}{\partial r} + \left(\widehat{v_{rki}} - \widehat{v_{rk}} \right) \Gamma_k \\
&+ M_{irk} - \left(\frac{\partial \alpha_k}{\partial r} \overline{\overline{\tau_{rrki}}} + \frac{1}{r} \frac{\partial \alpha_k}{\partial \theta} \overline{\overline{\tau_{\theta rki}}} + \frac{\partial \alpha_k}{\partial z} \overline{\overline{\tau_{zrki}}} \right)
\end{aligned}
$$

θ-component

$$
\begin{aligned}
&\alpha_k \overline{\overline{\rho_k}} \left(\frac{\partial \widehat{v_{\theta k}}}{\partial t} + \widehat{v_{rk}} \frac{\partial \widehat{v_{\theta k}}}{\partial r} + \frac{\widehat{v_{\theta k}}}{r} \frac{\partial \widehat{v_{\theta k}}}{\partial \theta} + \frac{\widehat{v_{rk}} \widehat{v_{\theta k}}}{r} + \widehat{v_{zk}} \frac{\partial \widehat{v_{\theta k}}}{\partial z} \right) \\
&= -\frac{\alpha_k}{r} \frac{\partial \overline{\overline{p_k}}}{\partial \theta} + \alpha_k \overline{\overline{\rho_k}} \widehat{g_{\theta k}} + \left\{ \frac{1}{r^2} \frac{\partial}{\partial r} r^2 \alpha_k \left(\overline{\overline{\tau_{r\theta k}}} + \tau_{r\theta k}^T \right) \right.
\end{aligned}
$$

$$+\frac{1}{r}\frac{\partial}{\partial\theta}\alpha_k\left(\overline{\overline{\tau_{\theta\theta k}}}+\tau_{\theta\theta k}^T\right)+\frac{\partial}{\partial z}\alpha_k\left(\overline{\overline{\tau_{\theta zk}}}+\tau_{\theta zk}^T\right)\Bigg\}$$

$$+\left(\overline{\overline{p_{ki}}}-\overline{\overline{p_k}}\right)\frac{1}{r}\frac{\partial\alpha_k}{\partial\theta}+\left(\widehat{v_{\theta ki}}-\widehat{v_{\theta k}}\right)\Gamma_k+M_{i\theta k} \tag{9-21}$$

$$-\left(\frac{\partial\alpha_k}{\partial r}\overline{\overline{\tau_{r\theta ki}}}+\frac{1}{r}\frac{\partial\alpha_k}{\partial\theta}\overline{\overline{\tau_{\theta\theta ki}}}+\frac{\partial\alpha_k}{\partial z}\overline{\overline{\tau_{z\theta ki}}}\right)$$

z-component

$$\alpha_k\overline{\overline{\rho_k}}\left(\frac{\partial\widehat{v_{zk}}}{\partial t}+\widehat{v_{rk}}\frac{\partial\widehat{v_{zk}}}{\partial r}+\frac{\widehat{v_{\theta k}}}{r}\frac{\partial\widehat{v_{zk}}}{\partial\theta}+\widehat{v_{zk}}\frac{\partial\widehat{v_{zk}}}{\partial z}\right)$$

$$=-\alpha_k\frac{\partial\overline{\overline{p_k}}}{\partial z}+\alpha_k\overline{\overline{\rho_k}}\widehat{g_{zk}}+\left\{\frac{1}{r}\frac{\partial}{\partial r}\left[r\alpha_k\left(\overline{\overline{\tau_{rzk}}}+\tau_{rzk}^T\right)\right]\right.$$

$$\left.+\frac{1}{r}\frac{\partial}{\partial\theta}\alpha_k\left(\overline{\overline{\tau_{\theta zk}}}+\tau_{\theta zk}^T\right)+\frac{\partial}{\partial z}\alpha_k\left(\overline{\overline{\tau_{zzk}}}+\tau_{zzk}^T\right)\right\} \tag{9-22}$$

$$+\left(\overline{\overline{p_{ki}}}-\overline{\overline{p_k}}\right)\frac{\partial\alpha_k}{\partial z}+\left(\widehat{v_{zki}}-\widehat{v_{zk}}\right)\Gamma_k+M_{izk}$$

$$-\left(\frac{\partial\alpha_k}{\partial r}\overline{\overline{\tau_{rzki}}}+\frac{1}{r}\frac{\partial\alpha_k}{\partial\theta}\overline{\overline{\tau_{\theta zki}}}+\frac{\partial\alpha_k}{\partial z}\overline{\overline{\tau_{zzki}}}\right).$$

For an axisymmetric flow, the terms with partial derivative with respect to θ drop from the equations. Furthermore, if the flow is free from the circulatory motion around the z-axis, then the velocity in θ-direction is also zero, thus the entire θ-components of the equation can be eliminated. For many practical problems of two-phase pipe flows, this is a sufficiently accurate model to be used.

Two-Fluid Model Energy Equations

The most fundamental form of the conservation of energy is expressed by considering the balance of the total energy. For the two-fluid model formulation, we have two *total energy equations* with the interfacial energy transfer condition. Thus from Eq.(5-32) we have

$$\frac{\partial}{\partial t}\left[\alpha_k\overline{\overline{\rho_k}}\left(\widehat{e_k}+\frac{\widehat{v_k}^2}{2}\right)\right]+\nabla\cdot\left[\alpha_k\overline{\overline{\rho_k}}\left(\widehat{e_k}+\frac{\widehat{v_k}^2}{2}\right)\widehat{\boldsymbol{v}_k}\right]$$
$$=-\nabla\cdot\left[\alpha_k\left(\overline{\overline{\boldsymbol{q}_k}}+\boldsymbol{q}_k^T\right)\right]+\nabla\cdot\left(\alpha_k\overline{\overline{\mathcal{T}_k}}\cdot\widehat{\boldsymbol{v}_k}\right) \tag{9-23}$$
$$+\alpha_k\overline{\overline{\rho_k}}\widehat{\boldsymbol{g}_k}\cdot\widehat{\boldsymbol{v}_k}+E_k.$$

Here we have assumed that $\boldsymbol{g}_k$ is constant, namely, $\boldsymbol{g}_k=\widehat{\boldsymbol{g}_k}$. The interfacial transfer condition (8-48) couples the energy transport processes of two phases, thus we have

$$\sum_{k=1}^{2}E_k=E_m \tag{9-24}$$

where

$$E_m=\overline{\overline{T}}_i\left(\frac{d\sigma}{dT}\right)\frac{D_ia_i}{Dt}+2\overline{\overline{H_{21}}}\,\overline{\overline{\sigma}}\frac{\partial\alpha_1}{\partial t}+E_m^H. \tag{9-25}$$

These relations show that the sum of the interfacial energy transfer terms E_k for each phase balances with the time rate of change of surface energy and the work done by the surface tension. We note here that the term given by E_m^H takes account for the effect of the changes in the mean curvature. If we use the detailed expression for E_k in terms of the interfacial variables given by Eq.(8-48), the total energy equation can be rewritten as

$$\frac{\partial}{\partial t}\left[\alpha_k\overline{\overline{\rho_k}}\left(\widehat{e_k}+\frac{\widehat{v_k}^2}{2}\right)\right]+\nabla\cdot\left[\alpha_k\overline{\overline{\rho_k}}\left(\widehat{e_k}+\frac{\widehat{v_k}^2}{2}\right)\widehat{\boldsymbol{v}_k}\right]$$
$$=-\nabla\cdot\left[\alpha_k\left(\overline{\overline{\boldsymbol{q}_k}}+\boldsymbol{q}_k^T\right)\right]+\nabla\cdot\left(\alpha_k\overline{\overline{\mathcal{T}_k}}\cdot\widehat{\boldsymbol{v}_k}\right)$$
$$+\alpha_k\overline{\overline{\rho_k}}\widehat{\boldsymbol{v}_k}\cdot\widehat{\boldsymbol{g}_k}+\left\{\Gamma_k\left(\widehat{h_{ki}}+\widehat{\boldsymbol{v}_{ki}}\cdot\widehat{\boldsymbol{v}_k}-\frac{\widehat{v_k}^2}{2}\right)+a_i\overline{\overline{q_{ki}''}}\right. \tag{9-26}$$
$$\left.-\overline{\overline{p_{ki}}}\frac{\partial\alpha_k}{\partial t}+\boldsymbol{M}_{ik}\cdot\widehat{\boldsymbol{v}_{ki}}-\left(\nabla\alpha_k\cdot\overline{\overline{\mathcal{C}_{ki}}}\right)\cdot\widehat{\boldsymbol{v}_{ki}}+W_{ki}^T\right\}.$$

By using the transformation on the convective derivative, Eq.(7-13), we can write

$$\alpha_k \overline{\overline{\rho_k}} \frac{D_k}{Dt}\left(\widehat{e_k} + \frac{\widehat{v_k}^2}{2}\right) = -\nabla\cdot\alpha_k\left(\overline{\overline{\boldsymbol{q}_k}} + \boldsymbol{q}_k^T\right) + \nabla\cdot\left(\alpha_k \overline{\overline{\mathcal{T}_k}}\cdot\widehat{\boldsymbol{v}_k}\right)$$

$$+\alpha_k \overline{\overline{\rho_k}}\widehat{\boldsymbol{v}_k}\cdot\widehat{\boldsymbol{g}_k} + \Gamma_k\left\{\left(\widehat{e_{ki}} - \widehat{e_k}\right) + \left(\widehat{\boldsymbol{v}_{ki}} - \widehat{\boldsymbol{v}_k}\right)\cdot\widehat{\boldsymbol{v}_k}\right\} + a_i \overline{\overline{q''_{ki}}} \tag{9-27}$$

$$-\overline{\overline{p_{ki}}}\left(\frac{\partial\alpha_k}{\partial t} - \frac{\Gamma_k}{\overline{\overline{\rho_{ki}}}}\right) + \boldsymbol{M}_{ik}\cdot\widehat{\boldsymbol{v}_{ki}} - \left(\nabla\alpha_k\cdot\overline{\overline{\mathcal{C}_{ki}}}\right)\cdot\widehat{\boldsymbol{v}_{ki}} + W_{ki}^T.$$

Equation (9-23) describes the transfer of energy seen from the observer at a fixed point, and Eq.(9-27) expresses the energy transfer by following the fluid with the barycentric velocity $\widehat{\boldsymbol{v}_k}$.

In many practical heat transfer problems, it is convenient to use the thermal energy equation instead of the total energy equation. This is particularly true for low-speed, two-phase flows with heat additions where the mechanical terms are insignificant in comparison with the high heat transfer rates. Thus, by recalling Eq.(5-38) and Eq.(5-39), *the thermal energy equation is* given by

$$\frac{\partial\alpha_k\overline{\overline{\rho_k}}\widehat{h_k}}{\partial t} + \nabla\cdot\left(\alpha_k\overline{\overline{\rho_k}}\widehat{h_k}\widehat{\boldsymbol{v}_k}\right) = -\nabla\cdot\alpha_k\left(\overline{\overline{\boldsymbol{q}_k}} + \boldsymbol{q}_k^T\right)$$

$$+\frac{D_k}{Dt}\left(\alpha_k\overline{\overline{p_k}}\right) - \widehat{\boldsymbol{v}_k}\cdot\nabla\cdot\left(\alpha_k\mathcal{C}_k^T\right) + \alpha_k\overline{\overline{\mathcal{C}_k}} : \nabla\widehat{\boldsymbol{v}_k} + \Lambda_k. \tag{9-28}$$

Substituting the expression for Λ_k of Eq.(8-40) into the above equation, we get

$$\frac{\partial}{\partial t}\left(\alpha_k\overline{\overline{\rho_k}}\widehat{h_k}\right) + \nabla\cdot\left(\alpha_k\overline{\overline{\rho_k}}\widehat{h_k}\widehat{\boldsymbol{v}_k}\right) = -\nabla\cdot\alpha_k\left(\overline{\overline{\boldsymbol{q}_k}} + \boldsymbol{q}_k^T\right)$$

$$+\frac{D_k}{Dt}\left(\alpha_k\overline{\overline{p_k}}\right) - \widehat{\boldsymbol{v}_k}\cdot\nabla\cdot\left(\alpha_k\mathcal{C}_k^T\right) + \alpha_k\overline{\overline{\mathcal{C}_k}} : \nabla\widehat{\boldsymbol{v}_k}$$

$$+\left(\Gamma_k\widehat{h_{ki}} + a_i\overline{\overline{q''_{ki}}}\right) - \overline{\overline{p_{ki}}}\frac{D_k\alpha_k}{Dt} + \boldsymbol{M}_{ik}\cdot\left(\widehat{\boldsymbol{v}_{ki}} - \widehat{\boldsymbol{v}_k}\right) \tag{9-29}$$

$$-\nabla\alpha_k\cdot\overline{\overline{\mathcal{C}_{ki}}}\cdot\left(\widehat{\boldsymbol{v}_{ki}} - \widehat{\boldsymbol{v}_k}\right) + W_{ki}^T.$$

This equation can also be transformed in terms of the convective derivatives as

$$\alpha_k \overline{\overline{\rho_k}} \frac{D_k \widehat{h_k}}{Dt} = -\nabla \cdot \alpha_k \left(\overline{\overline{\boldsymbol{q}_k}} + \boldsymbol{q}_k^T \right) - \widehat{\boldsymbol{v}_k} \cdot \nabla \cdot \left(\alpha_k \mathcal{T}_k^T \right)$$
$$+W_{ki}^T + \alpha_k \frac{D_k \overline{\overline{p_k}}}{Dt} + \alpha_k \overline{\overline{\mathcal{T}_k}} : \nabla \widehat{\boldsymbol{v}_k} + \Gamma_k \left(\widehat{h_{ki}} - \widehat{h_k} \right)$$
$$+a_i \overline{\overline{q_{ki}''}} + \left(\overline{\overline{p_k}} - \overline{\overline{p_{ki}}} \right) \frac{D_k \alpha_k}{Dt} + \boldsymbol{M}_{ik} \cdot \left(\widehat{\boldsymbol{v}_{ki}} - \widehat{\boldsymbol{v}_k} \right)$$
$$-\nabla \alpha_k \cdot \overline{\overline{\mathcal{T}_{ki}}} \cdot \left(\widehat{\boldsymbol{v}_{ki}} - \widehat{\boldsymbol{v}_k} \right) \tag{9-30}$$

which is the equation describing the exchanges of thermal energy as seen from the observer moving with the mass center of the kth-phase.

For simplicity we denote the turbulent energy source by Φ_k^T and the viscous dissipation term by Φ_k^μ, thus

$$\Phi_k^T \equiv -\widehat{\boldsymbol{v}_k} \cdot \nabla \cdot \left(\alpha_k \mathcal{T}_k^T \right) + W_{ki}^T \tag{9-31}$$

$$\Phi_k^\mu \equiv \alpha_k \overline{\overline{\mathcal{T}_k}} : \nabla \widehat{\boldsymbol{v}_k}. \tag{9-32}$$

Then Eq.(9-30) reduces to

$$\alpha_k \overline{\overline{\rho_k}} \frac{D_k \widehat{h_k}}{Dt} = -\nabla \cdot \alpha_k \left(\overline{\overline{\boldsymbol{q}_k}} + \boldsymbol{q}_k^T \right) + \alpha_k \frac{D_k \overline{\overline{p_k}}}{Dt} + \Phi_k^T + \Phi_k^\mu$$
$$+\Gamma_k \left(\widehat{h_{ki}} - \widehat{h_k} \right) + a_i \overline{\overline{q_{ki}''}} + \left(\overline{\overline{p_k}} - \overline{\overline{p_{ki}}} \right) \frac{D_k \alpha_k}{Dt}$$
$$+\boldsymbol{M}_{ik} \cdot \left(\widehat{\boldsymbol{v}_{ki}} - \widehat{\boldsymbol{v}_k} \right) - \nabla \alpha_k \cdot \overline{\overline{\mathcal{T}_{ki}}} \cdot \left(\widehat{\boldsymbol{v}_{ki}} - \widehat{\boldsymbol{v}_k} \right). \tag{9-33}$$

Now we expand the above thermal energy equation in two coordinate systems of practical importance. Thus, in the *rectangular coordinates* (x, y, z), Eq.(9-33) becomes

$$\alpha_k \overline{\overline{\rho_k}} \left(\frac{\partial \widehat{h_k}}{\partial t} + \widehat{v_{xk}} \frac{\partial \widehat{h_k}}{\partial x} + \widehat{v_{yk}} \frac{\partial \widehat{h_k}}{\partial y} + \widehat{v_{zk}} \frac{\partial \widehat{h_k}}{\partial z} \right)$$
$$= -\frac{\partial}{\partial x} \left[\alpha_k \left(\overline{\overline{q_{xk}}} + q_{xk}^T \right) \right] - \frac{\partial}{\partial y} \left[\alpha_k \left(\overline{\overline{q_{yk}}} + q_{yk}^T \right) \right]$$

$$-\frac{\partial}{\partial z}\left[\alpha_k\left(\overline{\overline{q_{zk}}}+q_{zk}^T\right)\right]+\alpha_k\left(\frac{\partial\overline{\overline{p_k}}}{\partial t}+\widehat{v_{xk}}\frac{\partial\overline{\overline{p_k}}}{\partial x}+\widehat{v_{yk}}\frac{\partial\overline{\overline{p_k}}}{\partial y}\right.$$

$$\left.+\widehat{v_{zk}}\frac{\partial\overline{\overline{p_k}}}{\partial z}\right)+\Phi_k^T+\Phi_k^\mu+\Gamma_k\left(\widehat{h_{ki}}-\widehat{h_k}\right)$$

$$+a_i\overline{\overline{q_{ki}''}}+\left(\overline{\overline{p_k}}-\overline{\overline{p_{ki}}}\right)\left(\frac{\partial\alpha_k}{\partial t}+\widehat{v_{xk}}\frac{\partial\alpha_k}{\partial x}+\widehat{v_{yk}}\frac{\partial\alpha_k}{\partial y}+\widehat{v_{zk}}\frac{\partial\alpha_k}{\partial z}\right)$$

$$+M_{ixk}\left(\widehat{v_{xki}}-\widehat{v_{xk}}\right)+M_{iyk}\left(\widehat{v_{yki}}-\widehat{v_{yk}}\right)+M_{izk}\left(\widehat{v_{zki}}-\widehat{v_{zk}}\right) \tag{9-34}$$

$$-\left(\frac{\partial\alpha_k}{\partial x}\overline{\overline{\tau_{xxki}}}+\frac{\partial\alpha_k}{\partial y}\overline{\overline{\tau_{yxki}}}+\frac{\partial\alpha_k}{\partial z}\overline{\overline{\tau_{zxki}}}\right)\left(\widehat{v_{xki}}-\widehat{v_{xk}}\right)$$

$$-\left(\frac{\partial\alpha_k}{\partial x}\overline{\overline{\tau_{xyki}}}+\frac{\partial\alpha_k}{\partial y}\overline{\overline{\tau_{yyki}}}+\frac{\partial\alpha_k}{\partial z}\overline{\overline{\tau_{zyki}}}\right)\left(\widehat{v_{yki}}-\widehat{v_{yk}}\right)$$

$$-\left(\frac{\partial\alpha_k}{\partial x}\overline{\overline{\tau_{xzki}}}+\frac{\partial\alpha_k}{\partial y}\overline{\overline{\tau_{yzki}}}+\frac{\partial\alpha_k}{\partial z}\overline{\overline{\tau_{zzki}}}\right)\left(\widehat{v_{zki}}-\widehat{v_{zk}}\right).$$

For a plane flow, the partial derivative with respect to x drops and the x-component of the velocity is zero.

In the *cylindrical coordinates* (r,θ,z), the thermal energy equation becomes

$$\alpha_k\overline{\overline{\rho_k}}\left(\frac{\partial\widehat{h_k}}{\partial t}+\widehat{v_{rk}}\frac{\partial\widehat{h_k}}{\partial r}+\frac{\widehat{v_{\theta k}}}{r}\frac{\partial\widehat{h_k}}{\partial\theta}+\widehat{v_{zk}}\frac{\partial\widehat{h_k}}{\partial z}\right)$$

$$=-\frac{1}{r}\frac{\partial}{\partial r}\left[r\alpha_k\left(\overline{\overline{q_{rk}}}+q_{rk}^T\right)\right]-\frac{1}{r}\frac{\partial}{\partial\theta}\left[\alpha_k\left(\overline{\overline{q_{\theta k}}}+q_{\theta k}^T\right)\right]$$

$$-\frac{\partial}{\partial z}\left[\alpha_k\left(\overline{\overline{q_{zk}}}+q_{zk}^T\right)\right]+\alpha_k\left(\frac{\partial\overline{\overline{p_k}}}{\partial t}+\widehat{v_{rk}}\frac{\partial\overline{\overline{p_k}}}{\partial r}+\frac{\widehat{v_{\theta k}}}{r}\frac{\partial\overline{\overline{p_k}}}{\partial\theta}\right. \tag{9-35}$$

$$\left.+\widehat{v_{zk}}\frac{\partial\overline{\overline{p_k}}}{\partial z}\right)+\Phi_k^T+\Phi_k^\mu+\Gamma_k\left(\widehat{h_{ki}}-\widehat{h_k}\right)$$

$$+a_i\overline{\overline{q_{ki}''}}+\left(\overline{\overline{p_k}}-\overline{\overline{p_{ki}}}\right)\left(\frac{\partial\alpha_k}{\partial t}+\widehat{v_{rk}}\frac{\partial\alpha_k}{\partial r}+\frac{\widehat{v_{\theta k}}}{r}\frac{\partial\alpha_k}{\partial\theta}+\widehat{v_{zk}}\frac{\partial\alpha_k}{\partial z}\right)$$

$$+M_{irk}\left(\widehat{v_{rki}}-\widehat{v_{rk}}\right)+M_{i\theta k}\left(\widehat{v_{\theta ki}}-\widehat{v_{\theta k}}\right)+M_{izk}\left(\widehat{v_{zki}}-\widehat{v_{zk}}\right)$$
$$-\left(\frac{\partial \alpha_k}{\partial r}\overline{\overline{\tau_{rrki}}}+\frac{1}{r}\frac{\partial \alpha_k}{\partial \theta}\overline{\overline{\tau_{\theta rki}}}+\frac{\partial \alpha_k}{\partial z}\overline{\overline{\tau_{zrki}}}\right)\left(\widehat{v_{rki}}-\widehat{v_{rk}}\right)$$
$$-\left(\frac{\partial \alpha_k}{\partial r}\overline{\overline{\tau_{r\theta ki}}}+\frac{1}{r}\frac{\partial \alpha_k}{\partial \theta}\overline{\overline{\tau_{\theta\theta ki}}}+\frac{\partial \alpha_k}{\partial z}\overline{\overline{\tau_{z\theta ki}}}\right)\left(\widehat{v_{\theta ki}}-\widehat{v_{\theta k}}\right)$$
$$-\left(\frac{\partial \alpha_k}{\partial r}\overline{\overline{\tau_{rzki}}}+\frac{1}{r}\frac{\partial \alpha_k}{\partial \theta}\overline{\overline{\tau_{\theta zki}}}+\frac{\partial \alpha_k}{\partial z}\overline{\overline{\tau_{zzki}}}\right)\left(\widehat{v_{zki}}-\widehat{v_{zk}}\right).$$

For an axisymmetric flow, the partial derivative with respect to θ drops from the equation. Moreover, if the flow is free from the circulatory motion around the z-axis, the θ-component of velocity is zero. This is a good approximation for many two-phase flows in pipes, particularly for vertical pipe flows.

It can be seen that both the total energy equation and the thermal energy equation are quite complicated in their full forms, and thus several simplifications are important for solving practical problems. We study several special cases below. If the heat transfer and phase changes dominate the energy exchanges, then we may neglect the terms arisen from the mechanical effects. Under this condition, Eq.(9-30) can be reduced to

$$\alpha_k\overline{\overline{\rho_k}}\frac{D_k\widehat{h_k}}{Dt}=-\nabla\cdot\alpha_k\left(\overline{\overline{\boldsymbol{q}_k}}+\boldsymbol{q}_k^T\right)+\Gamma_k\left(\widehat{h_{ki}}-\widehat{h_k}\right)+a_i\overline{\overline{q_k''}}. \quad (9\text{-}36)$$

The above equation suffices for many two-phase flow analyses except the problems of compressible wave propagations and/or at high speed flow conditions.

In the rectangular coordinates (x,y,z) Eq.(9-36) becomes

$$\alpha_k\overline{\overline{\rho_k}}\left(\frac{\partial\widehat{h_k}}{\partial t}+\widehat{v_{xk}}\frac{\partial\widehat{h_k}}{\partial x}+\widehat{v_{yk}}\frac{\partial\widehat{h_k}}{\partial y}+\widehat{v_{zk}}\frac{\partial\widehat{h_k}}{\partial z}\right)=\Gamma_k\left(\widehat{h_{ki}}-\widehat{h_k}\right)$$
$$+a_i\overline{\overline{q_{ki}''}}-\left\{\frac{\partial}{\partial x}\alpha_k\left(\overline{\overline{q_{xk}}}+q_{xk}^T\right)+\frac{\partial}{\partial y}\alpha_k\left(\overline{\overline{q_{yk}}}+q_{yk}^T\right)\right. \quad (9\text{-}37)$$
$$\left.+\frac{\partial}{\partial z}\alpha_k\left(\overline{\overline{q_{zk}}}+q_{zk}^T\right)\right\}.$$

If we use the cylindrical coordinates (r,θ,z) we have

$$\alpha_k \overline{\overline{\rho_k}} \left(\frac{\partial \widehat{h_k}}{\partial t} + \widehat{v_{rk}} \frac{\partial \widehat{h_k}}{\partial r} + \frac{\widehat{v_{\theta k}}}{r} \frac{\partial \widehat{h_k}}{\partial \theta} + \widehat{v_{zk}} \frac{\partial \widehat{h_k}}{\partial z} \right) = \Gamma_k \left(\widehat{h_{ki}} - \widehat{h_k} \right)$$

$$+ a_i \overline{\overline{q''_{ki}}} - \left\{ \frac{1}{r} \frac{\partial}{\partial r} r \alpha_k \left(\overline{\overline{q_{rk}}} + q^T_{rk} \right) + \frac{1}{r} \frac{\partial}{\partial \theta} \alpha_k \left(\overline{\overline{q_{\theta k}}} + q^T_{\theta k} \right) \right. \tag{9-38}$$

$$\left. + \frac{\partial}{\partial z} \alpha_k \left(\overline{\overline{q_{zk}}} + q^T_{zk} \right) \right\}.$$

Furthermore, if the flow is axisymmetric with negligible heat transfers in the axial direction it reduces to the following form

$$\alpha_k \overline{\overline{\rho_k}} \left(\frac{\partial \widehat{h_k}}{\partial t} + \widehat{v_{rk}} \frac{\partial \widehat{h_k}}{\partial r} + \widehat{v_{zk}} \frac{\partial \widehat{h_k}}{\partial z} \right) = \Gamma_k \left(\widehat{h_{ki}} - \widehat{h_k} \right) + a_i \overline{\overline{q''_{ki}}}$$

$$- \frac{1}{r} \frac{\partial}{\partial r} \left\{ r \alpha_k \left(\overline{\overline{q_{rk}}} + q^T_{rk} \right) \right\}. \tag{9-39}$$

It is a much simplified form of Eq.(9-35), yet the important heat transfer mechanisms are preserved in the above equation.

1.2 Two-fluid model constitutive laws

1.2.1 Entropy inequality

The general scheme of constructing the two-fluid model has been discussed at the beginning of this chapter. It is evident that the macroscopic field equations (9-1), (9-11) and (9-23) and the interfacial transfer conditions (9-2), (9-12) and (9-24) are insufficient to describe any particular system, since the number of the variables exceeds that of the available equations. Additional information which specifies the material and response characteristics of a particular group of materials is necessary. These are commonly called as constitutive equations, as explained in detail in Chapter 2.

The purpose of this section is to examine the necessary constitutive equations to close the system of equations. It is always possible to introduce more detailed mechanisms and variable to differentiate various effects of material and transfer mechanisms and then to complicate the set of equations. Consequently, we will discuss the most important aspect of the constitutive laws, namely the principle of the determinism, with *the simplest and the reasonably general set of the equations.* For this purpose we consider two sets of the macroscopic conservation equations of mass, momentum and energy given by Eqs.(9-1), (9-11) and (9-23) and the interfacial transfer conditions of mass, momentum and energy, Eqs.(9-2), (9-12) and (9-24).

In analogy with Chapter 2, we proceed to study entropy inequality in the macroscopic field. Thus, by applying the averaging procedures of Eqs.(5-8) and (5-10) to the inequality (2-23) and (2-85), we obtain

$$\begin{aligned}&\frac{\partial}{\partial t}\alpha_k\overline{\overline{\rho_k}}\widehat{s_k}+\nabla\cdot\left(\alpha_k\overline{\overline{\rho_k}}\widehat{s_k}\widehat{\boldsymbol{v}_k}\right)+\nabla\cdot\left\{\alpha_k\overline{\overline{\left(\frac{\boldsymbol{q}_k}{T_k}\right)}}+\alpha_k\overline{\overline{\rho_k s_k'\boldsymbol{v}_k'}}\right\}\\&+\frac{1}{\Delta t}\sum_j\frac{1}{v_{ni}}\left\{\dot{m}_k s_k+\boldsymbol{n}_k\cdot\left(\frac{\boldsymbol{q}_k}{T_k}\right)\right\}=\overline{\varDelta_k}\geq 0\end{aligned}\tag{9-40}$$

and

$$\begin{aligned}&\frac{1}{\Delta t}\sum_j\frac{1}{v_{ni}}\left\{\frac{d_s}{dt}s_a+s_a\nabla_s\cdot v_i-\sum_{k=1}^{2}\left[\dot{m}_k s_k+\boldsymbol{n}_k\cdot\left(\frac{\boldsymbol{q}_k}{T_k}\right)\right]\right\}\\&=\overline{\varDelta_a}\geq 0\end{aligned}\tag{9-41}$$

where we have taken the internal body heating $\dot{q}_k$ to be zero.

We recall that in Chapter 2 the interfacial entropy generation $\varDelta_a$ has been assumed to be zero in order to obtain simple boundary conditions at the interfaces. We follow exactly the same approach here, thus we have

$$\overline{\varDelta_a}=0.\tag{9-42}$$

Consequently, we obtain

$$\begin{aligned}&\overline{\overline{T_{1i}}}=\overline{\overline{T_i}}=\overline{\overline{T_{2i}}}\\&\widehat{\boldsymbol{v}_{t1}}=\widehat{\boldsymbol{v}_{ti}}=\widehat{\boldsymbol{v}_{t2}}\\&\sum_j a_{ij}\sum_{k=1}^{2}\frac{\dot{m}_k}{T_i}\left\{g_k+\frac{\left|\left(\boldsymbol{v}_k-\boldsymbol{v}_i\right)\right|^2}{2}-\frac{\tau_{nnk}}{\rho_k}\right\}=0.\end{aligned}\tag{9-43}$$

However, the last condition can be approximated for most practical problems by

$$\overline{\overline{p_{1i}}} - p^{sat}\left(\overline{\overline{T_i}}\right) = 2\overline{\overline{H_{21}}}\,\overline{\overline{\sigma}}\left(\frac{\overline{\overline{\rho_{1i}}}}{\overline{\overline{\rho_{2i}}} - \overline{\overline{\rho_{1i}}}}\right) \text{ or}$$
$$\overline{\overline{p_{2i}}} - p^{sat}\left(\overline{\overline{T_i}}\right) = 2\overline{\overline{H_{21}}}\,\overline{\overline{\sigma}}\left(\frac{\overline{\overline{\rho_{2i}}}}{\overline{\overline{\rho_{2i}}} - \overline{\overline{\rho_{1i}}}}\right) \tag{9-44}$$

which is the macroscopic form of Eq.(2-107).

The first and second conditions of Eq.(9-43) can be used to replace the fluid temperatures and the fluid tangential velocities by two parameters $\overline{\overline{T_i}}$ and $\widehat{\boldsymbol{v}_{ti}}$, whereas the *last condition remains very important in the macroscopic formulation to set the energy level of the interfaces.* Since at the lower reduced pressure, the density ratio is large and at the higher reduced pressure, surface tension effect is small, Eq.(9-44) may be approximated by

$$\overline{\overline{p_{gi}}} - p^{sat}\left(\overline{\overline{T_i}}\right) \doteq 0. \tag{9-45}$$

Thus, the vapor is almost always very close to the saturation condition at the interfaces. Equation (9-45) is simple enough and it is widely used even in the local instant formulation of two-phase flow problems. From the above discussion, it is seen that the result of Eq.(9-42) can be represented by a single equation (9-45) because the other conditions of Eq.(9-43) are satisfied by simply replacing the interfacial fluid temperatures and the tangential velocities by those of the interfaces, namely, $\overline{\overline{T_i}}$ and $\widehat{\boldsymbol{v}_{ti}}$.

Now we study the entropy inequality of each phase given by Eq.(9-40). If the fluctuations of the interfacial temperature are not important, then we have in analogy with Eq.(8-9) the following approximation

$$T_i \approx \overline{\overline{T_i}}. \tag{9-46}$$

Then Eq.(9-40) can be expressed with the interfacial macroscopic variables of the Section 1.1 of Chapter 8 as

$$\frac{\partial}{\partial t}\alpha_k\overline{\overline{\rho_k}}\widehat{s_k} + \nabla\cdot\left(\alpha_k\overline{\overline{\rho_k}}\widehat{s_k}\widehat{\boldsymbol{v}_k}\right) + \nabla\cdot\left[\alpha_k\overline{\overline{\left(\frac{\boldsymbol{q}_k}{T_k}\right)}} + \alpha_k\overline{\overline{\rho_k s_k'\boldsymbol{v}_k'}}\right]$$
$$-\left(\Gamma_k\widehat{s_{ki}} + a_i\overline{\overline{q_{ki}''}}\frac{1}{\overline{\overline{T_i}}}\right) = \overline{\Delta_k} \geq 0. \tag{9-47}$$

Although the above equation can be satisfied by the local instant formulation with the positive viscosity and the thermal conductivity, it still imposes some restrictions on the macroscopic constitutive equations. In other words, the exact form of $\overline{\Delta_k}$ which is obtainable from the local instant formulation satisfies Eq.(9-47), however it does not ensure that the left-hand side of the equation with various constitutive equations is always positive. It can be said that the formulation is consistent only if a set of constitutive equations with constraints imposed by the continuity, momentum and energy equations satisfy the inequality (9-47) trivially. Before we apply the above inequality, we will discuss one of the most important characteristics of the macroscopic model which appears only after the averaging.

1.2.2 Equation of state

It can be said that even if the original local instant formulation has simple linear constitutive laws with a standard equation of state as given in Chapter 2, the macroscopic model obtained by averaging may not have such simple constitutive equations. This is because the statistical effects of local instant fluctuations appear in the formulation. In general, these statistical effects depend not only on the present state in terms of the macroscopic variables, but also on the processes in which the present state has been reached. For example, fluid particles having the same values for the energy $\widehat{u_k}$ and the density $\overline{\overline{\rho_k}}$ can have entirely different values for the average temperature or the pressure. All these suggest that the macroscopic field has the characteristics of the materials with memory (Truesdell 1969). Thus the constitutive equations in general are given by the *functionals* of the past processes. This makes the analysis on the macroscopic constitutive equations extremely complicated and difficult. It is evident that the formulation will result in a set of coupled integro-differential equations. In order to avoid these difficulties, we have to make several assumptions at the expense of the accuracy of the model. We know that all materials show the characteristic of fading memory (Coleman and Noll, 1960). Thus, the importance of the effects of memory in a formulation depends on the ratio of the time span of the effective memory to the time constant of macroscopic processes.

Let us now examine the averaged equation of state corresponding to Eq.(2-24). We have

$$\frac{\partial \rho_k u_k}{\partial t} + \nabla \cdot \left(\rho_k u_k \boldsymbol{v}_k\right) \tag{9-48}$$

$$= T_k \left\{ \frac{\partial \rho_k s_k}{\partial t} + \nabla \cdot \left(\rho_k s_k \boldsymbol{v}_k \right) \right\} - p_k \nabla \cdot \boldsymbol{v}_k .$$

By averaging the above equation we obtain

$$\begin{aligned}
&\left\{ \alpha_k \overline{\overline{\rho_k}} \frac{D_k \widehat{u_k}}{Dt} + \nabla \cdot \left(\alpha_k \overline{\overline{\rho_k u_k' \boldsymbol{v}_k'}} \right) + \Gamma_k \left(\widehat{u_k} - \widehat{u_{ki}} \right) \right\} \\
&= \overline{\overline{T_k}} \left\{ \alpha_k \overline{\overline{\rho_k}} \frac{D_k \widehat{s_k}}{Dt} + \nabla \cdot \left(\alpha_k \overline{\overline{\rho_k s_k' \boldsymbol{v}_k'}} \right) + \Gamma_k \left(\widehat{s_k} - \widehat{s_{ki}} \right) \right\} \\
&+ \alpha_k \overline{\overline{T_k' \rho_k \left(\frac{\partial s_k}{\partial t} + \boldsymbol{v}_k \cdot \nabla s_k \right)}} \\
&- \left\{ -\alpha_k \overline{\overline{p_k}} \frac{1}{\overline{\overline{\rho_k}}} \frac{D_k \overline{\overline{\rho_k}}}{Dt} + \overline{\overline{p_k}} \nabla \cdot \left(\alpha_k \overline{\overline{\boldsymbol{v}_k'}} \right) + \Gamma_k \left(\frac{\overline{\overline{p_k}}}{\overline{\overline{\rho_k}}} - \frac{\overline{\overline{p_k}}}{\overline{\overline{\rho_{ki}}}} \right) \right\} \\
&- \alpha_k \overline{\overline{p_k' \frac{1}{\rho_k} \left(\frac{\partial \rho_k}{\partial t} + \boldsymbol{v}_k \cdot \nabla \rho_k \right)}} .
\end{aligned} \tag{9-49}$$

Here we have used the identities

$$\overline{A_k \frac{\partial B_k}{\partial t}} = \overline{\overline{A_k}} \frac{\partial}{\partial t} \alpha_k \overline{\overline{B_k}} - \overline{\overline{A_k}} \sum_j a_{ij} \boldsymbol{n}_k \cdot \boldsymbol{v}_i B_k + \alpha_k \overline{\overline{A_k' \left(\frac{\partial B_k}{\partial t} \right)}}$$

and (9-50)

$$\overline{A_k \nabla B_k} = \overline{\overline{A_k}} \nabla \alpha_k \overline{\overline{B_k}} + \overline{\overline{A_k}} \sum_j a_{ij} \boldsymbol{n}_k B_k + \alpha_k \overline{\overline{A_k' \nabla B_k}}$$

with

$$A_k' = A_k - \overline{\overline{A_k}} . \tag{9-51}$$

Equation (9-49) shows that, in general, we do not have simple equation of state in terms of averaged variables. The relation between the internal energy $\widehat{u_k}$, the entropy $\widehat{s_k}$ and the density $\overline{\overline{\rho_k}}$ is influenced by both the interfacial transfers and the statistical effects of the fluctuations of the variables.

A simple static relation between these mean values follows if the fluctuating components are sufficiently smaller than the macroscopic changes of these variables in question and thus the linear expansion of the equation of state is a good approximation. In this case we have

$$\widehat{u_k} = \widehat{u_k}\left(\widehat{s_k}, \overline{\overline{\rho_k}}\right) \doteq u_k\left(\widehat{s_k}, \overline{\overline{\rho_k}}\right) \tag{9-52}$$

with

$$T_k\left(\widehat{s_k}, \overline{\overline{\rho_k}}\right) = \frac{\partial u_k}{\partial s_k}\left(\widehat{s_k}, \overline{\overline{\rho_k}}\right) \doteq \overline{\overline{T}}_k \tag{9-53}$$

and

$$p_k\left(\widehat{s_k}, \overline{\overline{\rho_k}}\right) = -\frac{\partial u_k}{\partial\left(1/\rho_k\right)}\left(\widehat{s_k}, \overline{\overline{\rho_k}}\right) \doteq \overline{\overline{p_k}}. \tag{9-54}$$

These relations hold for a two-phase flow with each phase itself being in near equilibrium state in the time interval of Δt. Hereafter we assume that each phase obeys the static equation of state, Eq.(9-52) in the macroscopic field. It is a rather significant and practical assumption that enables us to construct the two-fluid model and its constitutive equations in parallel with the standard single-phase flow formulation. Under the above conditions, we have following relations in analogy with Eqs.(2-24), (2-25) and (2-26)

$$\widehat{u_k} = \widehat{u_k}\left(\widehat{s_k}, \overline{\overline{\rho_k}}\right)$$

$$\overline{\overline{T}}_k = \frac{\partial \widehat{u_k}}{\partial \widehat{s_k}}, \quad -\overline{\overline{p_k}} = \frac{\partial \widehat{u_k}}{\partial\left(1/\overline{\overline{\rho_k}}\right)}$$

$$\widehat{u_k} = \overline{\overline{T}}_k \widehat{s_k} - \frac{\overline{\overline{p_k}}}{\overline{\overline{\rho_k}}} + \widehat{g_k} \tag{9-55}$$

and

$$d\widehat{u_k} = \overline{\overline{T}}_k d\widehat{s_k} - \overline{\overline{p_k}}\, d\left(\frac{1}{\overline{\overline{\rho_k}}}\right).$$

The fundamental equation of state can also be represented by the combination of *the caloric and the thermal equations of state*, hence

$$\widehat{u_k} = \widehat{u_k}\left(\overline{\overline{T}}_k, \overline{\overline{\rho_k}}\right) \tag{9-56}$$

$$\overline{\overline{p_k}} = \overline{\overline{p_k}}\left(\overline{\overline{T}}_k, \overline{\overline{\rho}}_k\right) \tag{9-57}$$

or if we take the enthalpy as a variable, it becomes

$$\widehat{i}_k = \widehat{i}_k\left(\overline{\overline{T}}_k, \overline{\overline{p_k}}\right) \tag{9-58}$$

$$\overline{\overline{\rho_k}} = \overline{\overline{\rho_k}}\left(\overline{\overline{T}}_k, \overline{\overline{p_k}}\right). \tag{9-59}$$

In view of their great practical importance, we now study several thermodynamic second derivatives. The specific heats at constant pressure c_{pk} and at constant density c_{vk} are defined by

$$c_{pk} \equiv \left.\frac{\partial \widehat{i}_k}{\partial \overline{\overline{T}}_k}\right)_{\overline{\overline{p_k}}} = \overline{\overline{T}}_k \left.\frac{\partial \widehat{s}_k}{\partial \overline{\overline{T}}_k}\right)_{\overline{\overline{p_k}}} \tag{9-60}$$

and

$$c_{vk} \equiv \left.\frac{\partial \widehat{u_k}}{\partial \overline{\overline{T}}_k}\right)_{\overline{\overline{\rho_k}}} = \overline{\overline{T}}_k \left.\frac{\partial \widehat{s}_k}{\partial \overline{\overline{T}}_k}\right)_{\overline{\overline{\rho_k}}}. \tag{9-61}$$

Similarly, the thermal expansivity β_k and the isothermal compressibility κ_{Tk} are defined by

$$\beta_k \equiv -\frac{1}{\overline{\overline{\rho_k}}} \left.\frac{\partial \overline{\overline{\rho_k}}}{\partial \overline{\overline{T}}_k}\right)_{\overline{\overline{p_k}}} \tag{9-62}$$

$$\kappa_{Tk} \equiv \frac{1}{\overline{\overline{\rho_k}}} \frac{\partial \overline{\overline{\rho_k}}}{\partial \overline{\overline{p_k}}}\Bigg)_{\overline{\overline{T_k}}} = \frac{1}{\overline{\overline{\rho_k}}\left(a_{Tk}\right)^2} \tag{9-63}$$

where a_{Tk} is the isothermal sound velocity. Among these four derivatives we have the following identity

$$c_{pk} - c_{vk} = \frac{\overline{\overline{T_k}}\left(\beta_k\right)^2}{\kappa_{Tk}\overline{\overline{\rho_k}}}. \tag{9-64}$$

It is known that if Eq.(9-55) holds, then only three of the thermodynamic second derivatives are independent and others can be obtained from these three. Let us introduce the ratio of the specific heat

$$\gamma_k \equiv \frac{c_{pk}}{c_{vk}} \tag{9-65}$$

and the isentropic compressibility κ_{sk}

$$\kappa_{sk} = \frac{1}{\overline{\overline{\rho_k}}} \frac{\partial \overline{\overline{\rho_k}}}{\partial \overline{\overline{p_k}}}\Bigg)_{\widehat{s_k}} = \frac{1}{\overline{\overline{\rho_k}}\left(a_{Sk}\right)^2} \tag{9-66}$$

where a_{Sk} is the isentropic sound velocity. Then we have

$$\left(a_{Tk}\right)^2 = \frac{\left(a_{Sk}\right)^2}{\gamma_k}. \tag{9-67}$$

It shows that the isentropic sound velocity is always larger than the isothermal sound velocity, since from the stability of the system $\kappa_{Tk} \geq 0$, thus $\gamma_k \geq 1$. The importance of the thermodynamic second derivatives or the thermal and the caloric equations of state are related to the possibilities of measurements. For example, the fluid pressure and temperature are relatively easy to measure, thus the equations of state in the form of Eqs.(9-58) and (9-59) can be constructed experimentally.

Saturation Condition

The classical saturation condition is given by

$$\begin{cases} \overline{\overline{p_1}} = \overline{\overline{p_2}} = p^{sat} \\ \overline{\overline{T_1}} = \overline{\overline{T_2}} = T^{sat} \end{cases} \tag{9-68}$$

then

$$\widehat{g_1}\left(T^{sat}, p^{sat}\right) = \widehat{g_2}\left(T^{sat}, p^{sat}\right) = g^{sat}. \tag{9-69}$$

Thus we have the relation

$$p^{sat} = p^{sat}\left(T^{sat}\right) \tag{9-70}$$

which is assumed to be identical to Eq.(2-99).

1.2.3 Determinism

In the present analysis, we have assumed the existence of the static equation of state, Eq.(9-55). From the principle of determinism, we should be able to predict the present state from the past history. The necessary condition is that the system of equations is closed, or the number of unknown being same as that of equations. We see that this condition is not satisfied by the field equations (9-1), (9-11) and (9-23), the interfacial conditions (9-2), (9-12) and (9-24) and the equations of state (9-55). Consequently, it is necessary to add several constitutive equations that express the transfer mechanisms of average molecular diffusion, turbulent transfer and interfacial exchanges.

By taking the thermal and caloric equations of state, Eqs.(9-56) and (9-57), the variables appearing in the two-fluid model formulation are:

1. Conservation of Mass $\alpha_k, \overline{\overline{\rho_k}}, \widehat{\boldsymbol{v}_k}, \Gamma_k;$
2. Conservation of Momentum $\overline{\overline{p_k}}, \overline{\overline{\mathcal{T}_k}}, \mathcal{T}_k^T, \widehat{\boldsymbol{g}_k}, \boldsymbol{M}_k, \boldsymbol{M}_m;$
3. Conservation of Energy $\widehat{e_k}, \overline{\overline{\boldsymbol{q}_k}}, \boldsymbol{q}_k^T, E_k, E_m;$
4. Equations of State $\widehat{u_k}, \overline{\overline{T_k}}, \overline{\overline{T_i}}.$

where $k = 1$ and 2. Hence, the total number of the variables is thirty three. For a properly set model we should have also the same number of equations. These can be classified into the following groups.

	Equations		*Number of Equations*
1)	Field equations		
	mass	Eq.(9-1)	2
	momentum	Eq.(9-11)	2
	energy	Eq.(9-23)	2
2)	Interfacial transfer conditions		
	mass	Eq.(9-2)	1
	momentum	Eq.(9-12)	1
	energy	Eq.(9-24)	1
3)	Axiom of continuity		
	$\alpha_1 = 1 - \alpha_2$		1
4)	Average molecular diffusion fluxes		
	viscous stress $\overline{\overline{\mathcal{T}_k}}$		2
	conduction heat transfer $\overline{\overline{\boldsymbol{q}_k}}$		2
5)	Turbulent fluxes		
	turbulent stress $\mathcal{T}_k^T$		2
	turbulent energy transfer $\boldsymbol{q}_k^T$		2
6)	Body force fields $\widehat{\boldsymbol{g}_k}$		2
7)	Interfacial transfers		
	mass Γ_1		1
	momentum $\boldsymbol{M}_1$		1
	energy E_1		1
8)	Interfacial sources		
	momentum $\boldsymbol{M}_m$		1
	energy E_m		1
9)	Equations of State		
	thermal equation of state		2
	caloric equation of state		2
10)	Turbulent kinetic energy		
	$\widehat{e_k} - \widehat{u_k}$		2
11)	Phase change condition specifying the interfacial temperature $\overline{\overline{T_i}}$		1

12) Mechanical condition at interface specifying the relation between $\overline{\overline{p_1}}$ and $\overline{\overline{p_2}}$ (Average normal momentum jump condition) 1

This shows that we also have thirty three equations, thus the formulation is consistent. However, it should be noted that the constitutive equations shown above are expressed in the most primitive forms, thus it is quite possible that these equations are coupled with each other through some additional parameters with the same number of supplemental constitutive equations. Furthermore, if one is to use the entropy inequality then Eq.(9-55) should be introduced in the formulation.

1.2.4 Average molecular diffusion fluxes

Viscous Stress Tensor

The constitutive equations for $\overline{\overline{\mathcal{T}_k}}$ and $\overline{\overline{\boldsymbol{q}_k}}$ can be studied by using the identity (9-50). For simplicity, we assume that the fluid is Newtonian and

$$\begin{cases} \rho_k \approx \overline{\overline{\rho_k}} \\ \mu_k \approx \overline{\overline{\mu_k}} \end{cases} \quad \text{in } t \in [\Delta t]_T. \tag{9-71}$$

Then we obtain from Eqs.(2-38) and (9-50)

$$\overline{\overline{\mathcal{T}_k}} = \overline{\overline{\mu_k}} \left\{ \left[\nabla \widehat{\boldsymbol{v}_k} + \left(\nabla \widehat{\boldsymbol{v}_k} \right)^+ \right] + \frac{1}{\alpha_k} \sum_j a_{ij} \left(\boldsymbol{n}_k \boldsymbol{v}_k' + \boldsymbol{v}_k' \boldsymbol{n}_k \right) \right\} \tag{9-72}$$

where $\boldsymbol{v}_k'$ is the fluctuating component of the kth-phase velocity with respect to $\widehat{\boldsymbol{v}_k}$. It is easy to see that the second part of the stress tensor becomes important when the difference between the interfacial fluid velocity and the mean velocity is large. Thus it takes account for the effects of the interfacial motions and the mass transfers on the average deformation. Let us define the *interfacial extra deformation tensor* by

$$\begin{aligned} \mathcal{D}_{ki} &\equiv \frac{1}{2\alpha_k} \sum_j a_{ij} \left\{ \boldsymbol{n}_k \left(\boldsymbol{v}_k - \widehat{\boldsymbol{v}_k} \right) + \left(\boldsymbol{v}_k - \widehat{\boldsymbol{v}_k} \right) \boldsymbol{n}_k \right\} \\ &= \frac{a_i}{2\alpha_k} \left\{ \overline{\overline{\boldsymbol{n}_k \left(\boldsymbol{v}_k - \widehat{\boldsymbol{v}_k} \right) + \left(\boldsymbol{v}_k - \widehat{\boldsymbol{v}_k} \right) \boldsymbol{n}_k}} \right\}. \end{aligned} \tag{9-73}$$

The *bulk deformation tensor* is given by

$$\mathcal{D}_{kb} \equiv \frac{1}{2}\left[\nabla\widehat{\boldsymbol{v}_k} + \left(\nabla\widehat{\boldsymbol{v}_k}\right)^+\right]. \tag{9-74}$$

Consequently, we have

$$\overline{\overline{\mathcal{T}_k}} = 2\overline{\overline{\mu_k}}\left(\mathcal{D}_{kb} + \mathcal{D}_{ki}\right). \tag{9-75}$$

If the effect of the extra deformation tensor is included in the formulation, then a constitutive equation specifying $\mathcal{D}_{ki}$ for each phase should be given. In general, it is considered to be quite complex due to various mechanisms affecting $\mathcal{D}_{ki}$, however, under *special conditions* it can be reduced to a simple form. For example, if phase c is a continuous phase in a dispersed flow and the motions of interfaces are quite regular with little effects from the phase changes, then Eq.(9-73) with Eq.(4-62) can be approximated by

$$\begin{aligned} &\mathcal{D}_{ci} \doteq -\frac{1}{2\alpha_c}\left\{\left(\nabla\alpha_c\right)\left(\widehat{\boldsymbol{v}_d} - \widehat{\boldsymbol{v}_c}\right) + \left(\widehat{\boldsymbol{v}_d} - \widehat{\boldsymbol{v}_c}\right)\left(\nabla\alpha_c\right)\right\} \\ &\text{and} \\ &\mathcal{D}_{di} \doteq 0. \end{aligned} \tag{9-76}$$

For a more general case, we may approximate Eq.(9-73) by

$$\mathcal{D}_{ci} = -\frac{a_c^i}{2\alpha_c}\left\{\left(\nabla\alpha_c\right)\left(\widehat{\boldsymbol{v}_d} - \widehat{\boldsymbol{v}_c}\right) + \left(\widehat{\boldsymbol{v}_d} - \widehat{\boldsymbol{v}_c}\right)\left(\nabla\alpha_c\right)\right\} \tag{9-77}$$

where a_c^i represents the mobility of phase c.

Conduction Heat Transfer

The average heat flux $\overline{\overline{\boldsymbol{q}_k}}$ for a fluid obeying Fourier's Law of Heat Conduction, Eq.(2-41), can be given by

$$\overline{\overline{\boldsymbol{q}_k}} = -\overline{\overline{K_k}}\left\{\nabla\overline{\overline{T_k}} + \frac{1}{\alpha_k}\sum_j a_{ij}\boldsymbol{n}_{kj}\left(T_k - \overline{\overline{T_k}}\right)_j\right\} \tag{9-78}$$

in which we have used the identity (9-50) and assumed

$$K_k \approx \overline{\overline{K_k}} \quad \text{in } t \in [\Delta t]_T . \tag{9-79}$$

Furthermore, if we assume the thermal equilibrium at the interface, $T_{ki} = T_i \doteq \overline{\overline{T_i}}$,

$$T_i \approx \overline{\overline{T_i}} \quad \text{at } t \in [\Delta t]_S \tag{9-80}$$

then Eq.(9-78) reduces to

$$\overline{\overline{q_k}} = -\overline{\overline{K_k}} \left\{ \nabla \overline{\overline{T_k}} - \frac{\nabla \alpha_k}{\alpha_k} \left(\overline{\overline{T_i}} - \overline{\overline{T_k}} \right) \right\} \tag{9-81}$$

where we have used Eq.(4-62). It is interesting to note here that the second term represents the heat flux due to the concentration gradient and somehow it resembles the Dufour effect in the single-phase mixtures (Hirschfelder et al., 1954).

1.2.5 Turbulent fluxes

Turbulent Stress Tensor - Mixing Length Model

The difficulties encountered in writing the constitutive equations for turbulent fluxes, even in a single-phase flow, are quite considerable. The essential problem in turbulent flow analyses is to formulate a closure scheme for the averaged field equations. There are two different methods that have been used extensively in studying the transport mechanisms of turbulent flows. The first approach is based on the phenomenological construction of the constitutive equations for the turbulent fluxes. It is best represented by the mixing-length hypothesis of Prandtl who proposed a turbulent model by analogy with the kinematic theory of gases.

The second method is to use more accurate dynamical equations describing the turbulent transports. This can be done by taking the higher moments of the momentum equation. In this way, the number of dynamical equations can be increased as desired. This set is not closed, however, because a turbulent correlation term that arises as an additional flux in the moment equation is always one order higher than the other terms. Thus, these equations can never be closed mathematically. Consequently, it is necessary to make some approximations and use only limited number of the dynamical equations. In contrast to the statistical theories based on higher moment equations, the phenomenological approach is simple because it supplies directly the turbulent stress.

For many engineering problems, the mixing length model still remains as the primary means to obtain a solution particularly for the wall-induced turbulence. Inclusion of coupled higher-order moment equations almost always requires extensive computer calculations, whereas in many cases even the integral method suffices for engineering requirements.

Even in a single-phase flow, the statistical theories for turbulent flows are not firmly established and the method very often involves a system of quite complicated equations. Consequently, we do not discuss their applications to two-phase flow systems except for dispersed two-phase flow in the Section 1.4 of Chapter 12. Because of its simplicity, we now study the phenomenological approach for the turbulent fluxes in the two-fluid model formulation.

Following the standard analysis on the stress tensor (Truesdell and Toupin, 1960; Aris, 1962; Slattery, 1972), we assume that the local turbulent stress $\mathcal{T}_k^T$ can be decided if we know the phase velocity at the point and the deformation of the phase around it. The above assumption satisfies the constitutive *principle of local action*. Furthermore, if we use the *principle of material frame indifference*, we arrive to the conclusion that the stress tensor depends only on the deformation tensor

$$\mathcal{D}_k \equiv \mathcal{D}_{kb} + \mathcal{D}_{ki} \tag{9-82}$$

where the bulk deformation tensor $\mathcal{D}_{kb}$ and the interfacial extra deformation tensor are given by Eqs.(9-74) and (9-73), respectively. The turbulent stress which is caused by the bulk deformation can be called as the shear-induced turbulence, whereas the one that is caused by the interfacial extra deformation may be called as the bubble-induced turbulence.

In reality, these two constitutive principles may not be fulfilled in a strict sense even in single-phase turbulent flows as it has been discussed by Lumley (1970). Since there are very few experimental facts to depend on, we consider the simple case when above assumptions is valid. Consequently, the most general form permitted under the conditions is

$$\mathcal{T}_k^T = a_{k0}\mathcal{I} + a_{k1}\mathcal{D}_k + a_{k2}\mathcal{D}_k \cdot \mathcal{D}_k \tag{9-83}$$

where the coefficients a_{k0}, a_{k1}, and a_{k2} are functions of the three invariants of the deformation tensor $\mathcal{D}_k$ given by $\operatorname{tr}\mathcal{D}_k$, $\mathcal{D}_k : \mathcal{D}_k$ and $\det\mathcal{D}_k$. These are, namely, the trace, double dot product on itself and determinant of $\mathcal{D}_k$, respectively. Hence, we have

$$a_{kn} = a_{kn}\left(\overline{\overline{\rho_k}}, \overline{\overline{\mu_k}}, \alpha_k, \ell, a_i, \overline{\overline{H_{21}}}, \mathrm{tr}\,\mathcal{D}_k, \mathcal{D}_k : \mathcal{D}_k, \det \mathcal{D}_k\right). \tag{9-84}$$

In addition to the three invariants, the arguments of the coefficients are: the fluid density $\overline{\overline{\rho_k}}$; viscosity $\overline{\overline{\mu_k}}$; void fraction α_k; the distance from the wall ℓ; the interfacial area concentration a_i; and the mean curvature $\overline{\overline{H_{21}}}$. The expression for the turbulent stress tensor given by Eq.(9-83) with Eq.(9-84) is still very complicated. However, if we use the mixing length hypothesis similar to the one made in a single-phase flow, the result reduces to a simple form.

First, we assume that the stress tensor of Eq.(9-83) depends only on the second term, which is the Newtonian assumption. Then we have

$$\mathcal{T}_k^T = a_{k1}\mathcal{D}_{kb} \equiv 2\mu_k^T \mathcal{D}_{kb} \tag{9-85}$$

where μ_k^T is the turbulent viscosity. Furthermore, here the coefficient a_{k1} is taken to be

$$a_{k1} = a_{k1}\left(\overline{\overline{\rho_k}}, \overline{\overline{\mu_k}}, \alpha_k, \ell, a_i, \overline{\overline{H_{21}}}, \mathcal{D}_{kb} : \mathcal{D}_{kb}\right). \tag{9-86}$$

It is noted here that the bulk deformation tensor $\mathcal{D}_{kb}$ is used in the place of the total deformation $\mathcal{D}_k$, since the mixing length model is for the shear-induced turbulence. Because of its significance, it is discussed in more detail in Chapter 12. Consequently, from the dimensional analysis, we define

$$2\mu_k^{T*} = \frac{a_{k1}}{\overline{\overline{\rho_k}}\ell^2\sqrt{2\mathcal{D}_{kb} : \mathcal{D}_{kb}}}. \tag{9-87}$$

Then the non-dimensional function μ_k^{T*} should depend on four groups as

$$\mu_k^{T*} = \mu_k^{T*}\left(\frac{\overline{\overline{\rho_k}}\ell^2\sqrt{2\mathcal{D}_{kb} : \mathcal{D}_{kb}}}{\overline{\overline{\mu_k}}}, \ell a_i, \frac{\overline{\overline{H_{21}}}}{a_i}, \alpha_k\right). \tag{9-88}$$

The final expression then becomes

$$\mathcal{T}_k^T = 2\left(\mu_k^{T*}\right)\overline{\overline{\rho_k}}\ell^2\sqrt{2\mathcal{D}_{kb} : \mathcal{D}_{kb}}\,\mathcal{D}_{kb}. \tag{9-89}$$

This is the corresponding mixing length model for the two-fluid model formulation. The turbulent stress given by Eq.(9-89) with Eq.(9-88) is sufficiently simple to be a realistic model.

In order to visualize the model, let us consider a very simple two-phase pipe flow. By taking a fully developed flow with no phase changes, we have

$$\left(\overline{\overline{\mathcal{T}_d}}+\mathcal{T}_d^T\right)_{rz}=\left\{\overline{\overline{\mu_d}}+\left(\mu_d^{T*}\right)\overline{\overline{\rho_d}}\left(R-r\right)^2\left|\frac{d\widehat{v_{zd}}}{dr}\right|\right\}\frac{d\widehat{v_{zd}}}{dr}$$

$$\left(\overline{\overline{\mathcal{T}_c}}+\mathcal{T}_c^T\right)_{rz}$$

$$=\left\{\overline{\overline{\mu_c}}+\left(\mu_c^{T*}\right)\overline{\overline{\rho_c}}\left(R-r\right)^2\left|\frac{d\widehat{v_{zc}}}{dr}-\frac{1}{\alpha_c}\frac{d\alpha_c}{dr}\left(\widehat{v_{zd}}-\widehat{v_{zc}}\right)\right|\right\} \tag{9-90}$$

$$\times\left[\frac{d\widehat{v_{zc}}}{dr}-\frac{1}{\alpha_c}\frac{d\alpha_c}{dr}\left(\widehat{v_{zd}}-\widehat{v_{zc}}\right)\right]$$

where the coefficient μ_k^{T*} is given by

$$\mu_d^{T*}=\mu_d^{T*}\left(\frac{\overline{\overline{\rho_d}}\left(R-r\right)^2\left|\dfrac{d\widehat{v_{zd}}}{dr}\right|}{\overline{\overline{\mu_d}}},a_i\left(R-r\right),\frac{\overline{\overline{H_{21}}}}{a_i},\alpha_d\right)$$

$$\mu_c^{T*}=\mu_c^{T*}\left(\frac{\overline{\overline{\rho_c}}\left(R-r\right)^2\left|\dfrac{d\widehat{v_{zc}}}{dr}-\dfrac{1}{\alpha_c}\dfrac{d\alpha_c}{dr}\left(\widehat{v_{zd}}-\widehat{v_{zc}}\right)\right|}{\overline{\overline{\mu_c}}},\right. \tag{9-91}$$

$$\left.a_i\left(R-r\right),\frac{\overline{\overline{H_{21}}}}{a_i},\alpha_c\right).$$

If we exclude the region very near to the wall, the first non-dimensional group, which is a local Reynolds number, may be dropped from the arguments of the function μ_k^{T*}. Thus in this case, the constitutive equation for the turbulent flux, Eq.(9-88), depends only on the static parameters that express the mean geometrical configurations at a point in a flow.

Turbulent Heat Transfer - Mixing Length Model

The turbulent energy flux has been defined by Eq.(5-46). As we can see from the equation, it consists of three parts, namely the turbulent transfers of internal energy, of kinetic energy and the work done by the turbulences. For many practical two-phase flow systems, the latter two effects have less significant roles than the first effect as in the case of a single-phase flow. Thus, we construct a turbulent heat flux model by considering mainly the effect of the thermal energy transport, namely, the first and the last terms of Eq.(5-46) which give *enthalpy transport*. In analogy with Eq.(9-81) we assume

$$\begin{aligned}\boldsymbol{q}_k^T &= -K_k^T\left\{\nabla\overline{\overline{T}}_k - \frac{\nabla\alpha_k}{\alpha_k}\left(\overline{\overline{T}}_i - \overline{\overline{T}}_k\right)\right\}\\ &= -\overline{\overline{\rho}}_k\kappa_k^T\left\{\nabla\overline{\overline{i}}_k - \frac{\nabla\alpha_k}{\alpha_k}\left(\overline{\overline{i}}_{ki} - \overline{\overline{i}}_k\right)\right\}\end{aligned} \tag{9-92}$$

where the turbulent energy transport coefficient is expressed by

$$K_k^T = K_k^T\left(\overline{\overline{\rho}}_k, \overline{\overline{K}}_k, \alpha_k, c_{pk}, \ell, a_i, \overline{\overline{H}}_{21}, \sqrt{2\mathcal{D}_{kb} : \mathcal{D}_{kb}}\right). \tag{9-93}$$

From the dimensional analysis, we introduce

$$K_k^{T*} \equiv \frac{K_k^T}{\overline{\overline{\rho}}_k c_{pk}\ell^2\sqrt{2\mathcal{D}_{kb} : \mathcal{D}_{kb}}}. \tag{9-94}$$

Here c_{pk} and ℓ are the specific heat and the distance from the wall or the mixing length, respectively. Then the non-dimensional parameter K_k^{T*} is a function of four similarity groups as

$$K_k^{T*} = K_k^{T*}\left(\frac{\overline{\overline{\rho}}_k c_{pk}\ell^2\sqrt{2\mathcal{D}_{kb} : \mathcal{D}_{kb}}}{\overline{\overline{K}}_k}, \ell a_i, \frac{\overline{\overline{H}}_{21}}{a_i}, \alpha_k\right). \tag{9-95}$$

Thus, the turbulent heat flux can be given by

$$\boldsymbol{q}_k^T = -K_k^{T*}\overline{\overline{\rho}}_k c_{pk}\ell^2\sqrt{2\mathcal{D}_{kb} : \mathcal{D}_{kb}}\left\{\nabla\overline{\overline{T}}_k - \frac{\nabla\alpha_k}{\alpha_k}\left(\overline{\overline{T}}_i - \overline{\overline{T}}_k\right)\right\}. \tag{9-96}$$

It can be seen from Eqs.(9-73) and (9-82) that if a two-phase system undergoes a changing of phases, then the second invariant of the deformation tensor can be quite complicated. This effect, due to the extra interfacial deformation tensor, promotes the heat transfers in two-phase flow systems.

1.2.6 Interfacial transfer constitutive laws

From the entropy inequality (9-47), thermal energy equation (9-28) and the equation of state (9-55), it can be shown that the entropy productions associated with the interfacial transfer of mass Γ_k, generalized drag force $\boldsymbol{M}_{ik}$ and heat transfer $\overline{\overline{a_i q_k''}}$ become

$$\Gamma_k \left\{ \widehat{i_{ki}} \left(\frac{1}{\overline{\overline{T}}_k} - \frac{1}{\overline{\overline{T}}_i} \right) - \left(\frac{\widehat{g_k}}{\overline{\overline{T}}_k} - \frac{\widehat{g_{ki}}}{\overline{\overline{T}}_i} \right) \right\} + \frac{\boldsymbol{M}_{ik} \cdot \left(\widehat{\boldsymbol{v}_{ki}} - \widehat{\boldsymbol{v}_k} \right)}{\overline{\overline{T}}_k}$$
$$+ \overline{\overline{a_i q_{ki}''}} \left(\frac{1}{\overline{\overline{T}}_k} - \frac{1}{\overline{\overline{T}}_i} \right) \geq 0. \tag{9-97}$$

Here we have based our analysis on the assumption that these effects satisfy the entropy inequality independently. The standard theory on irreversible thermodynamics (De Groot and Mazur, 1962) gives a simple method to obtain linear constitutive equations. For this purpose, first we should arrange the terms in the entropy inequality into suitable combinations of fluxes and potentials (the fluxes are expanded linearly in terms of the potentials). We should pay special attention here because we have two inequalities from Eq.(9-97) for each phase; The mass transfer term Γ_k and the generalized drag force $\boldsymbol{M}_{ik}$ should satisfy the jump condition (8-5) and (8-19). Since in many practical problems the order of magnitude of $\boldsymbol{M}_m^H$ and $\boldsymbol{M}_k^t$ are much smaller than the drag force itself, we may approximate $\sum_{k=1}^{2} \boldsymbol{M}_{ik} \approx 0$ in Eq.(8-19). By taking into account these effects, we have the following inequality

$$\Gamma_1 \left\{ \overline{\overline{T}}_1 \left[\widehat{i_{1i}} \left(\frac{1}{\overline{\overline{T}}_2} - \frac{1}{\overline{\overline{T}}_i} \right) - \left(\frac{\widehat{g_1}}{\overline{\overline{T}}_1} - \frac{\widehat{g_{1i}}}{\overline{\overline{T}}_i} \right) \right] \right. \tag{9-98}$$

$$
-\overline{\overline{T}}_2\left[\widehat{i_{2i}}\left(\frac{1}{\overline{\overline{T}}_2}-\frac{1}{\overline{\overline{T}}_i}\right)-\left(\frac{\widehat{g_2}}{\overline{\overline{T}}_2}-\frac{\widehat{g_{2i}}}{\overline{\overline{T}}_i}\right)\right]\Bigg\}
$$

$$
+\boldsymbol{M}_{i1}\cdot\left\{\left(\widehat{\boldsymbol{v}_2}-\widehat{\boldsymbol{v}_1}\right)+\left(\widehat{\boldsymbol{v}_{1i}}-\widehat{\boldsymbol{v}_{2i}}\right)\right\}+\sum_{k=1}^{2}a_i\overline{\overline{q''_{ki}}}\left(1-\frac{\overline{\overline{T}}_k}{\overline{\overline{T}}_i}\right)\geq 0.
$$

Furthermore, if we neglect the thrust forces due to mass transfer and the normal stresses at interfaces, then from Eqs.(2-104) and (9-44) we have

$$
\widehat{g_{1i}}-g^{sat}\left(\overline{\overline{T}}_i\right)=\widehat{g_{2i}}-g^{sat}\left(\overline{\overline{T}}_i\right)\doteq-\left(\frac{2\overline{\overline{H_{21}}}\,\overline{\overline{\sigma}}}{\overline{\overline{\rho_{1i}}}-\overline{\overline{\rho_{2i}}}}\right). \tag{9-99}
$$

And the total momentum flux at the interfaces given by Eq.(8-11) can be simplified to

$$
\boldsymbol{M}_k\doteq\overline{\overline{p_{ki}}}\nabla\alpha_k+\boldsymbol{M}_{ik}+\Gamma_k\widehat{\boldsymbol{v}_i}-\nabla\alpha_k\cdot\overline{\overline{\mathcal{T}_{ki}}}. \tag{9-100}
$$

Thus the pressures at the interfaces should be related by

$$
\overline{\overline{p_{1i}}}-\overline{\overline{p_{2i}}}=-2\overline{\overline{H_{21}}}\,\overline{\overline{\sigma}} \tag{9-101}
$$

which is automatically satisfied by Eq.(9-100) with Eqs.(8-22), (8-23) and (8-25) as the normal component of the interfacial momentum transfer condition.

In what follows, we assume that the effects of the differences between the phase mean values in the bulk fluid and at the interfaces are negligible for the densities and pressures, but not for the temperatures. Thus we take

$$
\overline{\overline{\rho_{ki}}}\doteq\overline{\overline{\rho_k}} \tag{9-102}
$$

$$
\overline{\overline{p_{ki}}}\doteq\overline{\overline{p_k}}\quad\text{(for most cases)}. \tag{9-103}
$$

Under these assumptions, simple linear constitutive equations for interfacial transfer terms may be put into the following forms

$$\Gamma_1 = b_1^{\Gamma}\left(\overline{\overline{T_i}} - \overline{\overline{T_1}}\right) - b_2^{\Gamma}\left(\overline{\overline{T_i}} - \overline{\overline{T_2}}\right) \tag{9-104}$$

$$\boldsymbol{M}_{i1} = b_1^M\left(\widehat{\boldsymbol{v}_2} - \widehat{\boldsymbol{v}_1}\right) \tag{9-105}$$

$$a_i\overline{\overline{q_k''}} = b_k^E\left(\overline{\overline{T_i}} - \overline{\overline{T_k}}\right)\ (k = 1 \text{ or } 2) \tag{9-106}$$

in which the transport coefficients b_k^{Γ}, b_1^M and b_k^E are considered to be positive scalars.

Interfacial Mass Transfer Term

We assume that the transfer coefficient b_k^{Γ} in Eq.(9-104) is a function of following parameters

$$b_k^{\Gamma} = b_k^{\Gamma}\left(\overline{\overline{\rho_1}}, \overline{\overline{\rho_2}}, \widehat{i_k} - \widehat{i_{ki}}, \overline{\overline{K_k}} + K_k^T, \widehat{i_{1i}} - \widehat{i_{2i}}, \overline{\overline{H_{21}}}, a_i, \alpha_k\right). \tag{9-107}$$

In order to simplify the above equation, we first introduce a non-dimensional parameter

$$b_k^{\Gamma*} \equiv \frac{b_k^{\Gamma}\left|\widehat{i_{1i}} - \widehat{i_{2i}}\right|}{\left(\overline{\overline{K_k}} + K_k^T\right)a_i^2}. \tag{9-108}$$

And the Jakob numbers are defined by

$$N_{J1} \equiv \frac{\overline{\overline{\rho_1}}\left(\widehat{i_1} - \widehat{i_{1i}}\right)}{\overline{\overline{\rho_2}}\left(\widehat{i_{2i}} - \widehat{i_{1i}}\right)} \tag{9-109}$$

$$N_{J2} \equiv \frac{\overline{\overline{\rho_2}}\left(\widehat{i_2} - \widehat{i_{2i}}\right)}{\overline{\overline{\rho_1}}\left(\widehat{i_{1i}} - \widehat{i_{2i}}\right)}. \tag{9-110}$$

Consequently, the interfacial mass transfer term can be rewritten as

$$\Gamma_1 = a_i^2 \left\{ b_1^{\Gamma*} \frac{\left(\overline{\overline{K_1}} + K_1^T\right)}{\left|\widehat{i_{1i}} - \widehat{i_{2i}}\right|} \left(\overline{\overline{T_i}} - \overline{\overline{T_1}}\right) \right.$$
$$\left. - b_2^{\Gamma*} \frac{\left(\overline{\overline{K_2}} + K_2^T\right)}{\left|\widehat{i_{1i}} - \widehat{i_{2i}}\right|} \left(\overline{\overline{T_i}} - \overline{\overline{T_2}}\right) \right\} \tag{9-111}$$

where the non-dimensional function $b_k^{\Gamma*}$ can be expressed by four similarity groups as

$$b_k^{\Gamma*} = b_k^{\Gamma*} \left(\frac{\overline{\overline{\rho_1}}}{\overline{\overline{\rho_2}}}, N_{Jk}, \frac{\overline{\overline{H_{21}}}}{a_i}, \alpha_k \right). \tag{9-112}$$

The Jakob number, defined by Eq.(9-109), is the scale of the available energy. It is known to be an important parameter in the analyses of bubble growth.

Now let us examine some special cases in which the constitutive equation for Γ_1 can be reduced to a simple form. In many practical engineering problems, we may assume that the vapor phase is in saturation condition, thus we may take

$$\overline{\overline{p_g}} = \overline{\overline{p_g}}\left(\overline{\overline{T_i}}\right) \quad \text{and} \quad \overline{\overline{T_g}} = \overline{\overline{T_i}}. \tag{9-113}$$

Under this condition Eq.(9-111) reduces to

$$\Gamma_g = -\Gamma_f = \frac{\left(\overline{\overline{K_f}} + K_f^T\right)}{\left(\widehat{i_{gi}} - \widehat{i_{fi}}\right)} \left(\overline{\overline{T_f}} - \overline{\overline{T_i}}\right) b_f^{\Gamma*} a_i^2. \tag{9-114}$$

For example, the analyses on the bubble growth in a laminar flow suggest that for such flow $b_f^{\Gamma*}$ can be approximated by

$$b_f^{\Gamma*} \doteq b_f^{\Gamma*} \left(N_{Jf}, \frac{\overline{\overline{H_{21}}}}{a_i} \right) = C\left(1 + \frac{2N_{Jf}}{\pi}\right) \frac{\overline{\overline{H_{21}}}}{a_i}. \tag{9-115}$$

Here, C is a parameter that takes into account the thickness of the boundary layer. It varies approximately from 1 to 0.6 as the size of bubble increases. The form of the function $b_k^{\Gamma*}$ for more general case should be obtained from the analyses on a single bubble dynamics as well as from experimental data.

Interfacial Drag Force

The general expression for $\boldsymbol{M}_{ik}$ has been postulated by Eq.(9-105). Now we further assume that the coefficient b_1^M depends on the following parameters.

$$b_1^M = b_1^M\left(a_i, \overline{\overline{H_{21}}}, \overline{\overline{\rho_k}}, \left|\widehat{\boldsymbol{v}_2} - \widehat{\boldsymbol{v}_1}\right|, \overline{\overline{\mu_k}} + \mu_k^T, \alpha_k, \Gamma_1\right). \tag{9-116}$$

Then from a dimensional analysis, we can rewrite Eq.(9-105) as

$$\boldsymbol{M}_{i1} = \left(\overline{\overline{\rho_1}} + \overline{\overline{\rho_2}}\right)\left|\widehat{\boldsymbol{v}_2} - \widehat{\boldsymbol{v}_1}\right|\left(\widehat{\boldsymbol{v}_2} - \widehat{\boldsymbol{v}_1}\right) b_1^{M*} a_i \tag{9-117}$$

where

$$b_1^{M*} \equiv \frac{b_1^M}{\left(\overline{\overline{\rho_1}} + \overline{\overline{\rho_2}}\right)\left|\widehat{\boldsymbol{v}_2} - \widehat{\boldsymbol{v}_1}\right| a_i}. \tag{9-118}$$

The dimensionless function b_1^{M*} depends on the following similarity groups

$$b_1^{M*} = b_1^{M*}\left(\frac{\overline{\overline{\rho_1}}}{\overline{\overline{\rho_2}}}, \frac{\overline{\overline{H_{21}}}}{a_i}, \alpha_1, N_{\mathrm{Re}1}^i, N_{\mathrm{Re}2}^i, N_{pch}^i\right). \tag{9-119}$$

Here we have defined the interfacial Reynolds number by

$$N_{\mathrm{Re}k}^i \equiv \frac{\overline{\overline{\rho_k}}\left|\widehat{\boldsymbol{v}_2} - \widehat{\boldsymbol{v}_1}\right|}{\left(\overline{\overline{\mu_k}} + \mu_k^T\right) a_i} \tag{9-120}$$

and the phase change effect number by

$$N_{pch}^i \equiv \frac{\Gamma_1}{\left(\overline{\overline{\rho_1}} + \overline{\overline{\rho_2}}\right)\left|\widehat{\boldsymbol{v}_2} - \widehat{\boldsymbol{v}_1}\right| a_i}. \tag{9-121}$$

The phase change effect number takes account for the mass transfer effect on the drag forces. If this number is large, then the mass transfer effect significantly alters the standard drag correlations. This is exemplified in the field of aerodynamics by variations in the drag forces induced by changes in the rates of boundary-layer suction or blowing.

For a dispersed flow regime, there are numerous researches on the drag forces. The analysis is relatively easy for a flow of a dilute suspension with constant diameter solid spherical particles. However, the problem becomes increasingly complex as the void fraction of the dispersed phase increases or as the wall effect becomes important. It is evident that the drag correlations should depend extensively on experimental data, for a flow with deformable interfaces, interfacial mass transfer and the turbulences.

Some of the results on the dispersed flow drag law can be found in Brodkey (1967) , Soo (1967), Wallis (1969), Schlichting (1979), Happel and Brenner (1965). We discuss important special cases in Section 1.4 of Chapter 9. More general and complete modeling and discussion are presented in Chapter 12.

Interfacial Heat Flux

The constitutive equation for heat transfer at the interface has been postulated by Eq.(9-106). First, we introduce a non-dimensional heat transfer coefficient b_1^E as

$$b_1^{E*} \equiv \frac{b_1^E}{\left(\overline{\overline{K_1}} + K_1^T\right) a_i^2}. \tag{9-122}$$

Then we have

$$a_i \overline{\overline{q_{1i}''}} = \left(\overline{\overline{K_1}} + K_1^T\right) b_1^{E*} \left(\overline{\overline{T_i}} - \overline{\overline{T_1}}\right) a_i^2. \tag{9-123}$$

It is expected that b_1^{E*} depends on the following parameters

$$b_1^{E*} = b_1^{E*}\left(N_{\mathrm{Pr}1}^T, N_{\mathrm{Re}1}^i, N_{pch}^i, \frac{\overline{\overline{H_{21}}}}{a_i}, \alpha_1, \frac{\overline{\overline{\rho_1}}}{\overline{\overline{\rho_2}}}\right) \tag{9-124}$$

where the Prandtl number is defined by

$$N_{\mathrm{Pr}k}^{T} = \frac{c_{pk}\left(\overline{\overline{\mu_k}} + \mu_k^T\right)}{\overline{\overline{K_k}} + K_k^T}. \tag{9-125}$$

The interfacial Reynolds number $N_{\mathrm{Re}k}^{i}$ and the phase change effect number N_{pch}^{i} have been given by Eqs.(9-120) and (9-121), respectively. It is interesting to note here that actually the non-dimensional parameter b_1^{E*} is an interfacial Nusselt number. Furthermore, if we give a constitutive equation for Γ_1, then it is sufficient to supply a constitutive law for only *one* of the interfacial heat transfer, $a_i \overline{\overline{q_1''}}$. However, it is also possible to give a constitutive relation for $a_i \overline{\overline{q_1''}}$ and for $a_i \overline{\overline{q_2''}}$, then it is equivalent that Γ_1 is known due to the macroscopic energy jump condition.

Interfacial Shear Stress

Since bubbles are dispersed in continuous phase shear layer, the interfacial shear stress is approximated by the shear stress in continuous phase. Thus, we have

$$\overline{\overline{\mathcal{T}_{ki}}} \doteq \overline{\overline{\mathcal{T}_c}}. \tag{9-126}$$

Interfacial Momentum Source

In the original momentum jump condition there are two distinct pieces of information; the normal jump and the tangential jump balances. Since we preserved this special characteristic in the averaged formulation, we obtained the drag force balance (8-19) in addition to the interfacial momentum transfer condition given by Eq.(9-12). Furthermore, by neglecting the mass thrust effect and using the assumption, Eq.(9-103), we obtain from Eq.(8-21)

$$\begin{aligned} \sum_{k=1}^{2} \boldsymbol{M}_k &= \sum_{k=1}^{2}\left(\overline{\overline{p_k}}\nabla\alpha_k + \boldsymbol{M}_k^n + \boldsymbol{M}_k^t - \nabla\alpha_k \cdot \overline{\overline{\mathcal{T}_{ki}}}\right) \\ &= \boldsymbol{M}_m = 2\overline{\overline{H_{21}}}\,\overline{\overline{\sigma}}\nabla\alpha_2 + \boldsymbol{M}_m^H. \end{aligned} \tag{9-127}$$

Thus, in view of Eqs.(8-19), (9-12) and (9-13), we have

$$\overline{\overline{p_{1i}}} - \overline{\overline{p_{2i}}} = -2\overline{\overline{H_{21}}}\,\overline{\overline{\sigma}} \tag{9-128}$$

and

$$\sum_{k=1}^{2} \boldsymbol{M}_{ik} = \sum_{k=1}^{2} \left(\boldsymbol{M}_k^n + \boldsymbol{M}_k^t \right) = \boldsymbol{M}_m^H. \tag{9-129}$$

Here we have the thermal equation of state for the interfaces

$$\overline{\overline{\sigma}} = \overline{\overline{\sigma}} \left(\overline{\overline{T_i}} \right). \tag{9-130}$$

Since the normal component of the interfacial momentum transfer equation (9-129) specifies the mechanical equilibrium condition between two phases, it is necessary to specify the mean curvature $\overline{\overline{H_{21}}}$ by a constitutive equation. A simple case is to assume that the interfacial geometries are completely irregular, thus we may take $\overline{\overline{H_{21}}} = 0$. For dispersed two-phase flow, the mean curvature is the inverse of the radius of a particle. If the fluid particle size is uniform, the radius is given by $3\alpha/a_i$. Therefore, the ratio of α and a_i is in general an important length scale. $6\alpha/a_i$ is known as the Sauter mean diameter, thus the mean curvature is essentially proportional to the inverse of the Sauter mean diameter.

The importance of the parameter a_i remains in the macroscopic formulation, however, since it represents the available area of contact between two phases. It is evident that the interfacial transports of mass, momentum and energy are significantly influenced by the surface area concentration per unit volume a_i. In general, the constitutive equations for $\overline{\overline{H_{21}}}$ and a_i are extremely complicated because these are the parameters that decide the local geometric configuration in the macroscopic field. It is evident that we may supply this information directly or indirectly.

The direct information means that we have a prior knowledge of the flow structures. Then it is not very difficult to obtain a relation for $\overline{\overline{H_{21}}}$ and a_i in terms of various variables and initial and boundary conditions. For example, this can be done easily for a bubbly or droplet flow without phase changes, coalescences or disintegration of bubbles (or droplets). It is also possible to give indirect information on the flow structures through the constitutive equations for $\overline{\overline{H_{21}}}$ and a_i in terms of various parameters such as α_k, $\overline{\overline{\sigma}}$, $\widehat{\boldsymbol{v}}_2 - \widehat{\boldsymbol{v}}_1$, $\overline{\overline{\mu_k}}$, μ_k^T, Γ_k etc. Then we may solve the whole set of equations to find a local geometrical configuration. This is difficult because the geometrical configuration has a long-lasting memory and it does not obey the principle of local action in most cases. This means that the initial conditions as well as the wall effects on the flow geometries are very important.

Because of the difficulties encountered in the general case, let us start our discussion on the above constitutive equations from a simple case. Now let us suppose that phase 2 is dispersed in phase 1. Then we may assume that the volume occupied by phase 2 in a total volume V can be given as a function of the mean curvature $\overline{\overline{H_{21}}}$. Thus we have

$$V_2 = F_{V2}\left(\overline{\overline{H_{21}}}\right) = \alpha_2 V. \tag{9-131}$$

The surface area of phase 2 in the volume V is given by

$$A_2 = F_{A2}\left(\overline{\overline{H_{21}}}\right) = a_i V. \tag{9-132}$$

Exactly the same argument can be carried out in the time domain, hence we have

$$\Delta t_2 = f_{V2}\left(\overline{\overline{H_{21}}}\right) = \alpha_2 \Delta t \tag{9-133}$$

and

$$\sum_j \frac{1}{v_{ni}} = f_{A2}\left(\overline{\overline{H_{21}}}\right) = a_i \Delta t. \tag{9-134}$$

Then we may assume that

$$\frac{f_{V2}\left(\overline{\overline{H_{21}}}\right)}{f_{A2}\left(\overline{\overline{H_{21}}}\right)} = \frac{\alpha_S}{a_i} \doteq \frac{F_{V2}\left(\overline{\overline{H_{21}}}\right)}{F_{A2}\left(\overline{\overline{H_{21}}}\right)} = \frac{1}{3C^i\overline{\overline{H_{21}}}} \tag{9-135}$$

where C^i is a factor to take into account for the shapes and sizes of dispersed phase. Thus we can write the constitutive equation for $\overline{\overline{H_{21}}}$ as

$$\overline{\overline{H_{21}}} = \frac{a_i}{3C^i\alpha_2}. \tag{9-136}$$

Here the factor C^i is 1 for fairly uniform spherical droplets or bubbles and it does not vary much unless the dispersed phase has quite elongated shapes. The relation given by Eq.(9-136) is a static or geometric relation and we may take a more general form given by

$$\overline{\overline{H_{21}}} = \overline{\overline{H_{21}}}\left(\alpha_2, a_i, \left|\nabla\alpha_2\right|\right)$$
or (9-137)
$$\frac{\overline{\overline{H_{21}}}}{a_i} = \overline{\overline{H_{21}}}^* = \overline{\overline{H_{21}}}^*\left(\alpha_2, \frac{\left|\nabla\alpha_2\right|}{a_i}\right).$$

We call this relation as the geometric equation of state. From Eq.(9-137) we see that the mean curvature depends on the void fraction, surface area concentration and the void fraction gradient.

However, we still should have one more essential constitutive equation for a_i. It is considered that the information on a_i in terms of other parameters is really a part of the solutions for a local instant formulation. The most general method to include a_i in the two-fluid formulation would be to introduce one more transport equation for the interfacial area concentration as

$$\frac{\partial a_i}{\partial t} + \nabla\cdot\left(a_i\widehat{\boldsymbol{v}_i}\right) = \phi_L. \tag{9-138}$$

With this method, the source term takes account for the bubble or droplet expansions or collapses, coalescences, disintegration and the interfacial instabilities. It is evident that the constitutive equations for $\widehat{\boldsymbol{v}_i}$ and ϕ_L should be supplied. The interfacial area transport equation is a fundamental equation describing the change of surface area between phases. Because of its significance, it is discussed in detail in Chapter 10. In some cases the balance equation (9-138) may be replaced by a simpler algebraic constitutive relation such as

$$a_i = a_i\left(\widehat{\boldsymbol{v}_2} - \widehat{\boldsymbol{v}_1}, \overline{\overline{\rho_k}}, \overline{\overline{\mu_k}}, \alpha_1, \left|\nabla\alpha_1\right|, \overline{\overline{\sigma}}, g\right). \tag{9-139}$$

The constitutive equation for $\boldsymbol{M}_m^H$ can be given for a dispersed flow by

$$\boldsymbol{M}_m^H \doteq \alpha_2\nabla\left(2\overline{\overline{H_{21}}}\,\overline{\overline{\sigma}}\right) \tag{9-140}$$

where phase 2 is the dispersed phase. However, in many practical problems the order of magnitude of this term is small in comparison with $\boldsymbol{M}_{i1}$ or $\boldsymbol{M}_{i2}$. In such a case, we may set $\boldsymbol{M}_m^H$ is to be zero.

Interfacial Energy Source

The interfacial energy source E_m is given by Eq.(8-45). It is clear that the inclusion of the interfacial thermal energy term, namely, the terms given by Eq.(8-45), complicates the formulation significantly. Except for very few cases, this term can be neglected with respect to the large-energy exchanges that involve the latent heat at the changing of phases. Consequently, we approximate Eq.(8-45) by

$$E_m = \overline{\overline{T}}_i \left(\frac{d\sigma}{dT}\right) \frac{D_i a_i}{Dt} + 2\overline{\overline{H_{21}}}\,\overline{\overline{\sigma}} \frac{\partial \alpha_1}{\partial t} + E_m^H \approx 0 \tag{9-141}$$

which does not require any additional constitutive equations.

In order to complete the model for the interfacial energy transfer condition given by Eq.(8-48), we should supply the constitutive equations for the turbulent kinetic energies, the difference between the mean velocity and the average interfacial phase velocity, and the interfacial turbulent flux from the drag force W_{ki}^T. As in two-phase flow with phase changes, the order of magnitude of these terms compared to the thermal terms is relatively small, therefore, we may assume

$$\widehat{h_{ki}} - \widehat{i_{ki}} = \frac{\overline{\left(v_{ki}'\right)^2}}{2} \doteq 0 \tag{9-142}$$

$$\widehat{\boldsymbol{v}_{ki}} - \widehat{\boldsymbol{v}_k} \doteq 0 \tag{9-143}$$

$$W_{ki}^T \doteq 0. \tag{9-144}$$

In analogy with Eq.(9-142), we take for the bulk phases

$$\widehat{h_k} \doteq \widehat{i_k}. \tag{9-145}$$

Then, Eq.(8-48) can be reduced to the following form

$$E_k \doteq \Gamma_k \left(\widehat{i_{ki}} + \frac{\widehat{v_k}^2}{2}\right) + a_i \overline{\overline{q_k''}} - \overline{\overline{p_k}} \frac{\partial \alpha_k}{\partial t} + \boldsymbol{M}_{ik} \cdot \widehat{\boldsymbol{v}_{ki}} - \nabla \alpha_k \cdot \overline{\overline{\mathcal{T}_{ki}}} \cdot \widehat{\boldsymbol{v}_{ki}} \tag{9-146}$$

where we have used Eq.(9-103). And the interfacial total energy transfer condition becomes

$$\sum_{k=1}^{2} E_k = E_m \approx 0. \tag{9-147}$$

Then, the thermal energy transfer condition (8-49) can be approximated by

$$\Lambda_k \doteq \left(\Gamma_k \widehat{i_{ki}} + a_i \overline{\overline{q''_{ki}}}\right) - \overline{\overline{p_k}} \frac{D_k \alpha_k}{Dt} \tag{9-148}$$

and

$$\begin{aligned}\sum_{k=1}^{2} \Lambda_k = & \overline{\overline{T_i}} \left(\frac{d\sigma}{dT}\right) \frac{D_i a_i}{Dt} + E_m^H - \sum_{k=1}^{2} \overline{\overline{p_k}} \frac{D_k \alpha_k}{Dt} \\ & - \Gamma_1 \left(\frac{\widehat{v_1}^2}{2} - \frac{\widehat{v_2}^2}{2}\right) - \sum_{k=1}^{2} \boldsymbol{M}_{ik} \cdot \widehat{\boldsymbol{v}_k} + \sum_{k=1}^{2} \nabla \alpha_k \cdot \overline{\overline{\mathcal{T}_{ki}}} \cdot \widehat{\boldsymbol{v}_{ki}}\end{aligned} \tag{9-149}$$

in which we have used Eq.(9-128) in order to eliminate the surface tension term. By combing Eq.(9-148) with Eq.(9-149) we get

$$\begin{aligned}\sum_{k=1}^{2} \left(\Gamma_k \widehat{i_{ki}} + a_i \overline{\overline{q''_{ki}}} + \Gamma_k \frac{\widehat{v_k}^2}{2}\right) = & \left\{\overline{\overline{T_i}} \left(\frac{d\sigma}{dT}\right) \frac{D_i a_i}{Dt} + E_m^H\right\} \\ & - \sum_{k=1}^{2} \boldsymbol{M}_{ik} \cdot \widehat{\boldsymbol{v}_{ki}} + \sum_{k=1}^{2} \nabla \alpha_k \cdot \overline{\overline{\mathcal{T}_{ki}}} \cdot \widehat{\boldsymbol{v}_{ki}}.\end{aligned} \tag{9-150}$$

The first group on the right-hand side of the above equation is the effect of the surface tension, the second group arises from the interfacial drag work, and the third term is related to the work done by interfacial shear. Thus, we may assume for relatively low speed flow that

$$\sum_{k=1}^{2} \left(\Gamma_k \widehat{i_{ki}} + a_i \overline{\overline{q''_{ki}}}\right) \approx 0 \tag{9-151}$$

which is a well-known relation for a local instant formulation. We have shown here the conditions under which we can apply this important and useful formula to the macroscopic two-phase flow problems.

1.3 Two-fluid model formulation

The most general case of the two-fluid model formulation has been discussed in the Section 1.2.3 of Chapter 9 in connection with the principle of determinism. We will now set up a realistic formulation by combining the results of the previous two sections. We already have made a number of assumptions on the interfacial variables and the constitutive equations, thus the present analysis is not a complete mapping of the microscopic field in terms of the local instant variables into the macroscopic field. Rather, it should be considered as an approximate theory based on various constitutive assumptions. The results presented in this section are simplified to an extent of being realistic, yet it is general enough for most engineering problems encountered in two-phase flow system analyses.

First, we list all the important assumptions that have been made to obtain the model.

- Fundamental hypothesis on smoothness of mean values — Section 1.3 in Chapter 4
- Existence of the equation of state — Eq.(9-55)
- Transport properties μ_k and K_k are constant in the interval of time average — Eq.(9-71), Eq.(9-79)
- Interfacial variables are approximated by

  $\rho_{ki} \approx \overline{\overline{\rho_{ki}}},\ \sigma \approx \overline{\overline{\sigma}},\ \dot{m}_k \approx \overline{\overline{\dot{m}_k}}$ — Eq.(8-9)

  $T_i \approx \overline{\overline{T_i}}$ — Eq.(9-80)

  $\overline{\overline{\rho_{ki}}} \approx \overline{\overline{\rho_k}}$ — Eq.(9-102)

  $\overline{\overline{p_{ki}}} \approx \overline{\overline{p_k}}$ — Eq.(9-103)
- Interfacial normal stress and thrust due to mass transfer are neglected — Eq.(9-127)
- Negligible turbulent kinetic energy or energy transfer — Eq.(9-145)
- Mechanical interaction terms in the interfacial energy transfer condition are neglected — Eq.(9-148)
- Uniform body force — Eq.(5-50)

Under these conditions we have the following field equations

The continuity equations from Eq.(9-1)

$$\frac{\partial}{\partial t}\left(\alpha_k\overline{\overline{\rho_k}}\right)+\nabla\cdot\left(\alpha_k\overline{\overline{\rho_k}}\widehat{\boldsymbol{v}_k}\right)=\Gamma_k\ \ (k\text{=1 and 2}) \tag{9-152}$$

The equations of motion from Eq.(9-15)

$$\alpha_k \overline{\overline{\rho_k}} \frac{D_k \widehat{\boldsymbol{v}_k}}{Dt} = -\alpha_k \nabla \overline{\overline{p_k}} + \nabla \cdot \left[\alpha_k \left(\overline{\overline{\mathcal{T}_k}} + \mathcal{T}_k^T\right)\right] + \alpha_k \overline{\overline{\rho_k}} \widehat{\boldsymbol{g}_k}$$
$$+\boldsymbol{M}_{ik} - \nabla \alpha_k \cdot \overline{\overline{\mathcal{T}_{ki}}} + \Gamma_k \left(\widehat{\boldsymbol{v}_{ki}} - \widehat{\boldsymbol{v}_k}\right) + \left(\overline{\overline{p_{ki}}} - \overline{\overline{p_k}}\right) \nabla \alpha_k \tag{9-153}$$
$(k=1 \text{ and } 2)$

It is noted that the last term in the above equation is retained though for most cases it is very small, because for some cases such as horizontal flow it can be important.

The equations of thermal energy from Eq.(9-30)

$$\alpha_k \overline{\overline{\rho_k}} \frac{D_k \widehat{i_k}}{Dt} = -\nabla \cdot \alpha_k \left(\overline{\overline{\boldsymbol{q}_k}} + \boldsymbol{q}_k^T\right) + \alpha_k \frac{D_k \overline{\overline{p_k}}}{Dt}$$
$$+\alpha_k \overline{\overline{\mathcal{T}_k}} : \nabla \widehat{\boldsymbol{v}_k} + \Gamma_k \left(\widehat{i_{ki}} - \widehat{i_k}\right) + a_i \overline{\overline{q''_{ki}}} \tag{9-154}$$
$$+\left(\boldsymbol{M}_{ik} - \nabla \alpha_k \cdot \overline{\overline{\mathcal{T}_{ki}}}\right) \cdot \left(\widehat{\boldsymbol{v}_{ki}} - \widehat{\boldsymbol{v}_k}\right) \quad (k=1 \text{ and } 2)$$

Here we have neglected the turbulent work term $\widehat{\boldsymbol{v}_k} \cdot \nabla \cdot \left(\alpha_k \mathcal{T}_k^T\right)$, since it is considered to contribute mainly for the turbulent kinetic energy changes which have been neglected in the formulation. These two sets of three balance equations describe the physical laws of conservation of mass, momentum and energy in the macroscopic field.

Two phases that are governed by their own field equations are coupled by three interfacial transfer conditions given below.

The interfacial mass transfer condition from Eq.(9-2)

$$\sum_{k=1}^{2} \Gamma_k = 0 \tag{9-155}$$

The interfacial momentum transfer condition from Eqs.(9-127) and (9-128)

$$\sum_{k=1}^{2} \left(\overline{\overline{p_k}} \nabla \alpha_k + \boldsymbol{M}_{ik} - \nabla \alpha_k \cdot \overline{\overline{\mathcal{T}_{ki}}}\right) = 2\overline{\overline{H_{21}}}\, \overline{\overline{\sigma}} \nabla \alpha_2 + \boldsymbol{M}_m^H \tag{9-156}$$

with the normal component satisfying

$$\overline{\overline{p_{1i}}} - \overline{\overline{p_{2i}}} = -2\overline{\overline{H_{21}}}\,\overline{\overline{\sigma}} \tag{9-157}$$

The interfacial thermal energy transfer condition from Eq.(9-151)

$$\sum_{k=1}^{2}\left(\Gamma_k \widehat{i_{ki}} + a_i \overline{\overline{q''_{ki}}}\right) = 0. \tag{9-158}$$

Then, from the axiom of continuity we have

$$\alpha_1 = 1 - \alpha_2. \tag{9-159}$$

The equation of state for each phase is given by Eq.(9-55) or by Eqs.(9-58) and (9-59), thus we have the *caloric equations of state*

$$\widehat{i_k} = \widehat{i_k}\left(\overline{\overline{T_k}}, \overline{\overline{p_k}}\right) \quad (k{=}1 \text{ and } 2) \tag{9-160}$$

and the *thermal equations of state*

$$\overline{\overline{\rho_k}} = \overline{\overline{\rho_k}}\left(\overline{\overline{T_k}}, \overline{\overline{p_k}}\right) \quad (k{=}1 \text{ and } 2) \tag{9-161}$$

whereas the *equation of state for the surface* is given by Eq.(9-130)

$$\overline{\overline{\sigma}} = \overline{\overline{\sigma}}\left(\overline{\overline{T_i}}\right). \tag{9-162}$$

The interfacial temperature is given by the phase change condition (9-44) as

$$\overline{\overline{p_2}} - p^{sat}\left(\overline{\overline{T_i}}\right) = 2\overline{\overline{H_{21}}}\,\overline{\overline{\sigma}}\left(\frac{\overline{\overline{\rho_2}}}{\overline{\overline{\rho_2}} - \overline{\overline{\rho_1}}}\right) \tag{9-163}$$

where

$$p^{sat} = p^{sat}\left(\overline{\overline{T_i}}\right) \tag{9-164}$$

is the classical saturation condition. For many practical cases we may approximate Eq.(9-163) by

$$\overline{\overline{p_g}} = p^{sat}\left(\overline{\overline{T_i}}\right) \tag{9-165}$$

where $\overline{\overline{p_g}}$ is the vapor phase pressure.

The constitutive equation for the *viscous stress* $\overline{\overline{\mathcal{T}_k}}$ is given by

$$\overline{\overline{\mathcal{T}_k}} = 2\overline{\overline{\mu_k}}\left(\mathcal{D}_{kb} + \mathcal{D}_{ki}\right) = 2\overline{\overline{\mu_k}}\mathcal{D}_k \ \ (k\text{=1 and 2}). \tag{9-166}$$

Here the bulk and interfacial extra deformation tensors $\mathcal{D}_{kb}$ and $\mathcal{D}_{ki}$ are given by Eqs.(9-74) and (9-77). Thus we have

$$\mathcal{D}_{kb} = \frac{1}{2}\left[\nabla\widehat{\boldsymbol{v}_k} + \left(\nabla\widehat{\boldsymbol{v}_k}\right)^{+}\right] \tag{9-167}$$

and

$$\mathcal{D}_{ki} \doteq -\frac{a_k^i}{2\alpha_k}\left\{\left(\nabla\alpha_k\right)\left(\widehat{\boldsymbol{v}_2} - \widehat{\boldsymbol{v}_1}\right) + \left(\widehat{\boldsymbol{v}_2} - \widehat{\boldsymbol{v}_1}\right)\left(\nabla\alpha_k\right)\right\}. \tag{9-168}$$

The coefficient a_k^i represents the mobility of the kth-phase.

The constitutive equation for the *turbulent stress* $\mathcal{T}_k^T$ has been obtained from the mixing length hypothesis of Eqs.(9-85), (9-86), (9-87), (9-88) and (9-89), thus

$$\begin{aligned} \mathcal{T}_k^T &= 2\mu_k^T\mathcal{D}_{kb} \\ &= 2\mu_k^{T*}\overline{\overline{\rho_k}}\ell^2\sqrt{2\left(\mathcal{D}_{kb} : \mathcal{D}_{kb}\right)}\mathcal{D}_{kb} \ \ (k\text{=1 and 2}). \end{aligned} \tag{9-169}$$

Here the non-dimensional turbulent viscosity μ_k^{T*} is a function of the following parameters

$$\mu_k^{T*} = \mu_k^{T*}\left(\frac{\overline{\overline{\rho_k}}\ell^2\sqrt{2\mathcal{D}_k : \mathcal{D}_k}}{\overline{\overline{\mu_k}}}, a_i\ell, \frac{\overline{\overline{H_{21}}}}{a_i}, \alpha_k\right). \tag{9-170}$$

We recall here that ℓ and a_i are the distances from the wall and the surface area concentration, respectively. If we exclude the region very near to the wall, then the first parameter may be dropped from the arguments of μ_k^{T*}.

The constitutive law for average *conduction heat flux* $\overline{\overline{\boldsymbol{q}_k}}$ is given by Eq.(9-81), thus we have

$$\overline{\overline{\boldsymbol{q}_k}} = -\overline{\overline{K_k}}\left\{\nabla\overline{\overline{T_k}} - \frac{\nabla\alpha_k}{\alpha_k}\left(\overline{\overline{T_i}} - \overline{\overline{T_k}}\right)\right\}. \tag{9-171}$$

It is interesting to note that the second term, due to the concentration gradient, represents the effect of the temperature difference between the bulk phase and the interfaces, namely, thermal non-equilibrium.

From the mixing length model for the *turbulent heat flux* $\boldsymbol{q}_k^T$, we have

$$\boldsymbol{q}_k^T = -K_k^{T*}\overline{\overline{\rho_k}}\, c_{pk}\ell^2\sqrt{2\boldsymbol{D}_{kb}:\boldsymbol{D}_{kb}}\left\{\nabla\overline{\overline{T_k}} - \frac{\nabla\alpha_k}{\alpha_k}\left(\overline{\overline{T_i}} - \overline{\overline{T_k}}\right)\right\} \tag{9-172}$$

where the non-dimensional conductivity K_k^{T*} is a function of the following parameters

$$K_k^{T*} = K_k^{T*}\left(\frac{\overline{\overline{\rho_k}}c_{pk}\ell^2\sqrt{2\boldsymbol{D}_{kb}:\boldsymbol{D}_{kb}}}{\overline{\overline{K_k}}}, a_i\ell, \frac{\overline{\overline{H_{21}}}}{a_i}, \alpha_k\right). \tag{9-173}$$

We note here that the first dimensionless group may be dropped from the argument of the function K_k^{T*} if we exclude the region very near to the wall.

The mass transfer term Γ_1 is given by Eqs.(9-111) and (9-112), thus we have

$$\begin{aligned}\Gamma_1 = &\frac{\left(\overline{\overline{K_1}} + K_1^T\right)a_i^2}{\left|\widehat{i_{1i}} - \widehat{i_{2i}}\right|}b_1^{\Gamma*}\left(\overline{\overline{T_i}} - \overline{\overline{T_1}}\right)\\ &-\frac{\left(\overline{\overline{K_2}} + K_2^T\right)a_i^2}{\left|\widehat{i_{1i}} - \widehat{i_{2i}}\right|}b_2^{\Gamma*}\left(\overline{\overline{T_i}} - \overline{\overline{T_2}}\right)\end{aligned} \tag{9-174}$$

in which the coefficient $b_k^{\Gamma*}$ depends on four parameters

$$b_k^{\Gamma*} = b_k^{\Gamma*}\left(\frac{\overline{\overline{\rho_1}}}{\overline{\overline{\rho_2}}}, N_{Jk}, \frac{\overline{\overline{H_{21}}}}{a_i}, \alpha_k\right). \tag{9-175}$$

The group denoted by N_{Jk} is the Jakob number defined by Eqs.(9-109) and (9-110) and it is the most important parameter on the phase changes. A simpler case of the above constitutive law has been discussed in the previous section.

In view of the interfacial momentum transfer condition, Eq.(9-156), we should supply two constitutive equations that specify the drag force $\boldsymbol{M}_{i1}$ and the effects of surface tension. The *interfacial drag force* has been given by Eqs.(9-117) and (9-119), thus we have

$$\boldsymbol{M}_{i1} = \left(\overline{\overline{\rho_1}} + \overline{\overline{\rho_2}}\right) a_i \left|\widehat{\boldsymbol{v}_2} - \widehat{\boldsymbol{v}_1}\right| \left(\widehat{\boldsymbol{v}_2} - \widehat{\boldsymbol{v}_1}\right) b_1^{M*} \tag{9-176}$$

where the coefficient b_1^{M*} is expected to be a function of the following parameters

$$b_1^{M*} = b_1^{M*}\left(\frac{\overline{\overline{\rho_1}}}{\overline{\overline{\rho_2}}}, \frac{\overline{\overline{H_{21}}}}{a_i}, \alpha_1, N_{\mathrm{Re}1}^i, N_{\mathrm{Re}2}^i, N_{pch}^i\right). \tag{9-177}$$

Here, the interfacial Reynolds number $N_{\mathrm{Re}k}^i$ and the phase change effect number N_{pch}^i are defined by Eqs.(9-120) and (9-121), respectively.

Furthermore, the geometrical equation of state Eq.(9-137), specifies the mean curvature of the interfaces

$$\frac{\overline{\overline{H_{21}}}}{a_i} = \overline{\overline{H_{21}}}^*\left(\alpha_2, \left|\nabla\alpha_2\right|, a_i\right) \tag{9-178}$$

The interfacial thermal energy transfer condition given by Eq.(9-158) requires the *constitutive equation for the heat transfer at the interfaces.* Thus, from Eqs.(9-123) and (9-124) we have

$$a_i \overline{\overline{q_{1i}''}} = \left(\overline{\overline{K_1}} + K_1^T\right) a_i^2 b_1^{E*} \left(\overline{T_i} - \overline{\overline{T_1}}\right) \tag{9-179}$$

where

$$b_1^{E*} = b_1^{E*}\left(N_{\mathrm{Pr}1}^T, N_{\mathrm{Re}1}^i, N_{pch}^i, \frac{\overline{\overline{H_{21}}}}{a_i}, \alpha_1, \frac{\overline{\overline{\rho_1}}}{\overline{\overline{\rho_2}}}\right). \tag{9-180}$$

The definition of the Prandtl number is given by Eq.(9-125). In many practical problems, the dispersed phase can be assumed to be in thermal equilibrium, then it follows that the constitutive equation (9-179) reduces to a trivial form $a_i \overline{\overline{q''_{di}}} = 0$.

Finally, we should have a constitutive equation for the surface area concentration a_i. In general it should have the form of the balance equation

$$\frac{\partial a_i}{\partial t} + \nabla \cdot \left(a_i \widehat{\boldsymbol{v}}_i \right) = \phi_L \tag{9-181}$$

where the source term ϕ_L is expressed by various parameters that have already appeared. It is expected that, in general, the constitutive equation for ϕ_L is quite difficult to obtain unless the flow geometry is very simple, namely, such as the bubbly or droplet flows. The average interface velocity $\widehat{\boldsymbol{v}}_i$ can be given approximately by

$$\widehat{\boldsymbol{v}}_i \doteq \widehat{\boldsymbol{v}}_d - \frac{\Gamma_d}{\overline{\overline{\rho_d}} a_i^2} \nabla \alpha_d \doteq \widehat{\boldsymbol{v}}_d \tag{9-182}$$

in which the subscript *d* denotes the dispersed phase.

The constitutive equation for $\boldsymbol{M}_m^H$ for a dispersed flow can be given by Eq. (9-140), thus we have

$$\boldsymbol{M}_m^H \doteq \alpha_d \nabla \left(2 \overline{\overline{H_{dc}}}\, \overline{\overline{\sigma}} \right). \tag{9-183}$$

The basic variables appearing in the two-fluid formulation are

$$\alpha_k, \overline{\overline{\rho_k}}, \widehat{\boldsymbol{v}}_k, \Gamma_k$$

$$\overline{\overline{p_k}}, \overline{\overline{\mathcal{T}_k}}, \mathcal{T}_k^T, \overline{\overline{\mathcal{T}_{ki}}}, \boldsymbol{M}_{ik}, \boldsymbol{M}_m^H, \overline{\overline{p_{ki}}}$$

$$\widehat{i}_k, \overline{\overline{\boldsymbol{q}_k}}, \boldsymbol{q}_k^T, a_i \overline{\overline{q''_k}}, \overline{\overline{T_k}},$$

$$\overline{\overline{T_i}}, \overline{\overline{H_{21}}}, \overline{\overline{\sigma}}, a_i, p^{sat}.$$

Thus, the total number of unknown is thirty six, and we have:

- Six balance equations — Eqs.(9-152), (9-153) and (9-154);
- Three interfacial conditions — Eqs.(9-155), (9-156) and (9-158);
- Mechanical condition at interfaces — Eq.(9-157);
- Chemical condition at interfaces (Phase change) — Eq.(9-163);
- Saturation condition — Eq.(9-164);

- Axiom of continuity Eq.(9-159);
- Two caloric equations of state Eq.(9-160);
- Two thermal equations of state Eq.(9-161);
- Surface equation of state Eq.(9-162);
- Two constitutive equations for $\overline{\overline{\mathcal{T}_k}}$ Eq.(9-166);
- Two constitutive equations for $\mathcal{T}_k^T$ Eq.(9-169);
- Two constitutive equations for $\overline{\overline{\mathcal{T}_{ki}}}$ Eq.(9-126);
- Two constitutive equations for $\overline{\overline{\boldsymbol{q}_k}}$ Eq.(9-171);
- Two constitutive equations for $\boldsymbol{q}_k^T$ Eq.(9-172);
- Phase change constitutive law Eq.(9-174);
  (or constitutive equation for $a_i \overline{\overline{q''_{2i}}}$ similar to Eq. (9-179))
- Drag constitutive law Eq.(9-176);
- Geometrical equation of state Eq.(9-178);
- Constitutive equation for $a_i \overline{\overline{q''_{1i}}}$ Eq.(9-179);
- Balance equation for surface area Eq.(9-181);
- Constitutive equation for $\boldsymbol{M}_m^H$ Eq.(9-183);
- Constitutive equation for $\overline{\overline{p_{ki}}}$ (for most cases) Eq.(9-103).

This shows that we have thirty six equations. Hence the total number of unknown and of equations is the same. Consequently, our description is consistent and complete in mathematical sense, although this does not guarantee the uniqueness of the solution of the model nor even the existence of the solution. These should be checked by solving various simple cases, then as we have discussed in Chapter 2, the results should be compared to the experimental data to verify and improve the model. It is very important to note that Γ_1, $\overline{\overline{q''_{1i}}}$ and $\overline{\overline{q''_{2i}}}$ are related by the interfacial thermal energy transfer condition that represents the macroscopic energy jump condition. Therefore, the constitutive relation for $\overline{\overline{q''_{2i}}}$ can be used in place of the phase change constitutive law, Eq. (9-174). Since the heat flux can be modeled more easily than the phase change, this is a much more practical approach.

1.4 Various special cases

Scaling Parameter

The general formulation of the two-fluid model has been given above. In the following analysis, we obtain some important *scaling parameters from the field equations.* Before going into the detailed study, we recall that dimensionless groups can be obtained from the conservation equations, boundary conditions and constitutive laws. The similarity of two different systems can only be discussed by including all these groups. This will be very difficult to accomplish in the model based on the two-fluid formulation due to the large number of unknowns involved and of the complexity of the constitutive equations. For such systems, the dimensional analysis is more important for obtaining scaling parameters of various effects in the field equations than for making the similarity analysis of the entire system. The order of magnitude analysis based on these scaling parameters frequently leads to a much-simplified formulation that can be solved to yield meaningful answers to various engineering problems. It should be noted, however, that under certain conditions smaller terms cannot be neglected from the formulation. Thus, for a system of coupled nonlinear differential equations, the order of magnitude analysis should be accepted as a general trend with exceptions. Consequently, since many complicated problems can be solved only approximately, it becomes necessary to check the results with experimental data.

In the following analysis, the subscript o denotes the reference parameters chosen to be constant. The characteristic length is L_o, whereas the time constant is τ_o. For most problems it is taken as the ratio of L_o to the velocity scale, however, for oscillating flows it can be the period of oscillations. Below, we define dimensionless parameters whose order of magnitude is considered to be 1.

$$
\begin{aligned}
&\rho_k^* = \frac{\overline{\overline{\rho_k}}}{\rho_{ko}}, \boldsymbol{v}_k^* = \frac{\widehat{\boldsymbol{v}_k}}{\boldsymbol{v}_{ko}}, t^* = \frac{t}{\tau_o}, \nabla^* = L_o\nabla, \\
&\Gamma_k^* = \frac{\Gamma_k}{\left|\Gamma_{ko}\right|}, p_k^* = \frac{\overline{\overline{p_k}} - p_o}{\Delta p_o}, \mu_k^* = \frac{\overline{\overline{\mu_k}} + \mu_k^T}{\mu_{ko}}, \\
&\boldsymbol{M}_{ik}^* = \frac{\boldsymbol{M}_{ik}}{a_{io}\left(\rho_{1o} + \rho_{2o}\right)\left(v_{2o} - v_{1o}\right)^2}, i_k^* = \frac{\widehat{i_k} - i_{ko}}{\Delta i_{ko}}, \\
&K_k^* = \frac{\overline{\overline{K_k}} + K_k^T}{K_{ko}}, q_{ki}''^* = \frac{\overline{\overline{q_{ki}''}}}{a_{io}K_{ko}\left(T_{io} - T_{ko}\right)}, \\
&H_{21}^* = \frac{\overline{\overline{H_{21}}}}{H_{21o}}, \sigma^* = \frac{\overline{\overline{\sigma}}}{\sigma_o}, \left(\overline{\overline{\mathcal{T}_k}} + \mathcal{T}_k^T\right)^* = \frac{\left(\overline{\overline{\mathcal{T}_k}} + \mathcal{T}_k^T\right)}{\mu_{ko} v_{ko} / L_o}, \\
&\left(\overline{\overline{\mathcal{T}_{ki}}}\right)^* = \frac{\overline{\overline{\mathcal{T}_{ki}}}}{\mu_{ko} v_{ko} / L_o}, \left(\overline{\overline{\boldsymbol{q}_k}} + \boldsymbol{q}_k^T\right)^* = \frac{\left(\overline{\overline{\boldsymbol{q}_k}} + \boldsymbol{q}_k^T\right)}{K_{ko}\Delta T_{ko} / L_o}, \\
&T_k^* = \frac{\overline{\overline{T_k}} - T_o}{\Delta T_{ko}} \approx \frac{\widehat{i_k} - i_{ko}}{\Delta i_{ko}}, a_i^* = \frac{a_i}{a_{io}},
\end{aligned}
\tag{9-184}
$$

$$\boldsymbol{M}_m^{H*} = \frac{\boldsymbol{M}_m^H}{a_{io}\left(\rho_{1o} + \rho_{2o}\right)\left(v_{2o} - v_{1o}\right)^2}$$

Substituting these new parameters into the field equations we obtain the following results.

Non-dimensional continuity equations from Eq.(9-152)

$$\frac{1}{\left(N_{sl}\right)_k}\frac{\partial \alpha_k \rho_k^*}{\partial t^*} + \nabla^* \cdot \left(\alpha_k \rho_k^* v_k^*\right) = \left(N_{pch}\right)_k \Gamma_k^* \tag{9-185}$$

Non-dimensional equations of motion from Eq.(9-153)

$$\begin{aligned}
&\alpha_k \rho_k^* \left\{\frac{1}{\left(N_{sl}\right)_k}\frac{\partial \boldsymbol{v}_k^*}{\partial t^*} + \boldsymbol{v}_k^* \cdot \nabla^* \boldsymbol{v}_k^*\right\} = -\left(N_{Eu}\right)_k \alpha_k \nabla^* p_k^* \\
&+ \frac{1}{\left(N_{\mathrm{Re}}\right)_k} \nabla^* \cdot \left[\alpha_k \left(\overline{\overline{\mathcal{T}}}_k + \mathcal{T}_k^T\right)^*\right] + \frac{1}{\left(N_{Fr}\right)_k} \alpha_k \rho_k^* \frac{\widehat{\boldsymbol{g}_k}}{|\boldsymbol{g}|} \\
&+ \left(N_{drag}\right)_k \boldsymbol{M}_{ik}^* - \frac{1}{\left(N_{\mathrm{Re}}\right)_k} \nabla^* \alpha_k \cdot \left(\overline{\overline{\mathcal{T}}}_{ki}\right)^* \\
&+ \left(N_{pch}\right)_k \Gamma_k^* \left(\widehat{\boldsymbol{v}}_{ki} - \widehat{\boldsymbol{v}}_k\right)^* + \left(N_{Eu}\right)_k \left(p_{ki}^* - p_k^*\right) \nabla^* \alpha_k
\end{aligned} \tag{9-186}$$

Non-dimensional thermal energy equations from Eq.(9-154)

$$\begin{aligned}
&\alpha_k \rho_k^* \left\{\frac{1}{\left(N_{sl}\right)_k}\frac{\partial i_k^*}{\partial t^*} + \boldsymbol{v}_k^* \cdot \nabla^* i_k^*\right\} \\
&= -\frac{1}{\left(N_{Pe}\right)_k} \nabla^* \cdot \alpha_k \left(\overline{\overline{\boldsymbol{q}}}_k + \boldsymbol{q}_k^T\right)^* \\
&+ \left(N_{Eu}\right)_k \left(N_{Ec}\right)_k \alpha_k \left\{\frac{1}{\left(N_{sl}\right)_k}\frac{\partial p_k^*}{\partial t^*} + \boldsymbol{v}_k^* \cdot \nabla^* p_k^*\right\} \\
&+ \frac{\left(N_{Ec}\right)_k}{\left(N_{\mathrm{Re}}\right)_k} \alpha_k \left(\overline{\overline{\mathcal{T}}}_k\right)^* : \nabla^* \boldsymbol{v}_k^* + \left(N_{pch}\right)_k \Gamma_k^* \left(i_{ki}^* - i_k^*\right)
\end{aligned} \tag{9-187}$$

$$+\left(N_q\right)_k a_i^* q_{ki}''^* + \left(N_{drag}\right)_k \left(N_{Ec}\right)_k \boldsymbol{M}_{ik}^* \cdot \left(\widehat{\boldsymbol{v}_{ki}} - \widehat{\boldsymbol{v}_k}\right)^*$$
$$-\frac{\left(N_{Ec}\right)_k}{\left(N_{\mathrm{Re}}\right)_k}\left\{\nabla^*\alpha_k \cdot \left(\overline{\overline{\mathcal{T}_{ki}}}\right)^*\right\} \cdot \left(\widehat{\boldsymbol{v}_{ki}} - \widehat{\boldsymbol{v}_k}\right)^*$$

Here we have introduced several scaling parameters defined as

$$\text{Strouhal number } \left(N_{sl}\right)_k \equiv \frac{\tau_o v_{ko}}{L_o}$$

$$\text{Phase change number } \left(N_{pch}\right)_k \equiv \frac{\left|\Gamma_{ko}\right| L_o}{\rho_{ko} v_{ko}}$$

$$\text{Euler number } \left(N_{Eu}\right)_k \equiv \frac{\Delta p_o}{\rho_{ko} v_{ko}^2}$$

$$\text{Reynolds number } \left(N_{\mathrm{Re}}\right)_k \equiv \frac{\rho_{ko} v_{ko} L_o}{\mu_{ko}}$$

$$\text{Froude number } \left(N_{Fr}\right)_k \equiv \frac{v_{ko}^2}{|g| L_o}$$

$$\text{Drag number } \left(N_{drag}\right)_k \equiv \frac{\left(\rho_{1o} + \rho_{2o}\right) L_o a_{io} \left(v_{2o} - v_{1o}\right)^2}{\rho_{ko} v_{ko}^2}$$

$$\text{Peclet number } \left(N_{Pe}\right)_k \equiv \frac{\rho_{ko} v_{ko} \Delta i_{ko} L_o}{K_{ko} \Delta T_{ko}}$$

$$\text{Eckert number } \left(N_{Ec}\right)_k \equiv \frac{v_{ko}^2}{\Delta i_{ko}}$$

Interface heating number (9-188)

$$\left(N_q\right)_k \equiv \frac{K_{ko}\left(T_{io} - T_{ko}\right) L_o a_{io}^2}{\rho_{ko} v_{ko} \Delta i_{ko}}$$

The first two parameters, the Strouhal and phase change numbers, are the kinematic groups. The Euler, Reynolds, Froude and drag numbers are the dynamic groups, since they are the ratios of various forces appearing in the momentum equations. Similarly, the Peclet, Eckert and interface heating numbers are the energy groups that scale various energy transfer mechanisms. From the definitions (9-188) and the forms of the non-dimensional field equations, the physical meanings of various scaling parameters are evident. In two-fluid model formulation, the phase change, drag and interfacial heating numbers are particularly important since they are the parameters to scale the effects of the interactions between two phases.

Before we discuss various special cases that can be obtained by considering limiting conditions in terms of the scaling parameters, let us study the non-dimensional form of the conditions of interfacial transfer, mechanical state between phases and phase change (or chemical state). Non-dimensional interfacial mass transfer condition can be obtained from Eqs.(9-155) and (9-184), thus

$$\sum_{k=1}^{2} \Gamma_k^* = 0. \tag{9-189}$$

From Eqs.(9-156), (9-157) and (9-184) the interfacial momentum transfer condition is given by

$$\sum_{k=1}^{2} \boldsymbol{M}_{ik}^* = \boldsymbol{M}_m^{H*} \tag{9-190}$$

and the condition of the mechanical state by

$$p_1^* - p_2^* = -2N_\sigma H_{21}^* \sigma^*. \tag{9-191}$$

Similarly, the phase change condition (9-163) becomes

$$p_2^* - p^{sat*} = 2N_\sigma \left(\frac{\rho_2^*}{\rho_2^* - \rho_1^* / N_\rho} \right) H_{21}^* \sigma^*. \tag{9-192}$$

Furthermore, the energy transfer condition (9-158) becomes

$$\Gamma_1^* \left[\left\{ (N_i)_1 i_{1i}^* - (N_i)_2 i_{2i}^* \right\} - 1 \right] + \sum_{k=1}^{2} \frac{(N_q)_k (N_i)_k}{(N_{pch})_k} a_i^* q_k''^* = 0. \tag{9-193}$$

In these equations, we have introduced the following scaling parameters

$$\text{Surface tension number } N_\sigma \equiv \frac{H_{21o}\sigma_o}{\Delta p_o} \tag{9-194}$$

$$\text{Density ratio } N_\rho \equiv \frac{\rho_{2o}}{\rho_{1o}} \tag{9-195}$$

$$\text{Converted enthalpy ratio } \left(N_i\right)_k \equiv \frac{\Delta i_{ko}}{i_{2o} - i_{1o}}. \tag{9-196}$$

By combining the surface tension and Euler numbers we obtain the Weber number as

$$\left(N_{We}\right)_k \equiv \frac{\rho_{ko} v_{ko}^2}{H_{21o}\sigma_o} = \frac{1}{N_\sigma \left(N_{Eu}\right)_k}. \tag{9-197}$$

This shows that we obtain two Weber numbers in the two-fluid formulation, thus using the surface tension number is more convenient.

The converted enthalpy ratio scales the phase enthalpy change to the latent heat. This number is normally small, if the pressure is well below the critical pressure. However, the most important simplifications can be obtained by studying Eq.(9-192). If the surface tension number *or* the density ratio is very small we have

$$\overline{\overline{p_2}} \approx p^{sat}\left(\overline{\overline{T_i}}\right) \quad \text{for} \begin{cases} N_\sigma << 1 \\ \text{or} \\ N_\rho << 1. \end{cases} \tag{9-198}$$

Then from Eq.(9-157) we get

$$\overline{\overline{p_1}} \approx -2\overline{\overline{H_{21}}}\,\overline{\overline{\sigma}} + p^{sat}\left(\overline{\overline{T_i}}\right). \tag{9-199}$$

The simplest case happens, if the surface tension number is small, then

$$\overline{\overline{p_1}} \doteq \overline{\overline{p_2}} \doteq p^{sat}\left(\overline{\overline{T_i}}\right) \quad \text{for } N_\sigma << 1 \tag{9-200}$$

which indicates that the two phases are in mechanical equilibrium.

Now we study some of the important special cases.

Flow without Phase Changes

If the flow is without phase changes, we can set

$$\left(N_{pch}\right)_k = 0 \tag{9-201}$$

Then all the terms weighted by the phase change number in the field equations drop out from the formulation. In this case it is usually more convenient to transform the thermal energy equation in terms of c_{pk} and $\overline{\overline{T}}_k$. Thus, from the caloric equation of state (9-58), we have

$$d\widehat{i_k} = c_{pk}\, d\overline{\overline{T}}_k + \frac{1}{\overline{\overline{\rho}}_k}\left(1 + \frac{\overline{\overline{T}}_k}{\overline{\overline{\rho}}_k}\left.\frac{\partial \overline{\overline{\rho}}_k}{\partial \overline{\overline{T}}_k}\right)_{\overline{\overline{p}}_k}\right) d\overline{\overline{p}}_k$$

or (9-202)

$$d\widehat{i_k} = c_{pk}\, d\overline{\overline{T}}_k + \frac{1}{\overline{\overline{\rho}}_k}\left(1 + \overline{\overline{T}}_k \beta_k\right) d\overline{\overline{p}}_k.$$

Consequently, we have the following set of field equations from Eqs.(9-152), (9-153) and (9-154)

$$\frac{\partial}{\partial t}\alpha_k \overline{\overline{\rho}}_k + \nabla \cdot \left(\alpha_k \overline{\overline{\rho}}_k \widehat{\boldsymbol{v}_k}\right) = 0 \tag{9-203}$$

$$\alpha_k \overline{\overline{\rho}}_k \frac{D_k \widehat{\boldsymbol{v}_k}}{Dt} = -\alpha_k \nabla \overline{\overline{p}}_k + \nabla \cdot \left\{\alpha_k \left(\overline{\overline{\mathcal{T}}}_k + \mathcal{T}_k^T\right)\right\} + \alpha_k \overline{\overline{\rho}}_k \widehat{\boldsymbol{g}_k}$$
$$+ \boldsymbol{M}_{ik} - \nabla \alpha_k \cdot \overline{\overline{\mathcal{T}}}_{ki} \tag{9-204}$$

and

$$\alpha_k \overline{\overline{\rho}}_k c_{pk} \frac{D_k \overline{\overline{T}}_k}{Dt} = -\nabla \cdot \alpha_k \left(\overline{\overline{\boldsymbol{q}}}_k + \boldsymbol{q}_k^T\right) - \alpha_k \frac{\overline{\overline{T}}_k}{\overline{\overline{\rho}}_k}\left.\frac{\partial \overline{\overline{\rho}}_k}{\partial \overline{\overline{T}}_k}\right)_{\overline{\overline{p}}_k} \frac{D_k \overline{\overline{p}}_k}{Dt}$$
$$+ \alpha_k \overline{\overline{\mathcal{T}}}_k : \nabla \widehat{\boldsymbol{v}_k} + a_i \overline{\overline{q_{ki}''}} + \left(\boldsymbol{M}_{ik} - \nabla \alpha_k \cdot \overline{\overline{\mathcal{T}}}_{ki}\right) \cdot \left(\widehat{\boldsymbol{v}_i} - \widehat{\boldsymbol{v}_k}\right). \tag{9-205}$$

Here we have substituted Eq.(9-202) into Eq.(9-154).

Now we can define the Prandtl number as

$$\left(N_{\mathrm{Pr}}\right)_k \equiv \frac{c_{pko}\mu_{ko}}{K_{ko}}. \tag{9-206}$$

And the Peclet number should be modified to

$$\left(N_{Pe}\right)_k \equiv \frac{\rho_{ko}v_{ko}c_{pko}L_o}{K_{ko}} = \left(N_{\mathrm{Re}}\right)_k\left(N_{\mathrm{Pr}}\right)_k. \tag{9-207}$$

Furthermore, we note that the second term on the right-hand side of Eq.(9-205) reduces to a simple form for an ideal gas or for an incompressible fluid

$$-\alpha_k\frac{\overline{\overline{T_k}}}{\overline{\overline{\rho_k}}}\left.\frac{\partial\overline{\overline{\rho_k}}}{\partial\overline{\overline{T}}_k}\right)_{\overline{\overline{p_k}}}\frac{D_k\overline{\overline{p_k}}}{Dt} = \begin{cases}\alpha_g\dfrac{D_g\overline{\overline{p_g}}}{Dt} & \text{(ideal gas)}\\ 0 & \text{(incompressible).}\end{cases} \tag{9-208}$$

A simple form of the energy equations of a practical importance can be used if the Eckert numbers are very small, or the heat transfer dominates the energy exchanges. Then we have

$$\alpha_k\overline{\overline{\rho_k}}c_{pk}\frac{D_k\overline{\overline{T_k}}}{Dt} \doteq -\nabla\cdot\left\{\alpha_k\left(\overline{\overline{\boldsymbol{q}_k}}+\boldsymbol{q}_k^T\right)\right\}+a_i\overline{\overline{q_{ki}''}} \tag{9-209}$$

where the compressibility effect and the viscous dissipation term have been dropped from Eq.(9-205). In addition, if the two phases are incompressible with the temperature independent transport properties, the energy equations can be decoupled from the continuity and the momentum equations.

Isothermal Flow with No Phase Changes

Under the condition, the entire energy equations may be dropped from the formulation. And we have

$$\overline{\overline{\rho_k}} = \overline{\overline{\rho_k}}\left(\overline{\overline{p_k}}\right) \tag{9-210}$$

thus the flow is called *barotropic*. Furthermore, if the change of pressure or the isothermal compressibility is small, the flow can be considered as *incompressible*. Then we have

$$\overline{\overline{\rho_k}} = \text{constant}. \tag{9-211}$$

Under this condition, the pressure $\overline{\overline{p_k}}$ is independent of the density $\overline{\overline{\rho_k}}$ and it represents the hydrodynamic pressure. Moreover if the effects of the viscous stresses can be neglected, then we have

$$\begin{aligned} &\frac{\partial \alpha_k}{\partial t} + \nabla \cdot \alpha_k \widehat{\boldsymbol{v}_k} = 0 \\ &\frac{D_k \widehat{\boldsymbol{v}_k}}{Dt} = -\frac{1}{\overline{\overline{\rho_k}}} \nabla \overline{\overline{p_k}} + \boldsymbol{g} + \frac{\boldsymbol{M}_{ik}}{\alpha_k \overline{\overline{\rho_k}}}. \end{aligned} \tag{9-212}$$

In addition, if the system has a fixed interfacial geometry, the formulation reduces to a simple form due to the geometrical constitutive laws. Equations (9-178) and (9-181), as well as the drag law Eq.(9-176), can be obtained without much difficulties. Some of the results on the dispersed flow regime given below can be applied for this case.

Dispersed Flow with Fluid Particles

In the following analysis, we use subscript c and d for the continuous and dispersed phases, respectively. Thus we set:

phase 1 $\rightarrow$ phase c; continuous phase;
phase 2 $\rightarrow$ phase d; dispersed phase.

For simplicity, we assume that the dispersed phase has spherical geometry with fairly uniform diameters at any point and time. Then, from the *geometrical equation of state* (9-136) or Eq.(9-137), we have

$$\overline{\overline{R_d}} = \frac{3C^i \alpha_d}{a_i} \tag{9-213}$$

where

$$\begin{cases} \overline{\overline{H_{dc}}} = \dfrac{1}{\overline{\overline{R_d}}} \\ C^i \doteq 1. \end{cases} \tag{9-214}$$

Thus $\overline{\overline{R_d}}$ can be considered as the mean radius of the fluid particles. The *volume balance equation* can be put into the following form

$$\frac{\partial}{\partial t}\left(\frac{36\pi\alpha_d^2}{a_i^3}\right)+\nabla\cdot\left(\frac{36\pi\alpha_d^2}{a_i^3}\widehat{\boldsymbol{v}}_d\right)=\left(V_d^+-V_d^-\right) \tag{9-215}$$

where

$$\frac{1}{n_d}=\frac{36\pi\alpha_d^2}{a_i^3} \tag{9-216}$$

is the free volume available per each fluid particle, where n_d is the particle number density. The right-hand side represents the volume source due to coalescences and the sink due to disintegrations of particles.

We demonstrate the derivation of Eq.(9-138) for a simple case without the source or sink terms. Thus, by considering a fluid particle of average properties, we can approximate

$$\frac{D_d}{Dt}\left(\frac{4}{3}\pi\overline{\overline{R_d}}^3\overline{\overline{\rho_d}}\right)\doteq\Gamma_d\frac{4\pi\overline{\overline{R_d}}^2}{a_i}. \tag{9-217}$$

By substituting Eq.(9-136) for $\overline{\overline{R_d}}$, then using the dispersed phase continuity equation (9-152) to eliminate Γ_d, we have

$$\frac{\partial}{\partial t}\left(\frac{36\pi\alpha_d^2}{a_i^3}\right)+\nabla\cdot\left(\frac{36\pi\alpha_d^2}{a_i^3}\widehat{\boldsymbol{v}}_d\right)=0. \tag{9-218}$$

We note here that if the particle diameters vary considerably then the coefficient C^i is not a constant. In this case, we should have an additional term due to the changes in C^i because the average surface area and volume of fluid particles are not exactly the same as those calculated from the mean diameter.

Now let us study the drag constitutive equation in the fluid particle systems. The well-known Stokes' Law was extended by Hadamard to creeping motion of a *spherical fluid particle* in an infinite Navier-Stokes fluid (Brodkey, 1967; Soo, 1967). Thus, the total force acting on a fluid particle is given by

$$F=6\pi\mu_c\left(v_{c\infty}-v_d\right)R_d\left\{\frac{2\mu_c+3\mu_d}{3\left(\mu_c+\mu_d\right)}\right\}. \tag{9-219}$$

Then we define the drag coefficient $C_{D\infty}$ by

$$C_{D\infty} = \frac{F}{\frac{1}{2}\rho_c\left(v_{c\infty} - v_d\right)^2 \pi R_d^2} \tag{9-220}$$

and the particle Reynolds number by

$$(Re)_d = \frac{\rho_c\left(v_{c\infty} - v_d\right) 2 R_d}{\mu_c}. \tag{9-221}$$

Here $v_{c\infty}$ and v_d are the undisturbed flow velocity and the particle velocity. From the above, we have

$$C_{D\infty} = \frac{24}{(Re)_d}\left[\frac{2\mu_c + 3\mu_d}{3\left(\mu_c + \mu_d\right)}\right]; \ (Re)_d < 1. \tag{9-222}$$

The drag law given by Hadamard is good up to a Reynolds number of about 1. For higher Reynolds numbers we have the results of Levich (1962) and Chao (1962) given by

$$C_{D\infty} = \frac{48}{(Re)_d}; \ (Re)_d < 100 \tag{9-223}$$

$$C_{D\infty} = \frac{32}{(Re)_d}\left\{1 + 2\frac{\mu_d}{\mu_c} - 0.314\frac{\left(1 + 4\mu_d/\mu_c\right)}{\sqrt{(Re)_d}}\right\} \tag{9-224}$$

respectively. We also note the review work done by Clift et al. (1978) in these connections. At still higher Reynolds numbers, the value of $C_{D\infty} \approx 0.44$ given by Newton can be used for droplets. For bubbles, however, the interfacial deformations lead to ellipsoidal or spherical cap bubbles.

By combining these results, we can set that the drag coefficient as a function of the Reynolds number $(Re)_d$ and the ratio of the two viscosities, thus

$$C_{D\infty} = C_{D\infty}\left((Re)_d, \frac{\mu_c}{\mu_d}\right). \tag{9-225}$$

The above relation summarizes the ideal cases of a single fluid particle in infinite media.

In general cases, we have postulated that the interfacial drag force can be given by the constitutive law having the forms of Eqs.(9-117) and (9-119). For a dispersed flow restricted by Eqs.(9-213), (9-214) and (9-218), we may simplify the general drag constitutive law by introducing a drag coefficient C_D defined by

$$C_D = \frac{|\boldsymbol{F}|}{\frac{1}{2}\overline{\overline{\rho_c}}\left|\widehat{\boldsymbol{v}_c} - \widehat{\boldsymbol{v}_d}\right|^2 \pi\left(\overline{\overline{R_d}}\right)^2}. \tag{9-226}$$

In view of Eqs.(9-221) and (9-226), we redefine the appropriate particle Reynolds number by

$$N_{\mathrm{Re}c}^i \equiv \frac{\overline{\overline{\rho_c}}\left|\widehat{\boldsymbol{v}_c} - \widehat{\boldsymbol{v}_d}\right| 2\overline{\overline{R_d}}}{\overline{\overline{\mu_c}}}. \tag{9-227}$$

And the Reynolds number for the dispersed phase is redundant if we use the viscosity ratio as a non-dimensional group.

Thus, in view of Eqs.(9-119) and (9-227), we postulate that the drag law can be given by

$$C_D = C_{D\infty}\left(N_{\mathrm{Re}c}^i, \frac{\overline{\overline{\mu_c}}}{\overline{\overline{\mu_d}}}\right) f^*\left(\frac{\overline{\overline{\rho_c}}}{\overline{\overline{\rho_d}}}, N_{pch}^i, \alpha_c\right) \tag{9-228}$$

where f^* is the correction factor which takes into account for the effects of other particles and the changes of phase. It can be said that if N_{pch}^i is large then the linear correction of Eq.(9-228) cannot be applied because of the rapid expansions or collapses of fluid particles. A detailed discussion of the drag force in multi-particle systems is given in Chapter 12.

Chapter 10

INTERFACIAL AREA TRANSPORT

The interfacial transfer terms are strongly related to the interfacial area and to the local transfer mechanisms, such as the degree of turbulence near the interfaces and the driving potential. Basically, the interfacial transport of mass, momentum and energy is proportional to the interfacial area concentration and to a driving force. This area concentration, defined as the interfacial area per unit volume of the mixture, characterizes the kinematic effects; therefore, it must be related to the structure of the two-phase flow. The driving forces for the interphase transport characterize the local transport mechanism and they must be modeled separately.

Since the interfacial transfer rates can be considered as the product of the interfacial flux and the available interfacial area, the modeling of the interfacial area concentration is essential. In two-phase flow analysis the void fraction and the interfacial area concentration represent the two fundamental first-order geometrical parameters. Therefore, they are closely related to two-phase flow regimes. However, the concept of the two-phase flow regimes is difficult to quantify mathematically at a local point, because it is often defined at the scale close to the system scale. This may indicate that the modeling of the changes of the interfacial area concentration directly by a transport equation, namely interfacial area transport equation. This is a better approach than the conventional method using the flow-regime transition criteria and regime-dependent constitutive relations for interfacial area concentration. This is particularly true for a three-dimensional formulation of multiphase flow.

In this chapter, the detailed derivation and the necessary constitutive relations of the interfacial area transport equation is presented to establish the dynamic closure relation for the interfacial area concentration in the two-fluid model. Accounting for the substantial differences in the transport mechanisms in small spherical and large cap bubbles, the two-group transport equation is derived.

1.1 Three-dimensional interfacial area transport equation

The Boltzmann transport equation describes the particle transport by an integro-differential equation of the particle-distribution function. Since the interfacial area of the fluid particle is closely related to the particle number, the interfacial area transport equation can be formulated based on the Boltzmann transport equation (Kocamustafaogullari and Ishii, 1995; Ishii and Kim, 2004).

Consider a system of fluid particles in a continuous medium, where the source and sink of the fluid particle exist due to the particle interactions such as the coalescence and disintegration. Let $f(V,\boldsymbol{x},\boldsymbol{v},t)$ be the particle number density distribution function per unit mixture and bubble volume. This is assumed to be continuous and specifies the probable number density of fluid particles moving with particle velocity $\boldsymbol{v}$, at a given time t, in a spatial range $\delta\boldsymbol{x}$ with its center-of-volume located at $\boldsymbol{x}$ with particle volumes between V and $V+\delta V$. Assuming that the change of particle velocity within the time interval t to $t+\delta t$ is small, the particle number density distribution function per unit mixture and bubble volume can be simplified to be $f(V,\boldsymbol{x},t)$. This assumption of a uniform particle velocity for a given particle size is practical for most two-phase flow. However, for neutron transport, the energy of neutrons spans over many orders of magnitudes and is the essence of the transport theory. Therefore, the velocity dependence cannot be neglected. Then, we can write for a two-phase flow system

$$f\left(V+\delta V,\boldsymbol{x}+\delta\boldsymbol{x},t+\delta t\right)\delta\mu - f\left(V,\boldsymbol{x},t\right)\delta\mu = \left(\sum_j S_j + S_{ph}\right)\delta\mu\delta t \tag{10-1}$$

where $\delta\mu$ is a volume element in μ space. In the right-hand side of the equation, the S_j and S_{ph} are the particle source and sink rates per unit mixture volume due to j-th particle interactions (such as the disintegration or coalescence and that due to phase change, respectively). Expanding the first term on the left-hand side of Eq.(10-1) in a Taylor series in δt and dividing it by $\delta\mu\delta t$, Eq. (10-1) reduces to

$$\frac{\partial f}{\partial t} + \nabla\cdot(f\boldsymbol{v}) + \frac{\partial}{\partial V}\left(f\frac{dV}{dt}\right) = \sum_j S_j + S_{ph} \tag{10-2}$$

which is analogous to the Boltzmann transport equation of particles with the distribution function $f(V,\boldsymbol{x},t)$. Here, d/dt denotes the substantial derivative. In the following sections, we present the detailed derivations on the transport equations for fluid particle number (n), volume fraction (α_g), and interfacial area concentration (a_i).

1.1.1 Number transport equation

In two-phase flow applications, the particle transport equation given by Eq.(10-2) is much too detailed to be employed in practice, therefore, more macroscopic formulation is desirable. This can be done by integrating Eq.(10-2) over the volume of all sizes of particles from V_{min} to V_{max} and applying the Leibnitz rule of integration. We obtain the particle number transport equation as

$$\frac{\partial n}{\partial t}+\nabla\cdot\left(n\boldsymbol{v}_{pm}\right)=\sum_j R_j+R_{ph} \tag{10-3}$$

where the distribution function for the bubbles of volume V_{min} and V_{max} are assumed to be approximately zero. Here, the left-hand side of the equation represents the time rate of change of the total particle number density and its convection. The two terms in the right-hand side represent the number source and sink rates due to particle interaction (such as particle disintegration or coalescence) and the number source rate due to the phase change, respectively.

In Eq.(10-3), the total number of particles of all sizes per unit-mixture volume and the number source and sink rates are defined respectively by

$$n(\boldsymbol{x},t)=\int_{V_{min}}^{V_{max}} f(V,\boldsymbol{x},t)dV \tag{10-4}$$

and

$$R(\boldsymbol{x},t)=\int_{V_{min}}^{V_{max}} S(V,\boldsymbol{x},t)dV. \tag{10-5}$$

Also, $\boldsymbol{v}_{pm}$ is the average local particle velocity weighted by the particle number and is defined by

$$\boldsymbol{v}_{pm}(\boldsymbol{x},t) \equiv \frac{\int_{V_{min}}^{V_{max}} f(V,\boldsymbol{x},t)\boldsymbol{v}(V,\boldsymbol{x},t)\,dV}{\int_{V_{min}}^{V_{max}} f(V,\boldsymbol{x},t)dV}. \tag{10-6}$$

1.1.2 Volume transport equation

The particle volume (or void fraction) transport equation can be obtained by multiplying Eq.(10-2) by particle volume V and integrating it over the volume of all sizes of particles. Then, considering that the two-phase flow of interest consists of the dispersed bubbles in a continuous liquid medium, the void fraction transport equation is given by

$$\begin{aligned} &\frac{\partial \alpha_g}{\partial t} + \nabla\cdot\left(\alpha_g \boldsymbol{v}_g\right) + \int_{V_{min}}^{V_{max}} V \frac{\partial\left(f\dot{V}\right)}{\partial V} dV \\ &= \int_{V_{min}}^{V_{max}} \left(\sum_j S_j V + S_{ph} V\right) dV \end{aligned} \tag{10-7}$$

where $\dot{V}$ denotes the time derivative of volume V. Here, the void fraction and the average center-of-volume velocity of the dispersed (or gas) phase are defined respectively by

$$\alpha_g(\boldsymbol{x},t) = \int_{V_{min}}^{V_{max}} f(V,\boldsymbol{x},t)V\,dV \tag{10-8}$$

and

$$\boldsymbol{v}_g(\boldsymbol{x},t) \equiv \frac{\int_{V_{min}}^{V_{max}} f(V,\boldsymbol{x},t)V\boldsymbol{v}(V,\boldsymbol{x},t)\,dV}{\int_{V_{min}}^{V_{max}} f(V,\boldsymbol{x},t)V\,dV}. \tag{10-9}$$

In Eq.(10-7), the third term on the left-hand side attributes to the change in the particle volume (by expansion or contraction) due to the change in the pressure along the flow field. To better represent this term, we assume

$$\frac{\dot{V}}{V} \neq f(V) \tag{10-10}$$

such that the time-rate of change in relative particle volume is assumed to be independent of its volume. If evaporation effect is small compared to the compressibility effect, the dominant contribution in the change in a particle volume attributes to the changes in the pressure. Therefore, this assumption is valid in most two-phase flow conditions. It should be noted, however, that the evaporation effect is not completely neglected in the average transport equation development as will be shown later in Eq.(10-14). Then, the third term on the left-hand side of Eq.(10-7) reduces to

$$\int_{V_{min}}^{V_{max}} V\frac{\partial\left(f\dot{V}\right)}{\partial V}dV \cong -\left(\frac{\dot{V}}{V}\right)\alpha_g\left(\boldsymbol{x},t\right). \tag{10-11}$$

Furthermore, since the mass transfer by the evaporation process is given by

$$\frac{d\rho_g V}{dt} = \frac{\left(\Gamma_g - \eta_{ph}\rho_g\right)V}{\alpha_g} \tag{10-12}$$

where Γ_g is the total rate of change of mass-per-unit mixture volume and η_{ph} is the rate of volume generated by nucleation source per unit mixture volume, defined by

$$\eta_{ph} \equiv \int_{V_{min}}^{V_{max}} S_{ph} V dV. \tag{10-13}$$

The volume source can be written as

$$\begin{aligned} \frac{1}{V}\frac{dV}{dt} &= \frac{1}{\rho_g}\left(\frac{\Gamma_g - \eta_{ph}\rho_g}{\alpha_g} - \frac{d\rho_g}{dt}\right) \\ &= \frac{1}{\alpha_g}\left\{\frac{\partial\alpha_g}{\partial t} + \nabla\cdot\left(\alpha_g\boldsymbol{v}_g\right) - \eta_{ph}\right\}. \end{aligned} \tag{10-14}$$

Thus, combining Eq.(10-14) with Eq.(10-11), and substituting them into Eq.(10-7), the final form of the void fraction transport equation is obtained as

$$\frac{\partial \alpha_g}{\partial t} + \nabla \cdot \left(\alpha_g \boldsymbol{v}_g\right) - \frac{\alpha_g}{\rho_g}\left(\frac{\Gamma_g - \eta_{ph}\rho_g}{\alpha_g} - \frac{d\rho_g}{dt}\right) = \int_{V_{min}}^{V_{max}}\left(\sum_j S_j V + S_{ph} V\right) dV \tag{10-15}$$

where the first two terms on the left-hand side of the equation represents the time rate of change and convection of α_g, and the rest of the terms represent the change rates in α_g due to volume change, particle interactions and phase change, respectively.

By rearranging Eq.(10-15), it is interesting to note that we have

$$\frac{1}{\rho_g}\left\{\frac{\partial \alpha_g \rho_g}{\partial t} + \nabla \cdot \left(\alpha_g \rho_g \boldsymbol{v}_g\right) - \Gamma_g\right\} = \int_{V_{min}}^{V_{max}} \sum_j S_j V dV. \tag{10-16}$$

Here, it is noted that conservation of mass requires

$$\frac{\partial \alpha \rho_g}{\partial t} + \nabla \cdot \left(\alpha \rho_g \boldsymbol{v}_g\right) - \Gamma_g = 0. \tag{10-17}$$

Therefore, from Eqs.(10-16) and (10-17), we obtain the identity

$$\int_{V_{min}}^{V_{max}} \sum_j S_j V dV = 0. \tag{10-18}$$

Equation (10-18) satisfies both the volume and mass conservation, simultaneously.

1.1.3 Interfacial area transport equation

The transport equation for the interfacial area concentration can be obtained through a similar approach applied in the previous formulations. Hence, multiplying Eq.(10-2) by the surface area of particles of volume V, $A_i(V)$, (which is independent of the coordinate system) and integrating it over the volume of all particles, we obtain

$$\frac{\partial a_i}{\partial t} + \nabla \cdot (a_i \boldsymbol{v}_i) - \left(\frac{\dot{V}}{V}\right) \int_{V_{min}}^{V_{max}} fV dA_i = \int_{V_{min}}^{V_{max}} \left(\sum_j S_j + S_{ph}\right) A_i dV \tag{10-19}$$

where the average a_i of all fluid particles of volumes between V_{min} and V_{max}, and the interfacial velocity are given respectively by

$$a_i(\boldsymbol{x}, t) = \int_{V_{min}}^{V_{max}} f(V, \boldsymbol{x}, t) A_i(V) dV \tag{10-20}$$

and

$$\boldsymbol{v}_i(\boldsymbol{x}, t) \equiv \frac{\int_{V_{min}}^{V_{max}} f(V, \boldsymbol{x}, t) A_i(V) \boldsymbol{v}(V, \boldsymbol{x}, t) dV}{\int_{V_{min}}^{V_{max}} f(V, \boldsymbol{x}, t) A_i(V) dV}. \tag{10-21}$$

Now, in view of furnishing the third term on the left-hand side of Eq.(10-19), we define the volume-equivalent diameter, D_e, and surface-equivalent diameter, D_s, of a fluid particle with surface area A_i and volume V, as

$$V \equiv \frac{\pi}{6} D_e^3 \text{ and } A_i \equiv \pi D_s^2. \tag{10-22}$$

Therefore, combining them with Eq.(10-19) and recalling the volume source, given by Eq.(10-14), the interfacial area transport equation can be obtained as

$$\frac{\partial a_i}{\partial t} + \nabla \cdot (a_i \boldsymbol{v}_i) - \frac{2}{3}\left(\frac{a_i}{\alpha_g}\right)\left\{\frac{\partial \alpha_g}{\partial t} + \nabla \cdot (\alpha_g \boldsymbol{v}_g) - \eta_{ph}\right\} = \int_{V_{min}}^{V_{max}} \left(\sum_j S_j + S_{ph}\right) A_i dV \tag{10-23}$$

where the third term on the left-hand side represents the change in the interfacial area concentration due to the particle volume change. In deriving Eq.(10-23), the ratio $\left(D_s/D_e\right)$ is assumed to be constant in view of simplifying the equation. While this approximation may not be appropriate

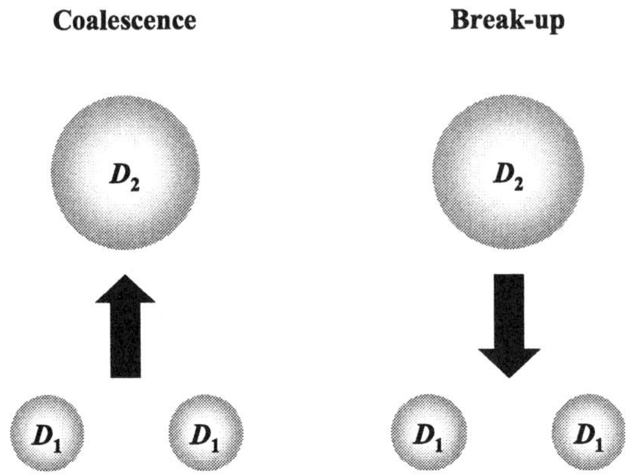

Figure 10-1. Illustration of fluid particle coalescence and disintegration process in view of ΔA_i (Ishii and Kim, 2004)

for the bubbles in distorted or slug shape, it is a good approximation for the spherical and cap bubbles. Essentially, this diameter ratio is a shape factor and for similar particle shapes this factor can be considered as constant.

To close the system of equations, the right-hand side of Eq.(10-23), which represents the source and sink rates of the interfacial area concentration, must be specified by the constitutive relations. In view of this, we define

$$\int_{V_{min}}^{V_{max}} \sum_j S_j dV = \sum_j R_j : \text{ particle number source and sink rate} \tag{10-24}$$

and

$$\int_{V_{min}}^{V_{max}} \sum_j S_j A_i dV = \sum_j \phi_j : \text{ source and sink rate for } a_i . \tag{10-25}$$

Furthermore, noting that ϕ_j can be expressed in terms of the change in surface area of a fluid particle after a certain particle interaction process, we can write

$$\phi_j = R_j \Delta A_i \tag{10-26}$$

where R_j can be mechanistically modeled for each interaction mechanism, and ΔA_i depends on the given interaction mechanism; such as the disintegration or coalescence processes.

In order to specify ΔA_i, consider a coalescence and a break-up processes as illustrated in Fig.10-1 for the spherical particles. Here, it was assumed that the given process is a binary process between the particles of same size. Then, since the total volume of the particles should be conserved, we can write

$$V_2 = 2V_1 \text{ or } D_2 = 2^{1/3} D_1 \tag{10-27}$$

where the subscripts 1 and 2 indicates the particles of smaller and bigger volumes, respectively. Hence, by assuming that the interaction process is a binary process, the change of surface area after one interaction process can be obtained for near spherical particles as

$$\Delta A_i = -0.413 A_i\text{: for a coalescence process} \tag{10-28}$$

and

$$\Delta A_i = +0.260 A_i\text{: for a break-up process} \tag{10-29}$$

where the minus and plus signs are used to indicate the reduction and gain of the surface area after one interaction process, respectively. Furthermore, recalling the definition given by Eq.(10-4), the particle number density n can be specified through a_i and α_g by

$$a_i = nA_i \text{ and } \alpha_g = nV \tag{10-30}$$

such that

$$n = \psi \frac{a_i^3}{\alpha_g^2} \tag{10-31}$$

with a shape factor defined by

$$\psi = \frac{1}{36\pi}\left(\frac{D_{Sm}}{D_e}\right)^3 \tag{10-32}$$

where the bubble Sauter mean diameter is given by

$$D_{Sm} = \frac{6\alpha_g}{a_i}. \tag{10-33}$$

Thus, combining these with Eq.(10-26), the surface source and sink rate, ϕ_j can be given by

$$\phi_j = \frac{1}{3\psi}\left(\frac{\alpha_g}{a_i}\right)^2 R_j. \tag{10-34}$$

Similarly, for the nucleation process, ϕ_{ph} can be given by

$$\phi_{ph} = \pi D_{bc}^2 R_{ph} \tag{10-35}$$

where D_{bc} is the critical bubble size. This should be determined depending on the given nucleation process; namely, the critical cavity size for the bulk boiling or condensation process, and the bubble departure size for the wall nucleation. For most two-phase flow, wall nucleation is the dominant mechanism.

After combining the constitutive relations given above and substituting them into Eq.(10-23), we obtain the interfacial area transport equation as

$$\frac{\partial a_i}{\partial t} + \nabla\cdot\left(a_i \boldsymbol{v}_i\right) = \frac{2}{3}\left(\frac{a_i}{\alpha_g}\right)\left\{\frac{\partial \alpha_g}{\partial t} + \nabla\cdot\left(\alpha_g \boldsymbol{v}_g\right) - \eta_{ph}\right\} + \frac{1}{3\psi}\left(\frac{\alpha_g}{a_i}\right)^2 \sum_j R_j + \pi D_{bc}^2 R_{ph} \tag{10-36}$$

where the left-hand side represents the time rate of change and convection of the interfacial area concentration. Each term on the right-hand side represents the rates of change in the interfacial area concentration due to the particle volume change caused by the change in pressure, various particle interactions and phase change, respectively. As can be seen in Eq.(10-36), R_j 's should be modeled independently, based on the given particle interaction mechanisms. Hence, the mechanistic models of the number source and sink rates for the coalescence and disintegration mechanisms, or those for the bubble nucleation and condensation phenomena, should be established as constitutive relations to solve the transport equation.

1.2 One-group interfacial area transport equation

When the transport phenomena of the fluid particles of interest do not vary significantly in a given two-phase flow system, and the particles remain similar in shape after the particle interactions, their characteristic transport phenomenon is similar and can be described by one transport equation. However, when fluid particles of various shapes and size present simultaneously, their transport mechanisms can be significantly different. In such cases, it may be necessary to employ multiple transport equations to describe the fluid particle transport.

In view of this, we first consider the two-phase flow system of the dispersed bubbles in a continuous liquid medium (namely, bubbly flow), where all the present bubbles can be categorized as *one group*. In such flow conditions, it is assumed that the bubbles are spherical in their shapes, and they are subject to the similar characteristic drag on their transport phenomena. Hence, accounting for the spherical shape in the one-group transport, ψ defined in Eq.(10-32) can be approximated by

$$\psi \approx \frac{1}{36\pi} = 8.85\times 10^{-3}: \quad \text{for dispersed bubbles} \tag{10-37}$$

because the bubble Sauter mean diameter is approximately equal to the volume-equivalent diameter. Furthermore, noting that critical bubble size due to nucleation is much smaller compared to the average bubble Sauter mean diameter, we may assume

$$\left(\frac{D_{bc}}{D_{Sm}}\right) \approx 0. \tag{10-38}$$

Also, since η_{ph} can be approximated as

$$\eta_{ph} \equiv \int_{V_{min}}^{V_{max}} S_{ph} V dV \approx R_{ph}\frac{\pi}{6}D_{bc}^{3} \tag{10-39}$$

the interfacial area transport equation for the dispersed bubbles, or the one-group interfacial area transport equation, is given by

$$\frac{\partial a_i}{\partial t} + \nabla \cdot \left(a_i \boldsymbol{v}_i\right) \approx \frac{2}{3}\left(\frac{a_i}{\alpha_g}\right)\left\{\frac{\partial \alpha_g}{\partial t} + \nabla \cdot \left(\alpha_g \boldsymbol{v}_g\right)\right\}$$
$$+\frac{1}{3\psi}\left(\frac{\alpha_g}{a_i}\right)^2 \sum_j R_j + \pi D_{bc}^2 R_{ph}. \tag{10-40}$$

The left-hand side represents the total rate of change in the interfacial area concentration, whereas the right-hand side represents the rates of change in the interfacial area concentration due to the change in particle volume, various particle interactions and phase change, respectively. It is noted that the effect of η_{ph} can be neglected because the departure size is smaller than the Sauter mean diameter.

1.3 Two-group interfacial area transport equation

In a gas-liquid two-phase flow system, a wide range of bubble shape and size exists depending on the given flow regime. Therefore, to develop the interfacial area transport equation describing the bubble transport in a wide range of two-phase flow regimes, the model must account for the differences in the transport characteristics of different types of bubbles. These variations in shape and size of bubbles cause substantial differences in their transport mechanisms due to the drag forces. Furthermore, the bubble interaction mechanisms in such flow conditions can be quite different compared to those in the one-group transport.

In most two-phase flow conditions, bubbles can be categorized into five types: spherical; distorted; cap; Taylor; and churn-turbulent bubbles. However, in view of their transport characteristics, they can be classified into two major groups, such that Group 1 includes the spherical and distorted bubbles, while Group 2 includes the cap, Taylor and churn-turbulent bubbles. Thus, in the present analysis, the approach employing two transport equations is given in describing the bubble transport over a wide range of two-phase flow conditions. That is, Group-1 transport equation describes the transport of spherical and distorted bubbles, and Group-2 transport equation describes the transport of cap, Taylor and churn-turbulent bubbles.

In the one-group formulation, the transport equation was averaged by the integration process over the volumes of all sizes of particles because the shape of the particles and their transport phenomena were assumed to be similar over the given range of particle volume. In the two-group formulation, however, the integration limit for each transport equation should be bounded by the bubble volume, by which the bubble group is determined. In view of this, we define V_c as the critical bubble volume given by $\pi D_{d,\max}/6$ with the maximum distorted bubble limit, $D_{d,max}$, specified by Ishii and Zuber (1979) as

$$D_{d,max} = 4\sqrt{\frac{\sigma}{g\Delta\rho}}: \quad \text{maximum distorted bubble limit} \quad (10\text{-}41)$$

over which the bubble becomes cap in shape and the drag effect starts to deviate from that on the smaller bubbles due to the large wake region. Therefore, the Group 1 bubbles exist in the range of V_{min} to V_c, whereas the Group 2 bubbles exist in the range of V_c to V_{max}.

1.3.1 Two-group particle number transport equation

The two-group particle number transport equation can be readily obtained by integrating Eq.(10-2) over the different ranges of integration limit bounded by V_c; namely, from V_{min} to V_c, for Group 1 and from V_c to V_{max} for Group 2. In two-group formulation, as in the one-group formulation, $f(V,\boldsymbol{x},t)$ describes the particle-number density distribution function per unit mixture and bubble volume. This is assumed to be continuous, specifying the probable number density of fluid particles moving at a velocity $\boldsymbol{v}$, at a given time t, in a spatial range $\delta\boldsymbol{x}$ with its center-of-volume located at $\boldsymbol{x}$ with particle volumes between V and $V+dV$. Then, the number transport equations for Group 1 and Group 2 are given by

$$\frac{\partial n_1}{\partial t} + \nabla\cdot\left(n_1\boldsymbol{v}_{pm1}\right) = -f_c V_c\left(\frac{\dot{V}}{V}\right) + \sum_j R_{j1} + R_{ph} \qquad (10\text{-}42)$$

and

$$\frac{\partial n_2}{\partial t} + \nabla\cdot\left(n_2\boldsymbol{v}_{pm2}\right) = f_c V_c\left(\frac{\dot{V}}{V}\right) + \sum_j R_{j2} \qquad (10\text{-}43)$$

where the subscripts 1 and 2 in the equations denote Group 1 and Group 2, respectively, and $\boldsymbol{v}_{pm1}$ and $\boldsymbol{v}_{pm2}$ are the average local particle velocity weighted by the particle number for each bubble group, such that they are defined by

$$\boldsymbol{v}_{pm1}(\boldsymbol{x},t) \equiv \frac{\int_{V_{min}}^{V_c} f(V,\boldsymbol{x},t)\boldsymbol{v}(V,\boldsymbol{x},t)\,dV}{\int_{V_{min}}^{V_c} f(V,\boldsymbol{x},t)dV}$$
$$\boldsymbol{v}_{pm2}(\boldsymbol{x},t) \equiv \frac{\int_{V_c}^{V_{max}} f(V,\boldsymbol{x},t)\boldsymbol{v}(V,\boldsymbol{x},t)\,dV}{\int_{V_c}^{V_{max}} f(V,\boldsymbol{x},t)dV}. \tag{10-44}$$

In Eqs.(10-42) and (10-43), the left-hand side of the equations represent the time rate of change and convection of fluid particle number for each bubble group. Each term on the right-hand side represents the rates of change of particle number through inter-group transfer by particle volume change, particle interaction and phase change for each bubble group. Here, it is interesting to note that in the two-group formulation, there are terms accounting for the inter-group transfer caused by particle volume change that did not appear in the one-group formulation. This is due to the fact that when two groups of bubbles present, the change in the particle volume in one group may serve as the number source in another due to the changes in bubble distribution function. These inter-group transfer terms disappear when the two equations are added together to obtain the total fluid particle number transport equation.

1.3.2 Two-group void fraction transport equation

The two-group void fraction transport equation can be obtained in a similar manner. Multiplying Eq.(10-2) by particle volume V, and integrating it over the specified limits for each group, we obtain

$$\frac{\partial \alpha_{g1}}{\partial t} + \nabla\cdot\left(\alpha_{g1}\boldsymbol{v}_{g1}\right) + \int_{V_{min}}^{V_c}\left\{V\frac{\partial}{\partial V}\left(f\frac{dV}{dt}\right)\right\}dV$$
$$= \int_{V_{min}}^{V_c}\left(\sum_j S_j + S_{ph}\right)VdV \tag{10-45}$$

and

$$\frac{\partial \alpha_{g2}}{\partial t} + \nabla \cdot \left(\alpha_{g2} \boldsymbol{v}_{g2}\right) + \int_{V_c}^{V_{max}} \left\{ V \frac{\partial}{\partial V} \left(f \frac{dV}{dt} \right) \right\} dV$$
$$= \int_{V_c}^{V_{max}} \left(\sum_j S_j + S_{ph} \right) V dV. \tag{10-46}$$

For Group 1 and Group 2, respectively, the third term on the left-hand side of the equations represents the rate of change in void fraction due to particle volume change. They are given by

$$\int_{V_{min}}^{V_c} \left\{ V \frac{\partial}{\partial V} \left(f \frac{dV}{dt} \right) \right\} dV = \left(\frac{\dot{V}}{V} \right) \left\{ -\alpha_{g1} + V_c \left(f_c V_c \right) \right\} \tag{10-47}$$

and

$$\int_{V_c}^{V_{max}} \left\{ V \frac{\partial}{\partial V} \left(f \frac{dV}{dt} \right) \right\} dV = \left(\frac{\dot{V}}{V} \right) \left\{ -\alpha_{g2} - V_c \left(f_c V_c \right) \right\} \tag{10-48}$$

for Group 1 and Group 2, respectively, where f_c is the distribution function of a bubble with critical volume V_c, or $f\left(V_c\right)$.

Here, the volume source $\left(\dot{V}/V\right)$ can be expressed by the total mass transfer rate as shown previously in the one-group formulation. However, due to the presence of the two groups of bubbles and their interactions, the rate of mass transfer between the two groups must be considered. Hence, by denoting the subscripts *ij* as the inter-group transfer from group *i* to group *j*, the volume sources for each group are given by

$$\frac{1}{V_1} \left(\frac{dV_1}{dt} \right) = \frac{1}{\rho_g} \left(\frac{\Gamma_{g1} - \Delta \dot{m}_{12}}{\alpha_{g1}} - \frac{d\rho_g}{dt} \right) - \frac{\eta_{ph}}{\alpha_{g1}}$$
$$= \frac{1}{\alpha_{g1}} \left\{ \frac{\partial \alpha_{g1}}{\partial t} + \nabla \cdot \left(\alpha_{g1} \boldsymbol{v}_{g1} \right) - \eta_{ph} \right\} \tag{10-49}$$

and

$$\frac{1}{V_2}\left(\frac{dV_2}{dt}\right)=\frac{1}{\rho_g}\left(\frac{\Gamma_{g2}+\Delta\dot{m}_{12}}{\alpha_{g2}}-\frac{d\rho_g}{dt}\right)$$
$$=\frac{1}{\alpha_{g2}}\left\{\frac{\partial\alpha_{g2}}{\partial t}+\nabla\cdot\left(\alpha_{g2}\boldsymbol{v}_{g2}\right)\right\} \tag{10-50}$$

for Group 1 and Group 2, respectively, where $\Delta\dot{m}_{12}$ represents the inter-group mass transfer rates from Group 1 to Group 2. The constitutive relation for the mass transfer between groups $\Delta\dot{m}_{12}$ will be discussed later. Hence, Eqs.(10-49) and (10-50) require the following identities

$$\frac{\partial\alpha_{g1}\rho_g}{\partial t}+\nabla\cdot\alpha_{g1}\rho_g\boldsymbol{v}_{g1}$$
$$=\Gamma_{g1}-\Delta\dot{m}_{12}:\quad \text{Mass Balance for Group 1} \tag{10-51}$$

and

$$\frac{\partial\alpha_{g2}\rho_g}{\partial t}+\nabla\cdot\alpha_{g2}\rho_g\boldsymbol{v}_{g2}$$
$$=\Gamma_{g2}+\Delta\dot{m}_{12}:\quad \text{Mass Balance for Group 2.} \tag{10-52}$$

Furthermore, by adding the two equations, we obtain the continuity equation for the gas phase as

$$\frac{\partial\alpha_g\rho_g}{\partial t}+\nabla\cdot\left(\alpha_g\rho_g\boldsymbol{v}_g\right)=\Gamma_g \tag{10-53}$$

with the following constitutive relations

$$\alpha_g=\alpha_{g1}+\alpha_{g2} \tag{10-54}$$

$$\Gamma_g=\Gamma_{g1}+\Gamma_{g2} \tag{10-55}$$

and

$$\boldsymbol{v}_g=\frac{\alpha_{g1}\boldsymbol{v}_{g1}+\alpha_{g2}\boldsymbol{v}_{g2}}{\alpha_{g1}+\alpha_{g2}}. \tag{10-56}$$

The term $f_c V_c$ on the right-hand side of Eqs.(10-47) and (10-48) represents the rate of change in the void fraction due to inter-group transfer, and it can be expressed in terms of other two-phase flow parameters. For convenience, the detailed discussion on this term will be presented in the following section. The two-group void fraction transport equation for each group is then given by

$$\frac{1}{\rho_g}\left\{\frac{\partial \alpha_{g1}\rho_g}{\partial t}+\nabla\cdot\left(\alpha_{g1}\rho_g \boldsymbol{v}_{g1}\right)-\Gamma_{g1}+\Delta\dot{m}_{12}\right\}$$
$$=-\left\{\frac{\partial \alpha_{g1}}{\partial t}+\nabla\cdot\left(\alpha_{g1}\boldsymbol{v}_{g1}\right)-\eta_{ph}\right\}\chi\left(\frac{D_{sc}}{D_{Sm1}}\right)^3 \tag{10-57}$$
$$+\int_{V_{\min}}^{V_c}\sum_j S_j V dV$$

and

$$\frac{1}{\rho_g}\left\{\frac{\partial \alpha_{g2}\rho_g}{\partial t}+\nabla\cdot\left(\alpha_{g2}\rho_g \boldsymbol{v}_{g_2}\right)-\Gamma_{g2}-\Delta\dot{m}_{12}\right\}$$
$$=\left\{\frac{\partial \alpha_{g1}}{\partial t}+\nabla\cdot\left(\alpha_{g1}\boldsymbol{v}_{g1}\right)-\eta_{ph}\right\}\chi\left(\frac{D_{sc}}{D_{Sm1}}\right)^3 \tag{10-58}$$
$$+\int_{V_c}^{V_{\max}}\sum_j S_j V dV$$

for Group 1 and Group 2, respectively. Here, D_{sc} is the critical bubble size for the group boundary with surface area and volume of A_{ic} and V_c. Also, χ is the coefficient accounting for the contribution from the inter-group transfer, which will be discussed in detail in the following section.

In Eqs.(10-57) and (10-58), the left-hand side of the equations represents the time rate of change and convection of void fraction for each group and the right-hand side represents the rates of change in the void fraction due to the volume change. This includes the inter-group transfer and various particle interactions. Furthermore, since the left-hand side of the equations corresponds to the continuity equations for each bubble group, it requires the following identities;

$$-\left\{\frac{\partial \alpha_{g1}}{\partial t}+\nabla\cdot\left(\alpha_{g1}\boldsymbol{v}_{g1}\right)-\eta_{ph}\right\}\chi\left(\frac{D_{sc}}{D_{Sm1}}\right)^3+\int_{V_{\min}}^{V_c}\sum_j S_j V dV \tag{10-59}$$
$$=0$$

and

$$\left\{\frac{\partial \alpha_{g1}}{\partial t}+\nabla\cdot\left(\alpha_{g1}\boldsymbol{v}_{g1}\right)-\eta_{ph}\right\}\chi\left(\frac{D_{sc}}{D_{Sm1}}\right)^3+\int_{V_c}^{V_{\max}}\sum_j S_j V dV \tag{10-60}$$
$$=0$$

for Group 1 and Group 2 respectively, which indicates the conservation in bubble volume. Here, the first terms in Eqs.(10-59) and (10-60) represent the inter-group transfer at the bubble group boundary, and the second terms represent sources and sinks due to various fluid particle interaction for the given bubble group.

1.3.3 Two-group interfacial area transport equation

Similarly, as in the one-group interfacial area transport equation formulation, multiply Eq.(10-2) by the surface area of particles with volume V, which is independent of the coordinate system. Then, after integrating it over the volume within which each bubble group is defined, we obtain

$$\begin{aligned}&\frac{\partial a_{i1}}{\partial t}+\nabla\cdot\left(a_{i1}\boldsymbol{v}_{i1}\right)+\int_{V_{\min}}^{V_c}\left\{A_i\frac{\partial}{\partial V}\left(f\frac{dV}{dt}\right)\right\}dV\\&=\int_{V_{\min}}^{V_c}\left(\sum_j S_j+S_{ph}\right)A_i dV\quad\text{for Group 1}\end{aligned} \tag{10-61}$$

and

$$\begin{aligned}&\frac{\partial a_{i2}}{\partial t}+\nabla\cdot\left(a_{i2}\boldsymbol{v}_{i2}\right)+\int_{V_c}^{V_{\max}}\left\{A_i\frac{\partial}{\partial V}\left(f\frac{dV}{dt}\right)\right\}dV\\&=\int_{V_c}^{V_{\max}}\sum_j S_j A_i dV\quad\text{for Group 2}\end{aligned} \tag{10-62}$$

where the average interfacial velocity for each bubble group is defined by

$$
\begin{aligned}
\boldsymbol{v}_{i1}(\boldsymbol{x},t) &\equiv \frac{\displaystyle\int_{V_{\min}}^{V_c} f(V,\boldsymbol{x},t) A_i \boldsymbol{v}(V,\boldsymbol{x},t)\,dV}{\displaystyle\int_{V_{\min}}^{V_c} f(V,\boldsymbol{x},t) A_i\,dV} \\
\boldsymbol{v}_{i2}(\boldsymbol{x},t) &\equiv \frac{\displaystyle\int_{V_c}^{V_{\max}} f(V,\boldsymbol{x},t) A_i \boldsymbol{v}(V,\boldsymbol{x},t)\,dV}{\displaystyle\int_{V_c}^{V_{\max}} f(V,\boldsymbol{x},t) A_i\,dV}.
\end{aligned}
\tag{10-63}
$$

In Eqs.(10-61) and (10-62), the third terms on the left-hand side of the equations represent the changes in the interfacial area concentration due to the particle volume change, such that

$$
\int_{V_{\min}}^{V_c} \left\{ A_i \frac{\partial}{\partial V}\left(f \frac{dV}{dt} \right) \right\} dV = \left(\frac{\dot{V}}{V} \right)\left(-\frac{2}{3} a_{i1} + A_{ic} f_c V_c \right) \tag{10-64}
$$

and

$$
\int_{V_c}^{V_{\max}} \left\{ A_i \frac{\partial}{\partial V}\left(f \frac{dV}{dt} \right) \right\} dV = \left(\frac{\dot{V}}{V} \right)\left(-\frac{2}{3} a_{i2} - A_{ic} f_c V_c \right). \tag{10-65}
$$

where $A_{ic} f_c V_c$ is attributed to the inter-group transfer as a result of bubble interactions between the two groups of bubbles. Hence, when $f_c \to 0$, there is no contribution due to the inter-group interaction. In reality, however, when two bubble groups present f_c is finite, and this inter-group transfer term plays an important role as sources or sinks of the interfacial area concentration for each group.

In order to incorporate the contributions from this inter-group transfer, the particle distribution function should be specified. However, the accurate mathematical description for the particle distribution function in two-phase flows with various sizes of bubbles requires the use of the original Boltzmann transport equation and statistical mechanics. For our purpose, we need to develop a simple integrated transport equation. Hence, in the present analysis, a linear profile or a uniform profile is assumed in the particle distribution for simplicity as shown in Fig.10-2. In this, V_{1p} is the peak bubble volume in Group 1 bubbles specifying the value f_1, and is defined by

$$
V_{1p} = \xi V_c \text{ where } \frac{V_{\min}}{V_c} < \xi < 1. \tag{10-66}
$$

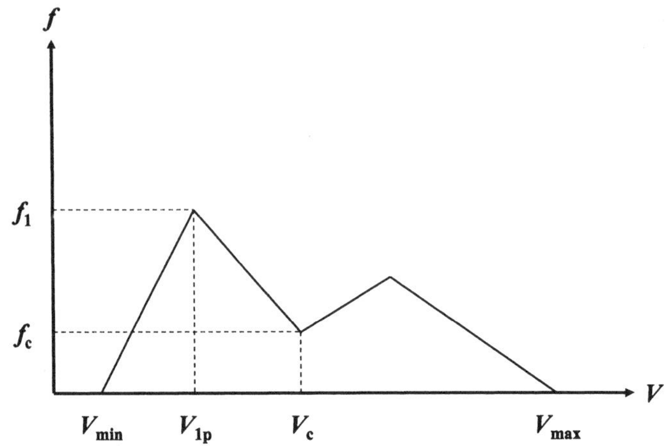

Figure 10-2. Linear approximation on profile of fluid particle distribution function (Ishii and Kim, 2004)

Then, the number density for Group 1 bubble can be expressed as

$$n_1 = \frac{1}{2} f_1 (V_c - V_{\min}) + \frac{1}{2} f_c V_c (1 - \xi) \tag{10-67}$$

which yields

$$f_c V_c A_{ic} = \left[\frac{2}{1 - \xi} - \frac{(V_c - V_{\min})}{n_1 (1 - \xi)} f_1 \right] n_1 A_{ic}. \tag{10-68}$$

Now, consider three limiting cases as shown in Figs.10-3(a) through 10-3(c) such that

Case I: $f = f_c = \text{constant}$, hence

$$n_1 = f_1 (V_c - V_{\min}) = f_c (V_c - V_{\min}) \tag{10-69}$$

Case II: $\xi \rightarrow V_{\min}/V_c$, or $V_{1p} \rightarrow V_{\min}$, hence

$$n_1 = \frac{1}{2} (f_1 + f_c)(V_c - V_{\min}) \tag{10-70}$$

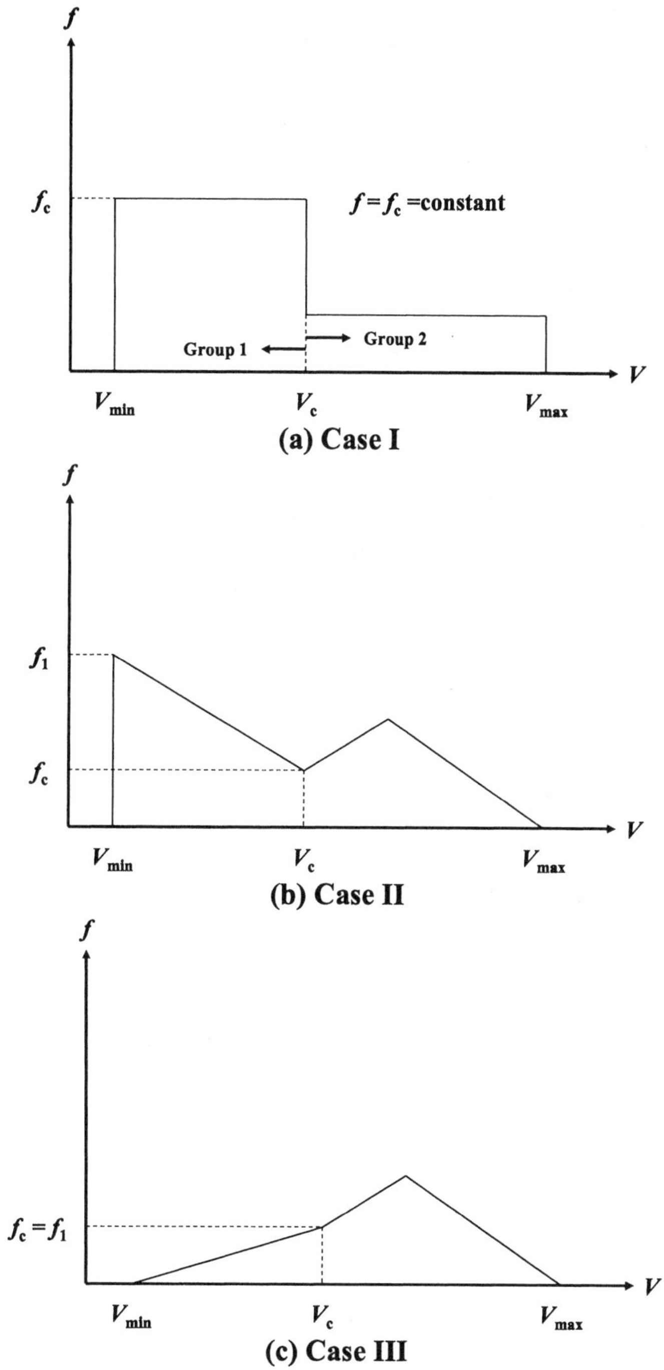

Figure 10-3. Limiting conditions for fluid Particle distribution (a) f=constant, (b) $\xi \to V_{min}/V_c$, (c) $\xi \to 1$ (Ishii and Kim, 2004)

and

Case III : $\xi \to 1$; $V_{1p} \to V_c$; $f_1 \to f_c$, hence

$$n_1 = \frac{1}{2} f_1 \left(V_c - V_{\min}\right) \cong \frac{1}{2} f_c \left(V_c - V_{\min}\right). \tag{10-71}$$

Therefore, assuming that $V_c >> V_{\min}$, and defining an arbitrary coefficient χ as

$$A_{ic} f_c V_c = \chi n_1 A_{ic} \tag{10-72}$$

we obtain, for each limiting case

$$\chi = \begin{cases} 1 & \text{for Case I} \\ 2 - \dfrac{V_c}{n_1} f_1 & \text{for Case II} \\ 2 & \text{for Case III} \end{cases} \tag{10-73}$$

and from Eq.(10-68)

$$\chi = \left\{ \frac{2}{1-\xi} - \frac{V_c}{n_1 (1-\xi)} f_1 \right\} \text{ in general.} \tag{10-74}$$

Thus, recalling that there is no inter-group transfer contribution when $f_c \to 0$, the constant χ should be bounded by

$$0 \le \chi \le 2. \tag{10-75}$$

In providing the analytical solution for χ, we have three unknowns with three equations, such that

$$\begin{aligned} n_1 &= f\left(f_1, f_c, V_{1p}\right) \\ \alpha_{g1} &= f\left(f_1, f_c, V_{1p}\right) \\ a_{i1} &= f\left(f_1, f_c, V_{1p}\right) \end{aligned} \tag{10-76}$$

where α_{g1} and a_{i1}, defined by Eqs.(10-8) and (10-20), can be obtained through experiment, and n_1 can be furnished by Eqs.(10-31) and (10-67).

Therefore, if necessary, the analytical solution for χ can be obtained. Furthermore, since the a_i can be written in terms of the average surface area of the particles A_i, and the particle number density, n as

$$a_{i1} = n_1 A_{i1} \tag{10-77}$$

and since $D_{s1} \approx D_{Sm1}$

$$\frac{A_{ic}}{A_{i1}} = \left(\frac{D_{sc}}{D_{s1}}\right)^2 \approx \left(\frac{D_{sc}}{D_{Sm1}}\right)^2 \tag{10-78}$$

we can rewrite Eqs.(10-64) and (10-65) as

$$\int_{V_{\min}}^{V_c} \left\{ A_i \frac{\partial}{\partial V}\left(f \frac{dV}{dt}\right)\right\} dV = \left(\frac{\dot{V}}{V}\right) a_{i1} \left\{ -\frac{2}{3} + \chi \left(\frac{D_{sc}}{D_{Sm1}}\right)^2 \right\} \tag{10-79}$$

and

$$\begin{aligned} &\int_{V_c}^{V_{\max}} \left\{ A_i \frac{\partial}{\partial V}\left(f \frac{dV}{dt}\right)\right\} dV \\ &= \left(\frac{\dot{V}}{V}\right) a_{i2} \left\{ -\frac{2}{3} - \chi \frac{a_{i1}}{a_{i2}} \left(\frac{D_{sc}}{D_{Sm1}}\right)^2 \right\} \end{aligned} \tag{10-80}$$

respectively.

Thus, combining above equations, the two-group interfacial area transport equation is given by

$$\begin{aligned} &\frac{\partial a_{i1}}{\partial t} + \nabla \cdot \left(a_{i1} \boldsymbol{v}_{i1}\right) = \frac{2}{3} \frac{a_{i1}}{\alpha_{g1}} \left\{ \frac{\partial \alpha_{g1}}{\partial t} + \nabla \cdot \left(\alpha_{g1} \boldsymbol{v}_{g1}\right) - \eta_{ph} \right\} \\ &-\chi \left(\frac{D_{sc}}{D_{Sm1}}\right)^2 \frac{a_{i1}}{\alpha_{g1}} \left\{ \frac{\partial \alpha_{g1}}{\partial t} + \nabla \cdot \left(\alpha_{g1} \boldsymbol{v}_{g1}\right) - \eta_{ph} \right\} \\ &+\int_{V_{\min}}^{V_c} \left(\sum_j S_j + S_{ph} \right) A_i dV \end{aligned} \tag{10-81}$$

$$\frac{\partial a_{i2}}{\partial t} + \nabla \cdot \left(a_{i2} \boldsymbol{v}_{i2} \right) = \frac{2}{3} \frac{a_{i2}}{\alpha_{g2}} \left\{ \frac{\partial \alpha_{g2}}{\partial t} + \nabla \cdot \left(\alpha_{g2} \boldsymbol{v}_{g2} \right) \right\}$$

$$+ \chi \left(\frac{D_{sc}}{D_{Sm1}} \right)^2 \frac{a_{i1}}{\alpha_{g1}} \left\{ \frac{\partial \alpha_{g1}}{\partial t} + \nabla \cdot \left(\alpha_{g1} \boldsymbol{v}_{g1} \right) - \eta_{ph} \right\} \qquad (10\text{-}82)$$

$$+ \int_{V_c}^{V_{\max}} \sum_j S_j A_i dV$$

for Group 1 and Group 2, respectively. In this, the left-hand sides of the equations represent the time-rate of change and convection of a_i for each group. Each term on the right-hand side represents the rate of change in a_i due to the particle volume change, inter-group transfer, various particle interactions and phase change for each group. The total interfacial area transport equation can be also obtained by adding the two equations, such that

$$\frac{\partial a_i}{\partial t} + \nabla \cdot \left(a_i \boldsymbol{v}_i \right) = \int_{V_{\min}}^{V_{\max}} \left(\sum_j S_j + S_{ph} \right) A_i dV$$

$$+ \sum_{k=1}^{2} \frac{2}{3} \frac{a_{ik}}{\alpha_{gk}} \left\{ \frac{\partial \alpha_{gk}}{\partial t} + \nabla \cdot \left(\alpha_{gk} \boldsymbol{v}_{gk} \right) - \eta_{ph} \right\} \qquad (10\text{-}83)$$

where the subscript k denotes the bubble group.

In this analysis, we demonstrate the inter-group transfer as a result of bubble interactions between the two groups of bubbles by assuming a liner profile or a uniform profile in the particle distribution. However, in two-phase flow applications, the uniform profile in the particle distribution may be assumed to be employed in practice.

1.3.4 Constitutive relations

In this section, the necessary constitutive relations in solving the interfacial area transport equation are summarized, and they are as follows:

$$\int_{V_{\min}}^{V_{\max}} \sum_j S_j dV = R_j: \quad \text{source and sink rate for particle number} \qquad (10\text{-}84)$$

$$\int_{V_{min}}^{V_{max}} \sum_j S_j A_i dV = \phi_j: \quad \text{source and sink rate for } a_i \tag{10-85}$$

$$\phi_{ph} = \pi D_{bc}^2 R_{ph}: \quad \text{source and sink rate for } a_i \text{ by phase-change} \tag{10-86}$$

$$n = \psi \frac{a_i^3}{\alpha_g^2} \text{ where } \psi = \frac{1}{36\pi}\left(\frac{D_{Sm}}{D_e}\right)^3 \tag{10-87}$$

$$D_{Sm} = \frac{6\alpha_g}{a_i} \tag{10-88}$$

$$\alpha_g = \alpha_{g1} + \alpha_{g2} \tag{10-89}$$

$$\Gamma_g = \Gamma_{g1} + \Gamma_{g2} \tag{10-90}$$

and

$$\boldsymbol{v}_g = \frac{\alpha_{g1}\boldsymbol{v}_{g1} + \alpha_{g2}\boldsymbol{v}_{g2}}{\alpha_{g1} + \alpha_{g2}}. \tag{10-91}$$

For the continuity equations, the net-mass transfer rate between Group 1 and Group 2 bubbles due to bubble interactions at steady state without phase change effect can be obtained from the modeling of the two-group bubble interactions as

$$\Delta\dot{m}_{12} = \rho_g \left\{ \sum_j \eta_{j,2} + \chi \left(\frac{D_{sc}}{D_{Sm1}}\right)^3 \nabla \cdot \left(\alpha_{g1} \boldsymbol{v}_{g1}\right) \right\}. \tag{10-92}$$

In this, $\eta_{j,2}$ is the net volume transfer from Group 1 bubbles to Group 2 bubbles due to the j-th interaction between the two groups of bubbles, such as bubble coalescence and disintegration.

Among the constitutive relations given above, the number source and sink rates defined by Eq.(10-84) should be established through mechanistic modeling of the major particle interactions that contribute to the change in the interfacial area concentration. Accounting for the wide range of gas-liquid two-phase flow, the major bubble interaction mechanisms that lead to the particle coalescence or disintegration can be summarized as follows:

- Random Collision (R_{RC}): coalescence through random collision driven by turbulent eddies;
- Wake Entrainment (R_{WE}): coalescence through collision due to acceleration of the following particle in the wake of the preceding particle;
- Turbulent Impact (R_{TI}): disintegration upon impact of turbulent eddies;
- Shearing-off (R_{SO}): shearing-off around the base rim of the cap bubble;
- Surface Instability (R_{SI}): break-up of large cap bubble due to surface instability;
- Rise Velocity (R_{RV}): collision due to the difference in the bubble rise velocity;
- Laminar Shear (R_{LS}): breakup due to the laminar shear in viscous fluid, and;
- Velocity Gradient (R_{VG}): collision due to the velocity gradient.

Chapter 11

CONSTITUTIVE MODELING OF INTERFACIAL AREA TRANSPORT

The two-fluid model is widely used in the current two-phase flow analysis codes, such as nuclear reactor safety analysis codes RELAP5, TRAC, and CATHARE. In the conventional model, the interfacial area concentration is given by empirical correlations. The correlations are based on two-phase flow regimes and regime-transition criteria that do not dynamically represent the changes in interfacial structure. There exist the following shortcomings caused by this static approach.

1. The flow-regime transition criteria are algebraic relations for steady-state, fully-developed flow. They do not fully reflect the true dynamic nature of changes in the interfacial structure. Hence, the effects of the entrance and developing flow cannot be taken into account correctly, nor can the gradual transition between regimes.
2. The method based on the flow-regime transition criteria is a two-step method that requires flow configuration transition criteria and interfacial area correlations for each flow configuration. The compound errors from the transition criteria and interfacial area correlations can be very significant.
3. The transition criteria and flow-regime dependent interfacial correlations are valid in limited parameter ranges for certain specific operational conditions and geometries. Most of them are obtained from simple air-water experiments and phenomenological models. Often the scale effects of geometry and fluid properties are not correctly taken into account. When applied to high-to-low pressure steam-water transients, these models may cause significant discrepancies, artificial discontinuities and numerical instability.

In Chapter 10, a physics-based approach, namely the interfacial area transport equation, was introduced to dynamically obtain the interfacial area concentration. In bubbly flow regime, bubbles may be assumed close to spherical in shape and with similar size, and thus a one-group interfacial area transport equation is sufficient to describe the interfacial area transport phenomena. However, in more generalized gas-liquid two-phase flows such as cap bubbly, slug and churn-turbulent flows, there exist bubbles with different sizes and shapes such as spherical, distorted, cap, slug, or churn-turbulent bubble. These variations in bubble size and shape substantially affect the bubble transport phenomena due to the differences in drag force and bubble interaction mechanisms. In developing the transport equation applicable to a wide range of two-phase flow, the differences in the shape and size of bubbles and in the characteristic transport phenomena should be accounted for. In view of this, the bubbles are categorized into two groups: spherical/distorted bubbles as Group 1 and cap/slug/churn-turbulent bubbles as Group 2. In Chapter 10, a general approach to treat bubbles in two groups was presented and the two-group interfacial area transport equation was formulated. In implementing the two-group interfacial area transport equation to the two-fluid model, some modifications of the conventional two-fluid model are required. This is mainly because the introduction of the two groups of bubbles requires two gas velocity fields while the conventional two-fluid model only provides one gas velocity through the momentum equation.

This chapter presents the modified two-fluid model that is ready to be implemented in the approach of the two-group interfacial area transport equation. Two momentum equations can be written for the two groups of bubbles, although it is not yet very practical to solve two gas momentum equations. However, for fully three-dimensional flow this may be necessary, whereas for one-dimensional flow a simplified approach is proposed. In this case, the momentum equation for the averaged velocity of the gas-phase is retained by combining the two gas momentum equations. Additional terms related to the velocity difference between Group-1 and Group-2 bubbles should be specified. This velocity difference can be estimated based on the simplified momentum equations for both Group-1 and Group-2 bubbles by accounting for the pressure gradient and general drag force. Furthermore, in one-dimensional simplification, a modified drift-flux model may be applied to solve for the velocity difference. In addition to this, this chapter demonstrates the modeling of sink and source terms in one-group and two-group interfacial area transport equations.

1.1 Modified two-fluid model for the two-group interfacial area transport equation

1.1.1 Conventional two-fluid model

As discussed in Chapter 9, a three-dimensional two-fluid model has been obtained by using temporal or statistical averaging. The model is expressed in terms of two sets of conservation equations governing the balance of mass, momentum and energy in each phase. However, since the averaged fields of one phase are not independent of the other phase, the interaction terms appear in the field equations as source terms. For most practical applications, the two-fluid model can be simplified to the following forms (Ishii, 1977; Ishii and Mishima, 1984) from Chapter 9.

Continuity equation for the gas phase

$$\frac{\partial\left(\alpha_g\rho_g\right)}{\partial t}+\nabla\cdot\left(\alpha_g\rho_g\boldsymbol{v}_g\right)=\Gamma_g \tag{11-1}$$

Continuity equation for the liquid phase

$$\frac{\partial\left[\left(1-\alpha_g\right)\rho_f\right]}{\partial t}+\nabla\cdot\left[\left(1-\alpha_g\right)\rho_f\boldsymbol{v}_f\right]=\Gamma_f \tag{11-2}$$

Momentum equation for the gas phase

$$\begin{aligned}&\frac{\partial\left(\alpha_g\rho_g\boldsymbol{v}_g\right)}{\partial t}+\nabla\cdot\left(\alpha_g\rho_g\boldsymbol{v}_g\boldsymbol{v}_g\right)=-\alpha_g\nabla p_g\\&+\nabla\cdot\left[\alpha_g\left(\mathcal{T}_g^{\mu}+\mathcal{T}_g^{T}\right)\right]+\alpha_g\rho_g\boldsymbol{g}\\&+\Gamma_g\boldsymbol{v}_{gi}+\boldsymbol{M}_{ig}-\nabla\alpha_g\cdot\mathcal{T}_{gi}+\left(p_{gi}-p_g\right)\nabla\alpha_g\end{aligned} \tag{11-3}$$

Momentum equation for the liquid phase

$$\begin{aligned}&\frac{\partial\left[\left(1-\alpha_g\right)\rho_f\boldsymbol{v}_f\right]}{\partial t}+\nabla\cdot\left[\left(1-\alpha_g\right)\rho_f\boldsymbol{v}_f\boldsymbol{v}_f\right]=-\left(1-\alpha_g\right)\nabla p_f\\&+\nabla\cdot\left[\left(1-\alpha_g\right)\left(\mathcal{T}_f^{\mu}+\mathcal{T}_f^{T}\right)\right]+\left(1-\alpha_g\right)\rho_f\boldsymbol{g}\\&+\Gamma_f\boldsymbol{v}_{fi}+\boldsymbol{M}_{if}-\nabla\left(1-\alpha_g\right)\cdot\mathcal{T}_{fi}+\left(p_{fi}-p_f\right)\nabla\left(1-\alpha_g\right)\end{aligned} \tag{11-4}$$

Thermal energy equation for the gas phase

$$\begin{aligned}
&\frac{\partial\left(\alpha_g \rho_g h_g\right)}{\partial t}+\nabla\cdot\left(\alpha_g \rho_g h_g \boldsymbol{v}_g\right)=-\nabla\cdot\left[\alpha_g\left(\boldsymbol{q}_g^C+\boldsymbol{q}_g^T\right)\right] \\
&+\alpha_g \frac{D_g p_g}{Dt}+\left(p_g-p_{gi}\right)\frac{D_g \alpha_g}{Dt}+\Gamma_g h_{gi}+a_i q_{gi}''+\phi_g
\end{aligned} \tag{11-5}$$

Thermal energy equation for the liquid phase

$$\begin{aligned}
&\frac{\partial\left[\left(1-\alpha_g\right)\rho_f h_f\right]}{\partial t}+\nabla\cdot\left[\left(1-\alpha_g\right)\rho_f \boldsymbol{v}_f h_f\right] \\
&=-\nabla\cdot\left[\left(1-\alpha_g\right)\left(\boldsymbol{q}_f^C+\boldsymbol{q}_f^T\right)\right]+\left(1-\alpha_g\right)\frac{D_f p_f}{Dt} \\
&+\left(p_f-p_{fi}\right)\frac{D_f\left(1-\alpha_g\right)}{Dt}+\Gamma_f h_{fi}+a_i q_{fi}''+\phi_f
\end{aligned} \tag{11-6}$$

Here, Γ_k, $\boldsymbol{M}_{ik}$, $\mathcal{T}_{ki}$, q_{ki}'' and ϕ_k are the mass generation, the generalized interfacial drag, the interfacial shear stress, the interfacial heat flux, and the dissipation, respectively. For simplicity, in the above equations the mathematical symbols of averaging are dropped, and $\overline{\overline{\mathcal{T}_k}}$, $\overline{\overline{\mathcal{T}_{ki}}}$ and $\overline{\overline{\boldsymbol{q}_k}}$ are represented by $\mathcal{T}_k^{\mu}$, $\mathcal{T}_{ki}$ and $\boldsymbol{q}_k^C$.

In Eqs.(11-1) to (11-6), the generation of mass per unit volume, the generalized drag force per unit volume, and the interfacial energy transfer per unit volume constitute the interfacial transfer terms. The jump conditions for the interfacial transfers are given as

$$\begin{cases}
\Gamma_g+\Gamma_f=0 \\
\boldsymbol{M}_{ig}+\boldsymbol{M}_{if}=0 \\
\left(a_i q_{gi}''+\Gamma_g h_{gi}\right)+\left(a_i q_{fi}''+\Gamma_f h_{fi}\right)=0.
\end{cases} \tag{11-7}$$

1.1.2 Two-group void fraction and interfacial area transport equations

The two-group void fraction transport equation is given by

$$\frac{\partial\left(\alpha_{gk}\rho_g\right)}{\partial t}+\nabla\cdot\left(\alpha_{gk}\rho_g\boldsymbol{v}_{gk}\right)=\Gamma_{gk}+(-1)^k\,\Delta\dot{m}_{12} \tag{11-8}$$

where $k=1$ and 2 for Groups 1 and 2, respectively. Γ_{gk} is the mass generation rate of Group-k bubbles due to phase change, and $\Delta\dot{m}_{12}$ represents the net inter-group mass transfer rate from Group-1 to Group-2 bubbles due to the bubble interactions and the hydrodynamic effect given by

$$\Delta\dot{m}_{12}=\rho_g\left[\sum_j\eta_{j,2}+\chi\left(D_{c1}^*\right)^3\left\{\frac{\partial\alpha_{g1}}{\partial t}+\nabla\cdot\left(\alpha_{g1}\boldsymbol{v}_{g1}\right)-\eta_{ph1}\right\}\right] \tag{11-9}$$

where $\eta_{j,2}$ and η_{phk} are the net inter-group void fraction transport from Group-1 to Group-2 bubbles and the source and sink term for the gas volume due to phase change, respectively. χ is the inter-group transfer coefficient and D_{c1}^* is the non-dimensional bubble diameter defined by

$$D_{c1}^*\equiv\frac{D_{crit}}{D_{Sm1}} \tag{11-10}$$

where D_{crit} is the volume-equivalent diameter of a bubble at the boundary between Groups 1 and 2.

The two-group interfacial area transport equation is given by

$$\begin{aligned}&\frac{\partial a_{i1}}{\partial t}+\nabla\cdot\left(a_{i1}\boldsymbol{v}_{gi1}\right)\\&=\left\{\frac{2}{3}-\chi\left(D_{c1}^*\right)^2\right\}\frac{a_{i1}}{\alpha_{g1}}\left[\frac{\partial\alpha_{g1}}{\partial t}+\nabla\cdot\left(\alpha_{g1}\boldsymbol{v}_{g1}\right)-\eta_{ph1}\right]\\&+\sum_j\phi_{j,1}+\phi_{ph1}\end{aligned} \tag{11-11}$$

$$\begin{aligned}&\frac{\partial a_{i2}}{\partial t}+\nabla\cdot\left(a_{i2}\boldsymbol{v}_{gi2}\right)=\frac{2}{3}\frac{a_{i2}}{\alpha_{g2}}\left[\frac{\partial\alpha_{g2}}{\partial t}+\nabla\cdot\left(\alpha_{g2}\boldsymbol{v}_{g2}\right)-\eta_{ph2}\right]\\&+\chi\left(D_{c1}^*\right)^2\frac{a_{i1}}{\alpha_{g1}}\left[\frac{\partial\alpha_{g1}}{\partial t}+\nabla\cdot\left(\alpha_{g1}\boldsymbol{v}_{g1}\right)-\eta_{ph1}\right]\\&+\sum_j\phi_{j,2}+\phi_{ph2}\end{aligned} \tag{11-12}$$

where $\phi_{j,k}$ and ϕ_{phk} are the source and sink terms for the interfacial area concentration due to bubble interactions for Group-k bubbles and phase change, respectively.

1.1.3 Modified two-fluid model

In what follows, the two-fluid model is modified for two-group interfacial area transport equation (Sun et al., 2003). The general form is given by the multi-field model for the gas phase. In general, the pressure and temperature for Group-1 and Group-2 bubbles can be assumed to be approximately the same. However, the velocities of two groups are not the same, therefore it is necessary to introduce two continuity and two momentum equations in principle. Based on the above assumption, the density of the gas phase is the same for Group-1 and Group-2 bubbles. This leads to the gas phase continuity equations as

Continuity equation for Group-1 bubbles

$$\frac{\partial\left(\alpha_{g1}\rho_g\right)}{\partial t}+\nabla\cdot\left(\alpha_{g1}\rho_g\boldsymbol{v}_{g1}\right)=\Gamma_{g1}-\Delta\dot{m}_{12} \tag{11-13}$$

Continuity equation for Group-2 bubbles

$$\frac{\partial\left(\alpha_{g2}\rho_g\right)}{\partial t}+\nabla\cdot\left(\alpha_{g2}\rho_g\boldsymbol{v}_{g2}\right)=\Gamma_{g2}+\Delta\dot{m}_{12}. \tag{11-14}$$

Here, $\Delta\dot{m}_{12}$ is the inter-group mass transfer due to hydrodynamic mechanisms. Furthermore, if the following identities are introduced,

$$\begin{cases}\alpha_{g1}+\alpha_{g2}=\alpha_g\\ \Gamma_{g1}+\Gamma_{g2}=\Gamma_g\\ \boldsymbol{v}_g=\dfrac{\left(\alpha_{g1}\boldsymbol{v}_{g1}+\alpha_{g2}\boldsymbol{v}_{g2}\right)}{\alpha_g}\end{cases} \tag{11-15}$$

then the summation of Eqs.(11-13) and (11-14) recovers the conventional continuity equation, i.e. Eq.(11-1). The continuity equation for the liquid phase remains the same as Eq.(11-2) with

$$\Gamma_f = -\Gamma_g = -\left(\Gamma_{g1} + \Gamma_{g2}\right). \tag{11-16}$$

The momentum equation is more complicated due to the introduction of the two groups of bubbles. Unlike the continuity equation for the gas phase, it is not desirable to have two momentum equations for Group-1 and Group-2 bubbles due to the complicated nature of the momentum equation at least for the one-dimensional formulation. If we assume Group-2 bubbles as the "third phases" in addition to the liquid phase and Group-1 bubbles and neglects the direct momentum interactions between the Group-1 and Group-2 bubbles, then two momentum equations may be written for both Group-1 and Group-2 bubbles as

$$\begin{aligned}&\frac{\partial\left(\alpha_{g1}\rho_g\boldsymbol{v}_{g1}\right)}{\partial t}+\nabla\cdot\left(\alpha_{g1}\rho_g\boldsymbol{v}_{g1}\boldsymbol{v}_{g1}\right)=-\alpha_{g1}\nabla p_{g1}\\&+\nabla\cdot\left[\alpha_1\left(\mathcal{T}_{g1}^{\mu}+\mathcal{T}_{g1}^{T}\right)\right]+\alpha_{g1}\rho_g\boldsymbol{g}+\left(\Gamma_{g1}-\Delta\dot{m}_{12}\right)\boldsymbol{v}_{gi1}\\&-\nabla\alpha_1\cdot\mathcal{T}_{gi1}+\boldsymbol{M}_{ig1}+\left(p_{gi1}-p_{g1}\right)\nabla\alpha_{g1}\end{aligned}\tag{11-17}$$

and

$$\begin{aligned}&\frac{\partial\left(\alpha_{g2}\rho_g\boldsymbol{v}_{g2}\right)}{\partial t}+\nabla\cdot\left(\alpha_{g2}\rho_g\boldsymbol{v}_{g2}\boldsymbol{v}_{g2}\right)=-\alpha_2\nabla p_{g2}\\&+\nabla\cdot\left[\alpha_2\left(\mathcal{T}_{g2}^{\mu}+\mathcal{T}_{g2}^{T}\right)\right]+\alpha_{g2}\rho_g\boldsymbol{g}+\left(\Gamma_{g2}+\Delta\dot{m}_{12}\right)\boldsymbol{v}_{gi2}\\&-\nabla\alpha_2\cdot\mathcal{T}_{gi2}+\boldsymbol{M}_{ig2}+\left(p_{gi2}-p_{g2}\right)\nabla\alpha_{g2}.\end{aligned}\tag{11-18}$$

Then, combining Eqs.(11-17) and (11-18) yields

$$\begin{aligned}&\frac{\partial\left(\alpha_g\rho_g\boldsymbol{v}_g\right)}{\partial t}+\nabla\cdot\left(\alpha_g\rho_g\boldsymbol{v}_g\boldsymbol{v}_g\right)\\&=-\nabla\cdot\left[\rho_g\frac{\alpha_{g1}\alpha_{g2}}{\alpha_g}\left(\boldsymbol{v}_{g1}-\boldsymbol{v}_{g2}\right)^2\right]-\left(\alpha_{g1}\nabla p_{g1}+\alpha_{g2}\nabla p_{g2}\right)\\&+\nabla\cdot\left[\alpha_{g1}\left(\mathcal{T}_{g1}^{\mu}+\mathcal{T}_{g1}^{T}\right)+\alpha_{g2}\left(\mathcal{T}_{g2}^{\mu}+\mathcal{T}_{g2}^{T}\right)\right]+\alpha_g\rho_g\boldsymbol{g}\\&+\left(p_{gi1}-p_{g1}\right)\nabla\alpha_{g1}+\left(p_{gi2}-p_{g2}\right)\nabla\alpha_{g2}\\&+\left[\left(\Gamma_{g1}-\Delta\dot{m}_{12}\right)\boldsymbol{v}_{gi1}+\left(\Gamma_{g2}+\Delta\dot{m}_{12}\right)\boldsymbol{v}_{gi2}\right]\end{aligned}\tag{11-19}$$

$$-\left(\nabla\alpha_1\cdot\mathcal{T}_{gi1}+\nabla\alpha_2\cdot\mathcal{T}_{gi2}\right)+\left(\boldsymbol{M}_{ig1}+\boldsymbol{M}_{ig2}\right)$$

with the definitions in Eq.(11-15). It is interesting to note that the first term on the right-hand side of Eq.(11-19) is an additional diffusion term due to the difference between the bubble velocities in different bubble groups.

However, Eq.(11-19) is too complicated to be applied in general applications. As mentioned earlier, for most of the practical applications, the pressure for the two groups of bubbles can be approximated as the same such that

$$p_{g1}\approx p_{g2}=p_g;\quad p_{gi1}\approx p_{gi2}=p_{gi}. \tag{11-20}$$

Furthermore, the interfacial shear for both groups of bubbles may be assumed to be very similar such that,

$$\mathcal{T}_{gi1}\approx\mathcal{T}_{gi2}=\mathcal{T}_{gi}. \tag{11-21}$$

We also have the following definition to further simplify Eq.(11-19)

$$\begin{cases}\mathcal{T}_g\equiv\dfrac{\alpha_{g1}\mathcal{T}_{g1}+\alpha_{g2}\mathcal{T}_{g2}}{\alpha_g}=\mathcal{T}_g^{\mu}+\mathcal{T}_g^{T}\\ \mathcal{T}_g^{\mu}\equiv\dfrac{\alpha_{g1}\mathcal{T}_{g1}^{\mu}+\alpha_{g2}\mathcal{T}_{g2}^{\mu}}{\alpha_g}\\ \mathcal{T}_g^{T}\equiv\dfrac{\alpha_{g1}\mathcal{T}_{g1}^{T}+\alpha_{g2}\mathcal{T}_{g2}^{T}}{\alpha_g}\\ \mathcal{T}_{g1}\equiv\mathcal{T}_{g1}^{\mu}+\mathcal{T}_{g1}^{T}\\ \mathcal{T}_{g2}\equiv\mathcal{T}_{g2}^{\mu}+\mathcal{T}_{g2}^{T}.\end{cases} \tag{11-22}$$

Therefore, Eq.(11-19) can be simplified as

$$\begin{aligned}&\frac{\partial\left(\alpha_g\rho_g\boldsymbol{v}_g\right)}{\partial t}+\nabla\cdot\left(\alpha_g\rho_g\boldsymbol{v}_g\boldsymbol{v}_g\right)=-\nabla\cdot\left[\rho_g\frac{\alpha_{g1}\alpha_{g2}}{\alpha_g}\left(\boldsymbol{v}_{g1}-\boldsymbol{v}_{g2}\right)^2\right]\\&-\alpha_g\nabla p_g+\nabla\cdot\left[\alpha_g\left(\mathcal{T}_g^{\mu}+\mathcal{T}_g^{T}\right)\right]+\alpha_g\rho_g\boldsymbol{g}\end{aligned} \tag{11-23}$$

$$+\left[\left(\Gamma_{g1}-\Delta\dot{m}_{12}\right)\boldsymbol{v}_{gi1}+\left(\Gamma_{g2}+\Delta\dot{m}_{12}\right)\boldsymbol{v}_{gi2}\right]-\nabla\alpha_g\cdot\mathcal{T}_{gi}$$
$$+\left(\boldsymbol{M}_{ig1}+\boldsymbol{M}_{ig2}\right)+\left(p_{gi}-p_g\right)\nabla\alpha_g.$$

It may be reasonable to assume that the averaged stresses in the bulk fluid and at the interface are approximately the same. Thus,

$$\mathcal{T}_{gi}\approx\left(\mathcal{T}_g^{\mu}+\mathcal{T}_g^{T}\right). \tag{11-24}$$

Then, Eq.(11-23) is further simplified as

$$\frac{\partial\left(\alpha_g\rho_g\boldsymbol{v}_g\right)}{\partial t}+\nabla\cdot\left(\alpha_g\rho_g\boldsymbol{v}_g\boldsymbol{v}_g\right)=-\nabla\cdot\left[\rho_g\frac{\alpha_{g1}\alpha_{g2}}{\alpha_g}\left(\boldsymbol{v}_{g1}-\boldsymbol{v}_{g2}\right)^2\right]$$
$$-\alpha_g\nabla p_g+\alpha_g\nabla\cdot\left(\mathcal{T}_g^{\mu}+\mathcal{T}_g^{T}\right)+\alpha_g\rho_g\boldsymbol{g} \tag{11-25}$$
$$+\left[\Gamma_{g1}\boldsymbol{v}_{g1}+\Gamma_{g2}\boldsymbol{v}_{g2}+\Delta\dot{m}_{12}\left(\boldsymbol{v}_{g2}-\boldsymbol{v}_{g1}\right)\right]+\left(\boldsymbol{M}_{ig1}+\boldsymbol{M}_{ig2}\right)$$
$$+\left(p_{gi}-p_g\right)\nabla\alpha_g.$$

The generalized interfacial drag terms, $\boldsymbol{M}_{ig1}$ and $\boldsymbol{M}_{ig2}$, should be individually modeled for Group-1 and Group-2 bubbles.

Furthermore, the momentum equation for the liquid phase has the same form as Eq.(11-3). Thus,

$$\frac{\partial\left[\left(1-\alpha_g\right)\rho_f\boldsymbol{v}_f\right]}{\partial t}+\nabla\cdot\left[\left(1-\alpha_g\right)\rho_f\boldsymbol{v}_f\boldsymbol{v}_f\right]=-\left(1-\alpha_g\right)\nabla p_f$$
$$+\nabla\cdot\left[\left(1-\alpha_g\right)\left(\mathcal{T}_f^{\mu}+\mathcal{T}_f^{T}\right)\right]+\left(1-\alpha_g\right)\rho_f\boldsymbol{g} \tag{11-26}$$
$$+\Gamma_f\boldsymbol{v}_{fi}+\boldsymbol{M}_{if}-\nabla\left(1-\alpha_g\right)\cdot\mathcal{T}_{fi}+\left(p_{fi}-p_f\right)\nabla\left(1-\alpha_g\right)$$

with

$$\boldsymbol{M}_{if}=-\boldsymbol{M}_{ig}=-\left(\boldsymbol{M}_{ig1}+\boldsymbol{M}_{ig2}\right). \tag{11-27}$$

It may be reasonable to assume that the averaged stresses in the liquid phase and at the interface are approximately the same. Thus,

$$\mathcal{T}_{fi} \approx \left(\mathcal{T}_f^{\mu} + \mathcal{T}_f^{T}\right). \tag{11-28}$$

Then, Eq.(11-26) is further simplified as

$$\begin{aligned}
&\frac{\partial\left[\left(1-\alpha_g\right)\rho_f \boldsymbol{v}_f\right]}{\partial t} + \nabla\cdot\left[\left(1-\alpha_g\right)\rho_f \boldsymbol{v}_f \boldsymbol{v}_f\right] = -\left(1-\alpha_g\right)\nabla p_f \\
&+\left(1-\alpha_g\right)\nabla\cdot\left(\mathcal{T}_f^{\mu} + \mathcal{T}_f^{T}\right) + \left(1-\alpha_g\right)\rho_f \boldsymbol{g} \\
&+\Gamma_f \boldsymbol{v}_{fi} + \boldsymbol{M}_{if} + \left(p_{fi} - p_f\right)\nabla\left(1-\alpha_g\right).
\end{aligned} \tag{11-29}$$

In the above derivation, it is assumed that the pressures and the temperatures for the two groups of bubbles are essentially the same. Then, similar to the momentum equation, the thermal energy equation for the gas phase can be expressed as

$$\begin{aligned}
&\frac{\partial\left(\alpha_g \rho_g h_g\right)}{\partial t} + \nabla\cdot\left(\alpha_g \rho_g h_g \boldsymbol{v}_g\right) = -\nabla\cdot\left[\alpha_g\left(\boldsymbol{q}_g^{C} + \boldsymbol{q}_g^{T}\right)\right] \\
&+\alpha_g \frac{D_g p_g}{Dt} + \Gamma_g h_{gi} + a_i q_{gi}'' + \phi_g
\end{aligned} \tag{11-30}$$

where the following definitions have been applied

$$\begin{cases}
\boldsymbol{q}_g^{C} = \dfrac{\alpha_{g1}\boldsymbol{q}_{g1}^{C} + \alpha_{g2}\boldsymbol{q}_{g2}^{C}}{\alpha_g} \\
\boldsymbol{q}_g^{T} = \dfrac{\alpha_{g1}\boldsymbol{q}_{g1}^{T} + \alpha_{g2}\boldsymbol{q}_{g2}^{T}}{\alpha_g} \\
q_{gi}'' = \dfrac{a_{i1}q_{gi1}'' + a_{i2}q_{gi2}''}{a_i} \\
a_i = a_{i1} + a_{i2}.
\end{cases} \tag{11-31}$$

The operator D_g/Dt is defined as

$$\frac{D_g}{Dt} \equiv \frac{\partial}{\partial t} + \frac{\alpha_{g1}\boldsymbol{v}_{g1} + \alpha_{g2}\boldsymbol{v}_{g2}}{\alpha_{g1} + \alpha_{g2}}\cdot\nabla = \frac{\partial}{\partial t} + \boldsymbol{v}_g\cdot\nabla. \tag{11-32}$$

Similarly, the thermal energy equation for the liquid phase is written as

$$\frac{\partial\left[\left(1-\alpha_g\right)\rho_f h_f\right]}{\partial t}+\nabla\cdot\left[\left(1-\alpha_g\right)\rho_f \boldsymbol{v}_f h_f\right]$$
$$=-\nabla\cdot\left[\left(1-\alpha_g\right)\left(\boldsymbol{q}_f^C+\boldsymbol{q}_f^T\right)\right]+\left(1-\alpha_g\right)\frac{D_f p_f}{Dt}+\Gamma_f h_{fi} \tag{11-33}$$
$$+a_i q_{fi}''+\phi_f$$

with the interfacial heat transfer at the liquid-phase side as

$$q_{fi}''=\frac{a_{i1}q_{fi1}''+a_{i2}q_{fi2}''}{a_i}. \tag{11-34}$$

Note that the following interfacial energetic condition should be satisfied

$$\left(a_i q_{gi}''+\Gamma_g h_{gi}\right)+\left(a_i q_{fi}''+\Gamma_f h_{fi}\right)=0. \tag{11-35}$$

In the above derivation, very complicated interfacial transfer terms are introduced. To solve the modified two-fluid model with the two-group interfacial area transport equation, various constitutive relations, interfacial transfer terms, and boundary conditions should be specified for the additional variables. These variables can be summarized as

$$\Gamma_{g1}, \Gamma_{g2}, \Gamma_f, \Delta\dot{m}_{12},$$
$$\mathcal{T}_g^{\mu}, \mathcal{T}_g^{T}, \mathcal{T}_f^{\mu}, \mathcal{T}_f^{T}, \boldsymbol{M}_{ig1}, \boldsymbol{M}_{ig2}, \boldsymbol{M}_{if}, \mathcal{T}_{gi}, \mathcal{T}_{fi}, \boldsymbol{v}_{gi1}, \boldsymbol{v}_{gi2}, \boldsymbol{v}_{fi}, \boldsymbol{g} \text{ and}$$
$$\boldsymbol{q}_g^C, \boldsymbol{q}_f^C, \boldsymbol{q}_g^T, \boldsymbol{q}_f^T, q_{gi}'', q_{fi}'', h_{gi}, h_{fi}.$$

1.1.4 Modeling of two gas velocity fields

For strongly one-dimensional flow, the introduction of two gas momentum equations may bring in unnecessary complications. In this case, the gas mixture momentum equation and an additional constitutive relation specifying the relative velocity between Group-1 and Group-2 gas velocities is sufficient. It is important to ensure that doing so will not over-specify the unknowns since the number of unknowns should equal the number of available equations.

The difference of the bubble velocities may be related to the local slip as

$$\boldsymbol{v}_{g1} - \boldsymbol{v}_{g2} = \left(\boldsymbol{v}_{g1} - \boldsymbol{v}_f\right) - \left(\boldsymbol{v}_{g2} - \boldsymbol{v}_f\right) = \boldsymbol{v}_{r1} - \boldsymbol{v}_{r2}. \tag{11-36}$$

To obtain the local relative velocity between the gas phase and liquid phase, a similar approach to Ishii (1977) may be taken along the drift-flux model formulation.

The momentum equation for Group-1 bubbles, i.e. Eq.(11-17) can be written in the following form by substituting the continuity equation and considering the assumptions of Eqs.(11-20) and (11-21) and $p_g = p_{gi}$.

$$\begin{aligned} &\alpha_{g1}\rho_g\left(\frac{\partial \boldsymbol{v}_{g1}}{\partial t} + \boldsymbol{v}_{g1}\cdot\nabla\boldsymbol{v}_{g1}\right) = -\alpha_{g1}\nabla p_{g1} + \alpha_{g1}\nabla\cdot\mathcal{T}_{g1} \\ &+\alpha_{g1}\rho_g\boldsymbol{g} + \left(\Gamma_{g1} - \Delta\dot{m}_{12}\right)\left(\boldsymbol{v}_{gi1} - \boldsymbol{v}_{g1}\right) + \boldsymbol{M}_{ig1} \end{aligned} \tag{11-37}$$

Similarly, we obtain the momentum equations for both Group-2 bubbles and the liquid phase as

$$\begin{aligned} &\alpha_{g2}\rho_g\left(\frac{\partial \boldsymbol{v}_{g2}}{\partial t} + \boldsymbol{v}_{g2}\cdot\nabla\boldsymbol{v}_{g2}\right) = -\alpha_{g2}\nabla p_{g2} + \alpha_{g2}\nabla\cdot\mathcal{T}_{g2} \\ &+\alpha_{g2}\rho_g\boldsymbol{g} + \left(\Gamma_{g2} + \Delta\dot{m}_{12}\right)\left(\boldsymbol{v}_{gi2} - \boldsymbol{v}_{g2}\right) + \boldsymbol{M}_{ig2} \end{aligned} \tag{11-38}$$

and

$$\begin{aligned} &\left(1-\alpha_g\right)\rho_f\left(\frac{\partial \boldsymbol{v}_f}{\partial t} + \boldsymbol{v}_f\cdot\nabla\boldsymbol{v}_f\right) = -\left(1-\alpha_g\right)\nabla p_f \\ &+\left(1-\alpha_g\right)\nabla\cdot\mathcal{T}_f + \left(1-\alpha_g\right)\rho_f\boldsymbol{g} + \Gamma_f\left(\boldsymbol{v}_{fi} - \boldsymbol{v}_f\right) + \boldsymbol{M}_{if}. \end{aligned} \tag{11-39}$$

To obtain the local relative velocity correlation, we consider a special condition such as steady-state condition without phase change and with negligible-transverse pressure gradient. Without phase change effect, the interfacial velocity and the phase velocity for each phase can be considered equal, i.e.

$$\boldsymbol{v}_{gi1} \approx \boldsymbol{v}_{g1};\quad \boldsymbol{v}_{gi2} \approx \boldsymbol{v}_{g2};\quad \boldsymbol{v}_{fi} \approx \boldsymbol{v}_f. \tag{11-40}$$

And the pressure for each phase may be approximated as

$$p_g \approx p_f \approx p_m. \tag{11-41}$$

Under these approximations, for a nearly one-dimensional flow, the above momentum equations can be expressed as

$$\boldsymbol{M}_{ig1} \approx \alpha_{g1}\nabla p_m - \alpha_{g1}\nabla\cdot\mathcal{T}_{g1} - \alpha_{g1}\rho_g\boldsymbol{g} \tag{11-42}$$

$$\boldsymbol{M}_{ig2} \approx \alpha_{g2}\nabla p_m - \alpha_{g2}\nabla\cdot\mathcal{T}_{g2} - \alpha_{g2}\rho_g\boldsymbol{g} \tag{11-43}$$

and

$$\boldsymbol{M}_{if} \approx \left(1-\alpha_g\right)\nabla p_m - \left(1-\alpha_g\right)\nabla\cdot\mathcal{T}_f - \left(1-\alpha_g\right)\rho_f\boldsymbol{g}. \tag{11-44}$$

From the interfacial force balance, i.e. Eq.(11-27), the summation of the above three equations yields

$$\nabla p_m - \boldsymbol{M}_{\tau m} - \rho_m\boldsymbol{g} \approx 0 \tag{11-45}$$

in which $\boldsymbol{M}_{\tau m}$ is the force associated with the mixture transverse stress gradient and given by

$$\begin{aligned}\boldsymbol{M}_{\tau m} &\equiv \left(\alpha_{g1}\nabla\cdot\mathcal{T}_{g1} + \alpha_{g2}\nabla\cdot\mathcal{T}_{g2}\right) + \left(1-\alpha_g\right)\nabla\cdot\mathcal{T}_f \\ &= \alpha_{g1}\boldsymbol{M}_{\tau g1} + \alpha_{g2}\boldsymbol{M}_{\tau g2} + \left(1-\alpha_g\right)\boldsymbol{M}_{\tau f}\end{aligned} \tag{11-46}$$

with

$$\boldsymbol{M}_{\tau g1} = \nabla\cdot\mathcal{T}_{g1};\quad \boldsymbol{M}_{\tau g2} = \nabla\cdot\mathcal{T}_{g2};\quad \boldsymbol{M}_{\tau f} = \nabla\cdot\mathcal{T}_f \tag{11-47}$$

while ρ_m is defined as

$$\rho_m \equiv \left(\alpha_{g1}+\alpha_{g2}\right)\rho_g + \left(1-\alpha_g\right)\rho_f = \alpha_g\rho_g + \left(1-\alpha_g\right)\rho_f. \tag{11-48}$$

Equation (11-45) also assumes that Eq.(11-28) is valid for most applications.

Furthermore, from Eq.(11-45), the gravitational force field may be replaced with the pressure field, which is an unknown in the momentum equation, as

$$\boldsymbol{g} \approx \frac{1}{\rho_m}\left(\nabla p_m - \boldsymbol{M}_{\tau m}\right). \tag{11-49}$$

This allows the approach to be applied in microgravity conditions. In steady state, the generalized interfacial drag force is approximated by neglecting the virtual mass force, the Basset force and non-drag force such as lift force as

$$\boldsymbol{M}_{ig1} \approx -\frac{3\alpha_{g1}}{8r_{d1}} C_{D1}\rho_f \boldsymbol{v}_{r1}\left|\boldsymbol{v}_{r1}\right| \tag{11-50}$$

where r_{d1} and C_{D1} are the drag radius and the drag coefficient of Group-1 bubble, and the relative velocity for Group-1 bubbles is defined as

$$\boldsymbol{v}_{r1} = \boldsymbol{v}_{g1} - \boldsymbol{v}_f. \tag{11-51}$$

Thus, in steady state, by using Eqs.(11-49) and (11-50), we can rewrite Eq.(11-42) as

$$\begin{aligned} &-\frac{3\alpha_{g1}}{8r_{d1}} C_{D1}\rho_f \boldsymbol{v}_{r1}\left|\boldsymbol{v}_{r1}\right| \\ &\approx \alpha_{g1}\left(1 - \frac{\rho_g}{\rho_m}\right)\nabla p_m + \alpha_{g1}\left(\frac{\rho_g}{\rho_m}\boldsymbol{M}_{\tau m} - \boldsymbol{M}_{\tau g1}\right) \end{aligned} \tag{11-52}$$

or in the following form

$$\boldsymbol{v}_{r1}\left|\boldsymbol{v}_{r1}\right| \approx -\frac{8r_{d1}}{3C_{D1}\rho_f}\left[\left(1 - \frac{\rho_g}{\rho_m}\right)\nabla p_m + \left(\frac{\rho_g}{\rho_m}\boldsymbol{M}_{\tau m} - \boldsymbol{M}_{\tau g1}\right)\right]. \tag{11-53}$$

Similarly, for Group-2 bubbles, we have the following formulation

$$\boldsymbol{v}_{r2}\left|\boldsymbol{v}_{r2}\right| \approx -\frac{8r_{d2}}{3C_{D2}\rho_f}\left[\left(1 - \frac{\rho_g}{\rho_m}\right)\nabla p_m + \left(\frac{\rho_g}{\rho_m}\boldsymbol{M}_{\tau m} - \boldsymbol{M}_{\tau g2}\right)\right] \tag{11-54}$$

where r_{d2} and C_{D2} are the drag radius and the drag coefficient of Group-2 bubbles, respectively, and the relative velocity for Group-2 bubbles is defined as

$$\boldsymbol{v}_{r2} = \boldsymbol{v}_{g2} - \boldsymbol{v}_f. \tag{11-55}$$

From Eqs.(11-53) and (11-54), together with Eq.(11-36), we can solve the local slip between the two groups of bubbles, i.e. $\boldsymbol{v}_{g1} - \boldsymbol{v}_{g2}$.

In the case of one-dimensional flows, the one-dimensional drift-flux model to be discussed in Chapter 14 can be utilized to specify the velocity difference. The one-dimensional drift-flux model is modified for two-group interfacial area transport equation as

$$\langle\langle v_{gk}\rangle\rangle = C_{0k}\langle j\rangle + \langle\langle V_{gjk}\rangle\rangle \tag{11-56}$$

where $\langle\langle v_{gk}\rangle\rangle$, C_{0k} and $\langle\langle V_{gjk}\rangle\rangle$ are the void fraction weighted mean gas velocity, the distribution parameter, and the void fraction weighted mean drift velocity of Group-k bubbles, respectively. Then, the velocity difference is given by

$$\begin{aligned} &\langle\langle v_{g1}\rangle\rangle - \langle\langle v_{g2}\rangle\rangle = \left(\langle\langle v_{g1}\rangle\rangle - \langle j\rangle\right) - \left(\langle\langle v_{g2}\rangle\rangle - \langle j\rangle\right) \\ &= \left[(C_{01}-1)\langle j\rangle + \langle\langle V_{gj1}\rangle\rangle\right] - \left[(C_{02}-1)\langle j\rangle + \langle\langle V_{gj2}\rangle\rangle\right]. \end{aligned} \tag{11-57}$$

The distribution parameters for both groups of bubbles should be obtained from experimental data for certain flow geometry. Furthermore, if we assume that the distribution parameters for both groups of bubbles are essentially similar for certain flows, then the following simplified form can be approximately obtained as

$$\langle\langle v_{g1}\rangle\rangle - \langle\langle v_{g2}\rangle\rangle \approx \langle\langle V_{gj1}\rangle\rangle - \langle\langle V_{gj2}\rangle\rangle. \tag{11-58}$$

1.2 Modeling of source and sink terms in one-group interfacial area transport equation

To model the integral source and sink terms in the interfacial area transport equation caused by particle coalescence and breakup, a general approach treats the bubbles in two groups: the spherical/distorted bubble group and the cap/slug bubble group, resulting in two interfacial area transport equations that involve the inner- and inter-group interactions as shown in Fig.11-1. As shown in Fig.11-2, the mechanisms of these interactions can be summarized in five categories: the coalescence due to random collisions driven by liquid turbulence; the coalescence due to wake

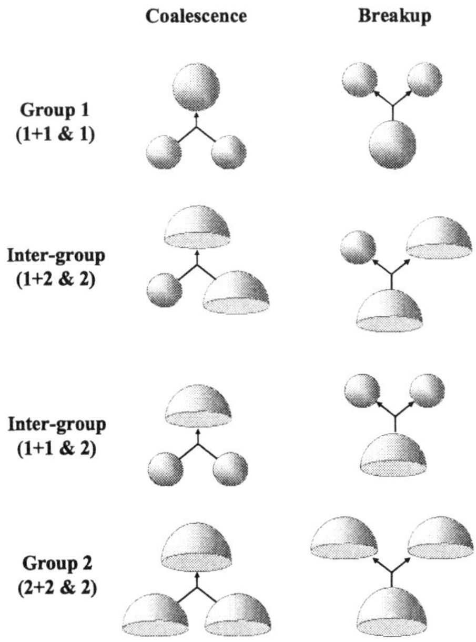

Figure 11-1. Classification of possible interactions of two-group bubbles (Hibiki and Ishii, 2000b)

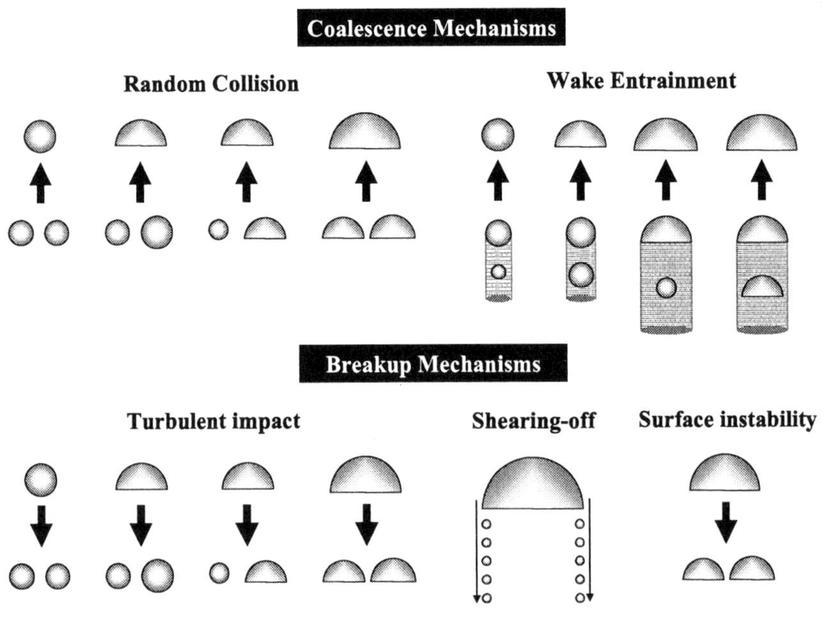

Figure 11-2. Schematic illustrations of two-group bubble interaction (Ishii et al., 2002)

entrainment; the breakup due to the impact of turbulent eddies; the shearing-off of small bubbles from cap/slug bubbles; and the breakup of large cap bubbles due to flow instability on the bubble surface (Kocamustafaogullari and Ishii, 1995; Wu et al., 1998). Some other mechanisms such as laminar-shearing induced coalescence (Friedlander, 1977) and the breakup due to velocity gradient (Taylor, 1934) are excluded because they are indirectly caused by the distributions of the flow parameters and void fraction, and the direct mechanisms still follow the above five categories.

In practice, when the void fraction of a two-phase bubbly flow is small, no cap or slug bubbles exist. The two-group interfacial area transport equation is then reduced to one group without the involvement of the interactions between the two groups as

$$\begin{aligned}&\frac{\partial a_i}{\partial t}+\nabla\cdot\left(a_i\boldsymbol{v}_i\right)\\&=\frac{2}{3}\left(\frac{a_i}{\alpha_g}\right)\left[\frac{\partial\alpha_g}{\partial t}+\nabla\cdot\left(\alpha_g\boldsymbol{v}_g\right)-\eta_{ph}\right]+\sum_j\phi_j+\phi_{ph}.\end{aligned} \tag{11-59}$$

In this section, some models of source and sink terms in one-group interfacial area transport equation are explained briefly.

1.2.1 Source and sink terms modeled by Wu et al. (1998)

Wu et al. (1998) considered three mechanisms of the interfacial area transport in an adiabatic bubbly flow, namely coalescence due to random collisions driven by liquid turbulence, coalescence due to wake entrainment, and breakup due to the impact of turbulent eddies. Then, Eq.(11-59) is further simplified as

$$\begin{aligned}&\frac{\partial a_i}{\partial t}+\nabla\cdot\left(a_i\boldsymbol{v}_i\right)\\&=\frac{2}{3}\left(\frac{a_i}{\alpha_g}\right)\left[\frac{\partial\alpha_g}{\partial t}+\nabla\cdot\left(\alpha_g\boldsymbol{v}_g\right)\right]+\left(\phi_{RC}+\phi_{WE}+\phi_{TI}\right).\end{aligned} \tag{11-60}$$

A. Bubble coalescence due to random collision

To model the bubble coalescence rate driven by turbulence in the continuous medium, the bubble random collision rate is of primary importance. These collisions are postulated to occur only between the neighboring bubbles because long-range interactions are driven by large

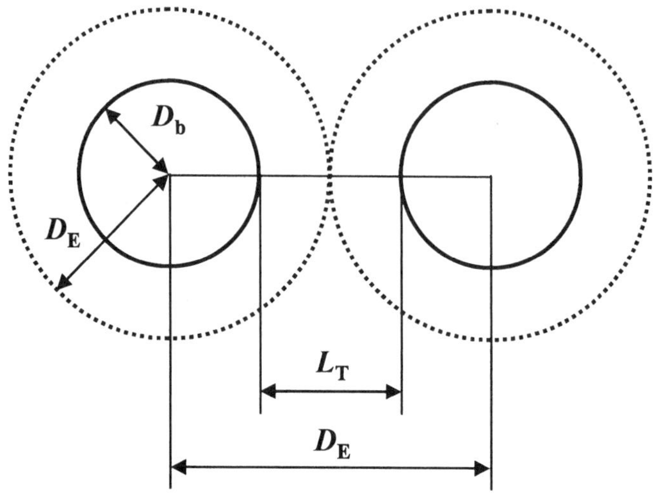

Figure 11-3. Geometric definitions of two approaching bubbles (Wu et al., 1998)

eddies that transport groups of bubbles without leading to significant relative motion (Prince and Blanch, 1990; Coulaloglou and Tavlarides, 1976). Between two neighboring spherical bubbles of the same size as shown in Fig.11-3, the time interval for one collision, Δt_C, is defined as

$$\Delta t_C = \frac{L_T}{u_t}. \tag{11-61}$$

Here, u_t is the root-mean-square approaching velocity of the two bubbles, and L_T represents the mean traveling distance between the two bubbles for one collision. This is approximated by

$$L_T \approx D_E - \delta D_b \propto \left(\frac{D_b}{\alpha_g^{1/3}} - \delta' D_b \right) = \frac{D_b}{\alpha_g^{1/3}} \left(1 - \delta' \alpha_g^{1/3} \right) \tag{11-62}$$

in which D_E is the effective diameter of the mixture volume that contains one bubble, and D_b is the bubble diameter. Since the bubble-traveling length for one collision varies from D_E to $\left(D_E - D_b\right)$, a factor δ is introduced in Eq.(11-62) to feature the averaged effect (whereas δ' is a collective parameter in considering the sign of proportionality between D_E and $D_b / \alpha_g^{1/3}$). For small void fraction, δ' plays a minor role due to the fact that D_E is much larger than D_b. However, it is important if the traveling length is comparable to the mean bubble size. When void fraction approaches the dense packing limit $\left(\alpha_g \cong \alpha_{g,max}\right)$, the mean traveling length should be zero, which leads to δ' equal to $\alpha_{g,max}^{-1/3}$. Using this asymptotic value as the approximation of δ', the mean traveling length is reduced to

$$L_T \propto \frac{D_b}{\alpha_g^{1/3}}\left\{1-\left(\frac{\alpha_g}{\alpha_{g,max}}\right)^{1/3}\right\}. \tag{11-63}$$

Accordingly, the collision frequency for two bubbles moving toward each other, f_{RC}, is given by

$$f_{RC} = \frac{1}{\Delta t_C} \propto \frac{u_t}{D_b}\alpha_g^{1/3}\left(\frac{\alpha_{g,max}^{1/3}}{\alpha_{g,max}^{1/3}-\alpha_g^{1/3}}\right). \tag{11-64}$$

Since the bubbles do not always move toward each other, however, a probability, P_C, for a bubble to move toward a neighboring bubble is considered here to modify the collision rate. By assuming a hexagonal close-packed structure, this probability is given by

$$P_C \sim \frac{D_b^2}{D_E^2} = \alpha_g^{2/3},\ \alpha_g < \alpha_{g,crit} \text{ and } P_C = 1,\ \alpha_g > \alpha_{g,crit}. \tag{11-65}$$

where $\alpha_{g,crit}$ is the critical void fraction when the center bubble cannot pass through the free space among the neighboring bubbles. In reality, the neighboring bubbles are in constant motion and the critical void fraction can be very close to the dense packing limit. This leads to

$$P_C \propto \left(\frac{\alpha_g}{\alpha_{g,max}}\right)^{2/3}. \tag{11-66}$$

Subsequently, the collision frequency for a mixture with bubble number density, n_b, is given by

$$f_{RC} \approx \frac{1}{\Delta t_C}P_c \propto u_t\frac{\alpha_g}{D_b}\left\{\frac{1}{\alpha_{g,max}^{1/3}\left(\alpha_{g,max}^{1/3}-\alpha_g^{1/3}\right)}\right\} \tag{11-67}$$

$$\propto n_b u_t D_b^2 \left\{ \frac{1}{\alpha_{g,max}^{1/3} \left(\alpha_{g,max}^{1/3} - \alpha_g^{1/3} \right)} \right\}.$$

The functional dependence of the above collision rate agrees with the model of Coulaloglou and Tavlarides (1976) proposed for a liquid-liquid droplet flow system, analogous to the particle collision model in an ideal gas. The difference is that the present model contains an extra term in the bracket, which covers the situation when the mean-free path of a bubble is comparable to the mean bubble size. Nevertheless, the model in the present form is still incomplete, since no matter how far away the neighboring bubble is located, the collision would occur as long as there is a finite approaching velocity. In actuality, when the mean distance is very large, no collision should be counted because the range of the relative motion for collisions between the neighboring bubbles is limited by the eddy size comparable to the bubble size. To consider this effect, the following modification factor is suggested for Eq.(11-67)

$$\left\{ 1 - \exp\left(-C_T \frac{L_t}{L_T} \right) \right\}. \tag{11-68}$$

where C_T and L_t are, respectively, an adjustable parameter depending on the properties of the fluid and the average size of the eddies that drive the neighboring bubbles together. These eddies are assumed to be on the same order of the mean bubble size because smaller eddies do not provide considerable bulk motion to a bubble. Larger eddies, however, transport groups of bubbles without inducing significant relative motion among the bubbles. Thereafter, the final form of the bubble collision frequency is given by

$$\begin{aligned} f_{RC} &\sim \left(u_t n_b D_b^2 \right) \left\{ \frac{1}{\alpha_{g,max}^{1/3} \left(\alpha_{g,max}^{1/3} - \alpha_g^{1/3} \right)} \right\} \\ &\times \left\{ 1 - \exp\left(-C_T \frac{\alpha_{max}^{1/3} \alpha^{1/3}}{\alpha_{max}^{1/3} - \alpha^{1/3}} \right) \right\}. \end{aligned} \tag{11-69}$$

For each collision, coalescence may not occur and thus a collision efficiency was suggested by many investigators (Oolman and Blanch,

1986b; Kirkpatrick and Lockett, 1974). The most popular model for the collision efficiency is the film thinning model (Kirkpatrick and Lockett, 1974). In this model, when the bubbles approach faster, they tend to bounce back without coalescence due to the limitation of the film-drainage rate governed by the surface tension. Mathematically, the coalescence rate decreases exponentially with respect to the turbulent fluctuating velocity, which is much stronger than the linear dependence of the collision rate, resulting in an overall decreasing trend of the coalescence rate as the turbulent fluctuation increases. Hence, a constant coalescence efficiency, λ_C, is employed in the model to depict the randomness of the coalescence phenomenon after each collision. Nevertheless, the constant coalescence efficiency is only an approximation and further efforts are needed to model the efficiency mechanistically. The mean-bubble fluctuation velocity, u_t, in Eq.(11-69) is proportional to the root-mean-square liquid fluctuating velocity difference between two points of length scale, D_b, and is given by $\varepsilon^{1/3}D_b^{1/3}$ where ε is the energy dissipation rate per unit mass of the continuous medium (Rotta, 1972). Thus, the decrease rate of the interfacial area concentration due to the bubble coalescence caused by random collisions, ϕ_{RC}, is given by

$$\begin{aligned}\phi_{RC} &= -\frac{1}{3\psi}\left(\frac{\alpha_g}{a_i}\right)^2 f_{RC} n_b \lambda_C \\ &= -\frac{\Gamma_{RC}\alpha_g^2\varepsilon^{1/3}}{D_b^{5/3}}\left\{\frac{1}{\alpha_{g,max}^{1/3}\left(\alpha_{g,max}^{1/3}-\alpha_g^{1/3}\right)}\right\} \\ &\quad \times\left\{1-\exp\left(-C_T\frac{\alpha_{g,max}^{1/3}\alpha_g^{1/3}}{\alpha_{g,max}^{1/3}-\alpha_g^{1/3}}\right)\right\}\end{aligned} \tag{11-70}$$

where Γ_{RC} is an adjustable parameter depending on the properties of the fluid, which is determined experimentally to be 0.016. The constants in Eq.(11-70) are set at $\alpha_{g,max}$=0.75 and C_T=3.

B. Bubble coalescence due to wake entrainment

When bubbles enter the wake region of a leading bubble, they will accelerate and may collide with the preceding one (Otake et al., 1977; Bilicki and Kestin, 1987; Stewart, 1995). For a spherical bubble with attached wake region in the liquid medium, the effective wake volume, V_W, in which the following bubbles may collide with the leading one, is defined as the projected bubble area multiplied by the effective length, L_W. The number of bubbles inside the effective volume, N_W, is then given by

$$N_W = V_W n_b \approx \frac{1}{4}\pi D_b^2 \left(L_W - \frac{D_b}{2} \right) n_b . \tag{11-71}$$

Assuming that the average time interval for a bubble in the wake region to catch up with the preceding bubble is Δt_W, the collision rate per unit mixture volume, R_{WE}, should satisfy

$$\begin{aligned} R_{WE} &\propto \frac{1}{2} n_b \frac{N_W}{\Delta t_W} \approx \frac{1}{8}\pi D_b^2 n_b^2 \left(\frac{L_W - D_b/2}{\Delta t_W} \right) \\ &\approx \frac{1}{8}\pi D_b^2 n_b^2 u_{rW} . \end{aligned} \tag{11-72}$$

where u_{rW} is the averaged relative velocity between the leading bubble and the bubble in the wake region. Schlichting (1979) gave the analytical expression of non-dimensionalized relative velocity as

$$\frac{v_{rW}}{v_r} \approx \left(\frac{C_D A}{\beta^2 y^2} \right)^{1/3} \tag{11-73}$$

where v_{rW}, v_r, C_D, A, β and y are: the relative velocity between the leading bubble and the bubble in the wake region; the relative velocity between the leading bubble and the liquid phase; the drag coefficient; the frontal area of the bubble; the ratio of the mixing length and the width of the wake; and the distance along the flow direction measured from the center of the leading bubble. The averaged relative velocity in the wake region, u_{rW}, is then obtained by integrating v_{rW} over the critical distance as

$$\begin{aligned} u_{rW} &\approx 3 v_{rW} \left(\frac{C_D \pi}{\beta^2} \right)^{1/3} \frac{1}{L_W / \left(D_b/2 \right) - 1} \left\{ \left(\frac{L_W}{D_b/2} \right)^{1/3} - 1 \right\} \\ &\approx F\left(\frac{L_W}{D_b} \right) C_D^{1/3} v_r \end{aligned} \tag{11-74}$$

where $F\left(L_W/D_b \right)$ is a function of L_W/D_b, since β is usually assumed to be constant (Schlichting, 1979). The exact form of $F\left(L_W/D_b \right)$ is not important since the effective bubble wake region may not be fully established. According to Tsuchiya et al. (1989), the wake length is roughly 5-7 times the bubble diameter in an air-water system, and thus L_W/D_b as well as $F(L_W/D_b)$ are treated as constants depending on the fluid properties. As long as their values obtained from experimental data fall into the range of L_W/D_b =5-7, the mechanism should be acceptable. Thus, the decrease rate of the interfacial area concentration due to the bubble coalescence caused by wake entrainment, ϕ_{WE}, is given by

$$\phi_{WE} = -\frac{1}{3\psi}\left(\frac{\alpha_g}{a_i}\right)^2 R_{WE}\lambda_C = -\frac{\Gamma_{WE} C_D^{1/3} \alpha_g^2 v_r}{D_b^2}. \tag{11-75}$$

where Γ_{WE} is an adjustable parameter mainly determined by the ratio of the effective wake length to the bubble size and the coalescence efficiency, which is determined experimentally to be 0.0076.

C. Bubble breakup due to turbulent impact

For binary bubble breakup due to the impact of turbulent eddies, the driving force comes from the inertial force, $F_{inertia}$, of the turbulent eddies in the continuous medium, while the holding force is the surface tension force, $F_{tension}$. To drive the daughter bubbles apart with a characteristic length of D_b within time interval Δt_B, a simple momentum balance approach gives the following relation.

$$\frac{\rho_f D_b^3 D_b}{\Delta t_B^2} \propto F_{inertia} - F_{tension} \tag{11-76}$$

Here, the inertia of the bubble is dominated by the virtual mass because of the large density ratio of the liquid and gas. Rearranging Eq.(11-76) leads to the following averaged bubble breakup frequency

$$f_{TI} \propto \frac{u_t}{D_b}\left(1 - \frac{We_{crit}}{We}\right)^{1/2}, \quad We \equiv \frac{\rho_f u_t^2 D_b}{\sigma} > We_{crit}. \tag{11-77}$$

The velocity, u_t, is assumed to be the root-mean-square velocity difference between two points of length D_b, which implies that only the eddies with sizes equivalent to the bubble size can break the bubble. We_{crit} is a collective constant, designated as a critical Weber number. The reported value of We_{crit} for bubble breakup varies in a wide range due to the resonance excitation of the turbulent fluctuation (Sevik and Park, 1973).

In a homogeneous turbulent flow, the probability for a bubble to collide with an eddy that has sufficient energy to break the bubble, namely the breakup efficiency, λ_B, is approximately (Coulaloglou and Tavlarrides, 1976)

$$\lambda_B \propto \exp\left(-\frac{u_{t,crit}^2}{u_t^2}\right) \tag{11-78}$$

where $u_{t,crit}^2$ is the critical mean-square fluctuation velocity obtained from We_{crit}. Finally, the increase rate of the interfacial area concentration due to the bubble breakup caused by turbulent impact, ϕ_{TI}, is given by

$$\begin{aligned}\phi_{TI} &= \frac{1}{3\psi}\left(\frac{\alpha_g}{a_i}\right)^2 f_{TI} n_b \lambda_B \\ &= \begin{cases} \dfrac{\Gamma_{TI}\alpha_g \varepsilon^{1/3}}{D_b^{5/3}}\left(1-\dfrac{We_{crit}}{We}\right)^{1/2} \exp\left(-\dfrac{We_{crit}}{We}\right), & We > We_{crit} \\ 0, \; We \le We_{crit}. \end{cases}\end{aligned} \tag{11-79}$$

The adjustable parameters Γ_{TI} and We_{crit} are determined experimentally to be 0.17 and 6.0, respectively. This expression differs from the previous models (Prince and Blanch, 1990) because the breakup rate equals zero when the Weber number is less than We_{crit}. This unique feature permits the decoupling of the bubble coalescence and breakup processes. At a low liquid flow rate with small void fraction, the turbulent fluctuation is small and thus no breakup would be counted. This allows the fine-tuning of the adjustable parameters in the coalescence terms, independent of the bubble breakage.

D. One-dimensional one-group model

The simplest form of the interfacial area transport equation is the one-dimensional formulation obtained by applying cross-sectional area averaging over Eq.(11-60)

$$\begin{aligned}&\frac{\partial \langle a_i \rangle}{\partial t} + \frac{\partial}{\partial z}\left(\langle a_i \rangle \langle\langle v_{i,z} \rangle\rangle_a\right) = \left(\langle \phi_{RC} \rangle + \langle \phi_{WE} \rangle + \langle \phi_{TI} \rangle\right) \\ &+ \frac{2}{3}\left(\frac{\langle a_i \rangle}{\langle \alpha_g \rangle}\right)\left[\frac{\partial \langle \alpha_g \rangle}{\partial t} + \frac{\partial}{\partial z}\left(\langle \alpha_g \rangle \langle\langle v_{g,z} \rangle\rangle\right)\right].\end{aligned} \tag{11-80}$$

Due to the uniform bubble size assumption, the area-averaged bubble interface velocity weighted by interfacial area concentration, $\langle\langle v_{i,z}\rangle\rangle_a$, is given by

$$\langle\langle v_{i,z}\rangle\rangle_a \equiv \frac{\langle a_i v_{i,z}\rangle}{\langle a_i\rangle} \approx \frac{\langle \alpha_g v_{g,z}\rangle}{\langle \alpha_g\rangle} \equiv \langle\langle v_{g,z}\rangle\rangle. \tag{11-81}$$

This is the same as the conventional area-averaged gas velocity weighted by void fraction, if the internal circulation in the bubble is neglected. The exact mathematical expressions for the area-averaged source and sink terms would involve many covariances that may further complicate the one-dimensional problem. However, since these local terms were originally obtained from a finite volume element of the mixture, the functional dependence of the area-averaged source and sink terms on the averaged parameters should be approximately the same if the hydraulic diameter of the flow path is considered as the length scale of the finite element. Therefore, Eqs.(11-70), (11-75) and (11-79), with the parameters averaged within the cross-sectional area, are still applicable for the area-averaged source and sink terms in Eq.(11-80).

In Eqs.(11-70), (11-75) and (11-79), the energy dissipation rate per unit mixture mass should be specified. In a complete two-fluid model, ε comes from its own constitutive relation such as the two-phase k - ε model (Lopez de Bertodano et al., 1994). For one-dimensional analysis, however, this term can be approximated by a simple algebraic equation as

$$\langle\varepsilon\rangle = \frac{f_{TW}}{2D_H}\langle v_m\rangle^3 \tag{11-82}$$

where v_m, D_H and f_{TW} are the mean mixture velocity, the hydraulic diameter of the flow path and the two-phase friction factor.

1.2.2 Source and sink terms modeled by Hibiki and Ishii (2000a)

Hibiki and Ishii (2000a; 2002c) discussed the contribution of wake entrainment to the interfacial area transport. Wake entrainment would play an important role in the bubbly-to-slug transition, slug and churn-turbulent flows. It may also be important for bubbly flow in a small diameter tube or for very low flow conditions as the lateral fluctuation of bubbles is small. However, for relatively high flow conditions, even bubbles captured in the wake region would easily leave the wake region due to liquid turbulence, resulting in a minor contribution of wake entrainment to the interfacial area transport. Thus, Hibiki and Ishii (2000a) dropped the wake entrainment term from the interfacial area transport equation in an adiabatic bubbly flow, and considered two terms of coalescence due to random collisions driven by liquid turbulence and breakup due to the impact of turbulent eddies.

A. Bubble coalescence due to random collision

The bubble coalescence is considered to occur due to the bubble random collision induced by turbulence in a liquid phase. For the estimation of bubble-bubble collision frequency, it is assumed that the movement of bubbles behaves like ideal gas molecules (Coulaloglou and Tavlarides, 1977). Following the kinetic theory of gases (Loeb, 1927), the bubble random collision frequency, f_{RC}, can be expressed by assuming the same velocity of bubbles, u_C, as a function of surface available to the collision, S_C, and volume available to the collision, U_C

$$f_{RC} = \frac{u_C S_C}{4U_C}. \tag{11-83}$$

Taking account of the excluded volume for bubbles, the surface and volume are given by

$$S_C = 4\pi\left(N_b - 1\right)D_b^2 \cong 4\pi N_b D_b^2 = V\frac{24\alpha_g}{D_b} \tag{11-84}$$

$$\begin{aligned} U_C &= V\left(1 - \beta_C \frac{2}{3}\pi n_b D_b^3\right) \\ &= 4\beta_C V\left(\alpha_{C,max} - \alpha_g\right),\ \alpha_{C,max} \equiv 1/4\beta_C \end{aligned} \tag{11-85}$$

where N_b, D_b, V, n_b and α_g denote the number of bubbles, the bubble diameter, the control volume, the bubble number density and the void fraction, respectively. The variable β_C (≤ 1) is introduced into the excluded volume in order to take account of the overlap of the excluded volume for high void fraction region. Although it may be a function of the void fraction, it is treated as a constant for simplicity. The distortion caused by this assumption will be adjusted by a tuning parameter in a final equation of the bubble coalescence rate as introduced later.

The mean fluctuation velocity difference between two points D_b in the inertial subrange of isotropic turbulence is given by (Hinze, 1959)

$$u_b = 1.4\left(\varepsilon D_b\right)^{1/3} \tag{11-86}$$

where ε denotes the energy dissipation. Taking account of the relative motion between bubbles, the average bubble velocity is given by

$$u_C = \gamma_C\left(\varepsilon D_b\right)^{1/3} \tag{11-87}$$

where γ_C is a constant.

The collision frequency will increase to infinity as the void fraction approaches to maximum void fraction, $\alpha_{C,max}$. Since 74.1 % of the volume is actually occupied by identical spheres close-packed according to a face-centered cubic lattice, $\alpha_{C,max}$ may be assumed to be 0.741. Finally, we obtain

$$f_{RC} = \frac{\gamma_C' \alpha_g \varepsilon^{1/3}}{D_b^{2/3}\left(\alpha_{C,max} - \alpha_g\right)} \tag{11-88}$$

where γ_C' is an adjustable valuable.

In order to obtain the bubble coalescence rate, it is necessary to determine a coalescence efficiency. Coulaloglou and Tavlarides (1977) gave an expression for the coalescence efficiency, λ_C, as a function of a time required for coalescence of bubbles, t_C, and a contact time for the two bubbles τ_C

$$\lambda_C = \exp\left(-\frac{t_C}{\tau_C}\right). \tag{11-89}$$

The time required for coalescence of bubbles was given by Oolman and Blanch (1986a; 1986b) for the thinning of the liquid film between bubbles of equal size as

$$t_C = \frac{1}{8}\sqrt{\frac{\rho_f D_b^3}{2\sigma}} \ln\frac{\delta_{init}}{\delta_{crit}} \tag{11-90}$$

where ρ_f, σ, δ_{init} and δ_{crit} are, respectively, the liquid density, interfacial tension, the initial film thickness, and the critical film thickness where rupture occurs. Levich (1962) derived the contact time in turbulent flows from dimensional consideration.

$$\tau_C = \frac{r_b^{2/3}}{\varepsilon^{1/3}} \tag{11-91}$$

where r_b is the bubble radius. Finally, we obtain

$$\lambda_C = \exp\left(-\frac{K_C \rho_f^{1/2} D_b^{5/6} \varepsilon^{1/3}}{\sigma^{1/2}}\right) \text{ where } K_C \equiv 2^{-17/6} \ln \frac{\delta_{init}}{\delta_{crit}}. \tag{11-92}$$

Kirkpatrick and Locket (1974) estimated the initial thickness of the film in air-water systems to be 1×10^{-4} m, whereas the final film thickness was typically taken as 1×10^{-8} m (Kim and Lee, 1987). Thus, the experimental coefficient, K_C, is determined to be 1.29 for an air-water system.

The decrease rate of the interfacial area concentration, ϕ_{RC}, is then expressed as

$$\begin{aligned} \phi_{RC} &= -\frac{1}{3\psi}\left(\frac{\alpha_g}{a_i}\right)^2 f_{RC} n_b \lambda_C \\ &= -\frac{\Gamma_{RC} \alpha_g^2 \varepsilon^{1/3}}{D_b^{5/3}\left(\alpha_{C,\max} - \alpha_g\right)} \exp\left(-\frac{K_C \rho_f^{1/2} D_b^{5/6} \varepsilon^{1/3}}{\sigma^{1/2}}\right). \end{aligned} \tag{11-93}$$

The adjustable variable, Γ_{RC}, would certainly be a function of the overlap of the excluded volume, the bubble deformation, and the bubble velocity distribution. However, the adjustable variable might be assumed to be a constant for simplicity and is determined experimentally to be 0.0314 for bubbly flow.

B. Bubble breakup due to turbulent impact

The bubble breakup is considered to occur due to the collision of the turbulent eddy with the bubble. For the estimation of bubble-eddy collision frequency, it is assumed that the movement of eddies and bubbles behaves like ideal gas molecules (Coulaloglou and Tavlarides, 1977). Furthermore, the following assumptions are made for the modeling of the bubble-eddy collision rate (Prince and Blanch, 1990): (i) the turbulence is isotropic; (ii) the eddy size D_e of interest lies in the inertial subrange; (iii) the eddy with the size from $c_e D_b$ to D_b can break up the bubble with the size of D_b, since larger eddies have the tendency to transport the bubble rather than to break it and smaller eddies do not have enough energy to break it. Azbel and

Athanasios (1983) developed the following expression for the number of eddies as a function of wave number.

$$\frac{dN_e(k_e)}{dk_e} = 0.1k_e^2 \tag{11-94}$$

where $N_e(k_e)$ denotes the number of eddies of wave number $k_e (= 2/D_e)$ per volume of fluid. Here, the number of eddies of wave number per volume of two-phase mixture, $n_e(k_e)$, is given by

$$n_e(k_e) = N_e(k_e)(1-\alpha_g). \tag{11-95}$$

Following the kinetic theory of gases (Loeb, 1927), the bubble-eddy random collision frequency, f_{TI}, can be expressed by assuming the same velocity of bubbles, u_B, as a function of the surface available to the collision, S_B, and the volume available to the collision U_B

$$f_{TI} = \frac{u_B S_B}{4U_B}. \tag{11-96}$$

Taking account of the excluded volume for the bubbles and eddies, the surface and volume are given by

$$\begin{aligned} S_B &= \frac{\int 4\pi(N_b - 1)\left(\frac{D_b}{2} + \frac{D_e}{2}\right)^2 dn_e}{\int dn_e} \\ &= 4\pi N_b D_b^2 \cdot F_S(c_e) = V\frac{24\alpha_g}{D_b}F_S(c_e) \end{aligned} \tag{11-97}$$

$$\begin{aligned} U_B &= V\left(1 - \frac{\beta_B \int \frac{2}{3}\pi n_b \left(\frac{D_b}{2} + \frac{D_e}{2}\right)^3 dn_e}{\int dn_e}\right) \\ &= V\left(1 - \beta_B \frac{2}{3}\pi n_b D_b^3 \cdot F_V(c_e)\right) \end{aligned} \tag{11-98}$$

$$= 4\beta_B F_V(c_e) V(\alpha_{B,max} - \alpha_g)$$

where $F_S(c_e)$ and $F_V(c_e)$ are functions of c_e defined by D_e/D_b. The variable β_B (≤ 1) is introduced into the excluded volume in order to take account of the overlap of the excluded volume for high void fraction region. Although it may be a function of the void fraction, it is treated as a constant for simplicity. The distortion caused by this assumption will be adjusted by a tuning parameter in a final equation of the bubble breakup rate as introduced later.

According to Kolmogorov's Law (Azbel, 1981), for the inertial subrange of the energy spectrum, the eddy velocity, u_e, is given as

$$u_e^2 = 8.2\left(\varepsilon/k_e\right)^{2/3} \quad \text{or} \quad u_e = 2.3\left(\varepsilon D_e\right)^{1/3}. \tag{11-99}$$

Here, taking account of the relative motion between bubble and eddy, the averaged relative velocity, u_B, can be expressed as

$$u_B = \gamma_B(c_e)\left(\varepsilon D_b\right)^{1/3} \tag{11-100}$$

where $\gamma_B(c_e)$ is a function of c_e. Finally, we obtain

$$f_B = \frac{\gamma_B'(c_e)\alpha_g \varepsilon^{1/3}}{D_b^{2/3}\left(\alpha_{B,max} - \alpha_g\right)} \tag{11-101}$$

where $\gamma_B'(c_e)$ and $\alpha_{B,max}$ are an adjustable variable depending on c_e and maximum allowable void fraction, respectively. The maximum allowable void fraction, $\alpha_{B,max}$, in Eq.(11-101) can approximately be taken at the same value as $\alpha_{C,max}$, namely, 0.741, if eddies with almost the same size of bubbles are assumed to break up the bubbles. Consequently, the functional form of the frequency of the bubble-eddy random collision, Eq.(11-101) looks similar to that of the frequency of the bubble-bubble random collision, Eq.(11-88).

In order to obtain the bubble breakup rate, it is necessary to determine a breakup efficiency, λ_B. The breakup efficiency is given in terms of the average energy of a single eddy, E_e, and the average energy required for bubble breakup, E_B, as (Prince and Blanch, 1990; Coulaloglou and Tavlarides, 1977; Tsouris and Tavlarides, 1994)

$$\lambda_B = \exp\left(-\frac{E_B}{E_e}\right). \tag{11-102}$$

For binary breakage, that is, the bubble breaks into two bubbles, the required energy, E_B, is simply calculated as the average value of the energy required for breakage into a small and a large daughter bubble as follows

$$E_B = \pi\sigma D_{b,\max}^2 + \pi\sigma D_{b,\min}^2 - \pi\sigma D_b^2. \tag{11-103}$$

The average values of the breakup energy for two extreme cases are calculated by averaging Eq.(11-103) from $D_{b,max} = D_b/2^{1/3}$ ($D_{b,min} = D_b/2^{1/3}$) to $D_{b,max} = D_b$ ($D_{b,min} = 0$) to be $0.200\pi\sigma D_b^2$ and by setting $D_{b,max} = D_{b,min} = D_b/2^{1/3}$ to be $0.260\pi\sigma D_b^2$. Thus, the breakup energy, E_B, is approximated to be $0.230\pi\sigma D_b^2$ by averaging the breakup energies for two extreme cases. It should be noted here that the relative difference between $E_B(=0.230\pi\sigma D_b^2)$ obtained by averaging Eq.(11-103) and $E_B(=0.260\pi\sigma D_b^2)$ assuming the binary breakage into two equal-size bubbles is about 13 %. Therefore, the assumption on the size of small and large daughter bubbles may not affect the estimation of E_B significantly.

The average energy of single eddies acting on the bubble breakup is simply calculated from

$$E_e = \frac{\int_{n_{e,\min}}^{n_{e,\max}} e\, dn_e}{\int_{n_{e,\min}}^{n_{e,\max}} dn_e} \tag{11-104}$$

where e is the energy of a single eddy given by

$$e = \frac{1}{2} m_e u_e^2. \tag{11-105}$$

In this, m_e is the mass per a single eddy. From Eqs.(11-94), (11-95), (11-99), (11-104) and (11-105), the average energy of single eddies acting on the bubble breakup is then given by

$$E_e = \frac{\int_{n_{e,\min}}^{n_{e,\max}} e dn_e}{\int_{n_{e,\min}}^{n_{e,\max}} dn_e} = \frac{0.546\pi\rho_f \varepsilon^{2/3}(1-\alpha_g)\int_{k_{e,\min}}^{k_{e,\max}} k_e^{-5/3} dk_e}{0.1(1-\alpha_g)\int_{k_{e,\min}}^{k_{e,\max}} k_e^2 dk_e}$$

$$= 1.93\pi\rho_f \varepsilon^{2/3} D_b^{11/3} \frac{1-c_e^{2/3}}{c_e^{-3}-1}. \tag{11-106}$$

Prince and Blanch (1990) set the minimum eddy size, which would not cause bubble breakup, at eddies smaller than 20 % of the bubble size, c_e^3=0.2. Thus, the average energy of single eddies is expressed by

$$E_e = 0.145\pi\rho_f \varepsilon^{2/3} D_b^{11/3}. \tag{11-107}$$

The final form of the breakup efficiency is then given by

$$\lambda_B = \exp\left(-\frac{K_B \sigma}{\rho_f D_b^{5/3} \varepsilon^{2/3}}\right) \tag{11-108}$$

where K_B is a constant to be 1.59 (=0.230/0.145).

The increase rate of the interfacial area concentration, ϕ_{TI}, is then expressed as

$$\phi_{TI} = \frac{1}{3\psi}\left(\frac{\alpha_g}{a_i}\right)^2 f_{TI} n_e \lambda_B$$

$$= \frac{\Gamma_B \alpha_g \left(1-\alpha_g\right)\varepsilon^{1/3}}{D_b^{5/3}\left(\alpha_{B,\max}-\alpha_g\right)} \exp\left(-\frac{K_B \sigma}{\rho_f D_b^{5/3} \varepsilon^{2/3}}\right) \tag{11-109}$$

where Γ_{TI} is an adjustable variable. The adjustable variable, Γ_B, would certainly be a function of the overlap of the excluded volume, the bubble deformation, the bubble velocity distribution, and the ratio of eddy size to bubble size. However, the adjustable variable might be assumed to be a constant for simplicity and is determined experimentally to be 0.0209 for bubbly flow.

It should be noted here that for one-dimensional analysis the energy dissipation rate per unit mass is simply obtained from the mechanical energy equation (Bello, 1968) as

$$\langle \varepsilon \rangle = \frac{\langle j \rangle}{\rho_m}\left(-\frac{dP}{dz}\right)_F \tag{11-110}$$

where j, ρ_m, P and z denote the mixture volumetric flux, the mixture density, the pressure, and the axial position from the test section inlet, respectively.

1.2.3 Source and sink terms modeled by Hibiki et al. (2001b)

Hibiki et al. (2001b) discussed the main mechanism of the interfacial area transport in a relatively small diameter tube at low liquid velocity where the bubble breakup is negligible. Here, a relatively small diameter tube is defined as a tube with a relatively high bubble size-to-pipe diameter ratio. In such a relatively small diameter tube, the radial bubble movement would be restricted due to the presence of the wall resulting in insignificant bubble random collision, whereas the bubbles are aligned along the flow direction resulting in significant wake entrainment. Thus, Hibiki et al. (2001b) developed the sink term due to the bubble coalescence considering the dependence of the bubble coalescence mechanism on the tube diameter.

A. Bubble coalescence due to random collision

The same model as in Hibiki and Ishii (2000a) was used.

$$\phi_{RC} = -\frac{\Gamma_{RC}\alpha_g^2\varepsilon^{1/3}}{D_b^{5/3}\left(\alpha_{C,\max}-\alpha_g\right)}\exp\left(-\frac{K_C\rho_f^{1/2}D_b^{5/6}\varepsilon^{1/3}}{\sigma^{1/2}}\right) \tag{11-111}$$

B. Bubble coalescence due to wake entrainment

The model was developed by modifying the model proposed by Wu et al. (1998).

$$\phi_{WE} = -\Gamma_{WE}C_D^{1/3}a_i^2v_r\exp\left(-\frac{K_C\rho_f^{1/2}D_b^{5/6}\varepsilon^{1/3}}{\sigma^{1/2}}\right) \tag{11-112}$$

where Γ_{WE} and K_C are, respectively, an adjustable parameter experimentally determined to be 0.082 and the experimental constant determined to be 1.29 for an air-water system.

C. Effect of tube size on interfacial area transport mechanism

The above simple consideration suggests that the major mechanism of the bubble coalescence in a relatively small diameter tube would be wake entrainment. However, experimental data (Hibiki and Ishii, 1999; Hibiki et al., 2001a) suggested that the bubble coalescence mechanism of bubbly flows in medium pipes (25.4 mm $\leq D \leq$ 50.8 mm) could successfully be modeled by considering the bubble random collision induced by liquid turbulence. Thus, the bubble coalescence mechanism is likely to be dependent on the ratio of bubble diameter to tube diameter, D_b/D. For example, a trailing bubble should certainly exist in a projected area of a leading bubble for D_b/D=0.5. Also, if the leading bubble rises in the center of the channel, the trailing bubble should certainly exist in the projected area of the leading bubble even for D_b/D=0.33. In a small diameter tube, since the radial bubble movement would be restricted due to the presence of the wall, the bubble coalescence due to bubble random collision is unlikely to occur. Thus, as the ratio of bubble diameter to tube diameter increases, the dominant bubble coalescence mechanism is expected to change from the bubble random collision to the wake entrainment. This suggests the following functional form of the sink term for bubbly flows in small and medium tubes.

$$\phi_C = \phi_{RC} \exp\left\{f\left(\frac{D_b}{D}\right)\right\} + \phi_{WE}\left[1 - \exp\left\{f\left(\frac{D_b}{D}\right)\right\}\right] \tag{11-113}$$

The function, $f\left(D_b/D\right)$, may be approximated based on experimental data as

$$f\left(\frac{D_b}{D}\right) = -1000\left(\frac{D_b}{D}\right)^5. \tag{11-114}$$

The interfacial area transport equation taking account of the tube size effect would be promising for predicting the interfacial area transport of bubbly flows in small and medium tubes.

1.3 Modeling of source and sink terms in two-group interfacial area transport equation

The interfacial structures in different flow regimes change dramatically. For cap bubbly, slug and churn-turbulent flows, bubbles are divided into two groups according to their geometrical and physical characteristics. The spherical and distorted bubbles are categorized as Group 1, and the cap, slug and churn-turbulent bubbles are categorized as Group 2. These two groups are subject to different coalescence/disintegration mechanisms. Therefore, a two-group interfacial area transport equation needs to be introduced and the bubble coalescence and breakup processes should be modeled properly. In this section, some two-group models are explained briefly.

1.3.1 Source and sink terms modeled by Hibiki and Ishii (2000b)

Hibiki and Ishii (2000b) developed the two-group interfacial area transport equation at adiabatic bubbly-to-slug transition flow in a moderate diameter tube and evaluated it using a vertical air-water flow data taken in a 50.8 mm-diameter tube (Hibiki et al., 2001a). In what follows, the classification of interfacial area transport mechanisms and the modeled source and sink terms are explained briefly.

A. Classification of interfacial area transport mechanisms

The boundary between Group-1 and Group-2 bubbles can be determined by (Ishii and Zuber, 1979)

$$D_{crit} = 4\sqrt{\frac{\sigma}{g\Delta\rho}} \tag{11-115}$$

where D_{crit} is the volumetric equivalent diameter of a bubble at the boundary between Group-1 and Group-2 bubbles. Equation (11-115) gives the value of about 10 mm for air-water system at atmospheric pressure.

To model the integral source and sink terms in two-group interfacial area transport equation caused by bubble coalescence and breakup, the possible combinations of bubble interactions can be classified into eight categories in terms of the belonging bubble group (see Fig.11-1): (1) the coalescence of bubbles (Group 1) into a bubble (Group 1); (2) the breakup of a bubble (Group 1) into bubbles (Group 1); (3) the coalescence of bubbles (Group 1 and 2) into a bubble (Group 2); (4) the breakup of a bubble (Group 2) into bubbles (Group 1 and 2); (5) the coalescence of bubbles (Group 1) into a bubble (Group 2); (6) the breakup of a bubble (Group 2) into bubbles (Group 1); (7) the coalescence of bubbles (Group 2) into a bubble (Group 2); and (8) the breakup of a bubble (Group 2) into bubbles (Group 2). As summarized in Table 11-1, Hibiki and Ishii (2000b) considered the three major bubble interactions: (1) the coalescence due to random collisions driven by turbulence; (2) the coalescence due to wake entrainment; and (3) the breakup upon the impact of turbulent eddies. They assumed that the bubble coalescence due to the shearing-off of cap or slug bubbles might be insignificant at bubbly-to-slug transition flow. They also assumed the bubble breakup due to surface instability could be neglected in a moderate

Table 11-1. List of intra- and inter-group interaction mechanisms in the model by Hibiki and Ishii (2000b)

Symbols	Mechanisms	Interaction	Parameters
$\phi_{RC}^{(1)}$	Random collision	(1)+(1)→(1)	$\Gamma_{RC,1}$=0.351, $K_{RC,1}$=0.258
$\phi_{WE}^{(12,2)}$	Wake entrainment	(1)+(2)→(1)	$\Gamma_{WE,12}$=24.9, $K_{WE,12}$=0.460
$\phi_{WE}^{(2)}$	Wake entrainment	(2)+(2)→(2)	$\Gamma_{WE,2}$=63.7, $K_{WE,2}$=0.258
$\phi_{TI}^{(1)}$	Turbulent impact	(1)→(1)+(1)	$\Gamma_{TI,1}$=1.12, $K_{TI,1}$=6.85
$\phi_{TI}^{(2)}$	Turbulent impact	(2)→(2)+(1)	$\Gamma_{TI,12}$=317, $K_{TI,12}$=11.7
$\phi_{TI}^{(2,12)}$	Turbulent impact	(2)→(2)+(2)	$\Gamma_{TI,2}$=4.26, $K_{TI,2}$=6.85

diameter tube where the tube size is smaller than the limit of the bubble breakup due to surface instability.

Hibiki and Ishii (2000b) developed the two-group model with the necessary inter-group coupling terms due to turbulent impact and wake entrainment as well as source and sink terms due to wake entrainment and turbulent impact in Group 2 (see Table 11-1). Here, some other mechanisms such as coalescence due to random collision between cap bubbles are excluded because the model is developed for bubbly-to-slug transition flow in a moderate diameter tube. In such a condition, cap and slug bubbles would rise around the tube center resulting in a minor role of random collision between cap bubbles. Two more mechanisms are also omitted in this model. They are interchange terms due to the complete breakup of a cap bubble (Group 2) into small bubbles (Group 1) and the coalescence of small bubbles (Group 1) into a cap bubble (Group 2). Since the ratio in diameter of cap bubbles to small bubbles is about 10 to 20 in the experimental conditions of the database (Hibiki et al., 2001a), these interchanges of bubbles are unlikely to occur. Eventually, six terms listed in Table 11-1 are considered as source and sink terms in the two-group interfacial area transport equations to be applied at the bubbly-to-slug transition flow in a moderate diameter tube.

B. Simplified two-group interfacial area transport equation

Here, an isothermal flow condition is assumed. In addition, the coefficient χ is neglected. It is due to the fact that there is very little portion of bubbles at the group boundary that could transfer to the other group simply due to expansion. Thus, two-group interfacial area transport equation is simplified as

$$\frac{\partial\left(\alpha_{g1}\rho_g\right)}{\partial t}+\nabla\cdot\left(\alpha_{g1}\rho_g\boldsymbol{v}_{g1}\right)=-\rho_g\left(\eta_{WE}^{(12,2)}+\eta_{TI}^{(2,12)}\right) \tag{11-116}$$

$$\frac{\partial a_{i1}}{\partial t}+\nabla\cdot\left(a_{i1}\boldsymbol{v}_{i1}\right)=\frac{2}{3}\frac{a_{i1}}{\alpha_{g1}}\left[\frac{\partial\alpha_{g1}}{\partial t}+\nabla\cdot\left(\alpha_{g1}\boldsymbol{v}_{g1}\right)\right] \tag{11-117}$$
$$+\phi_{RC}^{(1)}+\phi_{WE}^{(12,2)}+\phi_{TI}^{(1)}+\phi_{TI}^{(2,12)}$$

$$\frac{\partial a_{i2}}{\partial t}+\nabla\cdot\left(a_{i2}\boldsymbol{v}_{i2}\right)=\frac{2}{3}\frac{a_{i2}}{\alpha_{g2}}\left[\frac{\partial\alpha_{g2}}{\partial t}+\nabla\cdot\left(\alpha_{g2}\boldsymbol{v}_{g2}\right)\right] \tag{11-118}$$
$$+\phi_{WE}^{(2)}+\phi_{TI}^{(2)}.$$

C. Summary of modeled sink and source terms

The modeled sink and source terms are summarized as follows. In the one-dimensional formulation, all the two-phase parameters, such as α_g, a_i, and D_{Sm}, are area-averaged values. For simplicity, the $\langle\ \rangle$ signs standing for the area-average are omitted in the following formulations.

Bubble coalescence due to random collision

$$\phi_{RC}^{(1)}=-\frac{\Gamma_{RC,1}\alpha_{g1}^2\varepsilon^{1/3}}{D_{b,1}^{5/3}\left(\alpha_{C,\max}-\alpha_g\right)}\exp\left(-\frac{K_{RC,1}\rho_f^{1/2}D_{b,1}^{5/6}\varepsilon^{1/3}}{\sigma^{1/2}}\right) \tag{11-119}$$

Bubble coalescence due to wake entrainment

$$\phi_{WE}^{(12,2)}=-\frac{\Gamma_{WE,12}\alpha_{g1}\alpha_{g2}}{D_{b,1}D_{b,2}}\left(v_{g2}-v_f\right)$$
$$\times\exp\left\{-K_{WE,12}\sqrt[6]{\frac{\rho_f^3\varepsilon^2}{\sigma^3}\left(\frac{D_{b,1}D_{b,2}}{D_{b,1}+D_{b,2}}\right)^5}\right\} \tag{11-120}$$

$$\phi_{WE}^{(2)} = -\frac{\Gamma_{WE,2}\alpha_{g2}^{2}}{D_{b,2}^{2}}\left(v_{g2} - v_{f}\right)\exp\left\{-K_{WE,2}\sqrt[6]{\frac{D_{b,2}^{5}\rho_{f}^{3}\varepsilon^{2}}{\sigma^{3}}}\right\} \tag{11-121}$$

$$\eta_{WE}^{(12,2)} = \left(\frac{\alpha_{g1}}{a_{i,1}}\right)^{3}\frac{3\Gamma_{WE,12}\alpha_{g1}\alpha_{g2}}{D_{b,1}^{3}D_{b,2}}\left(v_{z,2} - v_{f}\right) \times\exp\left\{-K_{WE,12}\sqrt[6]{\frac{\rho_{f}^{3}\varepsilon^{2}}{\sigma^{3}}\left(\frac{D_{b,1}D_{b,2}}{D_{b,1} + D_{b,2}}\right)^{5}}\right\} \tag{11-122}$$

Bubble breakup due to turbulent impact

$$\phi_{TI}^{(1)} = \frac{\Gamma_{TI,1}\alpha_{g1}\left(1 - \alpha_{g}\right)\varepsilon^{1/3}}{D_{b,1}^{5/3}\left(\alpha_{TI,\max} - \alpha_{g}\right)}\exp\left(-\frac{K_{TI,1}\sigma}{\rho_{f}D_{b,1}^{5/3}\varepsilon^{2/3}}\right) \tag{11-123}$$

$$\phi_{TI}^{(2,12)} = \frac{\Gamma_{TI,12}\alpha_{g2}\left(1 - \alpha_{g}\right)\varepsilon^{1/3}}{D_{b,2}^{5/3}\left(\alpha_{TI,\max} - \alpha_{g}\right)} \times\exp\left(-\frac{K_{TI,12}\sigma\left\{\left(D_{b,2}^{3} - D_{b,1}^{3}\right)^{2/3} + \left(D_{b,1}^{2} - D_{b,2}^{2}\right)\right\}}{\rho_{f}D_{b,2}^{11/3}\varepsilon^{2/3}}\right) \tag{11-124}$$

$$\phi_{TI}^{(2)} = \frac{\Gamma_{TI,2}\alpha_{g2}\left(1 - \alpha_{g}\right)\varepsilon^{1/3}}{D_{b,2}^{5/3}\left(\alpha_{TI,\max} - \alpha_{g}\right)}\exp\left(-\frac{K_{TI,2}\sigma}{\rho_{f}D_{b,2}^{5/3}\varepsilon^{2/3}}\right) \tag{11-125}$$

$$\eta_{TI}^{(2,12)} = \left(\frac{\alpha_{g1}}{a_{i,1}}\right)^{3}\frac{3\Gamma_{TI,12}\alpha_{g2}\left(1 - \alpha_{g}\right)\varepsilon^{1/3}}{D_{b,2}^{11/3}\left(\alpha_{TI,max} - \alpha_{g}\right)} \times\exp\left(-\frac{K_{TI,12}\sigma\left\{\left(D_{b,2}^{3} - D_{b,1}^{3}\right)^{2/3} + \left(D_{b,1}^{2} - D_{b,2}^{2}\right)\right\}}{\rho_{f}D_{b,2}^{11/3}\varepsilon^{2/3}}\right) \tag{11-126}$$

The values of the coefficients in the source and sink terms are listed in Table 11-1.

1.3.2 Source and sink terms modeled by Fu and Ishii (2002a)

Fu and Ishii (2002a) developed the two-group interfacial area transport equation for bubbly flow, slug flow, and churn-turbulent flow in a moderate diameter tube and evaluated it using a vertical air-water flow data taken in a 50.8 mm-diameter tube. In what follows, the classification of interfacial area transport mechanisms and the modeled source and sink terms are explained briefly.

A. Classification of interfacial area transport mechanisms

Fu and Ishii (2002a) adopted five major bubble interactions: (1) the coalescence due to random collisions driven by turbulence; (2) the coalescence due to wake entrainment; (3) the breakup upon the impact of turbulent eddies; (4) the breakup due to shearing-off; and (5) the breakup of large-cap bubbles due to flow instability on the bubble surface. In view of the complexity for incorporating all source and sink terms into the interfacial area transport equation and the difficulty of experimental verification, they performed an analysis to simplify the interaction terms according to their nature and the order of magnitudes. It is verified from experiments that the majority of inter-group interactions is caused by the wake entrainment and the shearing-off of Group-1 bubbles to and from the Group-2 bubbles (Fu and Ishii, 2002a). In addition, the wake entrainment between Group-2 bubbles predominantly governs the Group-2 bubble number which significantly affects the flow structure and intensiveness of inter-group interactions. The Group-2 bubble disintegration due to surface instability is significantly enhanced by the high turbulent intensity and active eddy-bubble interaction in the wake region of the slug bubbles.

The random collision between Group-1 and Group-2 bubbles may be included into the wake entrainment of Group-1 into Group-2 bubbles due to the similar nature. In addition, the contribution from the collision of Group-1 bubble at the head of Group-2 bubble could be small due to the lower Group-1 bubble number density and lower turbulent intensity outside the wake region. Meanwhile, the random collision between Group-2 bubbles could also be negligible in a moderate diameter flow (2.5 cm $\leq D \leq$ 10 cm) because the predominant interaction is normally within a wake region of the leading bubble, and the coalescence mechanism can be treated as wake entrainment between Group-2 bubbles. Similarly, the turbulent disintegration that results in a generation of the Group-1 bubble from the Group-2 bubble can be seen as part of the shearing-off effect and might not

Table 11-2. List of intra- and inter-group interaction mechanisms in the model by Fu and Ishii (2002a; 2002b)

Symbols	Mechanisms	Interaction	Parameters
$\phi_{RC}^{(1)}$	Random collision	(1)+(1)→(1)	C_{RC}=0.0041, C_T=3.0
$\phi_{RC}^{(11,2)}$	Random collision	(1)+(1)→(2)	$\alpha_{g1,max}$=0.75
$\phi_{WE}^{(1)}$	Wake entrainment	(1)+(1)→(1)	C_{WE}=0.002, $C_{WE}^{(12,2)}$=0.015
$\phi_{WE}^{(11,2)}$	Wake entrainment	(1)+(1)→(2)	$C_{WE}^{(2)}$=10.0
$\phi_{WE}^{(12,2)}$	Wake entrainment	(1)+(2)→(2)	C_{TI}=0.0085, We_{crit}=6.0
$\phi_{WE}^{(2)}$	Wake entrainment	(2)+(2)→(2)	C_{SO}=0.031, γ_{SO}=0.032
$\phi_{TI}^{(1)}$	Turbulent impact	(1)→(1)+(1)	β_{SO}=1.6
$\phi_{TI}^{(2)}$	Turbulent impact	(2)→(2)+(2)	
$\phi_{SO}^{(2,12)}$	Shearing-off	(2)→(2)+(1)	

need to be modeled individually. Furthermore, the disintegration of Group-2 bubbles due to surface instability is considered to be very small and can be combined with the disintegration of Group-2 bubbles induced by turbulent impact. Eventually, nine terms listed in Table 11-2 are considered as sink and source terms in the two-group interfacial area transport equations to be applied at the bubbly, slug and churn-turbulent flows in a moderate diameter tube.

B. Simplified two-group interfacial area transport equation

Here, an isothermal flow condition is assumed. In addition, the coefficient χ is neglected. It is due to the fact that there is very little portion of bubbles at the group boundary that could transfer to the other group simply due to expansion. Thus, two-group interfacial area transport equation is simplified as

$$\frac{\partial\left(\alpha_{g1}\rho_g\right)}{\partial t}+\nabla\cdot\left(\alpha_{g1}\rho_g\boldsymbol{v}_{g1}\right)$$
$$=-\rho_g\left(\eta_{RC}^{(11,2)}+\eta_{WE}^{(11,2)}+\eta_{WE}^{(12,2)}+\eta_{SO}^{(2,12)}\right) \quad (11\text{-}127)$$

$$\frac{\partial a_{i1}}{\partial t}+\nabla\cdot\left(a_{i1}\boldsymbol{v}_{i1}\right)=\frac{2}{3}\frac{a_{i1}}{\alpha_{g1}}\left[\frac{\partial\alpha_{g1}}{\partial t}+\nabla\cdot\left(\alpha_{g1}\boldsymbol{v}_{g1}\right)\right]$$
$$+\phi_{RC}^{(1)}+\phi_{RC,1}^{(11,2)}+\phi_{WE}^{(1)}+\phi_{WE,1}^{(11,2)}+\phi_{WE,1}^{(12,2)}+\phi_{TI}^{(1)}+\phi_{SO,1}^{(2,12)} \tag{11-128}$$

$$\frac{\partial a_{i2}}{\partial t}+\nabla\cdot\left(a_{i2}\boldsymbol{v}_{i2}\right)=\frac{2}{3}\frac{a_{i2}}{\alpha_{g2}}\left[\frac{\partial\alpha_{g2}}{\partial t}+\nabla\cdot\left(\alpha_{g2}\boldsymbol{v}_{g2}\right)\right]$$
$$+\phi_{WE}^{(2)}+\phi_{SO,2}^{(2,12)}+\phi_{RC,2}^{(11,2)}+\phi_{TI}^{(2)}+\phi_{WE,2}^{(11,2)}+\phi_{WE,2}^{(12,2)}. \tag{11-129}$$

C. Assumed Group-2 bubble shape and bubble number density distribution

Group-2 bubbles consist of cap and slug bubbles. Bubble shapes are subject to the wall effects when the diameter ratio $D_{b,cross}/D$ exceeds certain limits, where $D_{b,cross}$ is the bubble cross sectional diameter, and D is the tube diameter. In addition, there are two important parameters that are considered crucial for determining the bubble shape. They are the viscosity number given by $N_{\mu f}=\mu_f/\left(\rho_f\sigma\sqrt{\sigma/g\Delta\rho}\right)^{1/2}$, and the length-scale ratio number given by $D^*=D/\sqrt{\sigma/g\Delta\rho}$. According to Clift et al. (1978), when $D_{b,cross}/D\leq 0.6$, the walls cause little deformation on the cap bubble shape as in an infinite medium. In this case, the shape of cap bubbles can be closely approximated as a segment of a sphere, and the wake angle is nearly 50°. When the diameter ratio $D_{b,cross}/D$ exceeds a value of about 0.6, the tube diameter becomes the controlling length governing the frontal shape of a bubble and then the bubble is called a slug bubble. The definitions of the geometrical parameters, including cross-sectional radius, a, the bubble height, h, and the wake angle, θ_w, are shown in Fig.11-4. It is shown (Clift et al., 1978) that the slug can be considered to be composed of two parts, a rounded nose region whose shape is independent of the slug length and a near-cylindrical section that is surrounded by an annular film of the liquid. It is also verified that for $N_{\mu f}\leq 0.032$ and $D^*>10$, the viscosity and surface tension forces are negligible and the bubble shape on the potential flow theory can be well applied. The air-water flow in a moderate diameter tube ($N_{\mu f}=2.36\times10^{-3}$ and $D^*\cong 19$ for a 50.8 mm-diameter tube) satisfies the above requirement. Therefore, the bubble shape can be predicted based on an application of the Bernoulli equation (Mishima and Ishii, 1984).

A simplified bubble number density distribution is given in Fig.11-5. It is assumed that all the bubble groups have flat number density distributions in the corresponding bubble volume range. The values of the distribution

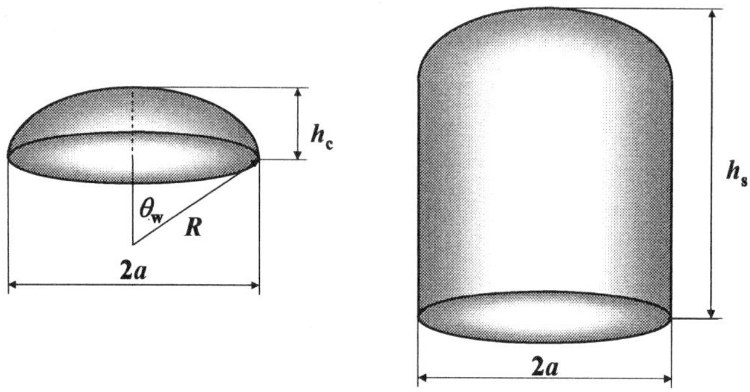

Figure 11-4. Definition of the geometrical parameters for cap and Taylor bubbles (Fu and Ishii, 2002a)

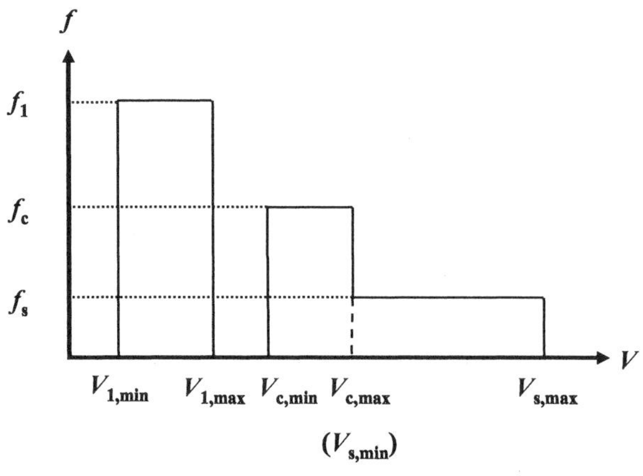

Figure 11-5. Illustration of the simplified bubble number density distribution (Fu and Ishii, 2002a)

function are denoted as f_1, f_c, and f_s for Group-1 bubbles, cap bubbles, and slug bubbles, respectively.

D. Summary of modeled source and sink terms

The modeled source and sink terms are summarized as follows. In the one-dimensional formulation, all the two-phase parameters such as α_g, a_i, and D_{Sm} are area-averaged values. For simplicity, the $\langle\ \rangle$ signs standing for the area-average are omitted in the following formulations.

Bubble coalescence due to random collision

$$\phi_{RC}^{(1)} = \left\langle \delta A_{i1}^{(11,1)} \right\rangle_R R_{RC}^{(1)} \tag{11-130}$$

$$\phi_{RC,1}^{(11,2)} = \left\langle \delta A_{i1}^{(11,2)} \right\rangle_R R_{RC}^{(1)} \tag{11-131}$$

$$\phi_{RC,2}^{(11,2)} = \left\langle \delta A_{i2}^{(11,2)} \right\rangle_R R_{RC}^{(1)} \tag{11-132}$$

$$\eta_{RC}^{(11,2)} = \left\langle \delta V^{(11,2)} \right\rangle_R R_{RC}^{(1)} \tag{11-133}$$

where

$$\begin{aligned} R_{RC}^{(1)} &= C_{RC} \left[\frac{u_t {n_{b1}}^2 {D_{Sm1}}^2}{\alpha_{g1,\max}^{1/3} \left(\alpha_{g1,\max}^{1/3} - {\alpha_{g1}}^{1/3} \right)} \right] \\ &\times \left[1 - \exp\left(-C_T \frac{\alpha_{g1,max}^{1/3} \alpha_{g1}^{1/3}}{\alpha_{g1,\max}^{1/3} - \alpha_{g1}^{1/3}} \right) \right] \end{aligned} \tag{11-134}$$

$$n_{b1} = \frac{1}{36\pi} \frac{a_{i1}^3}{\alpha_{g1}^2} \tag{11-135}$$

The turbulent fluctuation velocity (or root mean-squared velocity), u_t^2, is composed of the isotropic turbulence intensity, $u_{t,isot}^2$ and the wake turbulence intensity, $u_{t,wake}^2$, as

$$u_t^2 = u_{t,isot}^2 + u_{t,wake}^2 \tag{11-136}$$

$$u_{t,isot}^2 = \left(\varepsilon D_{Sm1} \right)^{2/3} \tag{11-137}$$

$$u_{t,wake}^2 = 0.056 C_g \left(\frac{D^3}{V_s^*} \right)^{1/2} \kappa_{fr} \tag{11-138}$$

where

$$C_g \equiv \sqrt{\frac{2g\Delta\rho}{\rho_f}} \tag{11-139}$$

$$V_s^* \equiv \frac{V_{s,min}}{V_{s,max}} \tag{11-140}$$

$$\kappa_{fr} = 1 - \exp\left(-\frac{C_{fr} V_s^{*1/2}}{D^{1/2}} \right). \tag{11-141}$$

Here, the constant C_{fr} is set at 1.8536. When $D_{Sm1} > 8.3338 \times 10^{-3}$ m,

$$\left\langle \delta A_{i1}^{(11,1)} \right\rangle_R = D_{Sm1}^2 \left[-3.142 D_{c1}^{*3} + 2.183 D_{c1}^{*5} - 0.395 D_{c1}^{*8} + 3.392 \left(0.579 D_{c1}^{*3} - 1 \right)^{8/3} \right] \tag{11-142}$$

$$\left\langle \delta A_{i1}^{(11,2)} \right\rangle_R = D_{Sm1}^2 \left[8.82 + 2.035 \left(0.579 D_{c1}^{*3} - 1 \right)^{8/3} - 5.428 D_{c1}^{*3} \right] \tag{11-143}$$

$$\left\langle \delta A_{i2}^{(11,2)} \right\rangle_R = D_{Sm1}^2 \left(6.462 - 2.182 D_{c1}^{*5} + 0.395 D_{c1}^{*8} \right) \tag{11-144}$$

$$\left\langle \delta V^{(11,2)} \right\rangle_R = D_{Sm1}^3 \xi \left(0.603 + 0.349 D_{c1}^{*3} \right) \tag{11-145}$$

$$\xi = 2\left(1 - 0.2894 D_{c1}^{*\,3}\right)^{2}. \tag{11-146}$$

Otherwise,

$$\left\langle \delta A_{i1}^{(11,1)} \right\rangle_{R} = 1.001 D_{Sm1}^{2} \tag{11-147}$$

and $\left\langle \delta A_{i1}^{(11,2)} \right\rangle_{R}$, $\left\langle \delta A_{i2}^{(11,2)} \right\rangle_{R}$, $\left\langle \delta V^{(11,2)} \right\rangle_{R}$ are all equal to zero.

Bubble coalescence due to wake entrainment

$$\phi_{WE}^{(1)} = \left\langle \delta A_{i1}^{(11,1)} \right\rangle_{R} R_{WE}^{(1)} \tag{11-148}$$

$$\phi_{WE,1}^{(11,2)} = \left\langle \delta A_{i1}^{(11,2)} \right\rangle_{R} R_{WE}^{(1)} \tag{11-149}$$

$$\phi_{WE,2}^{(11,2)} = \left\langle \delta A_{i2}^{(11,2)} \right\rangle_{R} R_{WE}^{(1)} \tag{11-150}$$

$$\phi_{WE,1}^{(12,2)} = -C_{WE}^{(12,2)} K_{WE,1}^{(12,2)} V_{s}^{*1/2} \frac{\alpha_{g1}\alpha_{g2}}{1-\alpha_{g2}} \kappa_{fr} D_{Sm1}^{-1} \tag{11-151}$$

$$\phi_{WE,2}^{(12,2)} = C_{WE}^{(12,2)} K_{WE,2}^{(12,2)} V_{s}^{*1/2} \frac{\alpha_{g1}\alpha_{g2}}{1-\alpha_{g2}} \kappa_{fr} \tag{11-152}$$

$$\phi_{WE}^{(2)} = -C_{WE}^{(2)} K_{WE}^{(2)} \alpha_{g2} \left[1 - \exp\left(\frac{-2331\alpha_{g2} V_{s}^{*2}}{D^{5}}\right)\right] \times \left[\exp\left(\frac{0.06 C_{l}\left(\alpha_{m2}/\alpha_{g2} - 1\right)}{V_{s}^{*}}\right) - 1\right]^{-1} \tag{11-153}$$

$$\eta_{WE}^{(11,2)} = \left\langle \delta V^{(11,2)} \right\rangle_R R_{WE}^{(1)} \tag{11-154}$$

$$\eta_{WE}^{(12,2)} = C_{WE}^{(12,2)} K_{WE}^{(12,2)} V_s^{*1/2} \frac{\alpha_{g1}\alpha_{g2}}{1-\alpha_{g2}} \kappa_{fr} \tag{11-155}$$

where

$$R_{WE}^{(1)} = C_{WE} C_D^{1/3} n_{b1}^2 D_{Sm1}^2 v_{r1} \tag{11-156}$$

$$v_{r1} \approx \sqrt{\frac{gD_{Sm1}}{3C_D}\frac{\Delta\rho}{\rho_f}} \tag{11-157}$$

$$C_D = \frac{2}{3} D_{Sm1} \sqrt{\frac{g\Delta\rho}{\sigma}} \left[\frac{1+17.67\left(1-\alpha_{g1}\right)^{2.6}}{18.67\left(1-\alpha_{g1}\right)^3} \right]^2 \tag{11-158}$$

$$K_{WE,1}^{(12,2)} = 3\pi C_g D^{1/2} \tag{11-159}$$

$$K_{WE,2}^{(12,2)} = 2\pi C_g D^{-1/2} \alpha_{m2}^{-1/2} \tag{11-160}$$

$$K_{WE}^{(2)} = 10.24 D^{3/2} \tag{11-161}$$

$$K_{WE}^{(12,2)} = 0.5\pi C_g D^{1/2} \tag{11-162}$$

$\left\langle \delta A_{i1}^{(11,2)} \right\rangle_R$, $\left\langle \delta A_{i1}^{(11,2)} \right\rangle_R$, $\left\langle \delta A_{i2}^{(11,2)} \right\rangle_R$, and $\left\langle \delta V^{(11,2)} \right\rangle_R$ are the same as those in Eqs.(11-142)-to-(11-145) and (11-147). The maximum cross-sectional void fraction of a slug bubble, α_{m2}, can be specified as 0.81 for most conditions and C_l is the adjustable parameter determined to be 0.1.

Bubble breakup due to turbulent impact

$$\phi_{TI}^{(1)} = \begin{cases} \dfrac{C_{TI}}{18}\left(\dfrac{u_t a_{i1}{}^2}{\alpha_{g1}}\right)\left(1-\dfrac{We_{crit}}{We^*}\right)^{1/2} \exp\left(-\dfrac{We_{crit}}{We^*}\right), \\ We^* > We_{crit} \\ 0, \ We^* \le We_{crit} \end{cases} \tag{11-163}$$

where We is the Weber number defined by

$$We^* \equiv \frac{\rho_f u_t^2 D_{Sm1}}{\sigma} \tag{11-164}$$

$$\phi_{TI}^{(2)} = C_{TI2} K_{TI}^{(2)} \alpha_{g2} \varepsilon^{1/3} V_s^* \left(\frac{1-\alpha_{g1}-\alpha_{g2}}{1-\alpha_{g2}}\right) \tag{11-165}$$

where

$$\begin{aligned} K_{TI}^{(2)} &= D^{-1}\left[1-\left(\frac{D_c}{\alpha_{2,\max}^{1/2} D}\right)^{5/3}\right] \\ &\times\left[14.38+1.57\alpha_{m2}^{-2/3}\left(\frac{D_{crit}}{D}\right)^{4/3} - 15.95\alpha_{m2}^{-1/6}\left(\frac{D_{crit}}{D}\right)^{1/3}\right] \end{aligned} \tag{11-166}$$

V_s^* can be determined by

$$D_{Sm2} = \frac{1.35D}{1+6.86V_s^* - 2.54V_s^{*2}}. \tag{11-167}$$

It is observed from experiments that for moderate and small diameter pipes, the $\phi_{TI}^{(2)}$ term is very small compared with other three mechanisms. Therefore, it could be neglected to simplify the equation for application (Fu and Ishii, 2002b).

Bubble breakup due to shearing-off

$$\phi_{SO,1}^{(2,1)} = C_{SO} K_{SO,1}^{(2,1)} \alpha_{g2} V_s^{*-4/5} \left(1 - 0.6535\kappa_{bl}\right) \xi_{SO} \kappa_{fr}^{\ 2} \tag{11-168}$$

$$\phi_{SO,2}^{(2,1)} = -C_{SO} K_{SO,2}^{(2,1)} \alpha_{g2} V_s^{*-1/5} \left(1 - 0.6474\kappa_{bl}\right) \kappa_{fr}^{4/5} \tag{11-169}$$

$$\eta_{SO}^{(2,1)} = C_{SO} K_{SO}^{(2,1)} \alpha_{g2} V_s^{*-1/5} \left(1 - 0.6474\kappa_{bl}\right) \kappa_{fr}^{4/5} \tag{11-170}$$

where

$$\xi_{SO} = \left[1 - \exp\left(-\gamma_{SO} \left(\frac{\alpha_{m2}}{\alpha_{m2} - \alpha_{g2}}\right)^{\beta_{SO}} \frac{We_{crit}}{We_1}\right)\right]^{-1} \tag{11-171}$$

$$\kappa_{bl} = \left(D^{-0.3} \alpha_{m2}^{-0.5} \nu_g^{0.2} C_g^{-0.2} V_s^{*-0.7} \kappa_{fr}^{-0.2}\right)^{1/7} \tag{11-172}$$

$$K_{SO,1}^{(2,1)} = 0.5755 C_g^{\ 2} \nu_g^{1/5} \left(\frac{\rho_f}{\sigma D}\right)^{3/5} \tag{11-173}$$

$$K_{SO,2}^{(2,1)} = 4.4332 \alpha_{g2} \nu_g^{1/5} D^{-9/5} \alpha_{m2}^{1/2} C_g^{\ 4/5} \tag{11-174}$$

$$K_{SO}^{(2,1)} = 1.1083 \nu_g^{1/5} D^{-4/5} C_g^{\ 4/5} \tag{11-175}$$

where ν_g is the kinematic viscosity of the gas phase. The values of the coefficients in the source and sink terms are listed in Table 11-2.

1.3.3 Source and sink terms modeled by Sun et al. (2004a)

Sun et al. (2004a) developed the two-group interfacial area transport equation for bubbly flow, cap-turbulent flow, and churn-turbulent flows in a confined channel and evaluated it using a vertical air-water flow data taken

Table 11-3. List of intra- and inter-group interaction mechanisms in the model by Sun et al. (2004a; 2004b)

Symbols	Mechanisms	Interaction	Parameters
$\phi_{RC}^{(1)}$	Random collision	(1)+(1)→(1)	$C_{RC}^{(1)}$=0.005, $C_{RC}^{(12,2)}$=0.005
$\phi_{RC}^{(11,2)}$	Random collision	(1)+(1)→(2)	$C_{RC}^{(2)}$=0.005 C_{RC1}=3.0
$\phi_{RC}^{(12,2)}$	Random collision	(1)+(2)→(2)	C_{RC2}=3.0
$\phi_{RC}^{(2)}$	Random collision	(2)+(2)→(2)	$C_{WE}^{(1)}$=0.002, $C_{WE}^{(12,2)}$=0.002
$\phi_{WE}^{(1)}$	Wake entrainment	(1)+(1)→(1)	$C_{WE}^{(2)}$=0.005
$\phi_{WE}^{(11,2)}$	Wake entrainment	(1)+(1)→(2)	$C_{TI}^{(1)}$=0.03, $C_{TI}^{(2)}$=0.02
$\phi_{WE}^{(12,2)}$	Wake entrainment	(1)+(2)→(2)	$We_{crit,TI1}$=6.5, $We_{crit,TI2}$=7.0
$\phi_{WE}^{(2)}$	Wake entrainment	(2)+(2)→(2)	C_{SO}=3.8×10^{-5}, C_d=4.80
$\phi_{TI}^{(1)}$	Turbulent impact	(1)→(1)+(1)	$We_{crit,SO}$=4500
$\phi_{TI}^{(2,11)}$	Turbulent impact	(2)→(1)+(1)	
$\phi_{TI}^{(2,12)}$	Turbulent impact	(2)→(1)+(2)	
$\phi_{TI}^{(2)}$	Turbulent impact	(2)→(2)+(2)	
$\phi_{SO}^{(2,12)}$	Shearing-off	(2)→(2)+(1)	
$\phi_{SI}^{(2)}$	Surface instability	(2)→(2)+(2)	

in a rectangular channel with the width, W, of 200 mm and the gap, G, of 10 mm. No stable slug flow regime was observed in the test section due to the large width of the test section (Sun et al., 2004a). In what follows, the classification of interfacial area transport mechanisms and the modeled source and sink terms are explained briefly.

A. Classification of interfacial area transport mechanisms

Sun et al. (2004a) adopted five major bubble interactions: (1) the coalescence due to random collisions driven by turbulence; (2) the coalescence due to wake entrainment; (3) the breakup upon the impact of turbulent eddies; (4) the breakup due to shearing-off; and (5) the breakup of large cap bubbles due to flow instability on the bubble surface. Fourteen terms listed in Table 11-3 are considered as source and sink terms in the two-group interfacial area transport equation to be applied at the bubbly, cap-turbulent and churn-turbulent flows in a confined channel.

B. Simplified two-group interfacial area transport equation

Here, an isothermal flow condition is assumed. Thus, two-group interfacial area transport equation is simplified as

$$\frac{\partial\left(\alpha_{g1}\rho_g\right)}{\partial t}+\nabla\cdot\left(\alpha_{g1}\rho_g\boldsymbol{v}_{g1}\right)=-\rho_g\left[\eta_{RC,2}^{(11,2)}+\eta_{RC,2}^{(12,2)}+\eta_{WE,2}^{(11,2)}\right.$$
$$\left.+\eta_{WE,2}^{(12,2)}+\eta_{SO,2}^{(2,12)}+\eta_{TI,2}^{(2,1)}+\chi\left(D_{c1}^*\right)^3\left\{\frac{\partial\alpha_{g1}}{\partial t}+\nabla\cdot\left(\alpha_{g1}\boldsymbol{v}_{g1}\right)\right\}\right] \quad (11\text{-}176)$$

$$\frac{\partial a_{i1}}{\partial t}+\nabla\cdot\left(a_{i1}\boldsymbol{v}_{i1}\right)$$
$$=\left\{\frac{2}{3}-\chi\left(D_{c1}^*\right)^2\right\}\frac{a_{i1}}{\alpha_{g1}}\left[\frac{\partial\alpha_{g1}}{\partial t}+\nabla\cdot\left(\alpha_{g1}\boldsymbol{v}_{g1}\right)\right] \quad (11\text{-}177)$$
$$+\phi_{RC}^{(1)}+\phi_{RC,1}^{(12,2)}+\phi_{WE}^{(1)}+\phi_{WE,1}^{(12,2)}+\phi_{TI}^{(1)}+\phi_{TI,1}^{(2,1)}+\phi_{SO,1}^{(2,12)}$$

$$\frac{\partial a_{i2}}{\partial t}+\nabla\cdot\left(a_{i2}\boldsymbol{v}_{i2}\right)=\frac{2}{3}\frac{a_{i2}}{\alpha_{g2}}\left[\frac{\partial\alpha_{g2}}{\partial t}+\nabla\cdot\left(\alpha_{g2}\boldsymbol{v}_{g2}\right)\right]$$
$$+\chi\left(D_{c1}^*\right)^2\frac{a_{i1}}{\alpha_{g1}}\left[\frac{\partial\alpha_{g1}}{\partial t}+\nabla\cdot\left(\alpha_{g1}\boldsymbol{v}_{g1}\right)\right]+\phi_{RC,2}^{(11,2)}+\phi_{RC,2}^{(12,2)} \quad (11\text{-}178)$$
$$+\phi_{RC}^{(2)}+\phi_{WE,2}^{(11,2)}+\phi_{WE,2}^{(12,2)}+\phi_{WE}^{(2)}+\phi_{TI,2}^{(2)}+\phi_{SO,2}^{(2,12)}+\phi_{SI}^{(2)}.$$

C. Assumed Group-2 bubble shape and bubble-number density distribution

In order to model source and sink terms analytically, the bubble shape should be simplified. For Group-1 bubbles, spherical shape can be assumed. For Group-2 bubbles, however, the unique geometry of the test section of interest should be accounted for. Since the boundary between Group-1 and Group-2 bubbles defined by Eq.(11-115) is approximately 10 mm in an adiabatic air-water system at atmospheric pressure, the cap bubbles are assumed to be sandwiched between the two parallel flat walls such that the cap bubbles have a thickness of G, as shown in Fig.11-6. Here, R and $2a$

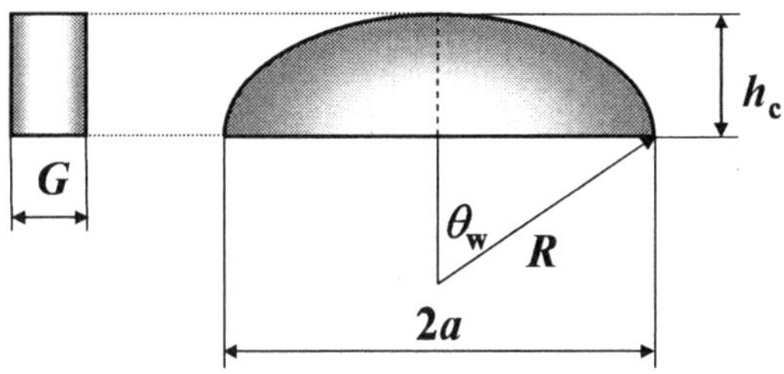

Figure 11-6. Definition of the geometrical parameters for a cap bubble (Sun et al., 2004a)

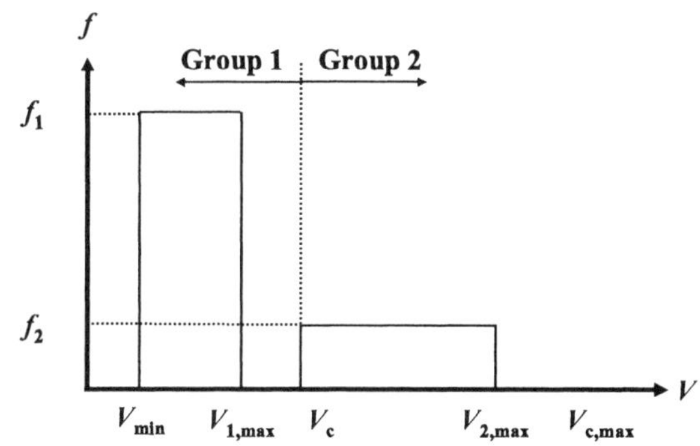

Figure 11-7. Illustration of the simplified bubble number density distribution (Sun et al., 2004)

are the radius of curvature and the base width of a cap bubble, and θ_w is the wake angle. A 50° wake angle is assumed as a reasonable approximation for all Group-2 bubbles in the flow conditions of interest in view of the wake angle correlation given by Clift et al. (1978). Furthermore, in view of the characteristic feature of the confined flows as well as the underlying physics, $2a$ is chosen as the characteristic length that determines the group boundary and the maximum stable bubble size.

A simplified bubble number density distribution is given in Fig.11-7. It is assumed that all the bubble groups have flat number density distributions in the corresponding bubble volume range. The values of the distribution function are denoted as f_1 and f_2 for Group-1 bubbles and Group-2 bubbles, respectively. In the figure, V_{min} is the volume for the minimum bubbles in the system, and $V_{c,max}$ is the volume of the maximum stable bubble, which corresponds to

$$D_{c,\max} = 40\sqrt{\frac{\sigma}{g\Delta\rho}}. \tag{11-179}$$

It should be noted, however, that bubbles of this size might not exist in the system and $D_{c,max}$ solely provides an upper limit for the maximum bubble size possible, beyond which the bubbles are assumed to disintegrate instantaneously. $V_{1,max}$ and $V_{2,max}$ are the maximum bubble volumes for Group 1 and Group 2, respectively, for a given flow condition by assuming the uniform bubble number density distribution. Furthermore, V_c is the critical bubble volume at the boundary for Group-1 and Group-2 bubbles, which corresponds to D_{crit} beyond which bubbles become cap-shaped and are categorized as Group-2 bubbles. The boundary between Group-1 and Group-2 bubbles can be determined for the narrow channel by

$$D_{crit} = 1.7G^{1/3}\left(\frac{\sigma}{g\Delta\rho}\right)^{1/3} \tag{11-180}$$

where D_{crit} is the volumetric equivalent diameter of a bubble at the boundary between Group-1 and Group-2 bubbles.

D. Summary of modeled source and sink terms

Bubble coalescence due to random collision

$$\phi_{RC}^{(1)} = -0.17C_{RC}^{(1)}\frac{\varepsilon^{1/3}\alpha_{g1}^{1/3}a_{i1}^{5/3}}{\alpha_{g1,max}^{1/3}\left(\alpha_{g1,max}^{1/3}-\alpha_{g1}^{1/3}\right)} \times\left[1-\exp\left(-C_{RC1}\frac{\alpha_{g1,max}^{1/3}\alpha_{g1}^{1/3}}{\alpha_{g1,max}^{1/3}-\alpha_{g1}^{1/3}}\right)\right] \tag{11-181}$$

$$\phi_{RC,1}^{(12,2)} = -4.85C_{RC}^{(12,2)}\varepsilon^{1/3}\frac{a_{i1}\alpha_{g1}^{2/3}\alpha_{g2}^{2}}{R_{2,max}^{2/3}} \times\left[1-\exp\left(-C_{RC1}\frac{\alpha_{g1,max}^{1/3}\alpha_{g1}^{1/3}}{\left(\alpha_{g1,max}^{1/3}-\alpha_{g1}^{1/3}\right)}\right)\right] \tag{11-182}$$

where $R_{2,max}$ is the radius of curvature of the maximum bubble in the system by assuming a uniform bubble number density distribution as

$$R_{2,max} \simeq 1.915 D_{Sm,2} \tag{11-183}$$

for the application of Sun et al. (2004a).

$$\begin{aligned}\phi_{RC,2}^{(11,2)} &= 0.68 C_{RC}^{(1)} \varepsilon^{1/3} \frac{\alpha_{g1}^{2} a_{i1}^{2/3}}{\alpha_{g1,max}^{2/3} G} \\ &\times \left[1 - \exp\left(-C_{RC1} \frac{\alpha_{g1,max}^{1/3} \alpha_{g1}^{1/3}}{\alpha_{g1,max}^{1/3} - \alpha_{g1}^{1/3}}\right)\right] \\ &\times \left[1 + 0.7 G^{7/6} \left(\frac{a_{i1}}{\alpha_{g1}}\right)^{1/2} \left(\frac{\sigma}{g\Delta\rho}\right)^{-1/3}\right] \left(1 - \frac{2}{3} D_{c1}^{*}\right) \\ &\text{for } D_{c1}^{*} < 1.5\end{aligned} \tag{11-184}$$

$$\begin{aligned}\phi_{RC,2}^{(12,2)} &= 13.6 C_{RC}^{(12,2)} \varepsilon^{1/3} \frac{\alpha_{g1}^{5/3} \alpha_{g2}^{2}}{R_{2,max}^{2/3} G} \left(1 + \frac{10.3 G}{R_{2,max}}\right) \\ &\times \left[1 - \exp\left(-C_{RC1} \frac{\alpha_{g1,max}^{1/3} \alpha_{g1}^{1/3}}{\left(\alpha_{g1,max}^{1/3} - \alpha_{g1}^{1/3}\right)}\right)\right]\end{aligned} \tag{11-185}$$

$$\begin{aligned}\phi_{RC}^{(2)} &= -13.6 C_{RC}^{(2)} \frac{\alpha_{g2}^{2} \varepsilon^{1/3}}{W^{2} G} R_{2,max}^{4/3} \left(1 - 2.0 R_{c}^{*2} + \frac{9.0 G}{R_{2,max}}\right) \\ &\times \left[1 - \exp\left(-C_{RC2} \alpha_{g2}^{1/2}\right)\right]\end{aligned} \tag{11-186}$$

where

$$R_{c}^{*} \equiv \frac{D_{crit}/2}{R_{2,max}} \tag{11-187}$$

$$\eta_{RC,2}^{(11,2)} = 3.4C_{RC}^{(1)} \frac{\varepsilon^{1/3}\alpha_{g1}^{2}a_{i1}^{2/3}}{\alpha_{g1,max}^{2/3}} \times\left[1-\exp\left(-C_{RC1}\frac{\alpha_{g1,max}^{1/3}\alpha_{g1}^{1/3}}{\alpha_{g1,max}^{1/3}-\alpha_{g1}^{1/3}}\right)\right]\left(1-\frac{2}{3}D_{c1}^{*}\right) \tag{11-188}$$

for $D_{c1}^{*} < 1.5$

$$\eta_{RC,2}^{(12,2)} = 4.85C_{RC}^{(12,2)}\varepsilon^{1/3}\frac{\alpha_{g1}^{5/3}\alpha_{g2}^{2}}{R_{2,max}^{2/3}} \times\left[1-\exp\left(-C_{RC1}\frac{\alpha_{g1,max}^{1/3}\alpha_{g1}^{1/3}}{\left(\alpha_{g1,max}^{1/3}-\alpha_{g1}^{1/3}\right)}\right)\right] \tag{11-189}$$

Bubble coalescence due to wake entrainment

$$\phi_{WE}^{(1)} = -0.27C_{WE}^{(1)}u_{r1}C_{D1}^{1/3}a_{i1}^{2} \tag{11-190}$$

$$\phi_{WE,1}^{(12,2)} = -4.35C_{WE}^{(12,2)}\sqrt{gC_{D2}G}\,\frac{a_{i1}\alpha_{g2}}{R_{2,max}} \tag{11-191}$$

$$\phi_{WE,2}^{(11,2)} = 1.08C_{WE}^{(11,2)}u_{r1}C_{D1}^{1/3}\frac{\alpha_{g1}a_{i1}}{G}\left(1-\frac{2}{3}D_{c1}^{*}\right) \times\left[1+0.7G^{7/6}\left(\frac{a_{i1}}{\alpha_{g1}}\right)^{1/2}\left(\frac{\sigma}{g\Delta\rho}\right)^{-1/3}\right] \tag{11-192}$$

for $D_{c1}^{*} < 1.5$

$$\phi_{WE,2}^{(12,2)} = 26.1C_{WE}^{(12,2)}\alpha_{g1}\alpha_{g2}\sqrt{\frac{gC_{D2}}{G}}\frac{1}{R_{2,max}}\left(1+4.31\frac{G}{R_{2,max}}\right) \tag{11-193}$$

$$\phi_{WE}^{(2)} = -15.9 C_{WE}^{(2)} \frac{\alpha_{g2}^{2}}{R_{2,max}^{2}} \sqrt{C_D g G} \left(1 + 0.51 R_c^*\right) \tag{11-194}$$

$$\eta_{WE,2}^{(11,2)} = 5.40 C_{WE}^{(11,2)} u_{r1} C_{D1}^{1/3} \alpha_{g1} a_{i1} \left(1 - \frac{2}{3} D_{c1}^*\right) \text{ for } D_{c1}^* < 1.5 \tag{11-195}$$

$$\eta_{WE,2}^{(12,2)} = 4.35 C_{WE}^{(12,2)} \sqrt{C_{D2} g G} \frac{\alpha_{g1}\alpha_{g2}}{R_{2,max}} \tag{11-196}$$

Bubble breakup due to turbulent impact

$$\phi_{TI}^{(1)} = \begin{cases} 0.12 C_{TI}^{(1)} \varepsilon^{1/3} \left(1 - \alpha_g\right) \dfrac{a_{i1}^{5/3}}{\alpha_{g1}^{2/3}} \exp\left(-\dfrac{We_{crit,TI1}}{We_1}\right) \\ \qquad \times \sqrt{1 - \dfrac{We_{crit,TI1}}{We_1}}, \quad We_1 > We_{crit,TI1} \\ 0, \quad We_1 \le We_{crit,TI1} \end{cases} \tag{11-197}$$

$$\begin{aligned} \phi_{TI,1}^{(2,1)} &= 2.71 C_{TI}^{(2)} \alpha_{g2} \left(1 - \alpha_g\right) \frac{\varepsilon^{1/3} G^{2/3} R_c^{*5/3} \left(1 - R_c^{*5/3}\right)}{R_{2,max}^{7/3}} \\ &\times \exp\left(-\frac{We_{crit,TI2}}{We_2}\right) \sqrt{1 - \frac{We_{crit,TI2}}{We_2}} \end{aligned} \tag{11-198}$$

$$\begin{aligned} \phi_{TI,2}^{(2)} &= 1.4 C_{TI}^{(2)} \alpha_{g2} \frac{\varepsilon^{1/3} G}{R_{2,max}^{8/3}} \left(1 - \alpha_g\right)\left(1 - 2R_c^*\right) \\ &\times \exp\left(-\frac{We_{crit,TI2}}{We_2}\right) \sqrt{1 - \frac{We_{crit,TI2}}{We_2}} \end{aligned} \tag{11-199}$$

$$\eta_{TI,2}^{(2,1)} = -0.34 C_{TI}^{(2)} \alpha_{g2} \left(1-\alpha_g\right) \frac{G\varepsilon^{1/3} R_c^{*7/3} \left(1 - R_c^{*5/3}\right)}{R_{2,max}^{5/3}} \\ \times \exp\left(-\frac{We_{crit,TI2}}{We_2}\right) \sqrt{1 - \frac{We_{crit,TI2}}{We_2}} \tag{11-200}$$

Bubble breakup due to shearing off

$$\phi_{SO,1}^{(2,12)} = 64.51 C_{SO} C_d^{\ 2} \frac{\alpha_{g2} v_{rb}}{G R_{2,max}} \left[1 - \left(\frac{We_{crit,SO}}{We_{2,max}}\right)^3\right] \tag{11-201}$$

where v_{rb} and $We_{crit,SO}$ are the relative velocity of the large bubble with respect to the liquid film near the cap bubble base and the critical Weber number, respectively, and $We_{2,max}$ is defined by

$$We_{2,max} \equiv \frac{2\rho_f v_{rb}^2 R_{2,max}}{\sigma} \tag{11-202}$$

In upward flow in round tubes, when a large cap or slug bubble rises, the liquid phase is pushed away and flows downward as a liquid film between the bubble side interface and the wall. However, in the flow channel considered by Sun et al. (2004a), when a large cap bubble rises, the liquid film between the bubble side interface and the wall may remain almost stagnant since more free space is available for the liquid phase in the width direction of the flow channel. This may be even truer when the cap bubble velocity is high and the shearing-off occurs. In view of this, the relative velocity of the cap bubble with respect to the liquid film around the bubble base, v_{rb}, may be estimated by the velocity of Group-2 bubbles in the main flow direction.

$$\phi_{SO,2}^{(2,12)} = -21.50 C_{SO} C_d^{\ 3} \left(\frac{\sigma}{\rho_f}\right)^{3/5} \frac{\alpha_{g2}}{v_{rb}^{1/5} G^{8/5} R_{2,max}} \\ \times \left[1 - \left(\frac{We_{crit,SO}}{We_{2,max}}\right)^3 + \frac{3.24 G}{R_{2,max}} \left\{1 - \left(\frac{We_{crit,SO}}{We_{2,max}}\right)^2\right\}\right] \tag{11-203}$$

$$\eta_{SO,2}^{(2,12)} = -10.75 C_{SO} C_d^{3} \left(\frac{\sigma}{\rho_f G} \right)^{3/5} \frac{\alpha_{g2}}{v_{rb}^{1/5} R_{m2}} \times \left\{ 1 - \left(\frac{We_{crit,SO}}{We_{m2}} \right)^{3} \right\} \tag{11-204}$$

Bubble breakup due to surface instability

$$\phi_{SI}^{(2)} = 1.25 \alpha_{g2}^{2} \left(\frac{\sigma}{g \Delta\rho} \right)^{-1} \left[C_{RC}^{(2)} \frac{\varepsilon^{1/3}}{W^2} \left(\frac{\sigma}{g \Delta\rho} \right)^{7/6} \times \left\{ 1 - \exp\left(-C_{RC2} \alpha_{g2}^{1/2} \right) \right\} + 2.3 \times 10^{-4} C_{WE}^{(2)} \sqrt{C_D g G} \right] \tag{11-205}$$

The values of the coefficients in the source and sink terms are listed in Table 11-3.

Inter-group transfer coefficient at group boundary

The inter-group transfer coefficient at group boundary is determined experimentally as

$$\chi = 4.44 \times 10^{-3} \left(\frac{D_{Sm1}}{D_{crit}} \right)^{0.36} \alpha_{g1}^{-1.35}. \tag{11-206}$$

This correlation is obtained based on the limited experimental database (Sun et al., 2004a). Nevertheless, in general cap-turbulent and churn-turbulent flow, the value of α_{g1} is usually between 0.05 and 0.40. Therefore, the correlation may be applicable to most of these flow conditions since the database by which the correlation is developed covers the similar range of α_{g1}.

Chapter 12

HYDRODYNAMIC CONSTITUTIVE RELATIONS FOR INTERFACIAL TRANSFER

In analyzing the interfacial force and relative motion between phases, consider first, the momentum equation for each phase. Under the assumption that both the average pressure and stress in the bulk fluid and at the interface are approximately the same, the k-phase momentum equation is given by

$$\alpha_k \overline{\overline{\rho_k}} \left(\frac{\partial \widehat{\boldsymbol{v}_k}}{\partial t} + \widehat{\boldsymbol{v}_k} \cdot \nabla \widehat{\boldsymbol{v}_k} \right) = -\alpha_k \nabla \overline{\overline{p_k}} + \nabla \cdot \left[\alpha_k \left(\overline{\overline{\mathcal{T}_k}} + \mathcal{T}_k^{T} \right) \right] + \alpha_k \overline{\overline{\rho_k}} \boldsymbol{g} + \boldsymbol{M}_{ik} + \left(\widehat{\boldsymbol{v}_{ki}} - \widehat{\boldsymbol{v}_k} \right) \Gamma_k - \nabla \alpha_k \cdot \overline{\overline{\mathcal{T}_{ki}}} \tag{12-1}$$

where $\overline{\overline{\mathcal{T}_k}}$, $\mathcal{T}_k^{T}$, $\overline{\overline{\mathcal{T}_{ki}}}$, and $\boldsymbol{M}_{ik}$ are the average viscous stress tensor, the average turbulent stress tensor, the interfacial shear stress, and the generalized interfacial drag force. The conservation of the mixture momentum requires

$$\sum_k \boldsymbol{M}_{ik} = 0 \tag{12-2}$$

which is the modified form of the average momentum-jump condition. Constitutive equations of the average turbulent stress tensor and the generalized interfacial drag force are required to analyze two-phase flows using the two-fluid model.

In a macroscopic two-phase flow analysis such as a one-dimensional two-phase flow analysis, the average turbulent stress term may be neglected except the wall shear contributions, whereas in a microscopic bubbly flow analysis, turbulence models such as mixing length model and k-ε model has been attempted to estimate the average turbulent stress term. However, due to the complexity of the two-phase flow turbulence, an accurate method to predict the turbulence in two-phase flow has not been established well.

In the two-fluid momentum equation, the most important term to be modeled by a constitutive relation is the generalized drag force $\boldsymbol{M}_{id}$ which specifies the interfacial surface forces. The simplest way to model this force is to formulate as the linear combination of various known interfacial forces as

$$\begin{aligned}\boldsymbol{M}_{id} &= \frac{\alpha_d}{B_d}\left(\boldsymbol{F}_d^D + \boldsymbol{F}_d^V + \boldsymbol{F}_d^B + \boldsymbol{F}_d^L + \boldsymbol{F}_d^W + \boldsymbol{F}_d^T\right)\\ &= \boldsymbol{M}_d^D + \boldsymbol{M}_d^V + \boldsymbol{M}_d^B + \boldsymbol{M}_d^L + \boldsymbol{M}_d^W + \boldsymbol{M}_d^T\end{aligned} \tag{12-3}$$

where B_d, $\boldsymbol{F}^D$, $\boldsymbol{F}^V$, $\boldsymbol{F}^B$, $\boldsymbol{F}^L$, $\boldsymbol{F}^W$, and $\boldsymbol{F}^T$ are the volume of a typical particle, the standard drag force, the virtual mass force, the Basset force, the lift force, the wall lift force and turbulent dispersion force for a typical single particle, respectively.

The significance of the various terms in the equation is as follows. The term on the left-hand side is the combined generalized interfacial drag force acting on the dispersed phase. The first term on the right-hand side is the skin and form drag under the steady-state condition. The second term is the force required to accelerate the apparent mass of the surrounding phase when the relative velocity changes. The third term, known as the Basset force, is the effect of the acceleration on the viscous drag and the boundary-layer development. The fourth term is the lift force normal to the relative velocity due to rotation of fluid. The fifth term is the wall lift force due to the velocity distribution change around particles near a wall. The last term is the turbulent dispersion force due to the concentration gradient. In a macroscopic two-phase flow analysis such as a one-dimensional two-phase flow analysis, forces except the standard drag force and the virtual mass force are not taken into account, whereas additional forces such as the lift force and the turbulent dispersion force are also considered in a microscopic analysis for three-dimensional flow.

In the present chapter, the constitutive equations for the interfacial transfer and the interfacial fluid mechanics of multiphase flows are discussed in detail following Ishii and Zuber (1979), Ishii and Chawla (1979) and Ishii and Mishima (1984). In the following discussion, symbols designating the time-average are omitted for simplicity except in the Section 1.4 of Chapter 12.

1.1 Transient forces in multiparticle system

The forms of the two transient terms are not firmly established. Because of their importance under transient conditions and for numerical-stability problems, further research in this area is required.

The Basset force is given by

$$\boldsymbol{F}_d^B = 6 r_d^2 \sqrt{\pi \rho_c \mu_m} \int_t \frac{D_d}{D\xi}\left(\boldsymbol{v}_c - \boldsymbol{v}_d\right) \frac{d\xi}{\sqrt{t-\xi}} \tag{12-4}$$

where μ_m is the mixture viscosity. This term represents the additional drag due to the development of boundary layer or viscous flow during a transient acceleration of particles. The derivative D_k/Dt is the convective derivative relative to velocity $\boldsymbol{v}_k$. The subscripts c and d stand for continuous phase and dispersed phase, respectively. The detailed expression for mixture viscosity is given in the Section 1.2 of Chapter 12. Due to its complicated time-integral form, the Basset force is not considered in a practical two-phase flow analysis. Some evaluation of this term for a higher Reynolds number is given by Clift et al. (1978).

Zuber (1964a) studied the effect of the concentration on the virtual mass force and obtained.

$$\frac{\alpha_d \boldsymbol{F}_d^V}{B_d} = -\frac{1}{2}\alpha_d \frac{1+2\alpha_d}{1-\alpha_d}\rho_c \frac{D_d}{Dt}\left(\boldsymbol{v}_d - \boldsymbol{v}_c\right). \tag{12-5}$$

Lahey et al. (1978) studied a necessary condition for the constitutive equation for the virtual mass term. From the requirement of the frame-indifference of the constitutive equation, they determined that the virtual mass force $\boldsymbol{F}_d^V$ should satisfy

$$\boldsymbol{F}_d^V \propto \left[\frac{D_d \boldsymbol{v}_d}{Dt} - \frac{D_c \boldsymbol{v}_c}{Dt} + (1-\lambda)\boldsymbol{v}_r \cdot \nabla \boldsymbol{v}_r\right]. \tag{12-6}$$

In view of Zuber's study (1964a) on the effect of concentration and the above frame-indifference condition, a new form for $\boldsymbol{F}_d^V$ is proposed here. Due to the acceleration of the particles relative to the fluid, the acceleration drag arises. This should be proportional to the induced mass $\rho_c \boldsymbol{B}_d^*$ and the frame-indifferent relative-acceleration vector. Hence,

$$\boldsymbol{F}_d^V = -\rho_c B_d^*\left[\frac{D_d\boldsymbol{v}_d}{Dt} - \frac{D_c\boldsymbol{v}_c}{Dt} + (1-\lambda)\boldsymbol{v}_r\cdot\nabla\boldsymbol{v}_r\right]. \tag{12-7}$$

The value of induced mass $\rho_c\boldsymbol{B}_d^*$ for a single particle in an infinite medium can be obtained from potential theory. Hence, the limiting value of $\boldsymbol{F}_d^V$ at $\alpha_d \to 0$ for a spherical particle is

$$\lim_{\alpha_d\to 0}\boldsymbol{F}_d^V = -\frac{1}{2}\rho_c B_d\frac{D_d(\boldsymbol{v}_d - \boldsymbol{v}_c)}{Dt}. \tag{12-8}$$

From this limit, it can be shown that

$$\lim_{\alpha_d\to 0} B_d^* = \frac{1}{2}B_d \tag{12-9}$$

and

$$\lim_{\alpha_d\to 0}\lambda = 2. \tag{12-10}$$

If λ is constant in Eq.(12-7), the value of λ should be 2.

The effect of the concentration on $\boldsymbol{B}_d^*$ can be taken into account by the method used by Zuber (1964a). Thus, from the solution for the induced mass for a sphere moving within an outer sphere, $\boldsymbol{B}_d^*$, may be approximated by

$$B_d^* \doteq \frac{1}{2}B_d\frac{1+2\alpha_d}{1-\alpha_d} \tag{12-11}$$

where α_d is the volumetric fraction of the dispersed phase. Under the assumption of λ=constant, the constitutive equation for the virtual mass force is obtained from Eqs.(12-7) and (12-11) as (Ishii and Mishima, 1984)

$$\frac{\alpha_d\boldsymbol{F}_d^V}{B_d} = -\frac{1}{2}\alpha_d\frac{1+2\alpha_d}{1-\alpha_d}\rho_c\left(\frac{D_d\boldsymbol{v}_r}{Dt} - \boldsymbol{v}_r\cdot\nabla\boldsymbol{v}_c\right). \tag{12-12}$$

The above equation indicates that the virtual mass force $\boldsymbol{F}_d^V$ per particle increases considerably with increasing particle concentration. This relation implies that the effect of concentration on dynamic coupling can be scaled by a factor of $(1+2\alpha_d)/(1-\alpha_d)$. Mokeyev (1977) used an electrohydrodynamic analog method to determine the velocity potential through an electric field potential and obtained an empirical function $B_d^*/B_d = 0.5 + 2.1\alpha_d$. The theoretical result of Eq.(12-12) compared favorably with this correlation.

A correlation for the virtual mass force in a slug flow can be developed from a simple potential flow analysis using a Bernoulli equation. First a cylindrical bubble of length L_b with diameter D_b in a tube of diameter D is considered, see Fig.12-1(a). Then the void fraction in a slug bubble section is given by

$$\alpha_b = \frac{D_b^2}{D^2} \tag{12-13}$$

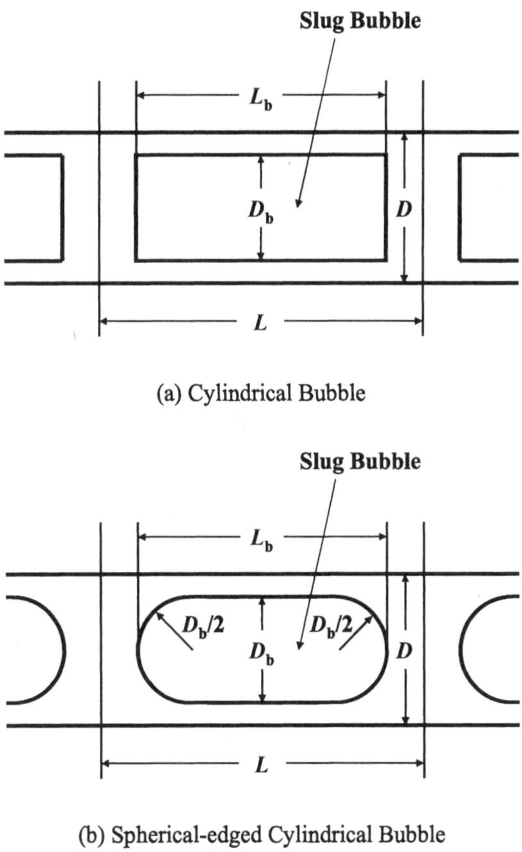

Figure 12-1. Slug-flow model for virtual-mass-force analysis (Ishii and Mishima, 1984)

and the average overall void fraction α_d by

$$\alpha_d = \frac{L_b}{L}\alpha_b \tag{12-14}$$

where L is the pitch. Now let the continuous phase accelerate with respect to a bubble. This will generate a pressure force acting on a bubble due to the acceleration along the film section. From a simple one-dimensional analysis, this force can be found as

$$\boldsymbol{F}_d^V = -\frac{\pi}{4}D_b^2 L_b \frac{\rho_c}{1-\alpha_b}\frac{\partial \boldsymbol{v}_r}{\partial t}. \tag{12-15}$$

However, the volume of a bubble is given by $B_d = \left(\pi/4\right) D_b^2 L_b$, thus the virtual mass force per unit volume becomes

$$\frac{\alpha_d \boldsymbol{F}_d^V}{B_d} = -\alpha_d \frac{\rho_c}{1-\alpha_d}\frac{\partial \boldsymbol{v}_r}{\partial t} \cong -5\alpha_d \rho_c \frac{\partial \boldsymbol{v}_r}{\partial t}. \tag{12-16}$$

Here the second form is obtained by approximating the void fraction in the slug bubble section by $\alpha_b \cong 0.8$.

The second case considered is a train of spherical-edged cylindrical bubbles, see Fig.12-1(b). Application of the Bernoulli equation

$$\frac{\partial \Phi}{\partial t} + \int \frac{dp}{\rho} + \Omega + \frac{v^2}{2} = \text{constant} \tag{12-17}$$

where Φ and Ω are respectively the velocity potential and the potential function to this geometry under a relative acceleration yields

$$\frac{\alpha_d \boldsymbol{F}_d^V}{B_d} = -5\left[0.66\alpha_d + 0.27\left(\frac{L_b - D_b}{L}\right)\right]\rho_c \frac{\partial \boldsymbol{v}_r}{\partial t} \tag{12-18}$$

where a simplification has been made on an approximation $\alpha_b \cong 0.8$ (Ishii and Mishima, 1981). For a limiting case of a train of spherical bubbles, $L_b = D_b$, the above equation reduces to

$$\frac{\alpha_d \boldsymbol{F}_d^V}{B_d} = -3.3\alpha_d \rho_c \frac{\partial \boldsymbol{v}_r}{\partial t}. \tag{12-19}$$

If $L_b >> D_b$, L_b/L can be approximated by α_d/α_b. Thus for long slug bubbles, Eq.(12-18) essentially converges to the simple solution given by Eq.(12-17). The virtual mass force for a slug flow given by Eq.(12-18) is expressed in terms of the relative acceleration in the absence of a large convective acceleration. However, if the convective acceleration can not be neglected, a special convective derivative in the form of Eq.(12-7) may be more appropriate. Thus for a general case,

$$\frac{\alpha_d \boldsymbol{F}_d^V}{B_d} = -5\left[0.66\alpha_d + 0.27\left(\frac{L_b - D_b}{L}\right)\right]\rho_c\left(\frac{D_d \boldsymbol{v}_r}{Dt} - \boldsymbol{v}_r \cdot \nabla \boldsymbol{v}_c\right). \tag{12-20}$$

This formula can also be applied to churn-turbulent flow.

Now the solutions for a dispersed flow, Eq.(12-12), and slug flow, Eq.(12-18), can be examined by introducing an induced mass coefficient C_M defined by

$$\frac{\alpha_d \boldsymbol{F}_d^V}{B_d} = -C_M \rho_c \left(\frac{D_d \boldsymbol{v}_r}{Dt} - \boldsymbol{v}_r \cdot \nabla \boldsymbol{v}_c\right) \tag{12-21}$$

where

$$C_M = \begin{cases} \dfrac{1}{2}\alpha_d \dfrac{1+2\alpha_d}{1-\alpha_d} & \text{(Bubbly flow)} \\ 5\alpha_d\left[0.66 + 0.34\left(\dfrac{1-D_b/L_b}{1-D_b/3L_b}\right)\right] & \text{(Slug flow).} \end{cases} \tag{12-22}$$

A plot of C_M against α_d is shown in Fig.12-2. The virtual mass force increases with an increasing void fraction of a dispersed phase due to stronger coupling between two phases. The intersection of the above two solutions occurs at the void fraction between 0.66 and 0.75. For a lower void fraction, the virtual mass force for a bubbly-flow is smaller than that for a slug-flow. This implies that the vapor phase has less resistance to an acceleration in a bubbly-flow configuration than a slug-flow configuration if $\alpha_d < 0.66$. This may also suggest that an accelerating slug flow has a tendency to disintegrate into a bubbly flow when $\alpha_d < 0.66$. For

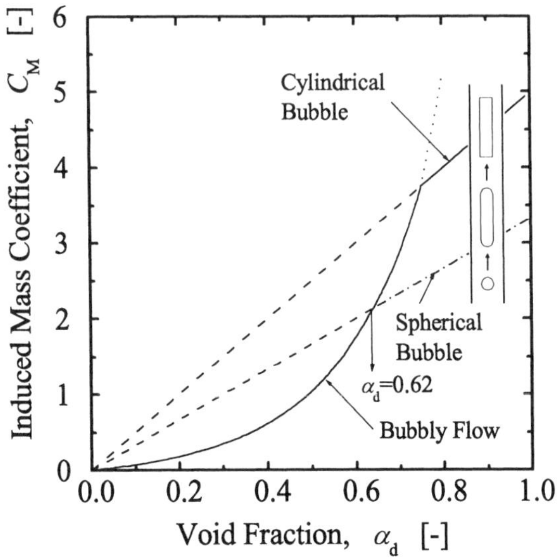

Figure 12-2. Virtual mass coefficient for dispersed and slug-flow regimes (Ishii and Mishima, 1984)

$\alpha_d > 0.66$ a slug flow should be quite stable even under a transient condition.

Due to a similarity in flow geometries, the virtual mass force for a churn-turbulent flow may be approximated by the solution for a slug flow given by Eq.(12-19). In a liquid-dispersed flow, the virtual mass force becomes considerably smaller than that in a vapor-dispersed flow. This decrease is caused by a change in the continuous phase density to be used in Eq.(12-12). By changing ρ_c from ρ_f to ρ_g, the virtual mass force for a droplet flow becomes insignificant. This also indicates that the virtual mass force should be reduced considerably in annular and annular dispersed flow.

1.2 Drag force in multiparticle system

The standard drag force acting on the particle under steady-state conditions can be given in terms of the drag coefficient C_D based on the relative velocity as

$$\boldsymbol{F}_d^D = -\frac{1}{2}C_D\rho_c\boldsymbol{v}_r\left|v_r\right|A_d \tag{12-23}$$

where A_d is the projected area of a typical particle and $\boldsymbol{v}_r$ is the relative velocity given by $\boldsymbol{v}_r = \boldsymbol{v}_d - \boldsymbol{v}_c$. Then $\boldsymbol{F}_d^D$ is related to the interfacial drag force by

$$\boldsymbol{F}_d^D = \frac{B_d \boldsymbol{M}_d^D}{\alpha_d}. \tag{12-24}$$

Hence, the portion of $\boldsymbol{M}_{id}$ represented by the drag force becomes

$$\frac{\alpha_d \boldsymbol{F}_d^D}{B_d} = -\left(\alpha_d \frac{A_d}{B_d}\right)\frac{C_D}{2}\rho_c \boldsymbol{v}_r \left|\boldsymbol{v}_r\right|. \tag{12-25}$$

In what follows, a constitutive relation for the drag coefficient C_D in dispersed two-phase flows will be explained in detail starting from a single-particle system.

1.2.1 Single-particle drag coefficient

Motion of the single solid particles, drops, or bubbles in an infinite medium has been studied extensively in the past, see for example, Peebles and Garber (1953), Harmathy (1960) and Wallis (1974). In what follows we summarize these results in simple forms useful for the development of the drag correlation in multiparticle systems (Ishii and Chawla, 1979).

By denoting the relative velocity of a single particle in an infinite medium by $v_{r\infty} = v_d - v_{c\infty}$, the drag coefficient is defined by

$$C_{D\infty} = -2F_D \Big/ \left\{\rho_c v_{r\infty} \left|v_{r\infty}\right| \pi r_d^2\right\} \tag{12-26}$$

where F_D is the drag force and r_d is the radius of a particle. To calculate the drag force F_D in terms of the relative velocity, we should determine a constitutive relation for $C_{D\infty}$ independent of Eq.(12-26). For a single-particle drag correlation, two similarity parameters are important. They are the particle Reynolds number and the viscosity number

$$N_{\text{Re}\infty} \equiv \frac{2 r_d \rho_c \left|v_{r\infty}\right|}{\mu_c} \tag{12-27}$$

and

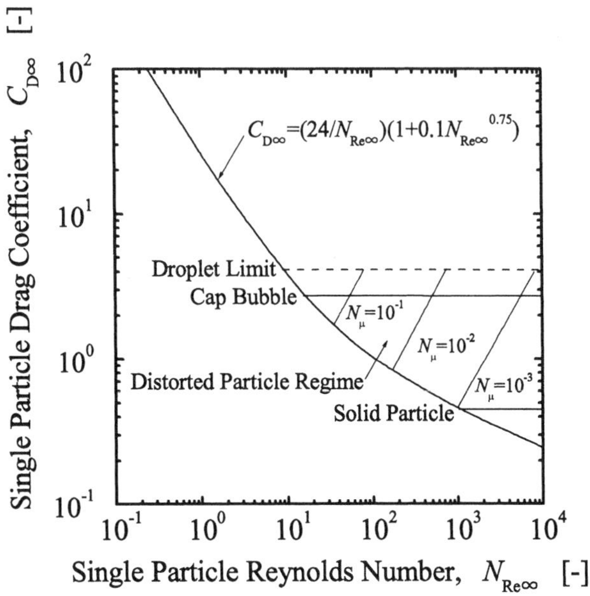

Figure 12-3. Single-particle drag coefficient (Ishii and Chawla, 1979)

$$N_\mu \equiv \frac{\mu_c}{\left(\rho_c\sigma\sqrt{\dfrac{\sigma}{g\Delta\rho}}\right)^{1/2}}. \tag{12-28}$$

Extensive studies on the single particle drag show that for most cases, the drag coefficient is a function of the Reynolds number (see Fig.12-3). However, the exact functional form depends on whether the particle is a solid particle, drop, or bubble. Briefly, for a solid-spherical-particle system, we have the viscous regime, in which the Reynolds-number dependence of $C_{D\infty}$ is pronounced, and Newton's regime, in which $C_{D\infty}$ is independent of $N_{Re\infty}$. In case of a clean fluid sphere in the viscous regime, $C_{D\infty}$ can be reduced up to 33 %, in comparison with the value predicted by the correlation for solid particles. This is explained by the internal circulation within the fluid particles. However, slight amounts of impurities are sufficient to eliminate this drag reduction. Therefore, for most practical applications, the drag law in a fluid- particle system may be approximated by that for a solid-particle system up to a certain particle size. Beyond this point, both the distortion of a particle shape and the irregular motions become pronounced. In this distorted-particle regime, $C_{D\infty}$ does not depend on the viscosity, but increases linearly with the radius of a particle. Because of the hydrodynamical instability, there is an upper limit on $C_{D\infty}$, and the particle reaches the cap bubble condition or the maximum droplet size. These regimes for the drag coefficient can be seen in Fig.12-3.

For a viscous regime, the function C_D is given by an empirical correlation as

$$C_D = \frac{24}{N_{\mathrm{Re}\infty}}\left(1 + 0.1 N_{\mathrm{Re}\infty}^{0.75}\right). \tag{12-29}$$

When the Reynolds number is small ($N_{\mathrm{Re}\infty} < 1$), the above correlation essentially reduces to the well-known Stokes drag law, $C_D = 24/N_{\mathrm{Re}\infty}$. The correlation for the viscous regime indicates that the dependence of the drag coefficient on the Reynolds number decreases with increasing values of the Reynolds number.

In solid particles, the drag coefficient becomes essentially constant at approximately

$$C_{D\infty} = 0.45 \text{ for } N_{\mathrm{Re}\infty} \geq 1000. \tag{12-30}$$

This Newton's regime holds up to $N_{\mathrm{Re}\infty} = 2 \times 10^5$. Beyond this Reynolds number the boundary layer separation point moves from the front side to back side of a particle due to the transition of the boundary layer from laminar to turbulent. This results in sharp drop in the drag coefficient.

For fluid particles such as drops or bubbles, we have a flow regime characterized by the distortion of particle shapes and the irregular motions. In this distorted particle regime, the experimental data show that terminal velocity is independent of the particle size (see Fig.12-4). In Fig.12-4, dimensionless terminal velocity, $v_{r\infty}^*$, and reduced radius, r_d^*, are defined by $|v_{r\infty}|\left(\rho_c^2/\mu_c g\Delta\rho\right)^{1/3}$ and $r_d\left(\rho_c g\Delta\rho/\mu_c^2\right)^{1/3}$, respectively. From this, it can be seen that the drag coefficient $C_{D\infty}$ does not depend on the viscosity, but it should be proportional to the radius of the particle (Harmathy, 1960). Physically, this indicates that the drag force is governed by distortion and swerving motion of the particle, and change of the particle shape is toward an increase in the effective cross section. Therefore, $C_{D\infty}$ may be scaled by the mean radius of the particle rather than the Reynolds number (Harmathy, 1960). Then we have

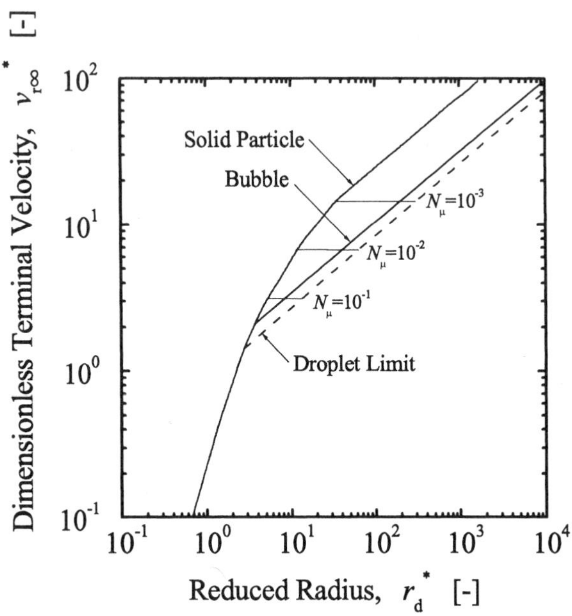

Figure 12-4. Terminal velocity for single-particle system (Ishii and Chawla, 1979)

$$C_{D\infty} = \frac{4}{3} r_d \sqrt{\frac{g\Delta\rho}{\sigma}} \text{ for } N_\mu \geq 36\sqrt{2}\left(1 + 0.1 N_{\mathrm{Re}\infty}^{0.75}\right) / N_{\mathrm{Re}\infty}^2 . \tag{12-31}$$

Here, the fluid particle size based on the terminal velocity is used.

Therefore, the flow regime transition between the viscous flow and the distorted particles flow can be given in terms of the viscosity number as shown in Figs.12-3 and 12-4. However, since in this regime the terminal velocity can be uniquely related to properties, Eq.(12-31) can be generalized in terms of the terminal velocity or the Reynolds number as

$$C_{D\infty} = \frac{\sqrt{2}}{3} N_\mu N_{\mathrm{Re}\infty} . \tag{12-32}$$

As the size of a bubble increases further, the bubble becomes spherical-cap shaped, and the drag coefficient reaches a constant value of

$$C_{D\infty} = \frac{8}{3}. \tag{12-33}$$

The transition from the distorted-bubble regime to the spherical-cap bubble regime occurs for

$$r_d = 2\sqrt{\frac{\sigma}{g\Delta\rho}}. \tag{12-34}$$

For a liquid drop, the drag coefficient may increase further according to Eq.(12-31). However, eventually a droplet becomes unstable and disintegrates into small drops. This limit can be given by the well-known Weber number criterion. By introducing the Weber number, $We \equiv 2\rho_g v_r^2 r_d / \sigma$, where v_r is the relative velocity, we can give the stability criterion approximately as $We \simeq 12$. Since the terminal velocity corresponding to Eq.(12-31) is $v_{r\infty} = \sqrt{2}\left(g\sigma\Delta\rho/\rho_g^2\right)^{1/4}$, the maximum possible drop radius is

$$r_{d\max} = 3\sqrt{\frac{\sigma}{g\Delta\rho}} \tag{12-35}$$

which corresponds to the maximum drag coefficient $C_D = 4$ for droplets. If the stability of a drop interface is governed by the Taylor instability, the characteristic drop radius is given by

$$r_{d\max} = \sqrt{\frac{\sigma}{g\Delta\rho}} \tag{12-36}$$

which may be a more practical upper limit of the drop size. It is also noted that in highly turbulent flow (Hinze, 1959) or under pressure shock conditions (Dinh et al., 2003; Theofanous et al., 2004) the stability limit Weber number can be much smaller than 12.

The cap bubble maintains a certain regular shape with the wake angle of about 50°, however there is also a maximum stable cap bubble diameter (Grace et al., 1978; Clift et al., 1978; Kocamstafaogullari et al., 1984; Miller, 1993). This instability is shown in Fig.12-5. Kocamustafaogullari et al. (1984) used the stability analysis based on the Kelvin-Helmholtz instability along the cap bubble surface. By comparing the surface wave residence time and the time for the wave amplitude to grow to the order of the magnitude of the bubble size, the stability criterion has been obtained. For most practical

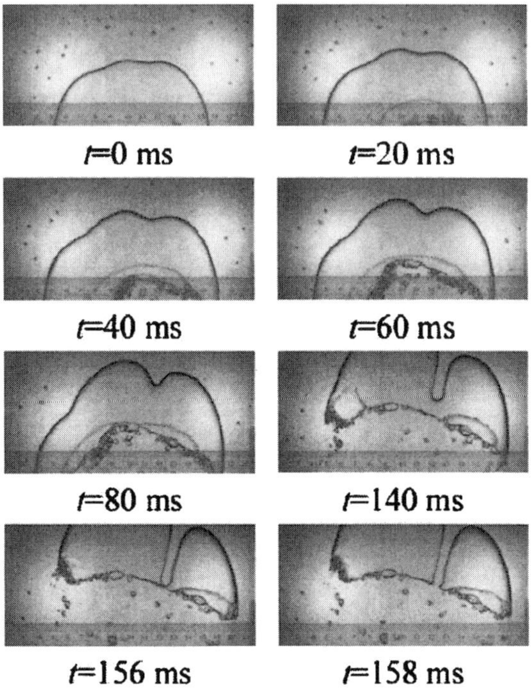

Figure 12-5. Large bubble breakup due to instability (500 frames/s)

cases, this stability limit can be approximated by

$$r_{cap,\max} = 20\sqrt{\frac{\sigma}{g\Delta\rho}}. \tag{12-37}$$

This result is significant because it defines the boundary between the smaller pipe where slug flow is possible and the larger pipe where slug flow cannot be formed due to the instability of the Taylor bubbles. For a pipe diameter $D < 2r_{cap,\max}\left(= 40\sqrt{\sigma/g\Delta\rho}\right)$ a stable slug flow can be formed. However for $D >> 2r_{cap,\max}$ bubbly flow is followed by cap-turbulent flow where multiple interacting cap bubbles exist at higher gas flux.

Using the above drag coefficient, we can obtain the terminal velocity in infinite media by balancing the pressure, gravity, and drag forces. The results are summarized in Fig.12-4 for various particles and flow regimes.

1.2.2 Drag coefficient for dispersed two-phase flow

A. Effects of Particles and Flow Regimes

In the preceding section it has been shown that the drag correlation for a single-particle system depends not only on the flow regimes but also on the nature of the particles; namely, solid particle, drop or bubble. Therefore, for a multiparticle system, these differences are also expected to play central roles in determining the drag correlation. In the present study, the multiparticle drag correlation is developed in parallel with the single-particle system by considering the following flow regimes

$$\text{Solid-particle system}\begin{cases}\text{Viscous regime}\\ \text{Newton's regime}\end{cases}$$

$$\text{Fluid-particle system}\begin{cases}\text{Viscous regime}\\ \text{(Undistorted-particle regime)}\\ \text{Distorted-particle regime}\\ \text{Churn-turbulent-flow regime}\\ \text{Slug-flow regime}\end{cases}$$

In the viscous regime, distortions of fluid particles are negligible. Therefore, for this regime, solid- and fluid-particle systems are considered together. Although small differences exist between these two systems due to the surface flow, for most cases these differences can be neglected (Clift et al., 1978). The other flow regimes are analyzed separately because of significant differences in the flow around the particles and the motions of the interfaces.

B. Viscous Regime (Undistorted-Particle Regime)

This regime is characterized by the strong effect of viscosity on the particle motion. For a fluid-particle system, this regime occurs only when particle shapes are not distorted due to interfacial instabilities or turbulent fluid motion. To develop a multiparticle drag correlation, several similarity hypotheses are introduced. First, it is assumed that the drag coefficient in the viscous regime can be given as a function of the particle Reynolds number. Thus,

$$C_D = C_D\left(N_{\mathrm{Re}}\right) \tag{12-38}$$

where the Reynolds number is defined in terms of the mixture viscosity μ_m as

$$N_{\mathrm{Re}} \equiv \frac{2\rho_c |v_r| r_d}{\mu_m}. \tag{12-39}$$

The introduction of the drag coefficient, Eq.(12-23), and the use of Eq. (12-38) are based on the assumption that the resistance to particle motion in a two-phase mixture can be evaluated by considering the local resistance to the shearing caused by the relative motion between the representative particle and the surrounding fluid. The effect of the other particles on the drag force arises from the resistance of the particle to the deformation of the flow field. Since the particles are more rigid than the fluid against deformations, the particles will impose a system of forces that will react upon the fluid. As a result of additional stresses, the original particles see an increase in the resistance to its motion, which appears to it as arising from an increase of viscosity.

Consequently, in analyzing the motion of the suspended particles, mixture viscosity should be used (Burgers, 1941; Zuber, 1964a). It is expected that the mixture viscosity is a function of concentration, fluid viscosity and particle viscosity. The viscosity of the dispersed phase takes account of the mobility of the interface and is the measure of the resistance to the particle-material motion along the interface. The effect of the particle collisions may be indirectly reflected in the mixture viscosity through the void fraction. Furthermore, for a fluid-particle system, the surface tension should have an effect on the particle collisions and coalescences. This is particularly important in determining the flow-regime transitions.

In the present analysis, we extend the linear correlation (Taylor, 1932) for the mixture viscosity for fluid particles along the power relation (Roscoe, 1952) for solid particles based on the maximum packing α_{dm}. Taylor's viscosity model for a fluid-particle system is given by

$$\frac{\mu_m}{\mu_c} = 1 + 2.5\alpha_d \frac{\mu_d + 0.4\mu_c}{\mu_d + \mu_c} \tag{12-40}$$

which is applicable only for $\alpha_d << 1$. The simple power-law viscosity model for a solid-particle system is given by

$$\frac{\mu_m}{\mu_c}=\left(1-\frac{\alpha_d}{\alpha_{dm}}\right)^{-2.5\alpha_{dm}}. \tag{12-41}$$

This shows that the viscosity of the mixture increases rapidly near the maximum packing. Note also that linear expansion of Eq.(12-41) at small α_d is $\mu_m/\mu_c=1+2.5\alpha_d$, which is similar to Eq.(12-40). The maximum packing α_{dm} for solid-particle systems ranges from 0.5 to 0.74. However, $\alpha_{dm}=0.62$ suffices for most of the practical cases. For a bubbly flow, theoretical α_{dm} can be much higher because of the deformation of bubbles. In the absence of turbulent motions and particle coalescences, the void fraction in a fluid-particle system can be as high as 0.95. By taking α_{dm} to be unity, we can include these foam or dense packing regimes in the analysis. Therefore, for fluid-particle systems, we take $\alpha_{dm}=1$. Combining the above two expressions produces the following model for both a solid-particle system and a fluid-particle system at all concentrations

$$\frac{\mu_m}{\mu_c}=\left(1-\frac{\alpha_d}{\alpha_{dm}}\right)^{-2.5\alpha_{dm}(\mu_d+0.4\mu_c)/(\mu_d+\mu_c)}. \tag{12-42}$$

Figure 12-6 compares this mixture-viscosity model to the various existing models for solid-particle systems (Eilers, 1941; Roscoe, 1952; Brinkman, 1952; Frankel and Acrivos, 1967; Landel et al., 1965; Thomas, 1965). Note that, for a solid particle system, μ_d approaches ∞. Thus, if we take the limit of Eq.(12-42), the viscosity-ratio term becomes unity and the correlation reduces to the power law, Eq.(12-41). By including the effect of viscosity of the dispersed phase in the correlation, this model has the advantage over the conventional correlations, because it is not limited to particulate flows, but can also be applied to droplet and bubble flows.

Using the recommended values for maximum packing, we can approximate the mixture viscosity by a simple power law given by

$$\mu_m/\mu_c=(1-\alpha_d)^{-n} \tag{12-43}$$

$$\text{where}\quad n=\begin{cases}1 & \text{Bubbly flow}\\ 1.75 & \text{Drops in liquid}\\ 2.5 & \text{Drops in gas, particulate flow.}\end{cases} \tag{12-44}$$

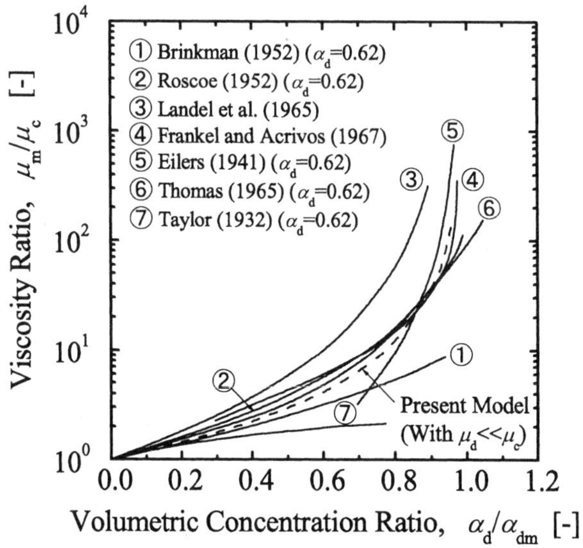

Figure 12-6. Comparison of present mixture-viscosity model and existing models for solid-particle system (Ishii and Chawla, 1979)

The expression for the solid-particle system is applicable only up to a moderate value of α_d. These relations are shown in Fig.12-7.

The second similarity hypothesis introduced in the analysis is that, in the viscous regime, a complete similarity exists between a single-particle system and a multi-particle system. Therefore, the multiparticle drag coefficient C_D has exactly the same functional form in terms of N_{Re} as $C_{D\infty}$ in terms of $N_{\text{Re}\infty}$ given by Eq.(12-29). Then $C_D = C_{D\infty}\left(N_{\text{Re}}\right)$ or

$$C_D = \frac{24}{N_{\text{Re}}}\left(1 + 0.1N_{\text{Re}}^{0.75}\right). \tag{12-45}$$

The relation given by Eq.(12-45) is shown in Fig.12-8. This correlation indicates that the drag coefficient increases with an increasing volumetric concentration α_d. This trend is clearly shown in Fig.12-9 for a solid-particle system by comparing single- and multi-particle systems.

The similarity criterion given by $C_D\left(N_{\text{Re}}\right) = C_{D\infty}\left(N_{\text{Re}}\right)$ with the Reynolds number based on mixture viscosity is first introduced for solid-particles system in the Stokes regime (Hawksley, 1951; Zuber, 1964a). Note that the present model is not limited to a solid-particle system or to the

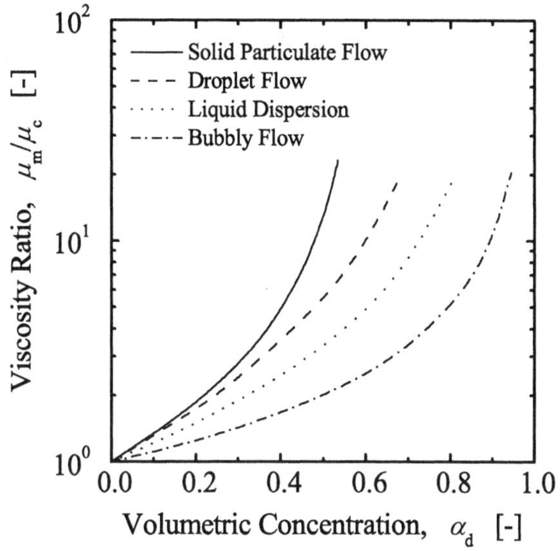

Figure 12-7. Mixture viscosity for various systems (Ishii and Chawla, 1979)

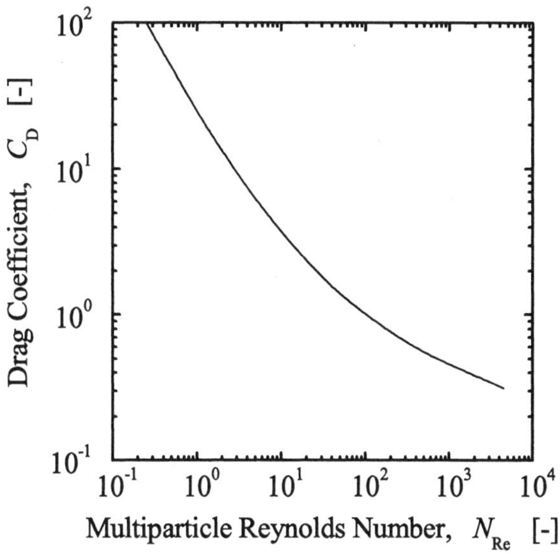

Figure 12-8. Drag coefficient in viscous regime (Ishii and Chawla, 1979)

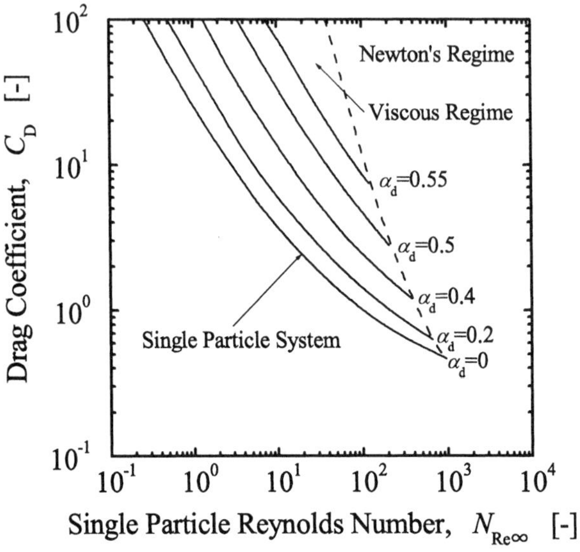

Figure 12-9. Effect of concentration on drag coefficient in viscous regime (Ishii and Chawla, 1979)

Stokes regime, however, because of the use of the generalized drag law and the applicability of the mixture-viscosity model to fluid-particle systems.

C. Newton's Regime

In Newton's regime, a vortex system develops behind a particle and its departure creates a large wake region. The drag force is mainly determined by the eddies generated by a separation of the flow. Hence, for a single-particle system, the drag force is approximately proportional to the inertia force and the drag coefficient can be considered constant.

For a multiparticle system, the drag coefficient in Newton's regime is assumed not to depend on the Reynolds number but on the void fraction. The effect of the other particles should be through α_d. Hence,

$$C_D = 0.45E\left(\alpha_d\right). \tag{12-46}$$

The function $E\left(\alpha_d\right)$ can be obtained by considering a special case of the terminal velocity in an infinite medium. From a force balance between gravity, pressure, and drag forces, we have

$$v_r\left|v_r\right| = \frac{8}{3}\frac{r_d}{C_D\rho_c}\left(\rho_c - \rho_d\right)g\left(1-\alpha_d\right). \tag{12-47}$$

For a single-particle system, this reduces to

$$v_{r\infty}\left|v_{r\infty}\right| = \frac{8}{3}\frac{r_d}{C_{D\infty}\rho_c}\left(\rho_c - \rho_d\right)g. \tag{12-48}$$

By comparing a multiparticle system to a single-particles system having the same particles size, we have

$$\frac{v_r}{v_{r\infty}} = \sqrt{\frac{C_{D\infty}\left(N_{\mathrm{Re}\infty}\right)\left(1-\alpha_d\right)}{C_D\left(N_{\mathrm{Re}}\right)}}. \tag{12-49}$$

Since Reynolds numbers can be a function of the velocities, Eq.(12-49) is an implicit equation for the terminal velocity v_r. If we consider the viscous-regime drag laws given by Eqs.(12-29) and (12-45), Eq.(12-49) becomes

$$\frac{v_r}{v_{r\infty}} = \frac{\mu_c}{\mu_m}\left(1-\alpha_d\right)\frac{1+0.1N_{\mathrm{Re}\infty}^{0.75}}{1+0.1N_{\mathrm{Re}}^{0.75}}. \tag{12-50}$$

The limiting case of $r_d \to 0$ (or $N_{\mathrm{Re}}, N_{\mathrm{Re}\infty} \to 0$) is

$$\lim_{r_d\to 0}\frac{v_r}{v_{r\infty}} = \frac{\mu_c}{\mu_m}\left(1-\alpha_d\right). \tag{12-51}$$

For $r_d \to \infty$ (or $N_{\mathrm{Re}}, N_{\mathrm{Re}\infty} \to \infty$),

$$\lim_{r_d\to \infty}\frac{v_r}{v_{r\infty}} = \left(\frac{\mu_c}{\mu_m}\right)^{1/7}\left(1-\alpha_d\right)^{4/7}. \tag{12-52}$$

By interpolating between these limits in view of Eq.(12-50), we obtain an approximate explicit solution for v_r given by

$$\frac{v_r}{v_{r\infty}} = \left(1-\alpha_d\right)\frac{\mu_c}{\mu_m}\frac{1+0.1N_{\mathrm{Re}\infty}^{0.75}}{1+0.1N_{\mathrm{Re}\infty}^{0.75}\left[\sqrt{1-\alpha_d}\,\mu_c/\mu_m\right]^{6/7}}. \tag{12-53}$$

Since the terminal velocity $v_{r\infty}$ is uniquely related to the Reynolds number by Eq.(12-48), however, $N_{\mathrm{Re}\infty}$ can be replaced by a radius of a particles. Thus,

$$\frac{v_r}{v_{r\infty}} \simeq (1-\alpha_d)\frac{\mu_c}{\mu_m}\frac{1+\psi(r_d^*)}{1+\psi(r_d^*)\left[\sqrt{1-\alpha_d}\,\mu_c/\mu_m\right]^{6/7}} \tag{12-54}$$

where

$$\begin{cases} r_d^* = r_d\left(\rho_c g\Delta\rho/\mu_c^2\right)^{1/3} \\ \text{and} \\ \psi(r_d^*) = 0.55\left[\left(1+0.08r_d^{*3}\right)^{4/7}-1\right]^{0.75}. \end{cases} \tag{12-55}$$

For a single-particles system, the transition from the viscous regime to Newton's regime occurs at $r_d^* = 34.65$ (or $N_{\mathrm{Re}\infty} \simeq 990$). At this particles size, Eq.(12-46) reduces to

$$\frac{v_r}{v_{r\infty}} = (1-\alpha_d)\frac{\mu_c}{\mu_m}\frac{18.67}{1+17.67\left[\sqrt{1-\alpha_d}\,\mu_c/\mu_m\right]^{6/7}}. \tag{12-56}$$

This equation is valid up to the transition from the viscous regime to Newton's regime in a multiparticle system. Therefore, at this transition point the drag-coefficient ratio can be calculated from Eqs.(12-49) and (12-56) as

$$C_D = C_{D\infty}\left(\frac{1+17.67\left(\sqrt{1-\alpha_d}\,\mu_c/\mu_m\right)^{6/7}}{18.67\sqrt{1-\alpha_d}\,\mu_c/\mu_m}\right)^2 \tag{12-57}$$

where $C_{D\infty} = 0.45$ at $r_d^* = 34.65$. In view of Eq.(12-42) with $\alpha_{dm} = 0.62$, we obtain

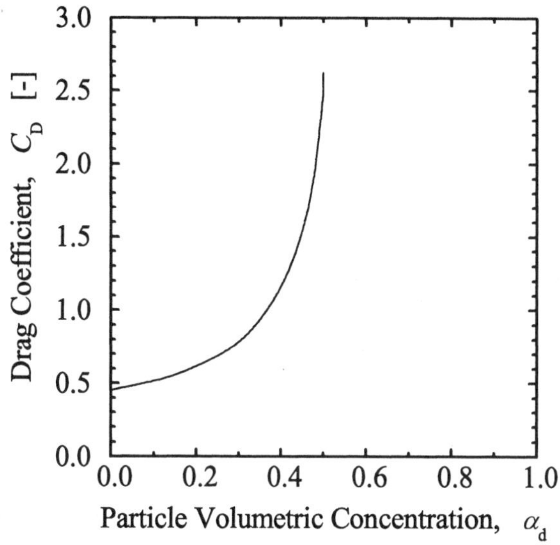

Figure 12-10. Drag coefficient for Newton's regime (Ishii and Chawla, 1979)

$$C_D = 0.45\left(\frac{1+17.67\left[f\left(\alpha_d\right)\right]^{6/7}}{18.67 f\left(\alpha_d\right)}\right)^2$$
$$\text{with } f\left(\alpha_d\right) \equiv \left(1-\alpha_d\right)^{0.5}\left(1-\frac{\alpha_d}{0.62}\right)^{1.55}. \tag{12-58}$$

The increase of the drag coefficient with increasing volumetric concentration α_d is shown in Fig.12-10 by plotting Eq.(12-58). This implies that the equilibrium relative velocity generally decreases with increases in the concentration due to stronger coupling between phases.

D. Distorted-fluid-particle Regime

In the distorted-fluid-particle regime, the single particle drag coefficient depends only on the particle radius and fluid properties and not on the velocity or the viscosity, namely, $C_{D\infty} = \left(4/3\right) r_d \sqrt{g\Delta\rho/\sigma}$, as discussed by Harmathy (1960). Thus, for a particle of a fixed diameter, $C_{D\infty}$ becomes constant. In considering the drag coefficient for a multiparticle system with the same radius, we must take into account the restrictions imposed by the existence of other particles on the flow field. Therefore, C_D is expected to be different from $C_{D\infty}$, in this regime.

Because of the strong contribution of the turbulent eddies on the wake region, a particle sees the increased drag due to other particles in essentially similar ways as in the Newton's regime for a solid-particle system where $C_{D\infty}$ is also constant under a wake-dominated flow condition. Hence, we postulate that, regardless of the differences in $C_{D\infty}$ in these regimes, the effect of increased drag in the distorted-fluid-particle regime can be predicted by an expression similar to that in the Newton's regime.

Under this assumption, Eq.(12-57) can be used with a proper expression for $C_{D\infty}$ given by Eq. (12-43). Thus,

$$C_D = \frac{\sqrt{2}}{3} N_\mu N_{\mathrm{Re}\infty} \left(\frac{1+17.67\left(\sqrt{1-\alpha_d}\mu_c/\mu_m\right)^{6/7}}{18.67\sqrt{1-\alpha_d}\mu_c/\mu_m} \right)^2. \tag{12-59}$$

In view of the approximation given by Eq.(12-43), the above correlation reduces to

$$C_D = \frac{\sqrt{2}}{3} N_\mu N_{\mathrm{Re}\infty} \left(\frac{1+17.67\left(1-\alpha_d\right)^{6(n+0.5)/7}}{18.67\left(1-\alpha_d\right)^{n+0.5}} \right)^2 \tag{12-60}$$

where n is given by Eq.(12-44). Thus, for a bubbly flow ($n = 1$),

$$C_D = \frac{\sqrt{2}}{3} N_\mu N_{\mathrm{Re}\infty} \left(\frac{1+17.67\left(1-\alpha_d\right)^{1.3}}{18.67\left(1-\alpha_d\right)^{1.5}} \right)^2. \tag{12-61}$$

For a droplet-liquid flow ($n = 1.75$)

$$C_D = \frac{\sqrt{2}}{3} N_\mu N_{\mathrm{Re}\infty} \left(\frac{1+17.67\left(1-\alpha_d\right)^{1.9}}{18.67\left(1-\alpha_d\right)^{2.3}} \right)^2. \tag{12-62}$$

For a droplet-gas flow ($n = 2.5$)

$$C_D = \frac{\sqrt{2}}{3} N_\mu N_{\mathrm{Re}\infty} \left(\frac{1+17.67\left(1-\alpha_d\right)^{2.6}}{18.67\left(1-\alpha_d\right)^{3}} \right)^2. \tag{12-63}$$

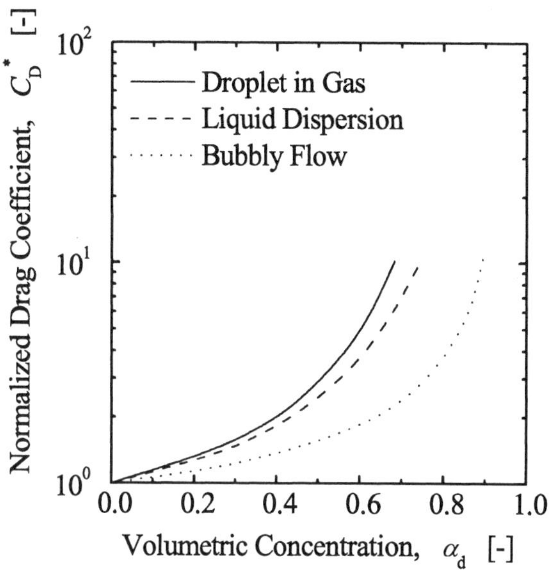

Figure 12-11. Drag coefficient in distorted-particle regime (Ishii and Chawla, 1979)

The above three correlations are shown in Fig.12-11. The form of the correlations indicates that the momentum coupling between phases increases with increasing particle concentration as in the case for Newton's regime.

E. Churn-turbulent-flow Regime

As the radius of the fluid particle is further increased, the wake and bubble boundary layer can overlap due to the formation of large wake regions. In other words, a particle can influence both the surrounding fluid and other particles directly. Hence, the entrainment of a particle in a wake of other particles becomes possible. This flow regime is known as the churn-turbulent flow regime and is commonly observed in bubbly flows. In the existence of sufficient turbulent motions in the continuous phase, the transition from the distorted-particle regime to the churn-turbulent flow regime occurs at the particle concentration around 0.3. This criterion for the transition can be applied to most forced-convection two-phase flows. In a batch process, however, detailed coalescence mechanisms and surface contaminations become important in determining the transition criterion.

In the churn-turbulent flow regime, a typical particle moves with respect to the average volumetric flux j rather than the average velocity of a continuous phase due to the hydrodynamic conditions discussed above.

Hence, the reference velocity in the definitions of the drag coefficient and the drag-similarity law should be the drift velocity rather than the relative velocity. Hence, the drag force should be given by

$$F_D = -\frac{1}{2} C'_D \rho_c V_{dj} \left| V_{dj} \right| \pi r_d^2. \tag{12-64}$$

Here the drift velocity V_{dj} is the relative velocity of the dispersed phase with respect to the center-of-volume velocity of a mixture. It can also be related to the true relative velocity between phases by

$$V_{dj} = v_d - j = (1 - \alpha_d) v_r \tag{12-65}$$

where the total flux j (center-of-volume velocity) is given by

$$j = \alpha_d v_d + (1 - \alpha_d) v_c. \tag{12-66}$$

In a churn-turbulent-flow regime, some particles should have reached the distortion limit corresponding to the cap-bubble transition or the droplet disintegration. This limit can be given as an extension of the Weber number criterion (Wallis, 1969) by using the drift velocity as a reference velocity in the following form

$$\frac{2\rho_c V_{dj}^2 r_d}{\sigma} = \begin{cases} 8 & \text{(bubble)} \\ 12 & \text{(droplet)}. \end{cases} \tag{12-67}$$

Due to the entrainment of particles in the wake of other larger particles and the coalescence and disintegration caused by the turbulence, the average motion of the dispersed phase is mainly governed by those particles that satisfy the Weber-number criterion. Hence, the effective drag coefficient is given by $C'_D = 8/3$. If we recast the above drag-force expression based on the drift velocity to the conventional one based on the relative velocity, we obtain

$$F_D = -\frac{8}{3}(1 - \alpha_d)^2 \frac{\rho_c v_r \left| v_r \right| \pi r_d^2}{2} \tag{12-68}$$

where the reference r_d for the drag-force expression is given by $r_d = (4 \text{ or } 6)\sigma / \left[\rho_c v_r^2 (1 - \alpha_d)^2\right]$, because these particles govern the relative

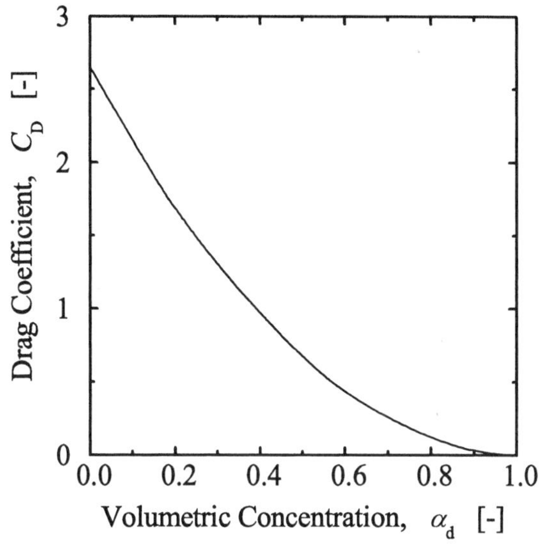

Figure 12-12. Drag coefficient for churn-turbulent flow (Ishii and Chawla, 1979)

motion. The above equation implies that the apparent drag coefficient based on the true relative velocity between phases should be given by

$$C_D = \frac{8}{3}(1-\alpha_d)^2 . \tag{12-69}$$

The form of Eq.(12-69) indicates that the drag coefficient decreases as the particle volumetric concentration increases, as shown in Fig.12-12. Therefore, the effect of α_d on C_D in the churn-turbulent-flow regime is opposite that in the other flow regimes. This peculiar trend can be explained by the effect of the entrainment of other particles behind a wake of larger particles. This entrainment promotes the channeling of the dispersed phase without increasing the drag force. As the volumetric concentration increases, the interaction among particles increases in the direction of reducing the drag force.

F. Slug-flow Regime

One of the limiting cases of the dispersed two-phase flow in a confined channel is a slug flow. When the volume of a bubble becomes very large, the shape of the bubble is significantly deformed to fit the channel geometry. The diameters of the bubbles become nearly that of the pipe with a thin liquid film separating the bubbles from the wall. The bubbles have an elongated bullet form with a cap-shaped nose. The motion of these bubbles in relatively inviscid fluids can be studied by using a potential flow analysis around a nose of a single bubble. Hence, Dumitrescu (1943) analytically obtained the rise velocity to be

$$v_{r\infty} = 0.35\sqrt{gD\Delta\rho/\rho_c} \tag{12-70}$$

where D is the hydraulic diameter. This result is also in good agreement with the experimental data of Dumitrescu (1943) and White and Beardmore (1962).

In a flowing system with chains of bubbles, the effect of the concentration and velocity profile should be considered. In general, the core velocity is higher than the cross-sectional area-averaged velocity due to the velocity profile. Therefore, the relative velocity based on the average velocities is larger than the local relative velocity in the core. This effect, known as the distribution-parameter effect, was studied extensively by Bankoff (1960), Zuber and Findlay (1965), and Ishii (1977) among others. When the average velocities are used, the results of Nicklin et al. (1962) and Neal (1963) show that

$$v_d - \langle j \rangle = 0.2\langle j \rangle + 0.35\sqrt{gD\Delta\rho/\rho_c}. \tag{12-71}$$

Here the left-hand side is the drift velocity of a bubble, namely, $\overline{V_{dj}} = v_d - \langle j \rangle$. The above equation can be rewritten as

$$v_d - j_{core} = 0.35\sqrt{gD\Delta\rho/\rho_c} \tag{12-72}$$

where $j_{core} \equiv 1.2\langle j \rangle$ and j_{core} may be considered as the local total flux in the core. In this case, $v_d - j_{core}$ is the local drift velocity in the core. In view of the relation given by Eq.(12-65), the local relative velocity v_r should satisfy

$$(1-\alpha_d)v_r = 0.35\sqrt{gD\Delta\rho/\rho_c} \tag{12-73}$$

which agrees with Dumitrescu's result at $\alpha_d \to 0$.

By limiting our discussion to the local drag coefficient, we can recast the above semiempirical result into a correlation for a drag coefficient in the slug-flow regime. In view of Eqs.(12-73) and (12-47), we obtain

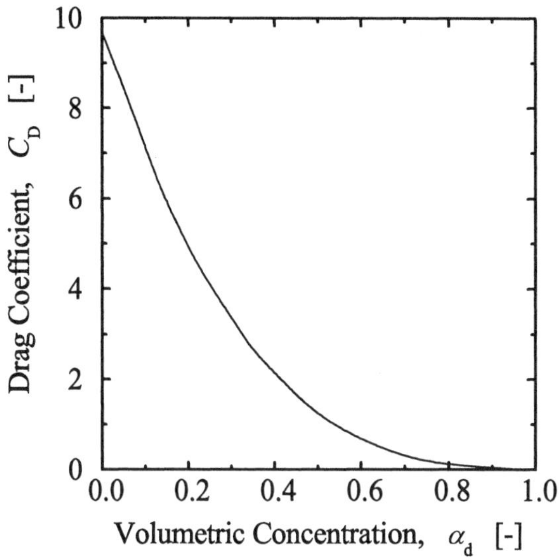

Figure 12-13. Drag coefficient for slug flow (Ishii and Chawla, 1979)

$$C_D = 10.9 \frac{2r_d}{D} \left(1 - \alpha_d\right)^3 . \tag{12-74}$$

For most practical applications, $2r_d / D$ can be approximated by 0.9. Then,

$$C_D \simeq 9.8 \left(1 - \alpha_d\right)^3 . \tag{12-75}$$

This correlation shows that the drag coefficient decreases with increase in the volumetric concentration, as shown in Fig.12-13. This clearly indicates the effect of the wake and channeling in the chains of bubbles in the slug-flow regime. Furthermore, C_D does not depend on the fluid properties. These two characteristics are similar to those of the churn-turbulent-flow regime. Table 12-1 summarizes the present drag coefficient in various flow regimes.

1.3 Other forces

In addition to the standard drag force and the virtual mass force, some forces such as the lift force and the turbulent dispersion forces are also

Table 12-1. Local drag coefficient in multiparticle system (Ishii and Chawla, 1979)

	Fluid Particle System			Solid Particle System
	Bubble in Liquid	Drop in Liquid	Drop in Gas	
Viscosity Model	$\frac{\mu_m}{\mu_c}=\left(1-\frac{\alpha_d}{\alpha_{dm}}\right)^{-2.5\alpha_{dm}\mu^*}$, $\mu^*\equiv\frac{\mu_d+0.4\mu_c}{\mu_d+\mu_c}$			
Max. Packing α_{dm}	~ 1	~ 1	$0.62\sim 1$	~ 0.62
μ^*	0.4	~ 0.7	1	1
μ_m/μ_c	$(1-\alpha_d)^{-1}$	$(1-\alpha_d)^{-1.75}$	$\sim(1-\alpha_d)^{-2.5}$	$\left(1-\frac{\alpha_d}{0.62}\right)^{-1.55}$
Stokes Regime C_D	$C_D=\frac{24}{N_{Re}}$ where $N_{Re}\equiv\frac{2r_d\rho_c v_r}{\mu_m}$			
Viscous Regime C_D	$C_D=\frac{24\left(1+0.1N_{Re}^{0.75}\right)}{N_{Re}}$			
Newton's Regime C_D	———			$C_D=0.45\left[\frac{1+17.67\{f(\alpha_d)\}^{6/7}}{18.67f(\alpha_d)}\right]^2$
Distorted Particle Regime C_D	$C_D=\frac{4}{3}r_d\sqrt{\frac{g\Delta\rho}{\sigma}}\left[\frac{1+17.67\{f(\alpha_d)\}^{6/7}}{18.67f(\alpha_d)}\right]^2$ $f(\alpha_d)=(1-\alpha_d)^{1.5}$	$(1-\alpha_d)^{2.25}$	$(1-\alpha_d)^3$	where $f(\alpha_d)=\sqrt{1-\alpha_d}\left(\frac{\mu_c}{\mu_m}\right)$
Churn-turbulent Flow Regime C_D	$C_D=\frac{8}{3}(1-\alpha_d)^2$			
Slug Flow C_D	$C_D=9.8(1-\alpha_d)^3$			

considered in a multidimensional two-phase flow analysis. As can be seen from Eq.(12-3), these forces are customarily added to the standard drag force and the virtual mass force linearly. In an actual two-phase flow, a wake behind a bubble may completely change the liquid turbulence structure, and thus the lift force may closely be coupled with the turbulence-induced force. Since the lift force and other lateral forces are small and may closely be coupled each other, it is difficult to identify each force experimentally. Therefore, it may be controversial that such forces can be added to the standard drag force and the virtual mass force linearly. In the present status of the development, constitutive equations for some forces are proposed

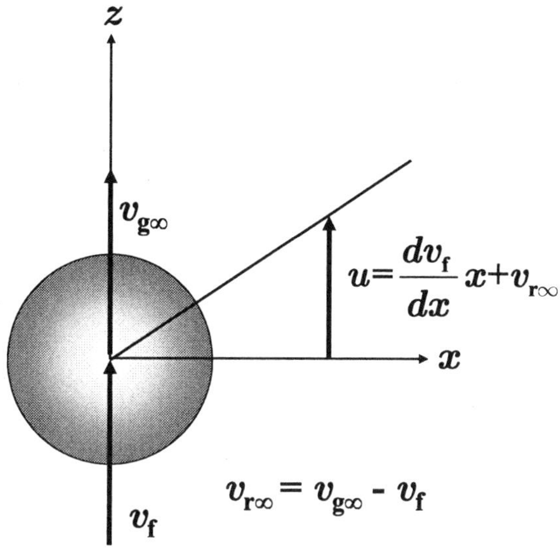

Figure 12-14. Schematic diagram of a particle in a shear flow

based on a speculation, and the applicable flow ranges of the constitutive equations are not given clearly. Thus, unlike the standard drag force, the constitutive equations for such lateral forces have not been well-developed. Nevertheless, these forces play an important role in predicting three-dimensional bubble distribution. In what follows, some constitutive equations for lift force and turbulent dispersion force, which are often used in a multidimensional two-phase flow analysis, are explained briefly. A review of lift force modeling can also be found in Akiyama and Aritomi (2002).

1.3.1 Lift force

Consider a single spherical particle moving through a very viscous liquid relative to a uniform simple shear, see Fig.12-14. Then, the particle experiences a lift force, $\boldsymbol{F}_{d\infty}^{LS-V}$, perpendicular to the flow direction as (Saffman, 1965)

$$\boldsymbol{F}_{d\infty}^{LS-V} = -6.46\mu_f \nu_f^{-1/2} v_{r\infty} r_d^2 \left|\frac{dv_f}{dx}\right|^{1/2} \operatorname{sgn}\left(\frac{dv_f}{dx}\right) \boldsymbol{e}_x. \tag{12-76}$$

For positive relative velocity and velocity gradient ($v_{r\infty} > 0$ and $dv_f/dx > 0$), the lift force pushes the particle towards the negative x direction.

Consider a single spherical particle placed in a weak shear flow of an inviscid flow. Then, the particle experiences a lift force, $\boldsymbol{F}_{d\infty}^{LS-I}$, as (Auton, 1987)

$$\boldsymbol{F}_{d\infty}^{LS-I} = -0.5\rho_f \frac{4}{3}\pi r_d^3 \boldsymbol{v}_{r\infty} \times \mathrm{rot}\boldsymbol{v}_f. \tag{12-77}$$

Mei and Klausner (1994) proposed an expression for the shear lift force at finite Reynolds number and finite shear by interpolating Saffman's result at small $N_{\mathrm{Re}\infty}$ (1965) and Auton's result at large $N_{\mathrm{Re}\infty}$ (1987). They also considered the extension of the lift force to a fluid sphere. The proposed lift force model is given by

$$\boldsymbol{F}_{d\infty}^{LS} = -f\left(G_{s\infty}, N_{\mathrm{Re}\infty}\right)\frac{1}{2}\pi r_d^2 \rho_f v_{r\infty}^2 \,\mathrm{sgn}\left(v_{r\infty}\frac{dv_f}{dx}\right)\boldsymbol{e}_x \tag{12-78}$$

$$f\left(G_{s\infty}, N_{\mathrm{Re}\infty}\right) = G_{s\infty}^{1/2}\left[\left\{\frac{1.72J\left(\sqrt{2G_{s\infty}/N_{\mathrm{Re}\infty}}\right)}{N_{\mathrm{Re}\infty}^{1/2}}\right\}^2 + \frac{16}{9}G_{s\infty}\right]^{1/2} \tag{12-79}$$

$$J\left(\sqrt{2G_{s\infty}/N_{\mathrm{Re}\infty}}\right) \approx 0.6765\left[1 + \tanh\left\{2.5\left(\log_{10}\sqrt{2G_{s\infty}/N_{\mathrm{Re}\infty}} + 0.191\right)\right\}\right] \times\left[0.667 + \tanh\left\{6\left(\sqrt{2G_{s\infty}/N_{\mathrm{Re}\infty}} - 0.32\right)\right\}\right] \tag{12-80}$$

$$G_{s\infty} \equiv \left|\frac{r_d}{v_{r\infty}}\frac{dv_f}{dx}\right|. \tag{12-81}$$

The Reynolds number of a single particle system, $N_{\mathrm{Re}\infty}$, is defined by Eq.(12-27).

In the 1980s and 1990s, extensive experiments were performed to identify important parameters to determine the lateral bubble migration characteristics. The experiments showed that relatively small and large bubbles tend to migrate toward a channel wall and center, respectively (Zun, 1988; Liu, 1993; Hibiki and Ishii, 1999). A numerical simulation of single bubbles in a Poiseuille flow (Tomiyama et al., 1993; 1995) suggested that the bubble migration toward the pipe center was related closely to a slanted wake behind a deformed bubble. Thus, it has been indicated that the bubble size and complex interaction between a bubble wake and a shear field around the bubble play an important role in the lateral bubble migration (Serizawa and Kataoka, 1988; 1994). Tomiyama et al. (2002) measured bubble trajectories of single air bubbles in simple shear flows of glycerol-water solutions to evaluate transverse lift force acting on single bubbles. Based on the experimental result, they assumed the lift force caused by the slanted wake has the same functional form as that of the shear-induced lift force, and proposed an empirical correlation of the lift coefficient.

Hibiki and Ishii proposed the correlation of the lift coefficient based on the shear-lift force model of Mei and Klausner (1994) and the concept of the lift force caused by slanted wake, $\boldsymbol{F}_{d\infty}^{LW}$, (Tomiyama et al., 2002) such as the functional form of $\boldsymbol{F}_{d\infty}^{LW}$ to be the same as that of $\boldsymbol{F}_{d\infty}^{LS}$. Thus, the net transverse lift force, $\boldsymbol{F}_{d\infty}^{LT}$, is given by

$$\begin{aligned} \boldsymbol{F}_{d\infty}^{LT} &= \boldsymbol{F}_{d\infty}^{LS} + \boldsymbol{F}_{d\infty}^{LW} \\ &= -f\left(G_{s\infty}, N_{\mathrm{Re}\infty}\right)\frac{1}{2}\pi r_d^2 \rho_f v_{r\infty}^2 \,\mathrm{sgn}\left(v_{r\infty}\frac{dv_f}{dx}\right)\boldsymbol{e}_x \\ &\quad -C_{LW} f\left(G_{s\infty}, N_{\mathrm{Re}\infty}\right)\frac{1}{2}\pi r_d^2 \rho_f v_{r\infty}^2 \,\mathrm{sgn}\left(v_{r\infty}\frac{dv_f}{dx}\right)\boldsymbol{e}_x \end{aligned} \tag{12-82}$$

where C_{LW} is the coefficient of the lift force caused by slanted wake. Here, we assume that the lift force caused by slanted wake pushes the particle towards the negative x direction for positive relative velocity. The coefficient was determined based on the data of Tomiyama et al. (2002) taken under the conditions of $-5.5 \le \log_{10} M \le -2.8$, $1.39 \le Eo \le 5.74$ and $0 \le \left|dv_f/dx\right| \le 8.3\ s^{-1}$, where M and Eo are the Morton number and the Eötvös number, respectively, as defined by

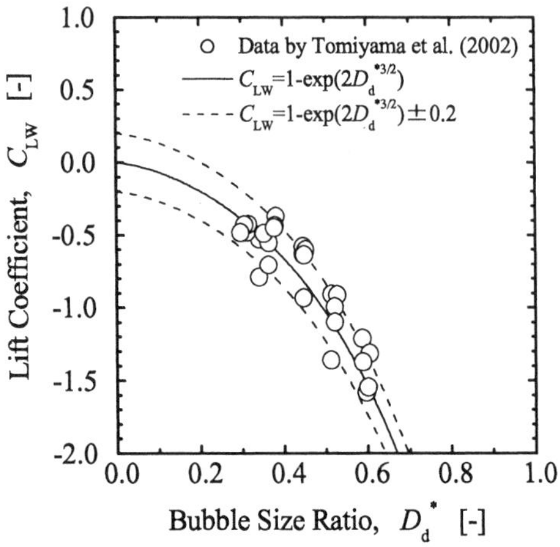

Figure 12-15. Dependence of lift coefficient on bubble size

$$M \equiv \frac{g\left(\rho_f - \rho_g\right)\mu_f^4}{\rho_f^2 \sigma^3} \tag{12-83}$$

$$Eo \equiv \frac{g\left(\rho_f - \rho_g\right)d_b^2}{\sigma}. \tag{12-84}$$

Then, as shown in Fig.12-15, the following correlation of the lift coefficient is proposed based on the data of Tomiyama et al. (2002) as

$$C_{LW} = 1 - \exp\left(2D_d^{*3/2}\right) \tag{12-85}$$

where D_d^* is the ratio of the bubble diameter to the bubble diameter at the distorted bubble limit as

$$D_d^* \equiv \frac{d_b}{4\sqrt{\dfrac{\sigma}{g\Delta\rho}}} = \frac{Eo^{1/2}}{4}. \tag{12-86}$$

We may extend the applicability of Eq.(12-82) in a single particle system to a multiparticle system by replacing $N_{\mathrm{Re}\infty}$ and $G_{s\infty}$ in Eq.(12-82) with N_{Re} and G_s, where the Reynolds number of a multiparticle system, N_{Re}, is defined by Eq.(12-39) and the non-dimensional velocity gradient, G_s, is defined by

$$G_s \equiv \left| \frac{r_d}{v_r} \frac{dv_f}{dx} \right|. \tag{12-87}$$

Thus, the net transverse lift force in multiparticle system, $\boldsymbol{M}_d^L$, is approximated as

$$\begin{aligned} \boldsymbol{M}_d^L &= \frac{\alpha_g}{B_d} \boldsymbol{F}_d^{LT} = -f\left(G_s, N_{\mathrm{Re}}\right) \frac{3\alpha_g \rho_f v_r^2}{8 r_d} \operatorname{sgn}\left(v_r \frac{dv_f}{dx}\right) \boldsymbol{e}_x \\ &-C_{LW} f\left(G_s, N_{\mathrm{Re}}\right) \frac{3\alpha_g \rho_f v_r^2}{8 r_d} \operatorname{sgn}\left(v_r \frac{dv_f}{dx}\right) \boldsymbol{e}_x. \end{aligned} \tag{12-88}$$

A reasonable agreement between the lift force calculated by Eq.(12-88) with Eq.(12-85) and air-water bubbly flow data (Wang et al., 1987) was obtained, which implies the lift force model, Eq.(12-88) with Eq.(12-85), to be promising to predict the net transverse lift force in multiparticle bubbly flow. Further efforts to examine the applicability of Eq.(12-88) with Eq.(12-85) to a multiparticle system should be made in a future study. As described above, the lift force is still poorly understood, and further experimental and numerical efforts are needed to understand the lift force (Sridhar and Katz, 1995; Ervin and Tryggvason, 1997; Loth et al., 1997).

1.3.2 Wall-lift (wall-lubrication) force

The wall-lift force $\boldsymbol{M}_d^W$ has been introduced and it is explained due to the velocity distribution change around particles near a wall (Antal et al., 1991). This force was used to predict the observed void profiles for cocurrent laminar upward and downward flows. This wall-lift force analogous to a lubrication force acts on a bubble near a wall and prevents the bubbles from touching the wall.

Consider the drainage of the liquid around a bubble moving in the vicinity of a wall. The no-slip condition at the wall should slow the drainage rate between the bubble and the wall, whereas the drainage rate should be increased on the opposite side. Thus, the asymmetrical drainage of the liquid around the bubble in the vicinity of the wall may be quite different from the symmetrical drainage of the liquid around the bubble in infinite liquid. As a consequence, the bubble experiences a hydrodynamic force, namely wall-lift force, which tends to move the bubble away from the wall. Antal et al. (1991) investigated the wall-lift force acting on a spherical bubble moving in a laminar flow analytically and numerically, and proposed the following functional form

$$\boldsymbol{M}_d^W = \frac{\alpha_g \rho_f \left|\boldsymbol{v}_\parallel\right|^2}{r_d}\left[C_{W1} + C_{W2}\left(\frac{r_d}{d_{bw}}\right)\right]\boldsymbol{n}_W \tag{12-89}$$

where $\boldsymbol{v}_\parallel = \left(\boldsymbol{v}_g - \boldsymbol{v}_f\right) - \left[\boldsymbol{n}_W \cdot \left(\boldsymbol{v}_g - \boldsymbol{v}_f\right)\right]\boldsymbol{n}_W$, $C_{W1} = -0.104 - 0.06 v_r$ and $C_{W2} = 0.147$. Here, d_{bw} and $\boldsymbol{n}_W$ are the distance between the bubble and the wall, and the unit outward normal vector on the surface of the wall, respectively. Equation (12-89) indicates that the direction of the wall-lift force is reversed at $y_0 = r_d / \left(0.707 + 0.408 v_r\right)$, and does not take into account the effect of the bubble deformation near the wall. Further experimental and analytical works should be required to establish the wall-lift force.

1.3.3 Turbulent dispersion force

The turbulent dispersion force $\boldsymbol{M}_d^T$ is due to the bubble motion produced by the turbulent energy of the liquid phase (Lahey et al., 1993). This force was introduced to compensate for the fact that the averaged two-phase continuity equations do not allow for a phasic diffusion effect. The turbulence dispersion force is driven by the void fraction gradient, and tends to flatten the void fraction distribution. On the analogy of the molecular dispersion force, the turbulent dispersion force is expressed by (Lahey et al., 1993)

$$\boldsymbol{M}_d^T = -C_T \rho_f k_f \nabla \alpha_g \tag{12-90}$$

where $C_T = 0.1$ and k_f is the total turbulent kinetic energy of the liquid phase. The applicable flow range of Eq.(12-90) is not clearly given.

1.4 Turbulence in multiparticle system

In a single-phase flow analysis, turbulence structure has been studied extensively, and several turbulence models have been proposed. The turbulence model is commonly classified into zero-equation model, one-equation model, two-equation model, and stress equation model. The large eddy simulation and direct numerical simulation of the Navier-Stokes equation are also possible. However, in a two-phase flow analysis, limited studies have been performed for turbulence modeling due to the complex nature of the two-phase flow turbulence. In what follows, some preliminary turbulence models in bubbly flow regime are explained briefly. In the models, the turbulent kinetic energy of gas phases is commonly neglected due to the large density difference between gas and liquid phases. To emphasize the time-average, the overbar is applied to a symbol. This subject is also reviewed in Akiyama and Aritomi (2002).

A. Zero-equation Model

The zero-equation turbulence model is commonly expressed as a model with no differential equation to determine the Reynolds stress. In what follows, a model proposed by Sato et al. (1981) will be explained briefly as an example of the zero-equation turbulence model.

Consider two-dimensional fully-developed bubbly flows such as a flow in a vertical pipe or between two parallel flat walls. The y- and z-axes are, respectively, normal and parallel to the main flow direction. This model assumes that (i) only liquid phase contributes to the momentum transfer, and (ii) there are two kinds of turbulences in the liquid phase independent of and dependent on bubble agitation. Then, the velocities in the y- and z-directions, $v_{f,y}$ and $v_{f,z}$ are expressed as

$$v_{f,y} = v'_{f,y} + v''_{f,y} \tag{12-91}$$

$$v_{f,z} = \overline{v_{f,z}} + v'_{f,z} + v''_{f,z} \tag{12-92}$$

where v'_f and v''_f are the liquid velocity fluctuations independent of and dependent on bubble agitation, respectively. $\overline{v_{f,z}}$ is the time-averaged liquid velocity in the z-direction. The turbulent stress for liquid phase, τ_f^T, can be expressed as

$$\tau_f^T = -\rho_f \overline{v'_{f,y} v'_{f,z}} - \rho_f \overline{v''_{f,y} v''_{f,z}} \tag{12-93}$$

where y is the normal-directional distance measured from the channel wall. Here, the eddy diffusivities ε' and ε'' are defined by

$$-\overline{v'_{f,y}v'_{f,z}} = \varepsilon' \frac{\partial \overline{v_{f,z}}}{\partial y} \text{ and } -\overline{v''_{f,y}v''_{f,z}} = \varepsilon'' \frac{\partial \overline{v_{f,z}}}{\partial y}. \tag{12-94}$$

Then, the turbulent stress, τ_f^T, is expressed as

$$\tau_f^T = \rho_f \left(\varepsilon' + \varepsilon''\right) \frac{\partial \overline{v_{f,z}}}{\partial y}. \tag{12-95}$$

Thus, the turbulent stress distribution can be calculated provided ε' and ε'' are given.

For fully developed turbulent bubbly flow, the eddy diffusivity independent of bubble agitation, ε', may be determined empirically based on the Prandtl's mixing length theory and the damping factor in the region close to the smooth wall as

$$\begin{aligned}\varepsilon' &= 0.4\left\{1 - \exp\left(-\frac{y^+}{16}\right)\right\}^2 \\ &\times \left\{1 - \frac{11}{6}\left(\frac{y^+}{R^+}\right) + \frac{4}{3}\left(\frac{y^+}{R^+}\right)^2 - \frac{1}{3}\left(\frac{y^+}{R^+}\right)^3\right\} \nu_f y^+.\end{aligned} \tag{12-96}$$

Here, $y^+ \equiv y v_f^* / \nu_f$ and $R^+ \equiv R v_f^* / \nu_f$ where v_f^* and R are, respectively, the friction velocity defined by $\sqrt{\tau_W / \rho_f}$ and the radius of a pipe or the half width of a channel with parallel flat walls. The eddy diffusivity that is dependent on bubble agitation, ε'', may be determined empirically based on the virtual kinematic viscosity of a free turbulent flow (such as a wake behind a solid body) and the damping factor in the region close to the smooth wall as

$$\varepsilon'' = 1.2\left\{1 - \exp\left(-\frac{y^+}{16}\right)\right\}^2 \alpha_g \left(\frac{d_B}{2}\right) v_{r\infty} \tag{12-97}$$

where d_B and $v_{r\infty}$ are the mean diameter of the bubbles given by Eq.(12-98) and the terminal velocity in the still liquid, respectively.

$$d_B = \begin{cases} 0 & 0\ \mu\text{m} < y < 20\ \mu\text{m} \\ 4y\left(\widehat{d_B} - y\right)/\widehat{d_B} & 20\ \mu\text{m} \le y \le \widehat{d_B}/2 \\ \widehat{d_B} & \widehat{d_B}/2 < y < R \end{cases} \tag{12-98}$$

where $\widehat{d_B}$ is the cross-sectional mean diameter of the bubbles. The turbulent stress is computed from the above equations and boundary conditions.

B. One-equation Model

The one-equation turbulence model is commonly expressed as a model with only one differential equation of turbulent kinetic energy conservation and constitutive equations for mixing length and other turbulent source terms. In what follows, a model proposed by Kataoka and Serizawa (1995) will be explained briefly as an example of the one-equation turbulence model.

Consider steady, fully developed adiabatic bubbly flows in a round tube. The y - and z-axes are, respectively, normal and parallel to the main flow direction. The turbulent stress, τ_f^T, is expressed in terms of the mixing length of two-phase flow, l_{TP}, and the turbulent velocity, v_t, as

$$\tau_f^T = \rho_f l_{TP} v_t \frac{d\overline{v_{f,z}}}{dy} \tag{12-99}$$

$$v_t \equiv \sqrt{\frac{\overline{\boldsymbol{v}_f' \cdot \boldsymbol{v}_f'}}{3}} \tag{12-100}$$

where $\boldsymbol{v}_f'$ is the liquid velocity fluctuation vector. Thus, the turbulent stress distribution can be calculated provided l_{TP} and v_t are given.

The mixing length of two-phase flow is assumed to be expressed by the linear superposition of the mixing length of the shear-induced turbulence in single-phase liquid flow, l_{SP}, and the mixing length due to the bubble-induced turbulence, l_B, as

$$l_{TP} = l_{SP} + l_B. \tag{12-101}$$

The mixing length of the shear-induced turbulence in single-phase liquid flow is given by

$$l_{SP} = 0.4y\left\{1-\exp\left(-\frac{yv_f^*}{26\nu_f}\right)\right\}. \tag{12-102}$$

The mixing length of the bubble-induced turbulence is given based on the mechanistic model in which the same volume of liquid is exchanged accompanying the bubble turbulent motion through the control surface as

$$l_B = \begin{cases} \frac{1}{3} d_B \alpha_g & \frac{3}{2} d_B \le y \le R \\ \frac{1}{6}\left\{d_B + \left(y - 0.5 d_B\right)\alpha_g\right\} & d_B \le y \le \frac{3}{2} d_B \\ \frac{1}{6}\left[d_B + \frac{4/3 - \left(y/d_B\right)}{2 - \left(4/3\right)\left(y/d_B\right)}\right]\alpha_g & 0 \le y \le d_B \end{cases} \tag{12-103}$$

where d_B is the bubble diameter.

The turbulence velocity, v_t, is calculated from the equation of the turbulent kinetic energy for liquid phase given by

$$\begin{aligned} &\frac{1}{R-y}\frac{\partial}{\partial y}\left\{(R-y)\left(1-\alpha_g\right)\left(\frac{\nu_f}{2}+\beta_2\sqrt{k}l_{TP}\right)\frac{\partial k}{\partial y}\right\} \\ &+\beta_1\sqrt{k}l_{TP}\left(1-\alpha_g\right)\left(\frac{\partial \overline{v_{f,z}}}{\partial y}\right)^2 - \gamma_1\left(1-\alpha_g\right)\frac{\left(\sqrt{k}\right)^3}{l_{TP}} \\ &-K_2\alpha_g\frac{\left(\sqrt{k}\right)^3}{d_B} + K_1\frac{3}{4d_B}\alpha_g C_D v_{r\infty}^3\left\{1-\exp\left(-\frac{yv_f^*}{26\nu_f}\right)\right\} \\ &-\nu_f\left(\frac{\partial\sqrt{k}}{\partial y}\right)^2 = 0 \end{aligned} \tag{12-104}$$

where k is the turbulent kinetic energy of liquid phase given by $k=\overline{v_f'\cdot v_f'}/2$. $\beta_1(=0.56)$, $\beta_2(=0.38)$, $\gamma_1=0.18$, $K_1(=0.075)$, and $K_2(=1.0)$ are coefficients. In this model, the turbulent velocity is determined by Eq.(12-100) with the assumption of equilateral turbulence as

$$v_t = \sqrt{\frac{2k}{3}}. \tag{12-105}$$

The significance of the various terms in the equation is as follows. The first, second, and third terms on the right-hand side represents the turbulence diffusion, turbulence generation due to shear, and the turbulence dissipation, respectively. The fourth and fifth terms represent the turbulence absorption due to small scale of interface and the turbulence generation due to bubble relative motion, respectively. The last term represents the compensation of numerical error very near to the wall. The turbulent stress is computed from the above equations and boundary conditions.

C. Two-equation Model

The two-equation turbulence model is commonly expressed as a model with two differential equations to determine the Reynolds stress. In what follows, k-ε model proposed by Lopez de Bertodano et al. (1994) will be explained briefly as an example of the two-equation turbulence model.

Consider steady, fully developed adiabatic dilute bubbly flows. The turbulent stress tensor, $\mathcal{T}_f^T$, is assumed to be expressed by the linear superposition of the shear-induced (SI) turbulent stress tensor, $\mathcal{T}_f^{SI}$, and the bubble-induced (BI) turbulent tensor, $\mathcal{T}_f^{BI}$, as

$$\mathcal{T}_f^T = \mathcal{T}_f^{SI} + \mathcal{T}_f^{BI}. \tag{12-106}$$

The shear-induced turbulent stress is computed by

$$\mathcal{T}_f^{SI} = \rho_f \nu_t \left\{ \nabla \overline{\boldsymbol{v}_f} + \left(\nabla \overline{\boldsymbol{v}_f} \right)^+ \right\} - \frac{2}{3} A \rho_f k^{SI} \tag{12-107}$$

where ν_t, A and k^{SI} are the turbulent kinematic viscosity, the turbulence anisotropy tensor and the turbulent kinetic energy due to the shear-induced turbulence, respectively. For the isotropic turbulence, $A = I$.

The bubble-induced turbulence is computed by

$$\mathcal{T}_f^{BI} = -\alpha_g \rho_f \left(\frac{1}{20} \overline{\boldsymbol{v}_r \boldsymbol{v}_r} + \frac{3}{20} \left| \overline{\boldsymbol{v}_r} \right|^2 I \right) \tag{12-108}$$

$$= -\alpha_g \rho_f \frac{1}{2} C_{vm} \left|\overline{\boldsymbol{v}_r}\right|^2 \begin{pmatrix} 4/5 & 0 & 0 \\ 0 & 3/5 & 0 \\ 0 & 0 & 3/5 \end{pmatrix}$$

where C_{vm} is the virtual volume coefficient, and the value for potential flow around a sphere is 1/2. The value of 2.0 and 1.2 are recommended for the low and high void fraction cases, respectively. The turbulent kinetic energy due to the bubble-induced turbulence, k^{BI}, can be obtained as

$$k^{BI} = \alpha_g \frac{1}{2} C_{vm} \left|\overline{\boldsymbol{v}_r}\right|^2. \tag{12-109}$$

Then, substituting Eqs.(12-108) and (12-107) into Eq.(12-106), yields

$$\boldsymbol{\tau}_f^T = \rho_f \nu_t \left\{ \nabla \overline{\boldsymbol{v}_f} + \left(\nabla \overline{\boldsymbol{v}_f} \right)^+ \right\} - \frac{2}{3} A \rho_f k^{SI} \\ -\alpha_g k^{BI} \begin{pmatrix} 4/5 & 0 & 0 \\ 0 & 3/5 & 0 \\ 0 & 0 & 3/5 \end{pmatrix}. \tag{12-110}$$

Thus, the shear stress distribution can be calculated provided ν_t and k^{SI} are given.

The turbulent kinematic viscosity is assumed to be expressed by the linear superposition of the turbulent kinematic viscosities due to the shear-induced turbulence, ν_t^{SI}, and the bubble-induced turbulence, ν_t^{BI} as

$$\nu_t = \nu_t^{SI} + \nu_t^{BI}. \tag{12-111}$$

The turbulent kinematic viscosities due to the shear-induced turbulence and the bubble-induced turbulence are given as Eqs.(12-112) and (12-113), respectively.

$$\nu_t^{SI} = 0.09 \frac{{k^{SI}}^2}{\varepsilon^{SI}} \tag{12-112}$$

where ε_{SI} is the dissipation of the shear-induced turbulence.

$$\nu_t^{BI} = 1.2\frac{d_B}{2}\alpha_g\left|\overline{\boldsymbol{v}_r}\right| \tag{12-113}$$

The turbulent kinetic energy and dissipation due to the shear-induced turbulence can be computed by the shear-induced turbulence kinetic energy transport equation given by Eq.(12-114) and the shear-induced turbulence dissipation rate transport equation given by Eq.(12-116).

$$\alpha_f\frac{Dk^{SI}}{Dt} = \nabla\cdot\left(\frac{\alpha_f\nu_t}{\sigma_k}\nabla k^{SI}\right) + \alpha_f\left(P^{SI} - \varepsilon^{SI}\right) \tag{12-114}$$

where $\sigma_k\,(=1.0)$ is a constant and P^{SI} is the production of the shear-induced turbulence given by

$$P^{SI} \equiv \nu_t\left\{\nabla\overline{\boldsymbol{v}_f} + \left(\nabla\overline{\boldsymbol{v}_f}\right)^{+}\right\} : \nabla\overline{\boldsymbol{v}_f} \tag{12-115}$$

$$\alpha_f\frac{D\varepsilon^{SI}}{Dt} = \nabla\cdot\left(\alpha_f\frac{\nu_t}{\sigma_\varepsilon}\nabla\varepsilon^{SI}\right) + \frac{\alpha_f\varepsilon^{SI}}{k^{SI}}\left(C_{\varepsilon1}P_{SI} - C_{\varepsilon2}\varepsilon^{SI}\right) \tag{12-116}$$

where $\sigma_\varepsilon\,(=1.3)$, $C_{\varepsilon1}\,(=1.44)$, and $C_{\varepsilon2}\,(=1.92)$ are constants. The turbulent stress is computed from the above equations and boundary conditions.

what?

what chat phrase slip?

Chapter 13

DRIFT-FLUX MODEL

The basic concept of the drift-flux model is to consider the mixture as a whole, rather than two phases separately. It is clear that the drift-flux model formulation will be simpler than the two-fluid model, however it requires some drastic constitutive assumptions causing some of the important characteristics of two-phase flow to be lost. However, it is exactly this simplicity of the drift-flux model that makes it very useful in many engineering applications. As it is the case with the analyses of two-phase flow *system dynamics*, information required in engineering problems is often the response of the total mixture and not of each constituent phase (Tong, 1965). Furthermore, detailed analyses on the local behavior of each phase can be carried out with less difficulty, *if* these mixtures responses are known.

Another important aspect of the drift-flux model is concerned with the scaling of systems that has direct applications in the planning and designing of two-phase flow experimental and engineering systems. The similarities of two different systems can be studied effectively by using the drift-flux model formulation and mixture properties. The most important aspect of the drift-flux model is the reduction in the total number of field and constitutive equations required in the formulation in comparison with the two-fluid model. The drift-flux model is expressed in terms of four field equations: the mixture continuity; momentum; energy equations; and the gas continuity equation.

It can be seen, therefore, that the drift-flux model follows the standard approach used to analyze the dynamics of a mixture of gases or of miscible liquids. It is generally accepted that the drift-flux model is appropriate to the mixture where the dynamics of two components are closely coupled. This suggests that the same argument may be used for the macroscopic two-phase flows. The usefulness of the drift-flux model in many practical engineering systems comes from the fact that even two-phase mixtures that are weakly coupled locally can be considered, because the relatively large axial dimension of the systems usually gives sufficient interaction times. The advantages of using the drift-flux model for the studies of system dynamics and instabilities caused by the *low velocity* wave propagation, namely, the void propagation, are demonstrated by Zuber (1967) and Ishii and Zuber (1970). However, there are some questions in applying the drift-flux model to the problems of acoustic wave propagations, choking phenomena and high frequency instabilities, as it has been discussed in detail by Bouré and Réocreux (1972), Bouré (1973) and Réocreux et al. (1973).

In the drift-flux model formulation we have only four field equations and, thus, one energy and one momentum equation have been eliminated from the original six field equations. Then, the relative motion and energy difference should be expressed by additional constitutive equations. In other words, the dynamic interaction relations are replaced by the constitutive laws. Furthermore, it is important to formulate the model based on the mixture center of mass in order to preserve the additive characteristic of the extensive variables, as explained in Chapter 4.

In this chapter, we develop a general formulation of the mixture model (Ishii, 1975) then discuss various special cases (Ishii, 1977) that are important in practical applications. Since we have carried out the detailed analysis on the field and constitutive equations for two-fluid model in Chapter 9, we recall and use these results for the establishment of the drift-flux model formulation whenever it is helpful. The following diagram summarizes the establishment of the drift-flux model formulation. Here we see the special importance of the kinematic, mechanical and thermal relations between the two phases. It is evident that the elimination of one of the two momentum equations from the formulation requires the kinematic relation between the phases, therefore, the relative velocity should be given by a constitutive law. Similarly, by using only the mixture energy equation for the balance of energies in a two-phase flow, thermal relation between the phases should be given.

1.1 Drift-flux model field equations

Formulation Based on the Center of Mass and Drift-flux Velocities

The most general forms of the four basic field equations for the drift-flux model have been given in the Section 1.3 of Chapter 5. In this section, first we put these equations into more realistic form by using some of the analysis on the constitutive equations for the two-fluid model. Then we discuss some appropriate simplifications which are important for practical applications. Here, we formulate the model based on the mixture continuity, momentum and thermal energy equations plus the continuity equation for one of the phases. These equations can be reduced to the following forms

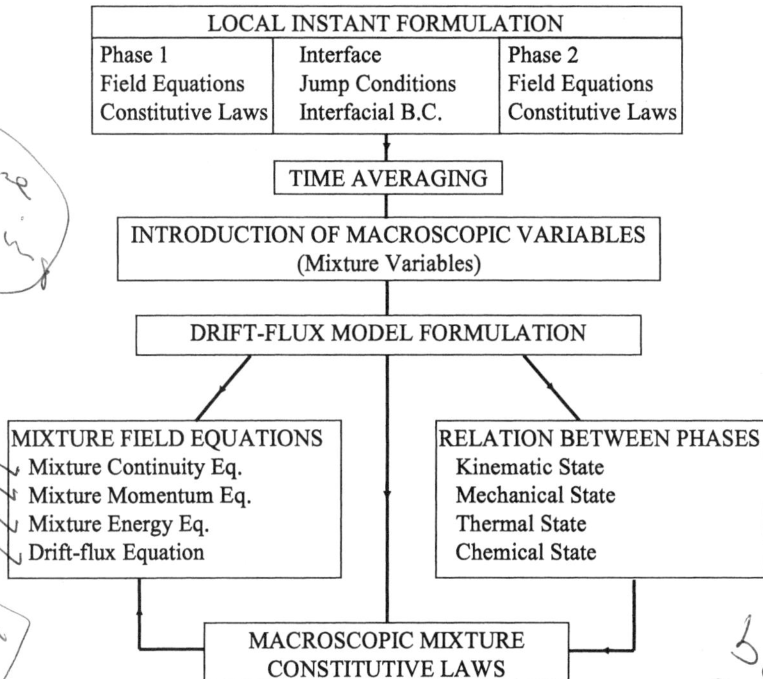

The mixture continuity equation from Eq.(5-40)

$$\frac{\partial \rho_m}{\partial t} + \nabla \cdot \left(\rho_m \boldsymbol{v}_m \right) = 0 \tag{13-1}$$

The continuity equation for phase 2 from Eq.(5-41)

$$\frac{\partial \alpha_2 \overline{\overline{\rho_2}}}{\partial t} + \nabla \cdot \left(\alpha_2 \overline{\overline{\rho_2}} \boldsymbol{v}_m \right) = \Gamma_2 - \nabla \cdot \left(\alpha_2 \overline{\overline{\rho_2}} \boldsymbol{V}_{2m} \right) \tag{13-2}$$

The mixture momentum equation from Eqs.(5-42) and (5-43)

$$\begin{aligned} &\frac{\partial \rho_m \boldsymbol{v}_m}{\partial t} + \nabla \cdot \left(\rho_m \boldsymbol{v}_m \boldsymbol{v}_m \right) = -\nabla p_m \\ &+\nabla \cdot \left[\overline{\overline{\mathcal{T}}} + \mathcal{T}^T - \sum_{k=1}^{2} \alpha_k \overline{\overline{\rho_k}} \boldsymbol{V}_{km} \boldsymbol{V}_{km} \right] + \rho_m \boldsymbol{g}_m + \boldsymbol{M}_m \end{aligned} \tag{13-3}$$

where we have from Eq.(9-127)

$$\boldsymbol{M}_m = 2\overline{\overline{H}}_{21}\,\overline{\overline{\sigma}}\nabla\alpha_2 + \boldsymbol{M}_m^H. \tag{13-4}$$

And in Eq.(13-3) the body force field has been taken as constant. The last term on the right-hand side of the mixture momentum equation represents the effects of the surface tension force on the mixture.

There are considerable difficulties obtaining an appropriate thermal energy equation for the mixture as we have discussed in the Section 1.3 of Chapter 5. It has been shown there that we have two different methods to obtain the equation. By adding the thermal energy equation for each phase, we obtained Eq.(5-53). Consequently, from Eqs.(9-154) and (9-150), we have

$$\begin{aligned}
&\frac{\partial \rho_m i_m}{\partial t} + \nabla\cdot\left(\rho_m i_m \boldsymbol{v}_m\right) = -\nabla\cdot\left(\overline{\boldsymbol{q}} + \boldsymbol{q}^T\right)\\
&-\nabla\cdot\left(\sum_k \alpha_k \overline{\overline{\rho_k}}\,\widehat{i_k}\boldsymbol{V}_{km}\right) + \sum_{k=1}^{2}\alpha_k \frac{D_k\overline{\overline{p_k}}}{Dt} + \sum_{k=1}^{2}\alpha_k \overline{\overline{\mathcal{T}_k}}:\nabla\widehat{\boldsymbol{v}_k}\\
&+\left(\overline{\overline{T}}_i \frac{d\sigma}{dT}\frac{D_i a_i}{Dt} + E_m^H\right)\\
&-\sum_{k=1}^{2}\left(\Gamma_k \frac{\widehat{v_k}^2}{2} - \boldsymbol{M}_{ik}\cdot\widehat{\boldsymbol{v}_{ki}} + \nabla\alpha_k\cdot\overline{\overline{\mathcal{T}_{ki}}}\cdot\widehat{\boldsymbol{v}_{ki}}\right).
\end{aligned} \tag{13-5}$$

From the definitions of the mixture properties and the interfacial momentum transfer condition, we have

$$\sum_{k=1}^{2}\alpha_k \frac{D_k\overline{\overline{p_k}}}{Dt} \doteq \frac{Dp_m}{Dt} + \sum_{k=1}^{2}\alpha_k \boldsymbol{V}_{km}\cdot\nabla\overline{\overline{p_k}} - 2\overline{\overline{H}}_{21}\,\overline{\overline{\sigma}}\frac{D\alpha_2}{Dt}. \tag{13-6}$$

For simplicity we define three different effects as follows

$$\Phi_m^\mu \equiv \sum_{k=1}^{2}\alpha_k \overline{\overline{\mathcal{T}_k}}:\nabla\widehat{\boldsymbol{v}_k} \tag{13-7}$$

$$\Phi_m^\sigma \equiv E_m^H - 2\overline{\overline{H}}_{21}\,\overline{\overline{\sigma}}\frac{D\alpha_2}{Dt} + \overline{\overline{T}}_i \frac{d\sigma}{dT}\frac{D_i a_i}{Dt} \tag{13-8}$$

and

$$\Phi_m^i \equiv \sum_{k=1}^{2}\left(-\Gamma_k \frac{\widehat{v_k}^2}{2} - \boldsymbol{M}_{ik}\cdot\widehat{\boldsymbol{v}_{ki}} + \nabla\alpha_k \cdot \overline{\overline{\boldsymbol{\mathcal{T}}_{ki}}}\cdot\widehat{\boldsymbol{v}_{ki}}\right). \tag{13-9}$$

Then *the thermal energy equation* (13-5) reduces to

$$\begin{aligned}\frac{\partial \rho_m i_m}{\partial t} + \nabla\cdot\left(\rho_m i_m \boldsymbol{v}_m\right) &= -\nabla\cdot\left(\overline{\boldsymbol{q}} + \boldsymbol{q}^T\right)\\ -\nabla\cdot\left(\sum_{k=1}^{2}\alpha_k\overline{\overline{\rho_k}}\,\widehat{i_k}\boldsymbol{V}_{km}\right) &+ \frac{Dp_m}{Dt} + \Phi_m^\mu + \Phi_m^\sigma + \Phi_m^i.\end{aligned} \tag{13-10}$$

The drift-flux model with the mixture thermal energy equation faces considerable difficulties when the last three terms in the above equation cannot be neglected. These terms arise from the viscous dissipation, work due to surface tension forces and interfacial mechanical energy transfer. This is particularly true, if the term Φ_m^i given by Eq.(13-9) has significant contributions to the thermal energy exchanges. It is evident that in this case the drift-flux model requires the constitutive equations for the relative velocity as well as for the interfacial mechanical energy transfer.

The alternative form of the thermal energy equation can be obtained from Eq.(5-55) by substituting Eqs.(9-141) and (9-127), then using the approximation given by (9-145). These two different forms of the thermal energy equation do not give an identical drift-flux model formulation, as it has been discussed in the Section 1.3 of Chapter 5. However, we use the relation given by Eq.(13-10), which has been obtained by adding the thermal energy equations for each phase, because it has a relatively simpler form than the one based on Eq.(5-55).

The above four field equations (13-1), (13-2), (13-3) and (13-10) state the balance laws which govern the *macroscopic mixture field*. They have been obtained by the time averaging applied to the two-phase flow systems with interfacial discontinuities. We note that mixture continuity, momentum and energy equations are somewhat similar to those of a single-phase flow. Actually, the mixture continuity equation has exactly the same form as that for the continuum without internal discontinuities. This has been done by using the properly defined mixture properties. The mixture momentum equation, however, has two additional terms that do not appear in a single-phase equation. One is the capillary force that takes into account the surface tension effects and can be considered as momentum source or sink. The other is a diffusion stress term, shown as the third stress on the right-hand side of Eq.(13-3). This term expresses the momentum diffusion due to the relative motion between two phases in addition to the molecular and turbulent diffusions that has been taken into account by the stress group $\left(\overline{\mathcal{T}} + \mathcal{T}^T\right)$. In the mixture thermal energy equation, we have three additional terms that do not appear in the single-phase flow equation. The second term on the right-hand side of Eq.(13-10) is an energy diffusion due to the transport of energy by relative motions of the phases with respect to the mixture center of mass. Recalling Eqs.(13-8) and (13-9), the terms given by Φ_m^σ and Φ_m^i represent the surface-tension effect and the contributions from the interfacial mechanical energy transfer, respectively. Under normal conditions, these two terms and energy dissipation term can be neglected almost always.

The mixture momentum and thermal energy equations given by Eqs.(13-3) and (13-10) describe the momentum and energy exchanges from the stationary observer. Thus, the convective fluxes with the mixture center of mass velocity and the additional diffusion fluxes defined with respect to the barycenter of mixture appear in the equations. These two equations can be transformed in terms of the convective derivative of Eq.(7-14) as follows

$$\rho_m \frac{D\boldsymbol{v}_m}{Dt} = -\nabla p_m + \nabla\cdot\left(\overline{\mathcal{T}} + \mathcal{T}^T\right) -\nabla\cdot\left(\sum_{k=1}^{2}\alpha_k\overline{\overline{\rho_k}}\boldsymbol{V}_{km}\boldsymbol{V}_{km}\right) + \rho_m \boldsymbol{g}_m + \boldsymbol{M}_m \tag{13-11}$$

and

$$\rho_m \frac{Di_m}{Dt} = -\nabla\cdot\left(\overline{\boldsymbol{q}} + \boldsymbol{q}^T\right) - \nabla\cdot\left(\sum_{k=1}^{2}\alpha_k\overline{\overline{\rho_k}}\widehat{i_k}\boldsymbol{V}_{km}\right) + \frac{Dp_m}{Dt} + \Phi_m^\mu + \Phi_m^\sigma + \Phi_m^i. \tag{13-12}$$

The above two equations described the transfers of momentum and energy as seen from the observer moving with the velocity $\boldsymbol{v}_m$. Because of its special form, Eq.(13-11) is called the equation of motion.

Field Equations in Several Coordinate Systems

In view of practical importance, we express these four field equations in two different coordinate systems. Since the derivatives are straightforward from the standard vector calculus (Aris, 1962; McConnell, 1957), we only list the results below.

In *rectangular coordinates* (x, y, z) we have for the conservation of mass of mixture

$$\frac{\partial \rho_m}{\partial t} + \frac{\partial}{\partial x}\left(\rho_m v_{xm}\right) + \frac{\partial}{\partial y}\left(\rho_m v_{ym}\right) + \frac{\partial}{\partial z}\left(\rho_m v_{zm}\right) = 0 \qquad (13\text{-}13)$$

for the drift-flux of mass of phase 2

$$\begin{aligned}
&\frac{\partial}{\partial t}\left(\alpha_2 \overline{\overline{\rho_2}}\right) + \left\{\frac{\partial}{\partial x}\left(\alpha_2 \overline{\overline{\rho_2}} v_{xm}\right) + \frac{\partial}{\partial y}\left(\alpha_2 \overline{\overline{\rho_2}} v_{ym}\right)\right. \\
&\left. + \frac{\partial}{\partial z}\left(\alpha_2 \overline{\overline{\rho_2}} v_{zm}\right)\right\} = \Gamma_2 - \left\{\frac{\partial}{\partial x}\left(\alpha_2 \overline{\overline{\rho_2}} V_{xm}\right)\right. \\
&\left. + \frac{\partial}{\partial y}\left(\alpha_2 \overline{\overline{\rho_2}} V_{ym}\right) + \frac{\partial}{\partial z}\left(\alpha_2 \overline{\overline{\rho_2}} V_{zm}\right)\right\}
\end{aligned} \qquad (13\text{-}14)$$

for the conservation of mixture momentum

x-component

$$\begin{aligned}
&\frac{\partial}{\partial t}\left(\rho_m v_{xm}\right) + \frac{\partial}{\partial x}\left(\rho_m v_{xm} v_{xm}\right) + \frac{\partial}{\partial y}\left(\rho_m v_{ym} v_{xm}\right) + \frac{\partial}{\partial z}\left(\rho_m v_{zm} v_{xm}\right) \\
&= -\frac{\partial p_m}{\partial x} + \rho_m g_{mx} + M_{xm} + \frac{\partial}{\partial x}\left(\overline{\tau_{xx}} + \tau_{xx}^T + \tau_{xx}^D\right) \\
&+ \frac{\partial}{\partial y}\left(\overline{\tau_{yx}} + \tau_{yx}^T + \tau_{yx}^D\right) + \frac{\partial}{\partial z}\left(\overline{\tau_{zx}} + \tau_{zx}^T + \tau_{zx}^D\right)
\end{aligned}$$

y-component

$$\begin{aligned}
&\frac{\partial}{\partial t}\left(\rho_m v_{ym}\right) + \frac{\partial}{\partial x}\left(\rho_m v_{xm} v_{ym}\right) + \frac{\partial}{\partial y}\left(\rho_m v_{ym} v_{ym}\right) \\
&+ \frac{\partial}{\partial z}\left(\rho_m v_{zm} v_{ym}\right) = -\frac{\partial p_m}{\partial y} + \rho_m g_{my} + M_{ym}
\end{aligned} \qquad (13\text{-}15)$$

$$+\frac{\partial}{\partial x}\left(\overline{\tau_{xy}}+\tau_{xy}^{T}+\tau_{xy}^{D}\right)+\frac{\partial}{\partial y}\left(\overline{\tau_{yy}}+\tau_{yy}^{T}+\tau_{yy}^{D}\right)$$
$$+\frac{\partial}{\partial z}\left(\overline{\tau_{zy}}+\tau_{zy}^{T}+\tau_{zy}^{D}\right)$$

z-component

$$\frac{\partial}{\partial t}\left(\rho_m v_{zm}\right)+\frac{\partial}{\partial x}\left(\rho_m v_{xm} v_{zm}\right)+\frac{\partial}{\partial y}\left(\rho_m v_{ym} v_{zm}\right)$$
$$+\frac{\partial}{\partial z}\left(\rho_m v_{zm} v_{zm}\right)=-\frac{\partial p_m}{\partial z}+\rho_m g_{mz}+M_{zm}$$
$$+\frac{\partial}{\partial x}\left(\overline{\tau_{xz}}+\tau_{xz}^{T}+\tau_{xz}^{D}\right)+\frac{\partial}{\partial y}\left(\overline{\tau_{yz}}+\tau_{yz}^{T}+\tau_{yz}^{D}\right)$$
$$+\frac{\partial}{\partial z}\left(\overline{\tau_{zz}}+\tau_{zz}^{T}+\tau_{zz}^{D}\right)$$

for the mixture thermal energy balance

$$\frac{\partial}{\partial t}\left(\rho_m i_m\right)+\frac{\partial}{\partial x}\left(\rho_m i_m v_{xm}\right)+\frac{\partial}{\partial y}\left(\rho_m i_m v_{ym}\right)+\frac{\partial}{\partial z}\left(\rho_m i_m v_{zm}\right)$$
$$=-\frac{\partial}{\partial x}\left(\overline{q_x}+q_x^T+\sum_{k=1}^{2}\alpha_k\overline{\overline{\rho_k}}\widehat{i_k}V_{xkm}\right)$$
$$-\frac{\partial}{\partial y}\left(\overline{q_y}+q_y^T+\sum_{k=1}^{2}\alpha_k\overline{\overline{\rho_k}}\widehat{i_k}V_{ykm}\right)$$
$$-\frac{\partial}{\partial z}\left(\overline{q_z}+q_z^T+\sum_{k=1}^{2}\alpha_k\overline{\overline{\rho_k}}\widehat{i_k}V_{zkm}\right) \tag{13-16}$$
$$+\left(\frac{\partial p_m}{\partial t}+v_{xm}\frac{\partial p_m}{\partial x}+v_{ym}\frac{\partial p_m}{\partial y}+v_{zm}\frac{\partial p_m}{\partial z}\right)$$
$$+\Phi_m^{\mu}+\Phi_m^{\sigma}+\Phi_m^{i}.$$

If the flow is restricted to two dimensions, then it is called a plane flow. In this case the partial derivative with respect to x as well as the x-component of the momentum equation can be dropped from the formulation.

Similarly in *cylindrical coordinates* (r,θ,z) we have for the conservation of mass of mixture

$$\frac{\partial}{\partial t}\rho_m + \frac{1}{r}\frac{\partial}{\partial r}\left(r\rho_m v_{rm}\right) + \frac{1}{r}\frac{\partial}{\partial \theta}\left(\rho_m v_{\theta m}\right) + \frac{\partial}{\partial z}\left(\rho_m v_{zm}\right) = 0 \tag{13-17}$$

for the drift-flux of mass of phase 2

$$\begin{aligned}&\frac{\partial}{\partial t}\left(\alpha_2\overline{\overline{\rho_2}}\right) + \frac{1}{r}\frac{\partial}{\partial r}\left(r\alpha_2\overline{\overline{\rho_2}}v_{r2}\right) + \frac{1}{r}\frac{\partial}{\partial \theta}\left(\alpha_2\overline{\overline{\rho_2}}v_{\theta 2}\right)\\ &+\frac{\partial}{\partial z}\left(\alpha_2\overline{\overline{\rho_2}}v_{z2}\right) = \Gamma_2 - \left\{\frac{1}{r}\frac{\partial}{\partial r}\left(r\alpha_2\overline{\overline{\rho_2}}V_{r2m}\right)\right.\\ &\left.+\frac{1}{r}\frac{\partial}{\partial \theta}\left(\alpha_2\overline{\overline{\rho_2}}V_{\theta 2m}\right) + \frac{\partial}{\partial z}\left(\alpha_2\overline{\overline{\rho_2}}V_{z2m}\right)\right\}\end{aligned} \tag{13-18}$$

for the conservation of mixture momentum

r-component

$$\begin{aligned}&\frac{\partial}{\partial t}\left(\rho_m v_{rm}\right) + \frac{1}{r}\frac{\partial}{\partial r}\left(r\rho_m v_{rm} v_{rm}\right) + \frac{1}{r}\frac{\partial}{\partial \theta}\left(\rho_m v_{rm} v_{\theta m}\right)\\ &-\frac{\rho_m v_{\theta m}^2}{r} + \frac{\partial}{\partial z}\left(\rho_m v_{rm} v_{zm}\right) = -\frac{\partial p_m}{\partial r} + \rho_m g_{mr} + M_{rm}\\ &+\frac{1}{r}\frac{\partial}{\partial r}\left\{r\left(\overline{\tau_{rr}} + \tau_{rr}^T + \tau_{rr}^D\right)\right\} + \frac{1}{r}\frac{\partial}{\partial \theta}\left(\overline{\tau_{r\theta}} + \tau_{r\theta}^T + \tau_{r\theta}^D\right)\\ &-\frac{1}{r}\left(\overline{\tau_{\theta\theta}} + \tau_{\theta\theta}^T + \tau_{\theta\theta}^D\right) + \frac{\partial}{\partial z}\left(\overline{\tau_{rz}} + \tau_{rz}^T + \tau_{rz}^D\right)\end{aligned}$$

θ-component

$$\begin{aligned}&\frac{\partial}{\partial t}\left(\rho_m v_{\theta m}\right) + \frac{1}{r^2}\frac{\partial}{\partial r}\left(r^2\rho_m v_{rm} v_{\theta m}\right) + \frac{1}{r}\frac{\partial}{\partial \theta}\left(\rho_m v_{\theta m} v_{\theta m}\right)\\ &+\frac{\partial}{\partial z}\left(\rho_m v_{\theta m} v_{zm}\right) = -\frac{\partial p_m}{\partial \theta} + \rho_m g_{m\theta} + M_{\theta m}\\ &+\frac{1}{r^2}\frac{\partial}{\partial r}\left\{r^2\left(\overline{\tau_{r\theta}} + \tau_{r\theta}^T + \tau_{r\theta}^D\right)\right\} + \frac{1}{r}\frac{\partial}{\partial \theta}\left(\overline{\tau_{\theta\theta}} + \tau_{\theta\theta}^T + \tau_{\theta\theta}^D\right)\end{aligned} \tag{13-19}$$

$$+\frac{\partial}{\partial z}\left(\overline{\tau_{\theta z}}+\tau_{\theta z}^{T}+\tau_{\theta z}^{D}\right)$$

z-component

$$\frac{\partial}{\partial t}\left(\rho_m v_{zm}\right)+\frac{1}{r}\frac{\partial}{\partial r}\left(r\rho_m v_{rm} v_{zm}\right)+\frac{1}{r}\frac{\partial}{\partial \theta}\left(\rho_m v_{\theta m} v_{zm}\right)$$
$$+\frac{\partial}{\partial z}\left(\rho_m v_{zm} v_{zm}\right)=-\frac{\partial p_m}{\partial z}+\rho_m g_{mz}+M_{zm}$$
$$+\frac{1}{r}\frac{\partial}{\partial r}\left\{r\left(\overline{\tau_{rz}}+\tau_{rz}^{T}+\tau_{rz}^{D}\right)\right\}+\frac{1}{r}\frac{\partial}{\partial \theta}\left(\overline{\tau_{\theta z}}+\tau_{\theta z}^{T}+\tau_{\theta z}^{D}\right)$$
$$+\frac{\partial}{\partial z}\left(\overline{\tau_{zz}}+\tau_{zz}^{T}+\tau_{zz}^{D}\right)$$

for the mixture thermal energy balance

$$\frac{\partial}{\partial t}\left(\rho_m i_m\right)+\frac{1}{r}\frac{\partial}{\partial r}\left(r\rho_m i_m v_{rm}\right)+\frac{1}{r}\frac{\partial}{\partial \theta}\left(\rho_m i_m v_{\theta m}\right)$$
$$+\frac{\partial}{\partial z}\left(\rho_m i_m v_{zm}\right)=-\left\{\frac{1}{r}\frac{\partial}{\partial r}\left[r\left(\overline{q_r}+q_r^{T}+\sum_{k=1}^{2}\alpha_k\overline{\overline{\rho_k}}\,\widehat{i_k}V_{rkm}\right)\right]\right.$$
$$+\frac{1}{r}\frac{\partial}{\partial \theta}\left(\overline{q_\theta}+q_\theta^{T}+\sum_{k=1}^{2}\alpha_k\overline{\overline{\rho_k}}\,\widehat{i_k}V_{\theta km}\right) \tag{13-20}$$
$$\left.+\frac{\partial}{\partial z}\left(\overline{q_z}+q_z^{T}+\sum_{k=1}^{2}\alpha_k\overline{\overline{\rho_k}}\,\widehat{i_k}V_{zkm}\right)\right\}+\Phi_m^{\mu}+\Phi_m^{\sigma}+\Phi_m^{i}$$
$$+\left(\frac{\partial p_m}{\partial t}+v_{rm}\frac{\partial p_m}{\partial r}+\frac{v_{\theta m}}{r}\frac{\partial p_m}{\partial \theta}+v_{zm}\frac{\partial p_m}{\partial z}\right).$$

If the flow is axisymmetric, the partial derivative with respect to θ drops from the equations. Furthermore if the flow is free from the circulatory motion around the z-axis, namely, the flow is restricted to two directions r and z, then $v_{\theta m}$ is zero, thus the momentum equation for θ direction can be eliminated.

1.2 Drift-flux (or mixture) model constitutive laws

It is evident that the drift-flux model based on the four field equations is an approximate theory of the two-fluid model. In order to complete the drift-flux model it is necessary to supply several constitutive laws for mixtures. We can consider two distinct approaches to accomplish our purpose. The first method is to start our analysis on the constitutive equations from the mixture field equations and the mixture entropy inequality, then to apply various constitutive axioms directly to the mixtures and independently of the two-fluid model. The second method is to obtain the necessary constitutive equations by the reduction from the two-fluid model formulation.

At first it seems to be more logical to follow the former approach, because it is a self-sufficient and independent formulation of the model for the mixtures. However, in reality it is confronted by great difficulties which cannot be overlooked lightly. The main problems arise from the fact that in general two phases are not in thermal equilibrium, thus it is not possible to introduce a mixture temperature. This suggests that we cannot expect the existence of a simple equation of state in terms of the macroscopic mixture properties.

It can be seen that the thermal non-equilibrium condition and the structures of the interfaces are the governing factors of the changes of phases. Furthermore, the kinematic and mechanical state between two phases is greatly influenced by the interfacial properties and structures. In order to bring these important effects into the drift-flux model formulation, it is simpler and more realistic to use reductions from the two-fluid model than the former approach. Consequently, in this section we develop our analysis on the mixture constitutive equations in parallel with the Section 1.2 of Chapter 9.

Principle of Determinism

The drift-flux model field equations have been given by Eqs.(13-1), (13-2), (13-3) and (13-10) which are not sufficient to describe the system completely. From the principle of determinism, it is necessary to supply additional constitutive equations that specify the response characteristics of certain group of macroscopic two-phase mixtures. In order to keep the thermal non-equilibrium effects in the drift-flux (or mixture) model formulation, we introduce a fundamental equation of state *for each phase*. Furthermore various mixture properties can be related to the properties of each phase through definitions of the Section 1.5 of Chapter 4.

By taking into account the above considerations, we have following variables appearing in the drift-flux model formulation.

1. Equation of State: $\overline{\overline{\rho_k}}, \overline{\overline{p_k}}, \overline{\overline{T}}_k, \widehat{i_k}, \overline{\overline{T}}_i, \overline{\overline{\sigma}};$
2. Conservation of Mass: $\rho_m, \boldsymbol{v}_m;$
3. Conservation of Momentum: $p_m, \boldsymbol{V}_{km}, \overline{\mathcal{T}}, \mathcal{T}^T, \boldsymbol{M}_m, \alpha_k;$
4. Conservation of Energy: $i_m, \overline{\boldsymbol{q}}, \boldsymbol{q}^T, \Phi_m^\mu, \Phi_m^\sigma, \Phi_m^i;$
5. Drift-flux Equation: $\Gamma_2;$

where k =1 and 2. Hence the total number of the variables is twenty seven. For a properly formulated drift-flux model we should have also the same number of equations. These can be classified into following groups.

	Equations		*Number of Equations*
1)	Field equations		
	mixture mass	Eq.(13-1)	1
	mixture momentum	Eq.(13-3)	1
	mixture energy	Eq.(13-10)	1
	drift-flux	Eq.(13-2)	1
2)	Axiom of continuity		
	$\alpha_1 = 1 - \alpha_2$	Eq.(4-13)	1
3)	Equation of state for phase		
	thermal equation of state	Eq.(9-56)	2
	caloric equation of state	Eq.(9-57)	2
	definition of ρ_m	Eq.(4-66)	1
	definition of i_m	Eq.(4-74)	1
	definition of p_m	Eq.(4-72)	1
4)	Equation of state for interfaces		
	$\overline{\overline{\sigma}} = \overline{\overline{\sigma}}\left(\overline{\overline{T}}_i\right)$	Eq.(9-130)	1
5)	Identity on the drift-flux velocities		
	$\sum_{k=1}^{2} \alpha_k \overline{\overline{\rho_k}} \boldsymbol{V}_{km} = 0$	Eq.(4-90)	1
6)	Kinematic constitutive equation for V_{2m}		1
7)	Mechanical state between two phases		
	$\overline{\overline{p_1}} - \overline{\overline{p_2}} = -2\overline{\overline{H}}_{21}\, \overline{\overline{\sigma}}$	Eq.(9-128)	1
8)	Thermal state between two phases		
	$\overline{\overline{T}}_1 - \overline{\overline{T}}_2 = 0$		1

9) Phase change condition

$$\overline{\overline{p_2}} - p^{sat}\left(\overline{\overline{T_i}}\right) = 2\overline{\overline{H_{21}}}\,\overline{\overline{\sigma}}\,\frac{\overline{\overline{\rho_2}}}{\overline{\overline{\rho_2}} - \overline{\overline{\rho_1}}} \qquad \text{Eq.(9-163)} \qquad 1$$

10) Mechanical constitutive equations

viscous stress $\overline{\mathcal{T}}$ 1
turbulent stress $\mathcal{T}^T$ 1
mixture momentum source $\boldsymbol{M}_m$ 1

11) Energetic constitutive equations

conduction heat flux $\overline{\boldsymbol{q}}$ 1
turbulent heat flux $\boldsymbol{q}^T$ 1
dissipation term Φ_m^{μ} 1
surface tension effect Φ_m^{σ} 1
mechanical energy effect Φ_m^{i} 1

12) Constitutive equation for phase change

mass generation Γ_2 1

This shows that we have also twenty seven equations, thus the present formulation is consistent. We note here that these field and constitutive equations are required from the principle of determinism, however they do not ensure the existence of a solution. It is very difficult to prove for our system that the problem is properly set, namely, the just setting (Truesdell and Toupin, 1960) because it concerns with the existence, uniqueness as well as the proper initial and boundary conditions. Usually, it can be checked only for very simplified classes of problems. Now we proceed to the detailed discussion of the above constitutive equations.

Equations of State and Mixture Properties

The axiom of continuity requires that an interface does not stay at a point for a finite time interval (see Section 1.4 of Chapter 5), thus we have

$$\alpha_1 + \alpha_2 = 1. \tag{13-21}$$

The mixture density ρ_m is defined by

$$\rho_m = \alpha_1\overline{\overline{\rho_1}} + \alpha_2\overline{\overline{\rho_2}} \tag{13-22}$$

with the thermal equation of state for each phase given by

$$\overline{\overline{\rho_1}} = \overline{\overline{\rho_1}}\left(\overline{\overline{T_1}}, \overline{\overline{p_1}}\right) \tag{13-23}$$

$$\overline{\overline{\rho_2}} = \overline{\overline{\rho_2}}\left(\overline{\overline{T_2}}, \overline{\overline{p_2}}\right). \tag{13-24}$$

And the mixture pressure is related to the phase pressures by

$$p_m = \alpha_1 \overline{\overline{p_1}} + \alpha_2 \overline{\overline{p_2}}. \tag{13-25}$$

The mixture density given by Eq.(13-22) with Eqs.(13-23) and (13-24) can be considered as the mixture thermal equation of state which has constraints imposed by the thermal, mechanical and the chemical relations between two phases.

The mixture enthalpy is defined by

$$i_m = \frac{\alpha_1 \overline{\overline{\rho_1}} \widehat{i_1} + \alpha_2 \overline{\overline{\rho_2}} \widehat{i_2}}{\rho_m} \tag{13-26}$$

with the caloric equation of state for each phase

$$\widehat{i_1} = \widehat{i_1}\left(\overline{\overline{T_1}}, \overline{\overline{p_1}}\right) \tag{13-27}$$

$$\widehat{i_2} = \widehat{i_2}\left(\overline{\overline{T_2}}, \overline{\overline{p_2}}\right). \tag{13-28}$$

By substituting Eqs.(13-27) and (13-28) into Eq.(13-26) we obtain the mixture caloric equation of state which shows the dependence of i_m on the temperatures, pressures and local void fractions. As in the case of mixture thermal equations of state, however, it has constraints imposed by the constitutive equations for the temperature difference and the pressure difference between two phases as well as by the phase change condition, Eq.(9-163).

The thermal equation of state for the interfaces in the macroscopic field can be approximated by

$$\overline{\overline{\sigma}} = \overline{\overline{\sigma}}\left(\overline{\overline{T_i}}\right) \tag{13-29}$$

which can be considered as the fundamental equation of state for the interfaces without surface mass.

Kinematic Constitutive Equation

As it has been explained in the previous section, we should supply a constitutive equation for the relative motion of phases. Since in the drift-flux model formulation we have eliminated one momentum equation, the kinematic constitutive equation stands as a relative equation of motion. It can be expected, however, that the *dynamic interactions* between two phases will be lost by replacing the momentum equation by a kinematic relation.

The diffusion velocity of each phase is related by an identity

$$\alpha_1\overline{\overline{\rho_1}}\boldsymbol{V}_{1m}+\alpha_2\overline{\overline{\rho_2}}\boldsymbol{V}_{2m}=0. \tag{13-30}$$

Thus, we should supply only one of the diffusion velocities by the kinematic constitutive equation. However, since the diffusion velocity $\boldsymbol{V}_{km}$ can be related to the relative velocity between phases or the drift velocities by the definitions in the Section 1.6 of Chapter 4 as

$$\begin{aligned}\boldsymbol{V}_{2m}&=-\frac{\alpha_1\overline{\overline{\rho_1}}}{\alpha_2\overline{\overline{\rho_2}}}\boldsymbol{V}_{1m}=-\frac{\alpha_1\overline{\overline{\rho_1}}}{\rho_m}\left(\widehat{\boldsymbol{v}_1}-\widehat{\boldsymbol{v}_2}\right)=\frac{\overline{\overline{\rho_1}}}{\rho_m}\boldsymbol{V}_{2j}\\&=-\frac{\alpha_1\overline{\overline{\rho_1}}}{\alpha_2\rho_m}\boldsymbol{V}_{1j}\end{aligned} \tag{13-31}$$

the constitutive equation can be given in terms of any of the above velocities.

The relative velocity between two phases depends upon the drag force acting at the interfaces as well as the interfacial geometry. Thus it can be expected that relative velocity will vary whenever the interfacial structure of the mixture changes. It has been shown in the Section 1.4 of Chapter 9 that, in a dispersed two-phase flow, the drag correlation should be expressed in terms of the drift velocity $\left(\boldsymbol{j}-\widehat{\boldsymbol{v}_d}\right)$ and the Reynolds number based on that velocity, Eqs.(9-223) and (9-224). This suggests that the kinematic constitutive equation on the relative motion between phases is best studied in terms of the drift velocity of the dispersed phase as it has been proposed by Zuber (1964b), Zuber et al. (1964) and Zuber and Staub (1966).

In view of the results obtained in the above references, it can be said that the drift velocity is a function of a terminal velocity $\boldsymbol{v}_\infty$ of a single particle in an infinite medium and the void fraction of a continuous phase. In order to take into account the drift effect due to the concentration gradient, we propose a linear constitutive law in the following form

$$V_{dj} = j - \widehat{v_d} = v_\infty (1-\alpha_d)^n - \frac{D_d^\alpha}{\alpha_d}\nabla\alpha_d \tag{13-32}$$

where D_d^α is the drift coefficient based on the void fraction α_d. The first term on the right-hand side takes into account for the *effect of gravity and forces which is usually the dominant part of the drift velocity.* A detailed analysis on this term in the bubbly flow regime has been made in Zuber et al. (1964), and Ishii (1977) also demonstrated that the constitutive equation of the drift velocity can be derived from the two-fluid model for various flow regimes. In the absence of the wall and under a steady-state condition without phase change, the multiparticle system in an infinite medium essentially reduces to a gravity and drag dominated one-dimensional flow, since the averaged void and velocity profiles become flat. Solving the momentum equations for each phase yields the relative velocity law. Thus, we use the results for drag correlations in Chapter 12.

For a viscous regime, the drift velocity can be given by

$$\begin{aligned} V_{dj} &\simeq 10.8\left(\frac{\mu_c g \Delta\rho}{\rho_c^2}\right)^{1/3} \frac{(1-\alpha_d)^{1.5} f(\alpha_d)}{r_d^*} \\ &\times \frac{\psi(r_d^*)^{4/3}\left\{1+\psi(r_d^*)\right\}}{1+\psi(r_d^*)\left\{f(\alpha_d)\right\}^{6/7}} \end{aligned} \tag{13-33}$$

where

$$f(\alpha_d) = (1-\alpha_d)^{1/2}\frac{\mu_c}{\mu_m} \tag{13-34}$$

and

$$\psi(r_d^*) = 0.55\left\{\left(1+0.08 r_d^{*3}\right)^{4/7} - 1\right\}^{0.75} \tag{13-35}$$

for the viscous regime. The non-dimensional radius, r_d^*, is defined by

$$r_d^* \equiv r_d\left(\frac{\rho_c g \Delta\rho}{\mu_c^2}\right)^{1/3}. \tag{13-36}$$

For a Newton's regime $\left(r_d^* \geq 34.65\right)$, the drift velocity can be given by

$$V_{dj} = 2.43\left(\frac{r_d g \Delta\rho}{\rho_c}\right)^{1/2}(1-\alpha_d)^{1.5} f(\alpha_d) \times \frac{18.67}{1+17.67\{f(\alpha_d)\}^{6/7}}. \tag{13-37}$$

For a distorted-fluid-particle regime, the drift velocity can be given by

$$V_{dj} \simeq \sqrt{2}\left(\frac{\sigma g \Delta\rho}{\rho_c^2}\right)^{1/4} \times \begin{cases} (1-\alpha_d)^{1.75} & \mu_c >> \mu_d \\ (1-\alpha_d)^{2} & \mu_c \simeq \mu_d \\ (1-\alpha_d)^{2.25} & \mu_d >> \mu_c. \end{cases} \tag{13-38}$$

The above criterion is applicable for $N_\mu \geq 0.11(1+\psi)/\psi^{8/3}$ where N_μ is the viscosity number given by Eq.(12-28).

For a churn-turbulent flow regime, the drift velocity can be given by

$$V_{dj} = \begin{Bmatrix} \sqrt{2} \\ \text{or } 1.57 \end{Bmatrix}\left(\frac{\sigma g \Delta\rho}{\rho_c^2}\right)^{1/4} \frac{\rho_c - \rho_d}{\Delta\rho}(1-\alpha_d)^{1/4} \simeq \sqrt{2}\left(\frac{\sigma g \Delta\rho}{\rho_c^2}\right)^{1/4} \frac{\rho_c - \rho_d}{\Delta\rho}. \tag{13-39}$$

In the exact expression for V_{dj}, the proportionality constant $\sqrt{2}$ is applicable for bubbly flows and 1.57 for droplet flows. However, in view of the uncertainty in predicting the drag coefficient, this difference as well as the effect of the void fraction may be neglected.

For a slug flow regime, the drift velocity can be given by

$$V_{dj} = 0.35\left(\frac{g \Delta\rho D}{\rho_c}\right)^{1/2} \tag{13-40}$$

where D is the diameter of the tube.

Figure 13-1 compares the present analytical result, Eq.(13-37), with the empirical correlation for solid-particles flow systems (Richardson and Zaki,

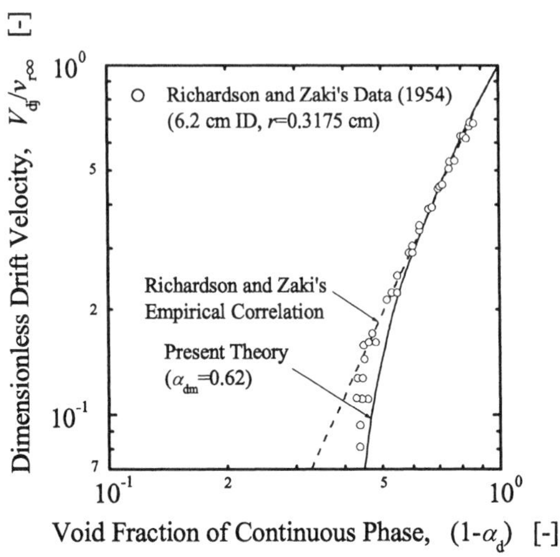

Figure 13-1. Comparison with experimental data for solid-particle system at high Reynolds number (Ishii, 1977)

1954). An agreement at relatively high void fraction of continuous phase is excellent. At very high values of α_d, Eq.(13-37) predicts much lower drift velocities than the Richardson-Zaki correlation. However, the original experimental data of Richardson and Zaki also indicate this trend, which is predicted by Eq.(13-37). Figure 13-2 shows the relative velocity in both the bubble and the droplet-liquid flow regimes. These data of Lackme (1973) clearly indicate the difference in the concentration dependence of the relative velocity between a bubbly flow and a droplet flow. These characteristics have been correctly predicted by the model. Figures 13-3 and 13-4 make further comparisons between the theoretical predictions and experimental data in both the batch and countercurrent bubbly flows and in a liquid-liquid dispersion system, respectively. The theoretical predictions agree with the data very well. Figure 13-5 compares the prediction for churn-turbulent flow and the experimental data of Yoshida and Akita (1965). The data were taken for an air-aqueous sodium sulfite solution system with various column diameters ranging from 7.7 to 60 cm. As can be seen from the figure, the theory underestimates the gas flux for smaller-column-diameter experiments. However, for larger column diameters, the agreement between the prediction and the data becomes increasingly satisfactory. This tendency can be easily explained by the two-dimensional effect due to the void and velocity profiles. The dispersed phase is locally transported with the local drift velocity with respect to local volumetric flux. Therefore, if more particles are

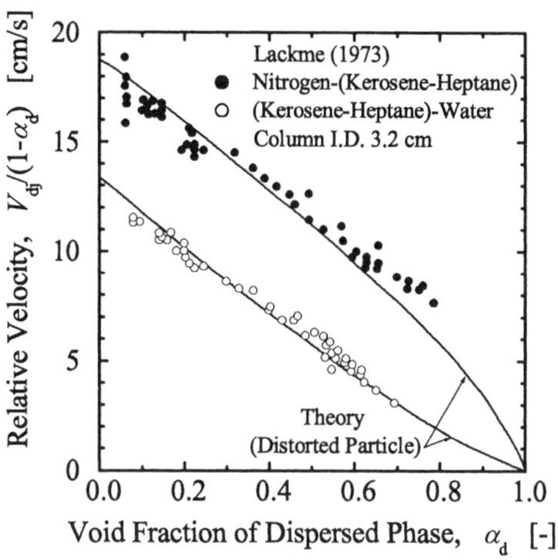

Figure 13-2. Difference between bubble system and droplet-dispersion system in distorted-particle regime (Ishii and Chawla, 1979)

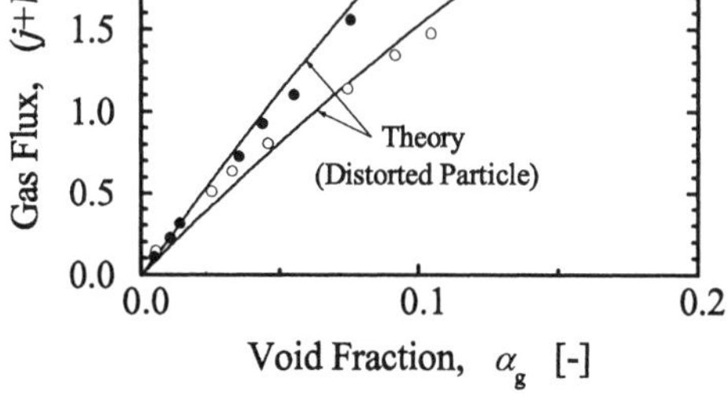

Figure 13-3. Comparison of gas volumetric flux with distorted-bubble-regime data in a flowing system (Ishii and Chawla, 1979)

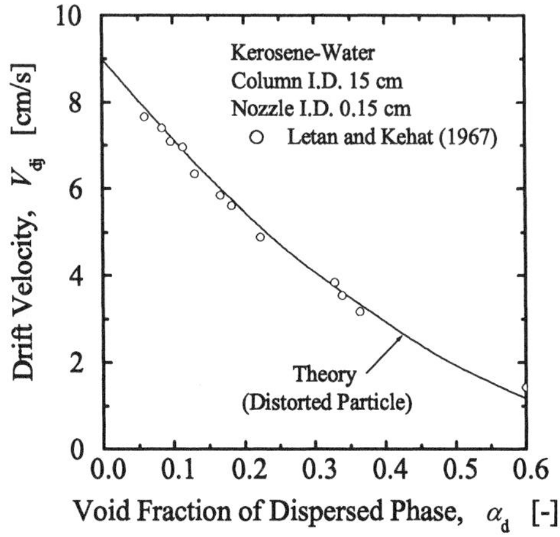

Figure 13-4. Comparison of predicted distorted-particle-regime drift velocity to data in a Kerosene-water system (Ishii and Chawla, 1979)

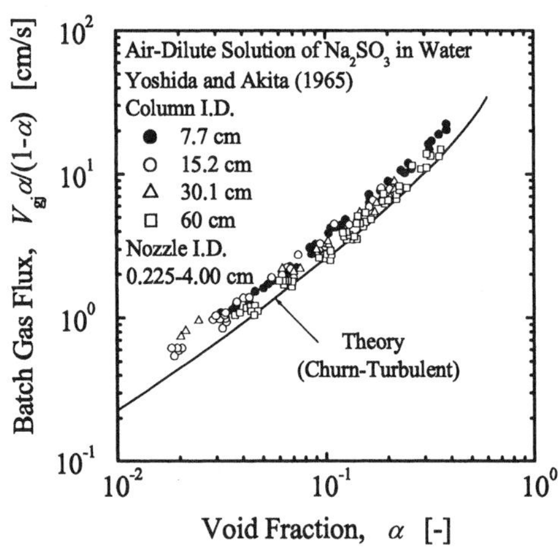

Figure 13-5. Comparison of predicted gas volumetric flux based on churn-flow-regime drift velocity to data (Ishii and Chawla, 1979)

concentrated in higher-flux regions, this will give a higher dispersed-phase flux than the case with uniform profiles. Then the mean gas volumetric flux should be somewhat higher than the prediction. For more detailed discussions on this point, see Zuber and Findlay (1965), Ishii (1977) and Werther (1974).

Thermodynamic State between Two Phases

As we have discussed in the preceding section it is necessary to specify the mechanical, thermal and chemical states between two phases. The simplified normal momentum jump condition gives the mechanical relation between two phases, thus we have

$$\overline{\overline{p_1}} - \overline{\overline{p_2}} = -2\overline{\overline{H_{21}}}\,\overline{\overline{\sigma}}. \tag{13-41}$$

As we see from the above equation, the pressure difference can be important only if the mean curvature $\overline{\overline{H_{21}}}$ is large, namely, for a bubbly or droplet flow with small fluid particle diameters. Consequently, in many practical engineering problems where the drift-flux model can be applied, the pressure difference between two phases can be neglected. Then we have

$$\overline{\overline{p_1}} \approx \overline{\overline{p_2}}. \tag{13-42}$$

The chemical state between phases decides the condition of phase changes and it is given from Eq.(9-163) as

$$\overline{\overline{p_2}} - p^{sat}\left(\overline{\overline{T_i}}\right) = 2\overline{\overline{H_{21}}}\,\overline{\overline{\sigma}}\left(\frac{\overline{\overline{\rho_2}}}{\overline{\overline{\rho_2}} - \overline{\overline{\rho_1}}}\right). \tag{13-43}$$

However, this can be approximated by

$$\overline{\overline{p_g}} \approx p^{sat}\left(\overline{\overline{T_i}}\right) \tag{13-44}$$

where $\overline{\overline{p_g}}$ denotes the vapor phase pressure. If the mechanical equilibrium condition (13-42) can be applied, then we have

$$\overline{\overline{p_1}} = \overline{\overline{p_2}} = p^{sat}\left(\overline{\overline{T_i}}\right). \tag{13-45}$$

The constitutive equation for the thermal state between two phases is one

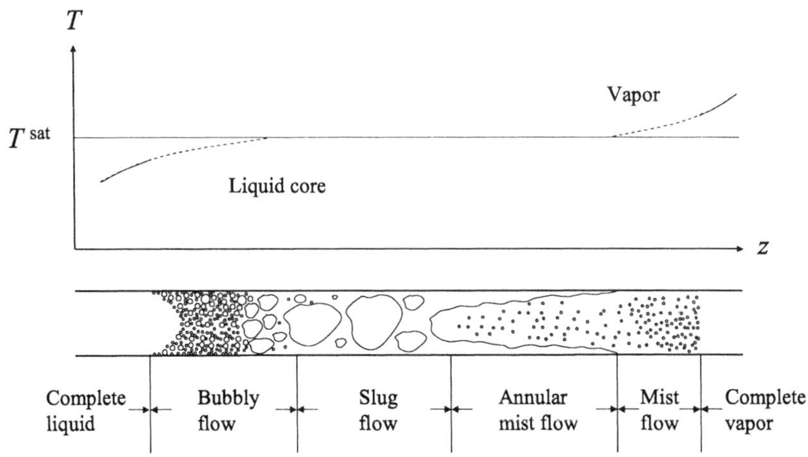

Figure 13-6. Axial temperature distribution (Ishii, 1975)

of the difficult constitutive equations to obtain, since this specifies the degree of thermal non-equilibrium. First we note that it can be replaced by the relation for $\overline{\overline{T_1}} - \overline{\overline{T_i}}$ *or* $\overline{\overline{T_2}} - \overline{\overline{T_i}}$. And if both relations are given, then the constitutive equation for the mass transfer Γ_2 becomes redundant. It is very important to realize that in many practical problems one of the phases is approximately in thermal equilibrium with the interfaces thus

$$\overline{\overline{T_k}} - \overline{\overline{T_i}} \approx 0 \quad \text{for } k = 1 \text{ or } 2 \tag{13-46}$$

For example in a boiling system we may assume that

$$\begin{aligned} \overline{\overline{T_g}} &= \overline{\overline{T_i}}; \quad \text{for bubbly and mixed flow} \\ \overline{\overline{T_f}} &= \overline{\overline{T_i}}; \quad \text{for droplet flow} \end{aligned} \tag{13-47}$$

as shown in the Fig.13-6.

Mechanical Constitutive Equations: $\overline{\mathcal{T}}$ *and* $\mathcal{T}^T$

The average viscous stress for each phase has been obtained in the Section 1.2 of Chapter 9, thus we have

$$\overline{\mathcal{T}} = \sum_{k=1}^{2} \left\{ \alpha_k \overline{\overline{\mu_k}} \left[\nabla \widehat{\boldsymbol{v}_k} + \left(\nabla \widehat{\boldsymbol{v}_k} \right)^+ \right] + 2\alpha_k \overline{\overline{\mu_k}} \mathcal{D}_{ki} \right\} \tag{13-48}$$

where the first term on the right-hand side has the standard form of the viscous stress based on the deformation tensor. The second term takes into account the effects of the interfaces, and the interfacial extra deformation tensor $\mathbb{D}_{ki}$ is defined by Eq.(9-73). By substituting the definition of the drift velocity we obtain

$$\begin{aligned}\overline{\overline{\mathscr{T}}} &= \left(\sum_{k=1}^{2}\alpha_k\overline{\overline{\mu_k}}\right)\left[\nabla \boldsymbol{v}_m + (\nabla \boldsymbol{v}_m)^+\right] \\ &+\sum_{k=1}^{2}\alpha_k\overline{\overline{\mu_k}}\left\{\left[\nabla \boldsymbol{V}_{km} + \left(\nabla \boldsymbol{V}_{km}\right)^+\right] + 2\mathbb{D}_{ki}\right\}.\end{aligned} \tag{13-49}$$

This shows that if the effects of the relative velocity and the interfacial deformation are small then the mixture viscosity can be given by $\sum_{k=1}^{2}\alpha_k\overline{\overline{\mu_k}}$.

For a dispersed two-phase flow, the interfacial extra deformation stress tensor can be put into a simple form, Eq.(9-76), if the mass transfer effects are not significant. In this case Eq.(13-48) can be reduced to the following form

$$\begin{aligned}\overline{\overline{\mathscr{T}}} &= \sum_{k=1}^{2}\alpha_k\overline{\overline{\mu_k}}\left[\nabla \boldsymbol{j} + (\nabla \boldsymbol{j})^+\right] \\ &+\alpha_d\left(\overline{\overline{\mu_d}} - \overline{\overline{\mu_c}}\right)\left[\nabla \boldsymbol{V}_{dj} + \left(\nabla \boldsymbol{V}_{dj}\right)^+\right]\end{aligned} \tag{13-50}$$

where $\boldsymbol{j}$ and $\boldsymbol{V}_{dj}$ are the volumetric flux of the mixture and the drift velocity of the dispersed phase, respectively. Furthermore we note that $\boldsymbol{j}$ can be related to $\boldsymbol{v}_m$ and $\boldsymbol{V}_{dj}$ by Eq.(4-93). The above result is important, since it shows that in a dispersed flow the mixture-deformation tensor should be based on the velocity of the volume center rather than that of the mass center. The mixture viscosity is then given by

$$\mu_m = \alpha_1\overline{\overline{\mu_1}} + \alpha_2\overline{\overline{\mu_2}}. \tag{13-51}$$

As we can see from Eq.(13-50), the mixture viscous stress has an additional term from the relative motion. The kinematic constitutive equation for $\boldsymbol{V}_{dj}$ shows that, in many cases, this term can be expressed as a function of the void fraction only.

Now let us consider the mixture turbulent stress $\mathscr{T}^T$ which appears in the mixture momentum equation (13-3). It is evident from the definition that $\mathscr{T}^T$ is the addition of the turbulent stresses for each phase $\alpha_k\mathscr{T}_k^T$. In the

Section 1.2 of Chapter 9 we applied the mixing-length theory to the two-fluid model formulation, thus we obtained the constitutive equation for $\mathcal{T}_k^T$. These expressions given by Eq.(9-89) are not appropriate for the drift-flux model formulation, however, since they are written in terms of the variables of each phase and not of the mixture.

We can consider two methods to obtain the mixture turbulent flux $\mathcal{T}^T$, namely the derivation of $\mathcal{T}^T$ from the ones for each phase by using the definitions of the mixture velocity and the drift velocity, or the establishment of the mixing length model in terms of the mixture properties in analogy with single-phase flows. If the two-phases are strongly coupled or the sizes of eddies are large in comparison with the characteristic dimension of a dispersed phase, the latter approach is justified. Then we have

$$\mathcal{T}^T = 2\mu_m^{T*}\rho_m \ell^2 \sqrt{2\mathcal{D}_m:\mathcal{D}_m}\,\mathcal{D}_m \tag{13-52}$$

where

$$\mathcal{D}_m = \frac{1}{2}\left[\nabla \widehat{\boldsymbol{v}_m} + \left(\nabla \widehat{\boldsymbol{v}_m}\right)^+\right]. \tag{13-53}$$

And the non-dimensional coefficient (μ_m^{T*}) corresponds to a mixing length constant.

For a dispersed flow, we may obtain a different expression for $\mathcal{T}^T$. If the dimension of a dispersed phase is comparative to that of turbulent eddies, the dominant part of the mixture stress is given by that of the *continuous phase*. Thus, we have

$$\mathcal{T}^T \doteq \alpha_c \mathcal{T}_c^T = 2\alpha_c \mu_c^{T*}\rho_c \ell^2 \sqrt{2\mathcal{D}_c:\mathcal{D}_c}\,\mathcal{D}_c \tag{13-54}$$

where we have used Eq.(9-89). The total deformation tensor of the continuous phase $\mathcal{D}_c$ is given by Eqs.(9-82), (9-74) and (9-76). Thus, we have

$$\mathcal{D}_c = \frac{1}{2}\left[\nabla \boldsymbol{j} + (\nabla \boldsymbol{j})^+\right] - \frac{1}{2}\left(\frac{\alpha_d}{1-\alpha_d}\right)\left[\nabla \boldsymbol{V}_{dj} + \left(\nabla \boldsymbol{V}_{dj}\right)^+\right]. \tag{13-55}$$

Here we used the definition (4-91) and the identity (4-95). For many practical cases the drift velocity of a dispersed phase can be taken as a function of α_d only, as we have discussed in connection with the kinematic constitutive equation. Further simplification can be made if we assume that the drift velocity is constant and, then, by combining Eqs.(13-50) and (13-54) the total mixture stress can be approximated by

$$\overline{\mathscr{T}} + \mathscr{T}^T + \mathscr{T}^D \doteq \left(\sum_{k=1}^{2} \alpha_k \overline{\overline{\mu_k}} + \alpha_c \mu_c^{T*} \overline{\overline{\rho_c}} \ell^2 \sqrt{2 D_c : D_c} \right) \left[\nabla \boldsymbol{j} + (\nabla \boldsymbol{j})^+ \right] - \frac{\alpha_d}{1-\alpha_d} \frac{\overline{\overline{\rho_c}}\,\overline{\overline{\rho_d}}}{\rho_m} \boldsymbol{V}_{dj} \boldsymbol{V}_{dj}. \tag{13-56}$$

For a fully developed pipe flow, we have

$$\left(\overline{\tau} + \tau^T + \tau^D \right)_{rz} = \left\{ \sum_{k=1}^{2} \alpha_k \overline{\overline{\mu_k}} + \alpha_c \mu_c^{T*} \overline{\overline{\rho_c}} (R-r)^2 \left| \frac{dj_z}{dr} \right| \right\} \frac{dj_z}{dr}. \tag{13-57}$$

The mixture momentum source $\boldsymbol{M}_m$ for a dispersed flow can be given by

$$\boldsymbol{M}_m = \nabla \left(2 \overline{\overline{H_{dc}}}\, \overline{\overline{\sigma}} \alpha_d \right). \tag{13-58}$$

And for a transitional flow we may take

$$\boldsymbol{M}_m \approx 0. \tag{13-59}$$

Energetic Constitutive Equations

The average conduction heat flux for each phase has been obtained in the Section 1.2 of Chapter 9, thus we have

$$\overline{\boldsymbol{q}} = -\sum_{k=1}^{2} \alpha_k \overline{\overline{K_k}} \left\{ \nabla \overline{\overline{T_k}} + \frac{1}{\alpha_k} \left(\overline{\overline{T_i}} - \overline{\overline{T_k}} \right) \left(-\nabla \alpha_k \right) \right\} \tag{13-60}$$

where the first term on the right-hand side has the standard significance of the conduction heat transfer due to the temperature gradient. The second term takes into account the effects of the interfaces. By rearranging the terms, the above equation reduces to

$$\overline{\boldsymbol{q}} = -\left(\sum_{k=1}^{2} \alpha_k \overline{\overline{K_k}}\right) \nabla \overline{\overline{T_i}} - \sum_{k=1}^{2} \overline{\overline{K_k}} \nabla \left[\alpha_k \left(\overline{\overline{T_k}} - \overline{\overline{T_i}}\right)\right]. \tag{13-61}$$

This form of the average conduction heat flux suggests that the concept of the mixture temperature can be represented by $\overline{\overline{T_i}}$ with the mixture conductivity given by $\sum_{k=1}^{2} \alpha_k \overline{\overline{K_k}}$. The second term on the right-hand side represents the effect of thermal non-equilibrium.

For many practical systems the effect of the pressure drop on the thermodynamic properties can be neglected, and as we have already discussed, the temperature of the dispersed phase can be approximated by the interfacial temperature. Thus by taking

$$\begin{aligned} &\nabla \overline{\overline{T_i}} \approx 0 \\ &\overline{\overline{T_d}} \approx \overline{\overline{T_i}} \end{aligned} \tag{13-62}$$

the constitutive equation for the heat flux becomes

$$\overline{\boldsymbol{q}} \doteq -\overline{\overline{K_c}} \nabla \left[\alpha_c \left(\overline{\overline{T_c}} - \overline{\overline{T_i}}\right)\right]. \tag{13-63}$$

Furthermore, the turbulent heat flux can be developed in parallel with the turbulent stress tensor. However, a special care should be taken here because the mixture temperature is not well defined. In view of the constitutive equation for the turbulent energy transfer for each phase, Eq.(9-92), we have

$$\boldsymbol{q}^T = \sum_{k=1}^{2} \alpha_k \boldsymbol{q}_k^T = -\sum_{k=1}^{2} \alpha_k K_k^T \left\{\nabla \overline{\overline{T_k}} - \frac{\nabla \alpha_k}{\alpha_k} \left(\overline{\overline{T_i}} - \overline{\overline{T_k}}\right)\right\} \tag{13-64}$$

where K_k^T is given by Eqs.(9-94) and (9-95).

For a dispersed flow, we may use the approximation given by Eq.(13-62), then we have

$$\boldsymbol{q}^T \doteq -K_c^T \nabla \left[\alpha_c \left(\overline{\overline{T_c}} - \overline{\overline{T_i}}\right)\right]. \tag{13-65}$$

In view of Eq.(9-94) we obtain

$$K_c^{T*} = \frac{K_c^T}{\overline{\overline{\rho_c}} c_{pc} \ell^2 \sqrt{2 D_c : D_c}}. \tag{13-66}$$

Consequently we get

$$\boldsymbol{q}^T = -K_c^{T*} \overline{\overline{\rho_c}} c_{pc} \ell^2 \sqrt{2 D_c : D_c} \nabla \left[\alpha_c \left(\overline{\overline{T_c}} - \overline{\overline{T_i}} \right) \right]. \tag{13-67}$$

Here the non-dimensional coefficient K_c^{T*} corresponds to a thermal mixing-length constant. We expect that it depends on the conductivity, surface area concentration, mean curvature and the void fraction as shown by Eq.(9-95).

For the terms represented by Φ_m^μ, Φ_m^σ, and Φ_m^i, we only note that if these effects due to the viscous dissipation, surface tension and mechanical energy interaction have to be included in the analyses, then they should be specified by three constitutive equations. It is evident from the definitions of these terms, Eqs.(13-7), (13-8) and (13-9), that such constitutive relations are expected to be quite complicated. This means that most of the advantages of using the thermal energy equation diminishes if these effects cannot be neglected.

Constitutive Equation for Phase Change

The constitutive equation for the mass transfer at the interface has been given by Eq.(9-111). Furthermore, we have noted that for a drift-flux model it is necessary to supply information on the thermal state between two phases. In a simplified form it can be given by Eq.(13-47) which is useful for most of the practical problems. Thus for a dispersed flow regime we have

$$\Gamma_c \doteq b_c^{\Gamma*} \frac{\left(\overline{\overline{K_c}} + K_c^T \right) a_i^2}{\left| \widehat{i_{di}} - \widehat{i_{ci}} \right|} \left(\overline{\overline{T_i}} - \overline{\overline{T_c}} \right) \tag{13-68}$$

where non-dimensional coefficient $b_c^{\Gamma*}$ is expected to depend on the following groups.

$$b_c^{\Gamma*} = b_c^{\Gamma*} \left(\frac{\overline{\overline{\rho_c}}}{\overline{\overline{\rho_d}}}, N_{Jc}, \frac{\overline{\overline{H_{dc}}}}{a_i}, \alpha_c \right) \tag{13-69}$$

Geometrical Constitutive Equations

If the effect of the mean curvature $\overline{\overline{H_{21}}}$ and the surface area concentration have to be included explicitly in the formulation, we should give two additional geometrical constitutive equations. In general these are given by Eqs.(9-137) and (9-138), however for a dispersed two-phase flow they can be simplified to Eqs.(9-213) and (9-215).

1.3 Drift-flux (or mixture) model formulation

The general case of the field and constitutive equations for the drift-flux model formulation has been discussed in the Sections 1.1 and 1.2 of Chapter 13. We have noted the importance of the mixture center of mass velocity as well as of the drift velocities in the formulation. It may be appropriate to call our present model the *drift-flux model* in order to emphasize that the effects of relative motions between two phases are taken into account by the drift velocities $\boldsymbol{V}_{kj}$.

1.3.1 Drift-flux model

In view of the definitions for $\boldsymbol{V}_{ki}$ and $\boldsymbol{V}_{km}$ given by Eqs.(4-91) and (4-89), respectively, the field equations for the drift-flux model can be given as follows:

The mixture continuity equation from Eq.(13-1)

$$\frac{\partial \rho_m}{\partial t} + \nabla \cdot \left(\rho_m \boldsymbol{v}_m\right) = 0 \tag{13-70}$$

The continuity equation for phase 2 from Eq.(13-2)

$$\frac{\partial \alpha_2 \overline{\overline{\rho_2}}}{\partial t} + \nabla \cdot \left(\alpha_2 \overline{\overline{\rho_2}} \boldsymbol{v}_m\right) = \Gamma_2 - \nabla \cdot \left(\frac{\alpha_2 \overline{\overline{\rho_1}}\, \overline{\overline{\rho_2}}}{\rho_m} \boldsymbol{V}_{2j}\right) \tag{13-71}$$

The mixture momentum equation from Eq.(13-3)

$$\begin{aligned} &\frac{\partial}{\partial t} \rho_m \boldsymbol{v}_m + \nabla \cdot \left(\rho_m \boldsymbol{v}_m \boldsymbol{v}_m\right) = -\nabla p_m + \nabla \cdot \left(\overline{\overline{\mathcal{T}}} + \mathcal{T}^T\right) \\ &-\nabla \cdot \left(\frac{\alpha_2}{1-\alpha_2} \frac{\overline{\overline{\rho_1}}\, \overline{\overline{\rho_2}}}{\rho_m} \boldsymbol{V}_{2j} \boldsymbol{V}_{2j}\right) + \rho_m \boldsymbol{g}_m + \boldsymbol{M}_m \end{aligned} \tag{13-72}$$

The mixture thermal energy equation from Eq.(13-10)

$$\begin{aligned}&\frac{\partial}{\partial t}\rho_m i_m+\nabla\cdot\left(\rho_m i_m \boldsymbol{v}_m\right)=-\nabla\cdot\left(\overline{\boldsymbol{q}}+\boldsymbol{q}^T\right)\\&-\nabla\cdot\left\{\alpha_2\frac{\overline{\overline{\rho_1}}\,\overline{\overline{\rho_2}}}{\rho_m}\boldsymbol{V}_{2j}\left(\widehat{i_2}-\widehat{i_1}\right)\right\}+\frac{Dp_m}{Dt}+\Phi_m^{\mu}+\Phi_m^{\sigma}+\Phi_m^{i}.\end{aligned}\tag{13-73}$$

Here we have formulated the model in terms of the mixture properties, void fraction α_2 and the drift velocity $\boldsymbol{V}_{2j}$. The model is most effective for a dispersed two-phase flow, since for this case the constitutive equations can be reduced to realistic forms as discussed in the Section 1.2 of Chapter 13.

It should be emphasized again that the drift-flux model is useful for the two-phase flow *system analyses*. This is particularly true if the motions of two phases are strongly coupled. Because of its simplicity the model can be used to make realistic similarity analyses as well as to solve many important engineering problems.

1.3.2 Scaling parameters

Let us denote the reference parameters by the subscript o. The characteristic length scale is L_o and the time scale is taken as the ratio of L_o to the velocity scale. Then we define the non-dimensional parameters whose order of magnitude is considered to be 1 as follows:

$$\begin{aligned}&\rho_m^*=\frac{\rho_m}{\rho_{mo}},\ \boldsymbol{v}_m^*=\frac{\boldsymbol{v}_m}{v_{mo}},\ t^*=\frac{t}{\tau_o}=\frac{t}{\left(L_o/v_{mo}\right)},\ \nabla^*=L_o\nabla\\&\rho_1^*=\frac{\overline{\overline{\rho_1}}}{\rho_{1o}},\ \rho_2^*=\frac{\overline{\overline{\rho_2}}}{\rho_{2o}},\ \boldsymbol{V}_{2j}^*=\frac{\boldsymbol{V}_{2j}}{V_{2jo}},\ \Gamma_2^*=\frac{\Gamma_2}{\Gamma_{2o}},\ p_m^*=\frac{p_m}{\rho_{mo}v_{mo}^2}\\&i_m^*=\frac{i_m-i_{1o}}{i_{2o}-i_{1o}},\ \Delta i_{12}^*=\frac{\widehat{i_2}-\widehat{i_1}}{i_{2o}-i_{1o}}\\&\left(\overline{\boldsymbol{\mathcal{T}}}+\boldsymbol{\mathcal{T}}^T\right)^*=\frac{\overline{\boldsymbol{\mathcal{T}}}+\boldsymbol{\mathcal{T}}^T}{\mu_{mo}v_{mo}/L_o},\ \boldsymbol{M}_m^*=\frac{\boldsymbol{M}_m}{2H_{21o}\sigma_o/L_o}\\&\left(\overline{\boldsymbol{q}}+\boldsymbol{q}^T\right)^*=\frac{\overline{\boldsymbol{q}}+\boldsymbol{q}^T}{K_{mo}\Delta T_o/L_o},\ \Phi_m^{\mu*}=\frac{\Phi_m^{\mu}}{\left(\mu_{mo}v_{mo}/L_o\right)\left(v_{mo}/L_o\right)}\end{aligned}\tag{13-74}$$

$$\Phi_m^{\sigma*} = \frac{\Phi_m^\sigma}{2H_{21}\sigma_o v_{mo} / L_o}, \Phi_m^{i*} = \frac{\Phi_m^i}{(\rho_{1o} - \rho_{2o}) v_{mo}^2 V_{2jo} / L_o}.$$

Substituting these new parameters into the field equations we obtain the following results:

Non-dimensional mixture continuity equation

$$\frac{\partial}{\partial t^*}\rho_m^* + \nabla^* \cdot \left(\rho_m^* \boldsymbol{v}_m^*\right) = 0 \tag{13-75}$$

Non-dimensional continuity equation for phase 2

$$\frac{\partial \alpha_2 \rho_2^*}{\partial t^*} + \nabla^* \cdot \left(\alpha_2 \rho_2^* \boldsymbol{v}_m^*\right) = N_{pch}\Gamma_2^* - N_D \nabla^* \cdot \left(\frac{\alpha_2 \rho_1^* \rho_2^*}{\rho_m^*} \boldsymbol{V}_{2j}^*\right) \tag{13-76}$$

Non-dimensional momentum equation for mixture

$$\begin{aligned}
&\frac{\partial}{\partial t^*}\left(\rho_m^* \boldsymbol{v}_m^*\right) + \nabla^* \cdot \left(\rho_m^* \boldsymbol{v}_m^* \boldsymbol{v}_m^*\right) = -\nabla^* p_m^* \\
&+ \frac{1}{N_{\mathrm{Re}}} \nabla^* \cdot \left(\overline{\mathcal{T}} + \mathcal{T}^T\right)^* + \frac{1}{N_{Fr}} \rho_m^* \frac{\boldsymbol{g}_m}{|\boldsymbol{g}_m|} \\
&- N_\rho N_D^2 \nabla^* \cdot \left\{\left(\frac{\alpha_2}{1-\alpha_2}\right) \frac{\rho_1^* \rho_2^*}{\rho_m^*} \boldsymbol{V}_{2j}^* \boldsymbol{V}_{2j}^*\right\} + N_\sigma \boldsymbol{M}_m^*
\end{aligned} \tag{13-77}$$

Non-dimensional thermal energy equation for mixture

$$\begin{aligned}
&\frac{\partial}{\partial t^*}\left(\rho_m^* i_m^*\right) + \nabla^* \cdot \left(\rho_m^* i_m^* \boldsymbol{v}_m^*\right) = -\frac{1}{N_{Pe}} \nabla^* \cdot \left(\overline{\boldsymbol{q}} + \boldsymbol{q}^T\right)^* \\
&- \frac{N_\rho}{(1-\alpha_2) + \alpha_2 N_\rho} N_D \nabla^* \cdot \left\{\frac{\alpha_2 \rho_1^* \rho_2^*}{\rho_m^*} \boldsymbol{V}_{2j}^* \Delta i_{12}^*\right\} \\
&+ N_{Ec}\left\{\frac{Dp_m^*}{Dt^*} + \frac{1}{N_{\mathrm{Re}}}\Phi_m^{\mu*} + N_\sigma \Phi_m^{\sigma*} + \left(N_\rho - 1\right) N_D \Phi_m^{i*}\right\}.
\end{aligned} \tag{13-78}$$

Here we have defined

Phase change number $N_{pch} \equiv \dfrac{\Gamma_{2o}L_o}{\rho_{2o}v_{mo}}$

Drift number $N_D \equiv \dfrac{\rho_{1o}V_{2jo}}{\rho_{mo}v_{mo}}$

Density ratio $N_\rho \equiv \dfrac{\rho_{2o}}{\rho_{1o}}$

Reynolds number $N_{\mathrm{Re}} \equiv \dfrac{\rho_{mo}v_{mo}L_o}{\mu_{mo}}$

Froude number $N_{Fr} \equiv \dfrac{v_{mo}^2}{|g_m|\,L_o}$

Surface number $N_\sigma \equiv \dfrac{2H_{21o}\sigma_o}{\rho_{mo}v_{mo}^2}$

Péclet number $N_{Pe} \equiv \dfrac{\rho_{mo}v_{mo}\Delta i_{12o}}{K_{mo}\Delta T_o}$

Eckert number $N_{Ec} \equiv \dfrac{v_{mo}^2}{\Delta i_{12o}}$. (13-79)

We note that these eight groups are *the scaling parameters for the mixtures* based on the drift-flux model formulation. These groups are analogous to the ones obtained for the one-dimensional, two-phase flow model in Ishii (1971). An exception is the Reynolds and Peclet numbers, however, since for the latter model they are replaced by the groups from the boundary conditions, namely, the friction and Stanton numbers.

The distinction between the *scaling parameters* and the *similarity groups* should be clearly made, since they are not synonymous. The similarity groups are obtainable from the field equations and the boundary and the initial conditions with all the constitutive equations specified. Consequently, the similarity criteria cannot be discussed in detail unless the system and the problem are clearly defined. The scaling parameter enables us to make various assumptions and approximations, because they give the order of magnitude of various terms appearing in the field equations.

The first three groups of Eq.(13-79), N_{pch}, N_D, and N_ρ are the kinematic groups. If the phase change number N_{pch} is much larger than the drift number N_D, the systems is controlled by the changes of phases. However, if $N_D >> N_{pch}$, then the system is controlled by the redistributions of phases. The number denoted by N_{Re}, N_{Fr}, and N_σ are dynamic groups, since they scale the various forces arisen in the mixture momentum equation, whereas the Peclet and Eckert numbers are the energy groups. We note the particular importance of the Eckert number N_{Ec}. It is evident that except for high speed flows, N_{Ec} is small, thus the terms given by Φ_m^μ, Φ_m^σ and Φ_m^i in the thermal energy equation can be neglected. As we have already discussed before, this is the single most important approximation which greatly simplifies the heat transfer problems.

From Eqs.(13-76), (13-77) and (13-78), we set that the drift transport of mass, momentum and energy are not weighted by the same scaling parameter. The phase drift and the enthalpy transport due to the relative motion between phases has approximately the same order N_D, thus these two terms should be treated under the same condition. However, the momentum drift are weighted by $N_\rho N_D^2$. Thus, depending on the magnitude of this scaling parameter, it is possible to neglect the drift-stress tensor independently of the other two effects.

We note that in the *scaling parameters*, we have introduced the mixture viscosity and the conductivity given by μ_m and K_m. These parameters need not to be precisely defined as long as the scaling parameters are concerned, because only their order of magnitude is important. We may choose them as the larger viscosity and conductivity among $\overline{\overline{\mu_k}}$ and $\overline{\overline{K_k}}$ or defined them by $\sum_{k=1}^{2}\alpha_{ko}\overline{\overline{\mu_{ko}}}$ and $\sum_{k=1}^{2}\alpha_{ko}\overline{\overline{K_{ko}}}$ where α_{ko} denotes the reference value of α_k, for example at the system boundary.

However, if the *similarity groups* are concerned, then the exact constitutive equations for the stress tensor and heat flux should be used. Thus, according to the forms and the variables appearing in the constitutive equations, the correct reference parameters should be chosen. Based on the order of magnitude analysis and the scaling parameters obtained above, we discuss some of the important special cases.

1.3.3 Homogeneous flow model

If the drift number N_D is much smaller than the phase change number N_{pch}, then the system is reaction (phase change) controlled and the drift or diffusion of mass is negligible in view of the continuity equation for phase 2. Furthermore, if the drift number N_D is much smaller than unity, then all of the drift terms and the interfacial mechanical energy transfer effect Φ_m^i can be neglected in the field equations. We may not drop the continuity equation for phase 2 from the formulation, however, since it takes into account for the thermal non-equilibrium effect as discussed by Zuber and Dougherty (1967) in connection with the one-dimensional model. Thus, we have the following four field equations for the general homogeneous flow model:

$$\frac{\partial \rho_m}{\partial t} + \nabla \cdot \left(\rho_m \boldsymbol{v}_m\right) = 0$$
$$\frac{\partial \alpha_2 \overline{\overline{\rho_2}}}{\partial t} + \nabla \cdot \left(\alpha_2 \overline{\overline{\rho_2}} \boldsymbol{v}_m\right) = \Gamma_2$$
$$\frac{\partial \rho_m \boldsymbol{v}_m}{\partial t} + \nabla \cdot \left(\rho_m \boldsymbol{v}_m \boldsymbol{v}_m\right) = -\nabla p_m + \nabla \cdot \left(\overline{\mathcal{C}} + \mathcal{C}^T\right) \tag{13-80}$$
$$+\rho_m \boldsymbol{g}_m + \boldsymbol{M}_m$$
$$\frac{\partial \rho_m i_m}{\partial t} + \nabla \cdot \left(\rho_m i_m \boldsymbol{v}_m\right) = -\nabla \cdot \left(\overline{\boldsymbol{q}} + \boldsymbol{q}^T\right) + \frac{Dp_m}{Dt}$$
$$+\Phi_m^\mu + \Phi_m^\sigma .$$

In this case, the mechanical constitutive equations for the dispersed two-phase flow can be reduced to the simple forms. Thus, from Eq.(13-50), we have

$$\overline{\mathcal{C}} = \left(\alpha_1 \overline{\overline{\mu_1}} + \alpha_2 \overline{\overline{\mu_2}}\right)\left[\nabla \boldsymbol{v}_m + \left(\nabla \boldsymbol{v}_m\right)^+\right]. \tag{13-81}$$

The turbulent stress is given by Eq.(13-52) as

$$\mathcal{C}^T = 2\mu_m^{T*} \rho_m \ell^2 \sqrt{2\mathcal{D}_m : \mathcal{D}_m}\, \mathcal{D}_m . \tag{13-82}$$

Furthermore, the constitutive equations for the heat fluxes are given by

$$\overline{\boldsymbol{q}} \doteq -\overline{\overline{K_c}} \nabla \left[\alpha_c \left(\overline{\overline{T_c}} - \overline{\overline{T_i}}\right)\right] \tag{13-83}$$

and

$$\boldsymbol{q}^T = -K_c^{T*} \overline{\overline{\rho_c}} c_{pc} \ell^2 \sqrt{2\mathcal{D}_m : \mathcal{D}_m}\, \nabla \left[\alpha_c \left(\overline{\overline{T_c}} - \overline{\overline{T_i}}\right)\right] \tag{13-84}$$

where the dispersed phase has been assumed to be in thermal equilibrium with interfaces, namely, Eq.(13-62). If the Eckert number N_{Ec} and the surface number N_σ are small, then the capillary force $\boldsymbol{M}_m$ in the mixture momentum equation, and the compressibility effect Dp_m/Dt, the dissipation term Φ_m^μ, and the surface tension effect Φ_m^σ in the energy equation can be dropped.

1.3.4 Density propagation model

For a number of systems of practical interest, particularly for systems at high-reduced pressures without large acoustic interactions, it is reasonable to assume that each phase is essentially incompressible. Then we have

$$\overline{\overline{\rho_k}} = \text{constant}. \tag{13-85}$$

Furthermore, we consider the case when the drift velocity of phase 2 is approximately a function of α_2 only. We recall that this is a valid assumption for many practical flow regimes as it has been discussed in the Section 1.3 of Chapter 7 and the Section 1.2 of Chapter 13. Thus we have

$$\boldsymbol{V}_{2j} \approx \boldsymbol{V}_{2j}(\alpha_2). \tag{13-86}$$

For simplicity, we also assume that the surface number N_σ and the Eckert number N_{Ec} are small such that

$$N_\sigma << 1 \text{ and } N_{Ec} << 1. \tag{13-87}$$

Then we obtain from Eqs.(7-31), (7-38), (13-72) and (13-73) the following field equations:

$$\begin{aligned}
&\nabla \cdot \boldsymbol{j} = \Gamma_2 \left(\frac{1}{\overline{\overline{\rho_2}}} - \frac{1}{\overline{\overline{\rho_1}}} \right) \\
&\frac{\partial \rho_m}{\partial t} + \boldsymbol{C}_K \cdot \nabla \rho_m = \frac{\rho_m}{\overline{\overline{\rho_1}}\,\overline{\overline{\rho_2}}} \Gamma_2 \left(\overline{\overline{\rho_2}} - \overline{\overline{\rho_1}} \right) \\
&\rho_m \frac{D\boldsymbol{v}_m}{Dt} = -\nabla p_m + \nabla \cdot \left(\overline{\mathcal{T}} + \mathcal{T}^T \right) \\
&-\nabla \cdot \left(\frac{\alpha_2}{1-\alpha_2} \frac{\overline{\overline{\rho_1}}\,\overline{\overline{\rho_2}}}{\rho_m} \boldsymbol{V}_{2j} \boldsymbol{V}_{2j} \right) + \rho_m \boldsymbol{g}_m
\end{aligned} \tag{13-88}$$

$$\rho_m \frac{Di_m}{Dt} = -\nabla \cdot \left(\overline{\boldsymbol{q}} + \boldsymbol{q}^T \right) - \nabla \cdot \left(\frac{\alpha_2 \overline{\overline{\rho_1}}\, \overline{\overline{\rho_2}}}{\rho_m} \boldsymbol{V}_{2j} \left(\widehat{i}_2 - \widehat{i}_1 \right) \right)$$

where $\boldsymbol{C}_K$ and $\boldsymbol{j}$ are given by

$$\boldsymbol{C}_K = \boldsymbol{j} + \frac{\partial}{\partial \alpha_2} \left(\alpha_2 \boldsymbol{V}_{2j} \right) \tag{13-89}$$

and

$$\boldsymbol{j} = \boldsymbol{v}_m + \frac{\alpha_2 \left(\overline{\overline{\rho_1}} - \overline{\overline{\rho_2}} \right)}{\rho_m} \boldsymbol{V}_{2j}. \tag{13-90}$$

The present density-wave propagation model is best suited for a dispersed (or mixed) two-phase flow regime. In this case, phase 2 should be taken as the dispersed phase. Then the constitutive equations for the stresses can be given by Eqs.(13-50) and (13-54), whereas the heat fluxes are given by Eqs.(13-63) and (13-67).

drift-flux model

Chapter 14

ONE-DIMENSIONAL DRIFT-FLUX MODEL

Two-phase flow always involves some relative motion of one phase with respect to the other; therefore, a two-phase-flow problem should be formulated in terms of two velocity fields. A general transient two-phase-flow problem can be formulated by using a two-fluid model or a drift-flux model, depending on the degree of the dynamic coupling between the phases. In the two-fluid model, each phase is considered separately; hence the model is formulated in terms of two sets of conservation equations governing the balance of mass, momentum, and energy of each phase. However, an introduction of two momentum equations in a formulation, as in the case of the two-fluid model, presents considerable difficulties due to mathematical complications and uncertainties in specifying interfacial interaction terms between two phases (Delhaye, 1968; Vernier and Delhaye, 1968; Bouré and Réocruex, 1972; Ishii, 1975). Numerical instabilities caused by improper choice of interfacial-interaction terms in the phase-momentum equations are common. Therefore, careful studies on the interfacial constitutive equations are required in the formulation of the two-fluid model. For example, it has been suggested (Réocruex, 1974) that the interaction terms should include first-order time and spatial derivatives under certain conditions.

These difficulties associated with a two-fluid model can be significantly reduced by formulating two-phase problems in terms of the drift-flux model (Zuber, 1967). In this model, the motion of the whole mixture is expressed by the mixture-momentum equation and the relative motion between phases is taken into account by a kinematic constitutive equation. Therefore, the basic concept of the drift-flux model is to consider the mixture as a whole rather than as two separated phases. The formulation of the drift-flux model based on the mixture balance equations is simpler than the two-fluid model based on the separate balance equations for each phase. The most important assumption associated with the drift-flux model is that the dynamics of two phases can be expressed by the mixture-momentum equation with the kinematic constitutive equation specifying the relative motion between phases. The use of the drift-flux model is appropriate when the motions of two phases are strongly coupled.

In the drift-flux model, the velocity fields are expressed in terms of the mixture center-of-mass velocity and the drift velocity of the vapor phase, which is the vapor velocity with respect to the volume center of the mixture. The effects of thermal non-equilibrium are accommodated in the drift-flux model by a constitutive equation for phase change that specifies the rate of mass transfer per unit volume. Since the rates of mass and momentum transfer at the interfaces depend on the structure of interface, these constitutive equations for the drift velocity and the vapor generation are functions of flow regimes (Zuber and Dougherty, 1967; Ishii et al., 1975).

The drift-flux model is an approximate formulation in comparison with the more rigorous two-fluid formulation. However, because of its simplicity and applicability to a wide range of two-phase-flow problems of practical interest, the drift-flux model is of considerable importance. Particularly, the one-dimensional drift-flux model obtained by averaging the local drift-flux formulation over the cross-sectional area is useful for complicated engineering problems involving fluid flow and heat transfer, since field equations can be reduced to quasi-one-dimensional forms. By area averaging, the information on changes of variables in the direction normal to the main flow within a channel is basically lost. Therefore, the transfer of momentum and energy between the wall and the fluid should be expressed by empirical correlations or by simplified models. In this chapter, we develop a general one-dimensional formulation of the drift-flux model, and discuss various special cases which are important in practical applications. For simplicity, mathematical symbols of time-averaging are dropped in the formulation in this chapter. The extensive review of this model is given by Ishii (1977).

1.1 Area average of three-dimensional drift-flux model

The three-dimensional form of the drift-flux model has been obtained by using the time or statistical averaging method. The result can be summarized as follows:

The mixture continuity equation from Eq.(13-70)

$$\frac{\partial \rho_m}{\partial t} + \nabla \cdot \left(\rho_m \boldsymbol{v}_m\right) = 0 \tag{14-1}$$

The continuity equation for dispersed phase from Eq.(13-71)

$$\frac{\partial \alpha_d \rho_d}{\partial t} + \nabla \cdot (\alpha_d \rho_d \boldsymbol{v}_m) = \Gamma_d - \nabla \cdot \left(\frac{\alpha_d \rho_d \rho_c}{\rho_m} \boldsymbol{V}_{dj} \right) \tag{14-2}$$

The mixture momentum equation from Eq.(13-72)

$$\begin{aligned} &\frac{\partial \rho_m \boldsymbol{v}_m}{\partial t} + \nabla \cdot (\rho_m \boldsymbol{v}_m \boldsymbol{v}_m) \\ &= -\nabla p_m + \nabla \cdot \left(\overline{\mathcal{T}} + \mathcal{T}^T - \frac{\alpha_d}{1-\alpha_d} \frac{\rho_d \rho_c}{\rho_m} \boldsymbol{V}_{dj} \boldsymbol{V}_{dj} \right) + \rho_m \boldsymbol{g} \end{aligned} \tag{14-3}$$

The mixture enthalpy-energy equation from Eq.(13-73)

$$\begin{aligned} &\frac{\partial \rho_m h_m}{\partial t} + \nabla \cdot (\rho_m h_m \boldsymbol{v}_m) \\ &= -\nabla \cdot \left[\overline{\boldsymbol{q}} + \boldsymbol{q}^T + \frac{\alpha_d \rho_d \rho_c}{\rho_m} (h_d - h_c) \boldsymbol{V}_{dj} \right] \\ &+ \frac{\partial p_m}{\partial t} + \left[\boldsymbol{v}_m + \frac{\alpha_d (\rho_c - \rho_d)}{\rho_m} \boldsymbol{V}_{dj} \right] \cdot \nabla p_m + \Phi_m^\mu . \end{aligned} \tag{14-4}$$

The rational approach to obtain a one-dimensional drift-flux model is to integrate the three-dimensional drift-flux model over a cross-sectional area and then to introduce proper mean values. A simple area average over the cross-sectional area, A, is defined by

$$\langle F \rangle = \frac{1}{A} \int_A F dA \tag{14-5}$$

and the void-fraction-weighted mean value is given by

$$\langle\langle F_k \rangle\rangle = \frac{\langle \alpha_k F_k \rangle}{\langle \alpha_k \rangle} . \tag{14-6}$$

In the subsequent analysis, the density of each phase ρ_d and ρ_c within any cross-sectional area is considered to be uniform, so that $\rho_k = \langle\langle \rho_k \rangle\rangle$. For most practical two-phase flow problems, this assumption is valid since the transverse pressure gradient within a channel is relatively small. The detailed analysis without this approximation appears in a reference (Ishii, 1971). Under the above simplifying assumption, the average mixture density is given by

$$\langle \rho_m \rangle \equiv \langle \alpha_d \rangle \rho_d + \left(1 - \langle \alpha_d \rangle\right) \rho_c. \tag{14-7}$$

The axial component of the weighted mean velocity of phase k is

$$\langle\langle v_k \rangle\rangle = \frac{\langle \alpha_k v_k \rangle}{\langle \alpha_k \rangle} = \frac{\langle j_k \rangle}{\langle \alpha_k \rangle} \tag{14-8}$$

where the scalar expression of the velocity corresponds to the axial component of the vector. Then the mixture velocity is defined by

$$\overline{v_m} \equiv \frac{\langle \rho_m v_m \rangle}{\langle \rho_m \rangle} = \frac{\langle \alpha_d \rangle \rho_d \langle\langle v_d \rangle\rangle + \left(1 - \langle \alpha_d \rangle\right) \rho_c \langle\langle v_c \rangle\rangle}{\langle \rho_m \rangle} \tag{14-9}$$

and the volumetric flux is given by

$$\langle j \rangle \equiv \langle j_d \rangle + \langle j_c \rangle = \langle \alpha_d \rangle \langle\langle v_d \rangle\rangle + \left(1 - \langle \alpha_d \rangle\right) \langle\langle v_c \rangle\rangle. \tag{14-10}$$

The mean mixture enthalpy also should be weighted by the density; thus,

$$\overline{h_m} \equiv \frac{\langle \rho_m h_m \rangle}{\langle \rho_m \rangle} = \frac{\langle \alpha_d \rangle \rho_d \langle\langle h_d \rangle\rangle + \left(1 - \langle \alpha_d \rangle\right) \rho_c \langle\langle h_c \rangle\rangle}{\langle \rho_m \rangle}. \tag{14-11}$$

The appropriate mean drift velocity is defined by

$$\overline{V_{dj}} \equiv \langle\langle v_d \rangle\rangle - \langle j \rangle = \left(1 - \langle \alpha_d \rangle\right) \left(\langle\langle v_d \rangle\rangle - \langle\langle v_c \rangle\rangle \right). \tag{14-12}$$

The experimental determination of the drift velocity is possible if the volume flow rate of each phase, Q_k, and the mean void fraction $\langle \alpha_d \rangle$ are measured. This is because Eq.(14-12) can be transformed into

$$\overline{V_{dj}} = \frac{\langle j_d \rangle}{\langle \alpha_d \rangle} - \left(\langle j_d \rangle + \langle j_c \rangle \right) \tag{14-13}$$

where $\langle j_k \rangle$ is given by $\langle j_k \rangle = Q_k / A$. Furthermore, the present definition of the drift velocity can also be used for annular two-phase flows. Under the definitions of various velocity fields we obtain several important relations, such as

$$\begin{cases} \langle\langle v_d \rangle\rangle = \overline{v_m} + \dfrac{\rho_c}{\langle \rho_m \rangle} \overline{V_{dj}} \\ \langle\langle v_c \rangle\rangle = \overline{v_m} - \dfrac{\langle \alpha_d \rangle}{1 - \langle \alpha_d \rangle} \dfrac{\rho_d}{\langle \rho_m \rangle} \overline{V_{dj}} \end{cases} \tag{14-14}$$

and

$$\langle j \rangle = \overline{v_m} + \frac{\langle \alpha_d \rangle (\rho_c - \rho_d)}{\langle \rho_m \rangle} \overline{V_{dj}}. \tag{14-15}$$

In the drift-flux formulation, a problem is solved for $\langle \alpha_d \rangle$ and $\overline{v_m}$ with a given constitutive relation for $\overline{V_{dj}}$. Thus, Eq.(14-14) can be used to recover a solution for the velocity of each phase after a problem is solved.

By area-averaging Eqs.(14-1)-(14-4) and using the various mean values, we obtain

Mixture Continuity Equation

$$\frac{\partial \langle \rho_m \rangle}{\partial t} + \frac{\partial}{\partial z}\left(\langle \rho_m \rangle \overline{v_m} \right) = 0 \tag{14-16}$$

Continuity Equation for Dispersed Phase

$$\frac{\partial \langle \alpha_d \rangle \rho_d}{\partial t} + \frac{\partial}{\partial z}\left(\langle \alpha_d \rangle \rho_d \overline{v_m} \right) = \langle \Gamma_d \rangle - \frac{\partial}{\partial z}\left(\frac{\langle \alpha_d \rangle \rho_d \rho_c}{\langle \rho_m \rangle} \overline{V_{dj}} \right) \tag{14-17}$$

Mixture Momentum Equation

$$\begin{aligned} &\frac{\partial \langle \rho_m \rangle \overline{v_m}}{\partial t} + \frac{\partial}{\partial z}\left(\langle \rho_m \rangle \overline{v_m}^2 \right) = -\frac{\partial}{\partial z} \langle p_m \rangle \\ &+ \frac{\partial}{\partial z} \left\langle \tau_{zz} + \tau_{zz}^T \right\rangle - \langle \rho_m \rangle g_z \end{aligned} \tag{14-18}$$

$$
-\frac{f_m}{2D}\langle\rho_m\rangle\overline{v_m}\left|\overline{v_m}\right|-\frac{\partial}{\partial z}\left[\frac{\langle\alpha_d\rangle\rho_d\rho_c}{\left(1-\langle\alpha_d\rangle\right)\langle\rho_m\rangle}\overline{V_{dj}}^2\right]
-\frac{\partial}{\partial z}\sum_{k=1}^{2}COV\left(\alpha_k\rho_k v_k v_k\right)
$$

Mixture Enthalpy-energy Equation

$$
\begin{aligned}
&\frac{\partial\langle\rho_m\rangle\overline{h_m}}{\partial t}+\frac{\partial}{\partial z}\left(\langle\rho_m\rangle\overline{h_m}\,\overline{v_m}\right)=-\frac{\partial}{\partial z}\left\langle\overline{q}+q^T\right\rangle+\frac{q_W''\xi_h}{A}\\
&-\frac{\partial}{\partial z}\frac{\langle\alpha_d\rangle\rho_d\rho_c}{\langle\rho_m\rangle}\Delta h_{dc}\overline{V_{dj}}-\frac{\partial}{\partial z}\left[\frac{\langle\alpha_d\rangle\rho_d\rho_c}{\langle\rho_m\rangle}\Delta h_{dc}\overline{V_{dj}}\right]\\
&-\frac{\partial}{\partial z}\sum_{k=1}^{2}COV\left(\alpha_k\rho_k h_k v_k\right)+\frac{\partial\langle p_m\rangle}{\partial t}\\
&+\left[\overline{v_m}+\frac{\langle\alpha_d\rangle\left(\rho_c-\rho_d\right)}{\langle\rho_m\rangle}\overline{V_{dj}}\right]\frac{\partial\langle p_m\rangle}{\partial z}+\left\langle\Phi_m^\mu\right\rangle.
\end{aligned}
\tag{14-19}
$$

Here, $\tau_{zz}+\tau_{zz}^T$ denotes the normal components of the stress tensor in the axial direction and Δh_{dc} is the enthalpy difference between phases; thus, $\Delta h_{dc}=\langle\langle h_d\rangle\rangle-\langle\langle h_c\rangle\rangle$. The covariance terms represent the difference between the average of a product and the product of the average of two variables such that $COV\left(\alpha_k\rho_k\psi_k v_k\right)\equiv\left\langle\alpha_k\rho_k\psi_k\left(v_k-\langle\langle v_k\rangle\rangle\right)\right\rangle$. If the profile of either ψ_k or v_k is flat, then the covariance term reduces to zero. The term represented by $f_m\langle\rho_m\rangle\overline{v_m}\left|\overline{v_m}\right|/(2D)$ in Eq.(14-18) is the two-phase frictional pressure drop. We note here that the effects of the mass, momentum, and energy diffusion associated with the relative motion between phases appear explicitly in the drift-flux formulation, since the convective terms on the left-hand side of the field equations are expressed in terms of the mixture velocity. These effects of diffusions in the present formulation are expressed in terms of the drift velocity of the dispersed phase V_{dj}. This may be formulated in a functional form as

$$
\overline{V_{dj}}=\overline{V_{dj}}\left(\langle\alpha_d\rangle,\langle p_m\rangle,g_z,\overline{v_m},\text{ etc.}\right).
\tag{14-20}
$$

To take into account the mass transfer across the interfaces, a constitutive equation for $\langle\Gamma_d\rangle$ should also be given. In a functional form, this phase-change constitutive equation may be written as

$$\langle\Gamma_d\rangle = \langle\Gamma_d\rangle\left(\langle\alpha_d\rangle, \langle p_m\rangle, \overline{\overline{v}}_m, \frac{\partial\langle p_m\rangle}{\partial t}, \text{etc.}\right). \tag{14-21}$$

The above formulation can be extended to non-dispersed two-phase flows, such as an annular flow, provided a proper constitutive relation for a drift velocity of one of the phases is given.

1.2 One-dimensional drift velocity

1.2.1 Dispersed two-phase flow

To obtain a kinematic constitutive equation for the one-dimensional drift-flux model, we must average the local drift velocity over the channel cross section. The constitutive relation for the local drift velocity V_{dj} in a confined channel was developed in the Section 1.2 of Chapter 13. Now we relate this to the mean drift velocity $\overline{\overline{V}}_{dj}$ defined by Eq.(14-12).

From Eqs.(14-6) and (14-12),

$$\overline{\overline{V}}_{dj} = \left\langle \frac{\alpha_d\left(j + V_{dj}\right)}{\langle\alpha_d\rangle} - j \right\rangle = \langle\langle V_{dj}\rangle\rangle + (C_0 - 1)\langle j\rangle \tag{14-22}$$

where

$$\langle\langle V_{dj}\rangle\rangle \equiv \frac{\langle\alpha_d V_{dj}\rangle}{\langle\alpha_d\rangle} \tag{14-23}$$

and

$$C_0 \equiv \frac{\langle\alpha_d j\rangle}{\langle\alpha_d\rangle\langle j\rangle}. \tag{14-24}$$

The second term on the right-hand side of Eq.(14-22) is a covariance between the concentration profile and the volumetric flux profile; thus it can also be expressed as $COV\left(\alpha_d j\right)/\langle\alpha_d\rangle$. The factor C_0, which has been used for bubbly or slug flows by several authors (Zuber and Findlay, 1965; Nicklin et al., 1962; Neal, 1963) is known as a distribution parameter. The inverse of this parameter was also used in the early work of Bankoff (1960). Physically, this effect arises from the fact that the dispersed phase is locally transported with the drift velocity V_{dj} with respect to local volumetric flux j and not to the average volumetric flux $\langle j \rangle$. For example, if the dispersed phase is more concentrated in the higher-flux region, then the mean transport of the dispersed phase is promoted by higher local j.

The value of C_0 can be determined from assumed profiles of the void fraction α_d and total volumetric flux j (Zuber and Findlay, 1965), or from experimental data (Zuber et al., 1967). By assuming power-law profiles in a pipe for j and α_d, we have

$$\begin{cases} \dfrac{j}{j_0} = 1 - \left(\dfrac{r}{R_W}\right)^m \\ \dfrac{\alpha_d - \alpha_{dW}}{\alpha_{d0} - \alpha_{dW}} = 1 - \left(\dfrac{r}{R_W}\right)^n \end{cases} \tag{14-25}$$

where j_0, α_{d0}, α_{dW}, r, and R_W are, respectively, the value of j and α at the center, the void fraction at the wall, radial distance, and the radius of a pipe. By substituting these profiles into the definition of C_0 given by Eq.(14-24), we obtain

$$C_0 = 1 + \frac{2}{m+n+2}\left(1 - \frac{\alpha_{dW}}{\langle \alpha_d \rangle}\right). \tag{14-26}$$

The distribution parameter based on the assumed profiles above is further discussed in a reference (Zuber et al., 1967).

Now Eq.(14-22) can be transformed to

$$\langle\langle v_d \rangle\rangle = \frac{\langle j_d \rangle}{\langle \alpha_d \rangle} = C_0 \langle j \rangle + \langle\langle V_{dj} \rangle\rangle \tag{14-27}$$

where $\langle\langle v_d \rangle\rangle$ and $\langle j \rangle$ are easily obtainable parameters in experiments, particularly under an adiabatic condition. Therefore, this equation suggests a plot of the mean velocity $\langle\langle v_d \rangle\rangle$ versus the average volumetric flux $\langle j \rangle$. If the concentration profile is uniform across the channel, then the value of the distribution parameter is equal to unity. In addition, if the effect of the local drift $\langle\langle V_{dj} \rangle\rangle$ is negligibly small, then the flow becomes essentially homogeneous. In this case, the relation between the mean velocity and flux reduces to a straight line through the origin at an angle of 45°. The deviation of the experimental data from this homogeneous flow line shows the magnitude of the drift of the dispersed phase with respect to the volume center of the mixture.

An important characteristic of such a plot is that, for two-phase flow regimes with fully developed void and velocity profiles, the data points cluster around a straight line (see Figs.14-1-14-3). This trend is particularly pronounced when the local drift velocity is constant or negligibly small. Hence, for a given flow regime, the value of the distribution parameter C_0 may be obtained from the slope of these lines, whereas the intercept of this line with the mean velocity axis can be interpreted as the weighed mean local drift velocity, $\langle\langle V_{dj} \rangle\rangle$. The extensive study by Zuber et al. (1967) shows that C_0 depends on pressure, channel geometry, and perhaps flow rate. An important effect of subcooled boiling and developing void profile on the distribution parameter has also been noted by Hancox and Nicoll (1972). Here, a simple correlation for the distribution parameter in bubbly-flow regime is presented based on study by Ishii (1977). First, by considering a fully developed bubbly flow, we assumed that C_0 depends on the density ratio ρ_g/ρ_f and on the Reynolds number based on liquid properties, GD/μ_f, where G, D, and μ_f are the total mass flow rate, hydraulic diameter, and the viscosity of the liquid, respectively. Hence,

$$C_0 = C_0\left(\frac{\rho_g}{\rho_f}, \frac{GD}{\mu_f}\right) \tag{14-28}$$

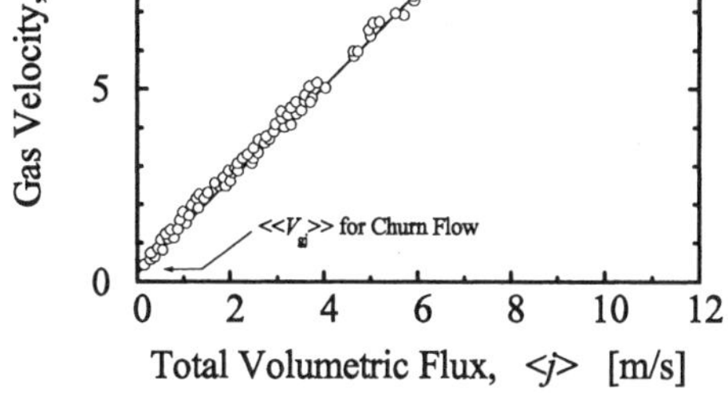

Figure 14-1. Fully developed air-water flow data (Ishii, 1977)

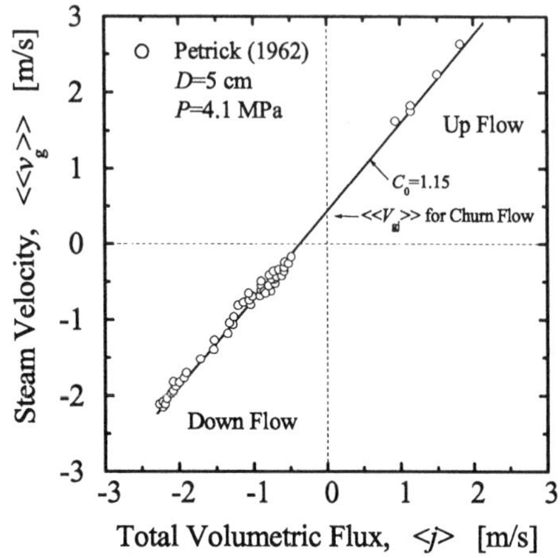

Figure 14-2. Experimental data for cocurrent upflow and cocurrent downflow of steam-water system (Ishii, 1977)

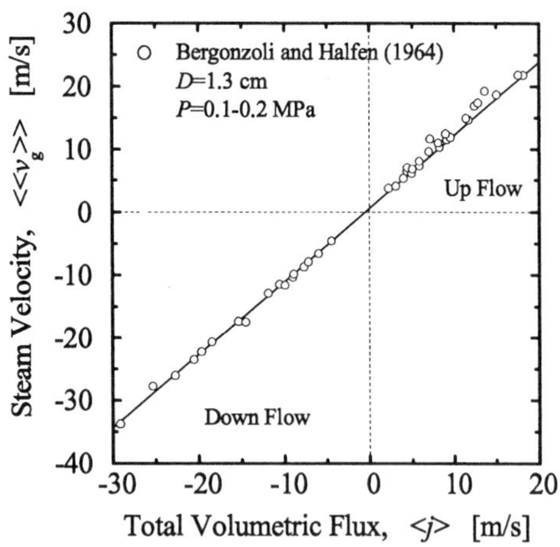

Figure 14-3. Experimental data for cocurrent upflow and cocurrent downflow of heated Santowax-R system (Ishii, 1977)

A single-phase turbulent-flow profile and the ratio of the maximum velocity to mean velocity give a theoretical limiting value of C_0 at $\alpha_d \to 0$ and $\rho_g/\rho_f \to 0$, since in this case all the bubbles should be concentrated at the central region. Thus from the experimental data of Nukuradse (1932) for a round tube, which gives the ratio of the maximum to mean velocity, we have

$$C_\infty = \lim \frac{\langle \alpha_d j \rangle}{\langle \alpha_d \rangle \langle j \rangle} = \frac{\langle \alpha_d \rangle j_0}{\langle \alpha_d \rangle \langle j \rangle} = 1.393 - 0.0155 \ln\left(\frac{GD}{\mu_f} \right) \tag{14-29}$$

as $\alpha_d \to 0$ and $\rho_g/\rho_f \to 0$. Furthermore, as the density ratio approaches unity, the distribution parameter C_0 should also become unity. Thus,

$$C_0 \to 1 \tag{14-30}$$

as $\rho_g/\rho_f \to 1$. Based on these limits and various experimental data in a fully developed flow, the distribution parameter can be given approximately by

$$C_0 = C_\infty - (C_\infty - 1)\sqrt{\rho_g/\rho_f} \tag{14-31}$$

where the density group scales the inertia effects of each phase in a transverse void distribution. Physically, Eq.(14-31) models the tendency of the lighter phase to migrate into a higher-velocity region, thus resulting in a higher void concentration in the central region (Bankoff, 1960). For a laminar flow, C_∞ is 2, but, due to the large velocity gradient, C_0 is very sensitive to $\langle \alpha_d \rangle$ at low void fractions.

Over a wide range of Reynolds number, GD/μ_f, Eq.(14-29) can be approximated by $C_\infty \cong 1.2$ for a flow in a round tube. Furthermore, for a rectangular channel, the experimental data show this value to be approximately 1.35. Thus, for a fully developed turbulent bubbly flow,

$$C_0 \cong \begin{cases} 1.2 - 0.2\sqrt{\rho_g/\rho_f}: & \text{round tube} \\ 1.35 - 0.35\sqrt{\rho_g/\rho_f}: & \text{rectangular channel.} \end{cases} \tag{14-32}$$

Figures 14-4 and 14-5 compare the above correlation with various experimental data. Each point in the figures represents anywhere from five to 150 data points. For example, the original experimental data of Smissaert (1963), shown in Fig.14-1, are presented by a single plot in Fig. 14-4. Each

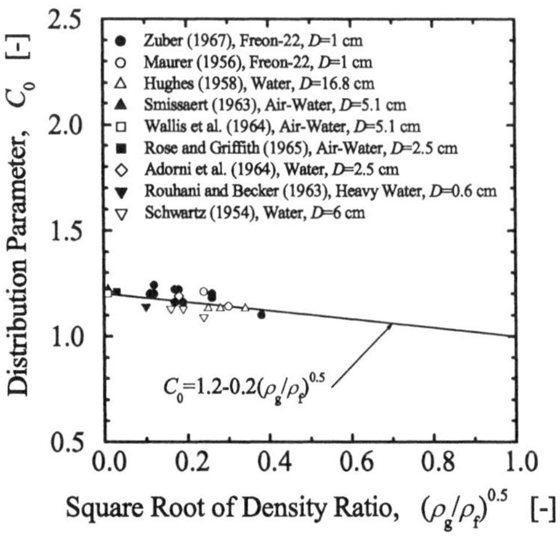

Figure 14-4. Distribution parameter for fully developed flow in a round tube (Ishii, 1977)

point in Fig.14-1 can be used to obtain a corresponding value for C_0 by using the existing correlation for the mean local drift velocity $\langle\langle V_{dj}\rangle\rangle$. However, in view of the strong linear relation between the mean velocity of the dispersed phase and the total flux, the average value of C_0 obtained by linear fitting has been used in Figs.14-4 and 14-5.

In the velocity-flux plane (see Figs.14-2 and 14-3), three operational modes can be easily identified. In the first quadrant, the flow is basically cocurrent upward; therefore both the liquid and vapor phases flow in an upward direction. In the second quadrant, the vapor phase is moving upward; however, there is a net downward flow of mixture. Consequently, the flow is countercurrent. The cocurrent downflow operation should appear in the third quadrant of the velocity-flux plane, as shown in Figs.14-2 and 14-3. These data indicate that the basic characteristic described by Eq.(14-27) is valid for both cocurrent up and down flows with an identical value for the distribution parameter, C_0. This fact demonstrates the usefulness of correlating the drift velocity in terms of the mean local drift velocity $\langle\langle V_{dj}\rangle\rangle$ and C_0.

In two-phase systems with heat addition, the change of void profiles from concave to convex can occur. The concave void-fraction profile is caused by the wall nucleation and delayed transverse migration of bubbles toward the center of a channel. Under these conditions, most of the bubbles are initially located near the nucleating wall, although even in adiabatic flow, small bubbles tend to accumulate near the wall region at low void fraction. The concave profile is particularly pronounced in the subcooled boiling regime,

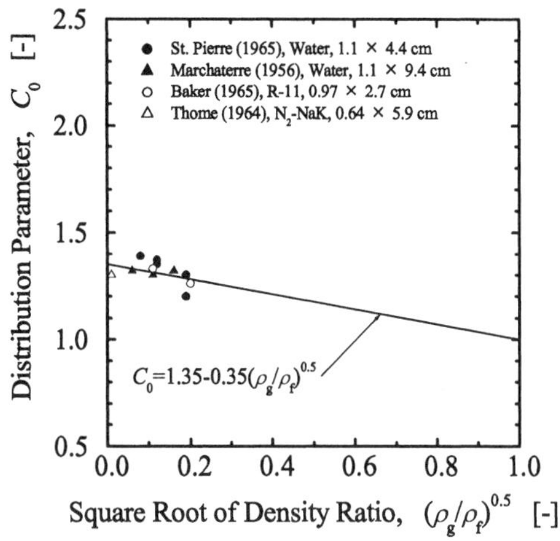

Figure 14-5. Distribution parameter for fully developed flow in a rectangular channel (Ishii, 1977)

because here only the wall-boundary layer is heated above the saturation temperature and the core liquid is subcooled. This temperature profile will induce collapses of migrating bubbles in the core region and resultant latent heat transport from the wall to the subcooled liquid. A similar concave profile can also be obtained by injecting gas into flowing liquid through a porous tube wall (Rose and Griffith, 1965).

In the region in which voids are still concentrated close to the wall, the mean velocity of vapor can be less than the mean velocity of liquid because the bulk of liquid moves with the high core velocity. However, as more and more vapor is generated along the channel, the void-fraction profile changes from concave to convex and becomes fully developed. For a flow with generation of void at the wall due to either nucleation or gas injection, the distribution parameter C_0 should have a near-zero value at the beginning of the two-phase flow region. This can be also seen from the definition of C_0 in Eq.(14-24). Hence, we have

$$\lim_{\langle\alpha_d\rangle\to 0} C_0 = \lim \frac{\langle\alpha_d j\rangle}{\langle\alpha_d\rangle\langle j\rangle} = \frac{\langle\alpha_d\rangle j_W}{\langle\alpha_d\rangle\langle j\rangle} = 0 \quad \text{for } \Gamma_g > 0. \qquad (14\text{-}33)$$

With the increase in the cross-sectional mean void fraction, the peak of the local void fraction moves from the near-wall region to the central region. This will lead to the increase in the value of C_0 as the void profile develops.

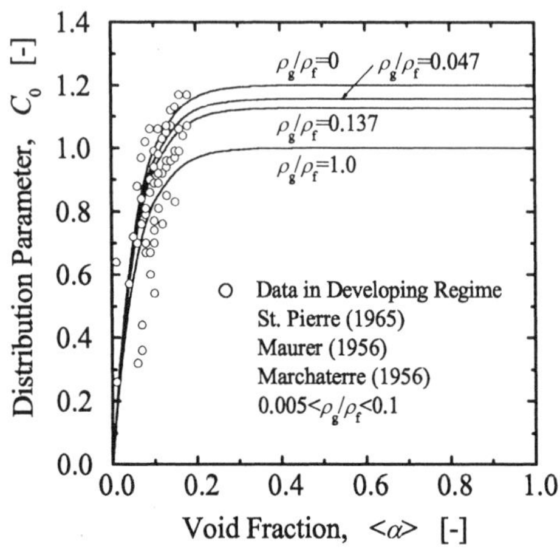

Figure 14-6. Distribution parameter in developing flow due to boiling (Data for the rectangular duct have been modified by a factor of 1.2/1.35 to obtain corresponding data for a round tube.) (Ishii, 1977)

In view of the basic characteristic described above and various experimental data (Zuber et al., 1967; Maurer, 1956; Pierre, 1965; Marchaterre, 1956) the following simple correlation is proposed (Ishii, 1977)

$$C_0 = \left[C_\infty - \left(C_\infty - 1\right)\sqrt{\rho_g/\rho_f}\right]\left(1 - e^{-18\langle\alpha_d\rangle}\right). \tag{14-34}$$

This expression indicates the significance of the developing void profile in the region given by $0 < \langle\alpha_d\rangle < 0.25$; beyond this region, the value of C_0 approaches rapidly to that for a fully developed flow (see Fig.14-6). Hence, for $\Gamma_g > 0$, we obtain

$$C_0 = \begin{cases} \left(1.2 - 0.2\sqrt{\rho_g/\rho_f}\right)\left(1\text{-}e^{-18\langle\alpha_d\rangle}\right): & \text{round tube} \\ \left(1.35 - 0.35\sqrt{\rho_g/\rho_f}\right)\left(1\text{-}e^{-18\langle\alpha_d\rangle}\right): & \text{rectangular channel.} \end{cases} \tag{14-35}$$

For most droplet or particulate flows in the turbulent regime, the volumetric flux profile is quite flat due to the turbulent mixing and particle slips near the wall, which increases the volumetric flux. The concentration of dispersed phase also tends to be uniform, except for weak peaking near the core of the flow. Because of these profiles for j and α_d, the value of the distribution parameter C_0 is expected to be close to unity $(1.0 \leq C_0 \leq 1.1)$. Thus, by assuming that the covariance terms are negligibly small for droplet or particulate flows, we have

$$\overline{V_{dj}} \cong \langle\langle V_{dj} \rangle\rangle \tag{14-36}$$

in which case the local slip becomes important.

The calculation of $\langle\langle V_{dj} \rangle\rangle$ based on the local constitutive equations is the integral transformation, Eq.(14-23); thus, it will require additional information on the void profile (Ishii, 1976). Since this profile is not known in general, we make the following simplifying approximations. The average drift velocity $\langle\langle V_{dj} \rangle\rangle$ due to the local slip can be predicted by the same expression as the local constitutive relations given in a reference (Ishii, 1976), provided the local void fraction α_d and the non-dimensional difference of the stress gradient are replaced by average values. These approximations are good for flows with a relatively flat void-fraction profile; also, they can be considered acceptable from the overall simplicity of the one-dimensional model.

For a fully developed vertical flow, the stress distribution in the fluid and in the dispersed phase should be similar; thus, the effect of shear gradient on the mean local drift velocity can be neglected. Under these conditions we obtain the following results.

Undistorted-particle Regime

$$\overline{V_{dj}} = (C_0 - 1)\langle j \rangle + \frac{10.8\mu_c}{\rho_c r_d}\frac{\mu_c}{\langle \mu_m \rangle}(1 - \langle \alpha_d \rangle)^2 \times \frac{\psi^{4/3}(1+\psi)}{1+\psi\left[\frac{\mu_c}{\langle \mu_m \rangle}(1-\langle \alpha_d \rangle)^{0.5}\right]^{6/7}} \frac{\rho_c - \rho_d}{\Delta\rho} \tag{14-37}$$

where $\psi(r_d^*) = 0.55\left[\left(1 + 0.08 r_d^{*3}\right)^{4/7} - 1\right]^{0.75}$ for $r_d^* < 34.65$ and $\psi(r_d^*) = 17.67$ for $r_d^* \geq 34.65$. The limiting case of the undistorted-particle regime is the Stokes regime in which the mean drift velocity reduces to

$$\overline{V_{dj}} = (C_0 - 1)\langle j\rangle + \frac{2}{9}r_d^2\frac{g\Delta\rho}{\mu_c}\left(1-\langle\alpha_d\rangle\right)^2\frac{\mu_c}{\langle\mu_m\rangle}\frac{\rho_c-\rho_d}{\Delta\rho} \tag{14-38}$$

Distorted-particle Regime $(1.75 \le n \le 2.25)$

$$\overline{V_{dj}} = (C_0 - 1)\langle j\rangle + \sqrt{2}\left(\frac{\sigma g\Delta\rho}{\rho_c^2}\right)^{1/4}\left(1-\langle\alpha_d\rangle\right)^n\frac{\rho_c-\rho_d}{\Delta\rho}. \tag{14-39}$$

Here the value of n depends on the viscosities as

$$\begin{aligned} n &= 1.75 \quad ; \quad \mu_d << \mu_c \\ n &= 2 \quad ; \quad \mu_d \cong \mu_c \\ n &= 2.25 \quad ; \quad \mu_d >> \mu_c \end{aligned} \tag{14-40}$$

Churn-turbulent-flow Regime

$$\overline{V_{dj}} = (C_0 - 1)\langle j\rangle + \sqrt{2}\left(\frac{\sigma g\Delta\rho}{\rho_c^2}\right)^{1/4}\frac{\rho_c-\rho_d}{\Delta\rho}. \tag{14-41}$$

Here the mean mixture viscosity (Ishii, 1976) is given by

$$\frac{\langle\mu_m\rangle}{\mu_c} = \left(1-\frac{\langle\alpha_d\rangle}{\alpha_{dm}}\right)^{-2.5\alpha_{dm}(\mu_d+0.4\mu_c)/(\mu_d+\mu_c)}. \tag{14-42}$$

The value of maximum packing, $\alpha_{dm} = 0.62$, is recommended for solid particle-fluid systems, although it can range from 0.5 to 0.74. However, for a bubbly flow, the theoretical value of α_{dm} can be much higher. If we consider the standard range of interest of void fraction in bubbly flow, α_{dm} may be approximated by $\alpha_{dm} = 1$. Hence, for a bubbly flow, the mixture viscosity becomes

$$\frac{\langle\mu_m\rangle}{\mu_c} = \frac{1}{1-\langle\alpha_d\rangle}. \tag{14-43}$$

However, for a particulate flow with a low particle concentration, namely, $\langle\alpha_d\rangle << 1$, $\langle\mu_m\rangle$ can be approximated by

$$\frac{\langle\mu_m\rangle}{\mu_c} = \left(1 - \langle\alpha_d\rangle\right)^{-2.6}. \tag{14-44}$$

In a horizontal flow with a complete suspension of the dispersed phase, the transverse mixing, which keeps the particles suspended, can significantly influence the stress gradient of each phase; thus, the stress gradient effect may not be neglected. However, in view of the present state of the art, the assumption $\langle\langle V_{dj}\rangle\rangle \cong 0$ may be used as a first-order approximation, particularly in high-flux flows. As explained at the end of this chapter, the actual local drift velocity depends also upon the pressure gradient due to friction and, therefore, in strict sense it is not zero even in horizontal flow.

For high-flux flows, the effect of the local drift $\langle\langle V_{dj}\rangle\rangle$ on the mean drift velocity is small in comparison with the covariance term $(C_0 - 1)\langle j\rangle$. Thus, by neglecting the former, we have

$$\overline{V_{dj}} = \frac{(C_0 - 1)\langle\rho_m\rangle\overline{v_m}}{\langle\rho_m\rangle - (C_0 - 1)\langle\alpha_d\rangle(\rho_c - \rho_d)}. \tag{14-45}$$

For bubbly flows, the above equation imposes a condition on applicable void-fraction ranges; thus, we should have $\langle\rho_m\rangle > (C_0 - 1)\langle\alpha_d\rangle(\rho_d - \rho_c)$. Here, a simple criterion for the boundary between the high- and low-flux flow can be obtained by taking the ratio of the total volumetric flux and the terminal velocity. If this ratio is more than 10, the flow can be considered a high-flux flow.

The other limiting case of the dispersed two-phase flow in a confined channel is slug flow. When the volume of a bubble is very large, the shape of the bubble is significantly deformed to fit the channel geometry. The diameters of the bubbles become approximately that of the pipe with a thin liquid film separating the bubbles from the wall. The bubbles have the bullet form with a cap-shaped nose. The motion of these bubbles in relatively inviscid fluids can be studied by using a potential flow analysis around a sphere (Dumitrescu, 1943), and the result is shown to agree with experimental data. Thus,

$$\overline{V_{dj}} = 0.2\langle j\rangle + 0.35\left(\frac{gD\Delta\rho}{\rho_c}\right)^{1/2} \tag{14-46}$$

which was originally proposed by Nicklin et al. (1962) and Neal (1963).

1.2.2 Annular two-phase flow

In annular two-phase flow, the relative motions between phases are governed by the interfacial geometry, the body-force field, and the interfacial momentum transfer. The constitutive equation for the vapor-drift velocity in annular two-phase flow has been developed by taking into account those macroscopic effects of the structured two-phase flows (Ishii et al., 1976). Assuming steady-state adiabatic two-phase annular flow with constant single-phase properties, we have the following one-dimensional momentum equations for each phase.

$$-\left(\frac{dp_m}{dz}+\rho_g g_z\right)=\frac{\tau_i P_i}{\langle\alpha_g\rangle A} \tag{14-47}$$

and

$$-\left(\frac{dp_m}{dz}+\rho_f g_z\right)=\frac{\tau_{wf} P_{wf}}{\left(1-\langle\alpha_g\rangle\right)A}-\frac{\tau_i P_i}{\left(1-\langle\alpha_g\rangle\right)A} \tag{14-48}$$

where τ_i, τ_{wf}, P_i, and P_{wf} are the interfacial shear, wall shear, interfacial wetted perimeter, and wall wetted perimeter, respectively. The hydraulic diameter and the ratio of wetted perimeters are defined by $D\equiv 4A/P_{wf}$ and $\xi\equiv P_i/P_{wf}$. By assuming that the film thickness δ is small compared with D, we have $4\delta/D\cong 1-\langle\alpha_g\rangle$. Furethermore, for an annular flow in a pipe, ξ reduces to $\sqrt{\alpha_g}$.

The wall shear can be expressed through the friction factor with a gravity-correction term by $\tau_{wf}=f_{wf}\rho_f\langle\langle v_f\rangle\rangle\left|\langle\langle v_f\rangle\rangle\right|/2-\Delta\rho g_z\delta/3$, where f_{wf} can be given by the standard friction-factor correlation: $f_{wf}=16/Re_f$ for laminar film flows and $f_{wf}=0.0791\,Re_f^{-0.25}$ for turbulent flows. Here the liquid-film Reynolds number is given by $Re_f=\rho_f\left|\langle j_f\rangle\right|D/\mu_f$. Similarly, the interfacial shear can be expressed as $\tau_i=f_i\rho_g\left|\overline{v_r}\right|\overline{v_r}/2$ with the interfacial friction factor given by $f_i=0.005\left[1+75\left(1-\langle\alpha_g\rangle\right)\right]$ for rough wavy films (Wallis, 1969).

By definition, the vapor-drift velocity is related to v_r, namely, $\overline{V_{gj}}=\left(1-\langle\alpha_g\rangle\right)\overline{v_r}$. Hence, by eliminating the pressure gradient from the momentum equations, we obtain for a laminar film

$$\overline{V_{gj}} = \pm\left[\frac{16\langle\alpha_g\rangle}{\rho_g f_i \xi}\left|\frac{\mu_f\langle j_f\rangle}{D} + \frac{\Delta\rho g_z D\left(1-\langle\alpha_g\rangle\right)^3}{48}\right|\right]^{1/2} \tag{14-49}$$

and for a turbulent film

$$\overline{V_{gj}} = \pm\left[\frac{\langle\alpha_g\rangle\left(1-\langle\alpha_g\rangle\right)^3 D}{\rho_g f_i \xi} \times\left|\frac{0.005\rho_f\langle j_f\rangle\left|\langle j_f\rangle\right|}{D\left(1-\langle\alpha_g\rangle\right)^3} + \frac{1}{3}\Delta\rho g_z\right|\right]^{1/2}. \tag{14-50}$$

Here, the negative root is taken when the term within the absolute signs becomes negative. The drift velocity in the form expressed by Eqs.(14-49) and (14-50) is convenient for use in analyzing steady-state adiabatic or thermal-equilibrium flows since, in these cases, the value of $\langle j_f\rangle$ can be easily obtained.

In a general drift-flux-model formulation, $\overline{V_{gj}}$ should be expressed in terms of the mixture velocity $\overline{v_m}$ rather than $\langle j_f\rangle$, as $\overline{v_m}$ is the velocity used in the formulation. From the definition, we have

$$\langle j_f\rangle = \left(1-\langle\alpha_g\rangle\right)\overline{v_m} - \frac{\langle\alpha_g\rangle\rho_g}{\langle\rho_m\rangle}\overline{V_{gj}}. \tag{14-51}$$

By substituting Eq.(14-51) into Eq.(14-49), we obtain for a laminar film

$$\overline{V_{gj}} = \pm\frac{8\mu_f\langle\alpha_g\rangle^2}{\langle\rho_m\rangle D f_i \xi}\left[-1+\left\{1+\frac{f_i D\langle\rho_m\rangle^2\left(1-\langle\alpha_g\rangle\right)\xi}{4\mu_f\langle\alpha_g\rangle^3\rho_g} \times\left|\overline{v_m} + \frac{\Delta\rho g_z D^2\left(1-\langle\alpha_g\rangle\right)^2}{48\mu_f}\right|\right\}^{1/2}\right] \tag{14-52}$$

which is valid for the laminar range given by

$$\frac{\left(1-\left\langle \alpha_g\right\rangle\right)\left\langle \rho_m\right\rangle \overline{v_m}-\left\langle \rho_m\right\rangle\left\langle j_f\right\rangle_{tr}}{\left\langle \alpha_g\right\rangle \rho_g}\leq \overline{V_{gj}}$$
$$\leq \frac{\left(1-\left\langle \alpha_g\right\rangle\right)\left\langle \rho_m\right\rangle \overline{v_m}+\left\langle \rho_m\right\rangle\left\langle j_f\right\rangle_{tr}}{\left\langle \alpha_g\right\rangle \rho_g}. \tag{14-53}$$

Here the laminar turbulent-transition volumetric flow is defined by $\left\langle j_f\right\rangle_{tr} = 3200\mu_f/\rho_f D$. The negative root of Eq.(14-52) applies when the term within the absolute signs becomes negative. It is easy to show that, for $\overline{V_{gj}} \leq \left(1-\left\langle \alpha_g\right\rangle\right)\left\langle \rho_m\right\rangle \overline{v_m}/\left(\left\langle \alpha_g\right\rangle \rho_g\right)$, the flow is cocurrent upward, whereas, for $\overline{V_{gj}}$ larger than the above limit, the liquid flow is downward. The solution for the case of turbulent film flow is somewhat more complicated. For convenience, let us introduce the following parameters.

$$\begin{cases} a \equiv \dfrac{f_i \xi \rho_g}{0.005\left\langle \alpha_g\right\rangle \rho_f\left(1-\left\langle \alpha_g\right\rangle\right)^2} \\ b \equiv \dfrac{\left\langle \alpha_g\right\rangle \rho_g}{\left\langle \rho_m\right\rangle\left(1-\left\langle \alpha_g\right\rangle\right)} \\ c \equiv \dfrac{\Delta\rho g_z D\left(1-\left\langle \alpha_g\right\rangle\right)}{0.015\rho_f} \end{cases} \tag{14-54}$$

Then, for upward liquid flow, we have

$$\overline{V_{gj}} = \begin{cases} \dfrac{-b\overline{v_m}+\left[a\overline{v_m}^2+\left(a-b^2\right)c\right]^{1/2}}{\left(a-b^2\right)} & \text{if } a-b^2 \neq 0 \\ \dfrac{\overline{v_m}^2+c}{2b\overline{v_m}} & \text{if } a-b^2 = 0 \end{cases} \tag{14-55}$$

which applies under the condition $\overline{v_m} \geq \sqrt{cb^2/a}$. However, in the transition regime given by $-\sqrt{c} \leq \overline{v_m} < \sqrt{cb^2/a}$, where the liquid film flow is downward with upward interfacial shear forces on the film, the vapor-drift velocity becomes

$$\overline{V_{gj}} = \frac{b\overline{v_m} + \left[-a\overline{v_m}^2 + \left(a + b^2\right)c\right]^{1/2}}{a + b^2}. \tag{14-56}$$

In the range of $\overline{v_m}$ given by $\overline{v_m} \leq -\sqrt{c}$,

$$\overline{V_{gj}} = \frac{-b\overline{v_m} - \left[a\overline{v_m}^2 - \left(a - b^2\right)c\right]^{1/2}}{a - b^2} \tag{14-57}$$

which applies to the cocurrent downward flow.

The above solution can be applied only if the following turbulent-flow criterion is satisfied.

$$\begin{cases} \overline{V_{gj}} \leq \dfrac{\left(1 - \langle\alpha_g\rangle\right)\langle\rho_m\rangle\overline{v_m} - \langle\rho_m\rangle\langle j_f\rangle_{tr}}{\langle\alpha_g\rangle\rho_g} \\ \text{or} \\ \overline{V_{gj}} \geq \dfrac{\left(1 - \langle\alpha_g\rangle\right)\langle\rho_m\rangle\overline{v_m} + \langle\rho_m\rangle\langle j_f\rangle_{tr}}{\langle\alpha_g\rangle\rho_g} \end{cases} \tag{14-58}$$

These results do not have a very simple form for a turbulent film. However, if the absolute value of the mixture velocity is large, so that the flow is essentially cocurrent and the gravity effect is small, then the turbulent solution can be approximated by the simple form

$$\overline{V_{gj}} = \frac{\left(1 - \langle\alpha_g\rangle\right)\overline{v_m}}{\dfrac{\langle\alpha_g\rangle\rho_g}{\langle\rho_m\rangle} + \left\{\dfrac{\xi\rho_g\left[1 + 75\left(1 - \langle\alpha_g\rangle\right)\right]}{\langle\alpha_g\rangle\rho_f}\right\}^{1/2}}. \tag{14-59}$$

Equation (14-59) for the drift velocity can be transformed to obtain the slip ratio v_g/v_f under the simplifying assumption that the average liquid velocity is much smaller than the vapor velocity. Then we have

$$\frac{\langle\langle v_g\rangle\rangle}{\langle\langle v_f\rangle\rangle}=\sqrt{\frac{\rho_f}{\rho_g}}\left[\frac{\sqrt{\langle\alpha_g\rangle}}{1+75\left(1-\langle\alpha_g\rangle\right)}\right]^{1/2} \tag{14-60}$$

for an annular flow in a pipe for which $\xi=\sqrt{\langle\alpha_g\rangle}$. The above expression for slip ratio is similar to that obtained by Fauske (1962), namely $\langle\langle v_g\rangle\rangle/\langle\langle v_f\rangle\rangle=\sqrt{\rho_f/\rho_g}$, which has no dependence on the void fraction. The factor that takes the void fraction into account in Eq.(14-60) varies roughly from 0.24 to 1 for the range $0.8<\langle\alpha_g\rangle<1$. Therefore, for a turbulent film, the Fauske correlation should give reasonably accurate results at high void fractions.

The drift-velocity correlation for the annular flow has been expressed in terms of the mixture velocity, since $\overline{v_m}$ is the basic variable in the formulation of the general drift-flux model. However, it is also interesting and important to resolve the expression for $\overline{V_{gj}}$ in terms of the total volumetric flux $\langle j\rangle$, since $\langle j\rangle$ was the variable used to correlate $\overline{V_{gj}}$ in dispersed two-phase flow regimes.

By considering the turbulent film-flow regime and using the definition $\langle j_f\rangle=\left(1-\langle\alpha_g\rangle\right)\langle j\rangle-\langle\alpha_g\rangle\overline{V_{gj}}$, we can resolve Eq.(14-50) for the mean drift velocity $\overline{V_{gj}}$. The result does not have a simple form; however, for most practical cases, it can be approximated by a linear function of $\langle j\rangle$.

$$\overline{V_{gj}}\cong\frac{1-\langle\alpha_g\rangle}{\langle\alpha_g\rangle+\left[\dfrac{1+75\left(1-\langle\alpha_g\rangle\right)}{\sqrt{\langle\alpha_g\rangle}}\dfrac{\rho_g}{\rho_f}\right]^{1/2}} \tag{14-61}$$

$$\times\left[\langle j\rangle+\sqrt{\frac{\Delta\rho g_z D\left(1-\langle\alpha_g\rangle\right)}{0.015\rho_f}}\right]$$

This expression may be further simplified for $\rho_g/\rho_f<<1$ as

$$\overline{V_{gj}}\cong\frac{1-\langle\alpha_g\rangle}{\langle\alpha_g\rangle+4\sqrt{\rho_g/\rho_f}}\left[\langle j\rangle+\sqrt{\frac{\Delta\rho g_z D\left(1-\langle\alpha_g\rangle\right)}{0.015\rho_f}}\right]. \tag{14-62}$$

From the comparison of Eq.(14-62) to Eq.(14-22), the apparent distribution parameter for annular flow becomes

$$C_0 \simeq 1+\frac{1-\langle\alpha_g\rangle}{\langle\alpha_g\rangle+4\sqrt{\rho_g/\rho_f}};\quad \left(\rho_g/\rho_f<<1\right). \tag{14-63}$$

This indicates that the apparent C_0 in annular flow should be close to unity.

1.2.3 Annular mist flow

As the gas velocity increases in the annular flow, the entrainment of liquid from the film to the gas-core flow takes place. Based on criteria developed for an onset of entrainment (Ishii and Grolmes, 1975), the critical gas velocity for a rough turbulent film flow can be given by

$$|\langle j_g\rangle|>\frac{\sigma}{\mu_f}\sqrt{\frac{\rho_f}{\rho_g}}\times\begin{cases}N_{\mu f}^{0.8} & \text{for } N_{\mu f}\le\dfrac{1}{15}\\ 0.1146 & \text{for } N_{\mu f}>\dfrac{1}{15}\end{cases} \tag{14-64}$$

where $N_{\mu f}\equiv\mu_f\Big/\left[\rho_f\sigma\sqrt{\sigma/g\Delta\rho}\right]^{1/2}$. However, in general, the vapor flux is much larger than the liquid flux in the annular-mist-flow regime. Then, for a weakly viscous fluid such as water or sodium, the above correlation may be replaced by

$$|\langle j_g\rangle|\simeq|\langle j\rangle|>\left(\frac{\Delta\rho g}{\rho_g^2}\right)^{1/4}N_{\mu f}^{-0.2}. \tag{14-65}$$

If Inequality (14-65) is satisfied, then the droplet entrainment into the gas-core flow should be considered; otherwise the correlation for annular flow, Eq.(14-62) can be applied.

The correlation for $\overline{V_{gj}}$ in annular mist flow can be readily developed by combining the previous results for a dispersed flow and pure annular flow. The area fraction of liquid entrained in the gas core from total liquid area at any cross section is denoted by E_d, and the cross-sectional-area-averaged void fraction by $\langle\alpha_g\rangle$. Then the film-area fraction is given by

$$1-\alpha_{core} = \frac{\text{liquid-film cross-sectional area}}{\text{total cross-sectional area}} \\ = \left(1-\langle\alpha_g\rangle\right)\left(1-E_d\right) \tag{14-66}$$

and the mean liquid-droplet fraction in the gas core alone is given by

$$\alpha_{drop} = \frac{\text{cross-sectional area of drops}}{\text{cross-sectional area of core}} \\ = \frac{\left(1-\langle\alpha_g\rangle\right)E_d}{1-\left(1-\langle\alpha_g\rangle\right)\left(1-E_d\right)}. \tag{14-67}$$

Consequently, α_{core} should be used in the annular-flow correlation, Eq.(14-62), to obtain the relative motion between the core and the film, whereas α_{drop} should be used in the dispersed-flow correlation to obtain a slip between droplets and gas-core flow.

By denoting the gas-core velocity, liquid-drop velocity, and film velocity by v_{gc}, v_{fc}, and v_{ff}, respectively, the total volumetric flux is given by

$$\langle j\rangle = \left[v_{gc}\left(1-\alpha_{drop}\right)+\alpha_{drop}v_{fc}\right]\alpha_{core} + v_{ff}\left(1-\alpha_{core}\right). \tag{14-68}$$

Furthermore, by denoting the total volumetric flux in the core based on the core area by j_{core}, we have from the annular correlation, Eq.(14-62),

$$j_{core} - \langle j\rangle \cong \frac{\left(1-\langle\alpha_g\rangle\right)\left(1-E_d\right)}{\langle\alpha_g\rangle + 4\sqrt{\rho_g/\rho_f}} \tag{14-69}$$

$$\times\left[\langle j\rangle + \sqrt{\frac{\Delta\rho g_z D\left(1-\langle\alpha_g\rangle\right)\left(1-E_d\right)}{0.015\rho_f}}\right].$$

From the dispersed-flow correlations, it can be shown that, for a distorted-droplet or churn-droplet flow regime, the drift velocity can be given approximately by

$$\langle\langle v_g \rangle\rangle - j_{core} = \sqrt{2}\left(\frac{\sigma g \Delta\rho}{\rho_g^2}\right)^{1/4} \frac{E_d\left(1-\langle\alpha_g\rangle\right)}{\langle\alpha_g\rangle + E_d\left(1-\langle\alpha_g\rangle\right)}. \tag{14-70}$$

Here we have used an approximation based on $\left(1-\langle\alpha_g\rangle\right) << 1$. However, depending on the core-gas velocity, the dispersed-flow drift-velocity correlation for a much smaller particle should be used. When the droplets are generated by the entrainment of liquid film, the following approximate form is suggested for an undistorted-particle regime outside the Stokes regime (Ishii, 1976).

$$\langle\langle v_g \rangle\rangle - j_{core} = 0.5 r_d \left[\frac{(g\Delta\rho)^2}{\mu_g \rho_g}\right]^{1/3} \frac{E_d\left(1-\langle\alpha_g\rangle\right)}{\langle\alpha_g\rangle + E_d\left(1-\langle\alpha_g\rangle\right)} \tag{14-71}$$

where the particle radius may be approximated from the Weber-number criterion at the shearing-off wave crests. Thus,

$$r_d \cong \frac{6\sigma}{\rho_g} \frac{1}{\langle j \rangle^2}. \tag{14-72}$$

The above relations apply only when the total volumetric flux is sufficiently high to induce fragmentations of the wave crests. Hence, Eq.(14-71) should be used when

$$|\langle j \rangle| > 1.456\left(\frac{\sigma g \Delta\rho}{\rho_g^2}\right)^{1/4} \left[\frac{\mu_g^2}{\rho_g \sigma \sqrt{\sigma/g\Delta\rho}}\right]^{-1/12}. \tag{14-73}$$

By combining the above results, we obtain

$$\overline{\overline{V}}_{gj} = \frac{\left(1-\langle\alpha_g\rangle\right)\left(1-E_d\right)}{\langle\alpha_g\rangle + 4\sqrt{\rho_g/\rho_f}} \times\left[\langle j \rangle + \sqrt{\frac{\Delta\rho g_z D\left(1-\langle\alpha_g\rangle\right)\left(1-E_d\right)}{0.015\rho_f}}\right] \tag{14-74}$$

$$+\frac{E_d\left(1-\langle\alpha_g\rangle\right)}{\langle\alpha_g\rangle+E_d\left(1-\langle\alpha_g\rangle\right)}\times\begin{cases}\sqrt{2}\left(\dfrac{\sigma g\Delta\rho}{\rho_g^2}\right)^{1/4}\\ \text{or}\\ \dfrac{3\sigma}{\rho_g}\left[\dfrac{\left(g\Delta\rho\right)^2}{\mu_g\rho_g}\right]^{1/3}\dfrac{1}{\langle j\rangle^2}\end{cases}$$

where the latter expression applies under the condition given by Eq.(14-73). If the radius of the particle is very small, then the essential contribution to the relative motion between phases comes from the first term of Eq.(14-74), and the core flow may be considered as a homogeneous dispersed flow. In such a case, Eq.(14-74) reduces to

$$\overline{V_{dj}}\cong\frac{\left(1-\langle\alpha_g\rangle\right)\left(1-E_d\right)}{\langle\alpha_g\rangle+4\sqrt{\rho_g/\rho_f}}\times\left(\langle j\rangle+\sqrt{\frac{\Delta\rho g_z D\left(1-\langle\alpha_g\rangle\right)\left(1-E_d\right)}{0.015\rho_f}}\right). \tag{14-75}$$

This expression shows a linear decrease of drift velocity in terms of entrained liquid fraction, which can be observed in various experimental data (Alia et al., 1965; Cravarolo et al., 1964).

1.3 Covariance of convective flux

In the one-dimensional drift-flux model, the momentum and energy convective fluxes have been divided into three terms: the mixture convective flux; the drift convective flux; and the covariance term, as can be seen from Eqs.(14-18) and (14-19). In other words, the convective flux of quantity ψ for the mixture can be written as

$$\frac{\partial}{\partial z}\left(\sum_{k=1}^{2}\langle\alpha_k\rho_k\psi_k v_k\rangle\right)=\frac{\partial}{\partial z}\left(\rho_m\overline{\psi_m}\overline{v_m}\right)+\frac{\partial}{\partial z}\left(\frac{\langle\alpha_d\rangle\rho_d\rho_c}{\langle\rho_m\rangle}\Delta\psi_{dc}\overline{V_{dj}}\right)+\frac{\partial}{\partial z}\sum_{k=1}^{2}COV\left(\alpha_k\rho_k\psi_k v_k\right) \tag{14-76}$$

where $\Delta\psi_{dc} \equiv \langle\langle\psi_d\rangle\rangle - \langle\langle\psi_c\rangle\rangle$ and $COV(\alpha_k\rho_k\psi_k v_k) \equiv \left\langle\alpha_k\rho_k\psi_k\left(v_k - \langle\langle v_k\rangle\rangle\right)\right\rangle$. Therefore, for the momentum flux, we have $\psi_k = v_k$ and $\Delta\psi_{dc} = \overline{V}_{dj}/\left(1-\langle\alpha_g\rangle\right)$. For the enthalpy flux, we have $\psi_k = h_k$ and $\Delta\psi_{dc} = \langle\langle h_d\rangle\rangle - \langle\langle h_c\rangle\rangle$, which is equivalent to the latent heat if phases are in thermal equilibrium.

To close the set of the governing equations, we must specify relations for these covariance terms. This can be done by introducing distribution parameters for the momentum and energy fluxes. If we define a distribution parameter for a flux as

$$C_{\psi k} \equiv \frac{\langle\alpha_k\psi_k v_k\rangle}{\langle\alpha_k\rangle\langle\langle\psi_k\rangle\rangle\langle\langle v_k\rangle\rangle} \tag{14-77}$$

the covariance term becomes

$$\begin{aligned} COV(\rho_k\alpha_k\psi_k v_k) &= \rho_k\left\langle\alpha_k\psi_k\left(v_k - \langle\langle v_k\rangle\rangle\right)\right\rangle \\ &= \left(C_{\psi k}-1\right)\rho_k\langle\alpha_k\rangle\langle\langle\psi_k\rangle\rangle\langle\langle v_k\rangle\rangle. \end{aligned} \tag{14-78}$$

For the momentum flux, the distribution parameter is defined by

$$C_{vk} \equiv \frac{\langle\alpha_k v_k^2\rangle}{\langle\alpha_k\rangle\langle\langle v_k\rangle\rangle^2}. \tag{14-79}$$

Physically, C_{vk} represents the effect of the void and momentum-flux profiles on the cross-sectional-area-averaged momentum flux of k phase. A quantitative study of C_{vk} can be made by considering a symmetric flow in a circular duct and introducing the power-law expressions in parallel with the analysis of C_0 in the Section 1.2.1 of Chapter 14. Hence we postulate that

$$\frac{\alpha_k - \alpha_{kw}}{\alpha_{k0} - \alpha_{kw}} = 1 - \left(\frac{r}{R_w}\right)^n \tag{14-80}$$

and

$$\frac{v_k}{v_{k0}} = 1 - \left(\frac{r}{R_w}\right)^m \tag{14-81}$$

where the subscripts 0 and w refer to the value at the centerline and at the wall of a tube.

For simplicity, it is assumed that the void and velocity profiles are similar; namely, $n = m$. This assumption is widely used in mass-transfer problems. It may also be reasonable for fully developed two-phase flows if one considers that the vapor flux and, hence, the void concentration greatly influence the velocity distributions. Under this assumption, it can be shown that

$$C_{vk} = \frac{\dfrac{n+2}{n+1}\left(\alpha_{kw} + \Delta\alpha_k \dfrac{3n}{3n+2}\right)\left(\alpha_{kw} + \Delta\alpha_k \dfrac{n}{n+2}\right)}{\left(\alpha_{kw} + \Delta\alpha_k \dfrac{n}{n+1}\right)^2} \tag{14-82}$$

where $\Delta\alpha_k = \alpha_{k0} - \alpha_{kw}$.

For a dispersed vapor phase, $\alpha_{gw} << \Delta\alpha_g$; hence,

$$C_{vg} \cong \frac{3n+3}{3n+2}. \tag{14-83}$$

However, from Eq.(14-26), the volumetric-flux-distribution parameter C_0 becomes

$$C_0 \cong \frac{n+2}{n+1}. \tag{14-84}$$

Therefore, in the standard range of n, the parameter C_{vg} can be given approximately by

$$C_{vg} \simeq 1 + 0.5\left(C_0 - 1\right). \tag{14-85}$$

For a liquid phase in a vapor-dispersed-flow regime, $\alpha_{fw} \cong 1$ and $\alpha_{f0} < 1$. Then from Eq.(14-82) it can be shown that, for a standard range of α_{f0} in the bubbly- and churn-flow regimes, C_{vf} can be approximated by

$$C_{vf} = 1 + 1.5\left(C_0 - 1\right). \tag{14-86}$$

For an annular flow, the momentum covariance term can also be calculated by using the standard velocity profiles for the vapor and liquid flows. Thus we obtain

$$C_{vk} \simeq \begin{cases} 1.02 & \text{(turbulent flow)} \\ 1.33 & \text{(laminar flow)}. \end{cases} \tag{14-87}$$

The above result for the individual phases can now be used to study the mixture covariance term. By defining the mixture-momentum-distribution parameter as

$$C_{vm} \equiv \frac{C_{vd}\rho_d\langle\alpha_d\rangle + C_{vc}\rho_c\langle\alpha_c\rangle}{\langle\rho_m\rangle} \tag{14-88}$$

the covariance term becomes

$$\begin{aligned}&\sum_{k=1}^{2} COV\left(\alpha_k\rho_k v_k^2\right)\\&= \left(C_{vm}-1\right)\left[\langle\rho_m\rangle\overline{\overline{v}}_m^{\,2} + \frac{\rho_c\rho_d\langle\alpha_d\rangle}{\left(1-\langle\alpha_d\rangle\right)\langle\rho_m\rangle}\overline{\overline{V}}_{dj}^{\,2}\right]\\&+\frac{2\rho_c\rho_d\langle\alpha_d\rangle}{\langle\rho_m\rangle}\left(C_{vd}-C_{vc}\right)\overline{\overline{v}}_m\overline{\overline{V}}_{dj}.\end{aligned} \tag{14-89}$$

In view of the above analysis, the order of magnitude of $\left(C_{vd}-C_{vc}\right)$ is the same as that of $\left(C_{vm}-1\right)$ or less; therefore, the last term on the right-hand side of Eq.(14-89) can be neglected for almost all cases. This term may be important only in the near critical regime and if $\overline{\overline{v}}_m \cong \overline{\overline{V}}_{dj}$. However, in general, $\overline{\overline{V}}_{dj}$ becomes insignificant as the density ratio approaches unity. Hence, under the above conditions, the convective term itself becomes relatively small. Consequently, even for this case, the term may be dropped. Thus we have

$$\begin{aligned}&\sum_{k=1}^{2} COV\left(\alpha_k\rho_k v_k^2\right)\\&\cong \left(C_{vm}-1\right)\left[\langle\rho_m\rangle\overline{\overline{v}}_m^{\,2} + \frac{\rho_c\rho_d\langle\alpha_d\rangle}{\left(1-\langle\alpha_d\rangle\right)\langle\rho_m\rangle}\overline{\overline{V}}_{dj}^{\,2}\right].\end{aligned} \tag{14-90}$$

The value of C_{vm} can be evaluated from Eq.(14-88) by using Eqs.(14-85) and (14-86) or Eq.(14-87). In the bubbly- and churn-flow regimes of practical importance, C_{vm} can be given approximately by

$$C_{vm} \cong 1 + 1.5\left(C_0 - 1\right). \tag{14-91}$$

However, in the near critical regime, C_{vm} depends also on the void fraction and the density ratio. Furthermore, at very low void fractions in a fully developed flow or in a developing flow, the value of C_{vm} should be reduced to the one for the single-phase flow given by Eq.(14-87). The effect of the development of the void profile into that given by the power law may be taken into account by a similar void-fraction correction term used in the correlation for C_0 in Eq.(14-34). By recalling that for a turbulent flow $C_{vm} = 1.0$ at $\alpha_d \to 0$, we obtain for a round tube

$$C_{vm} \cong 1 + 0.3\left(1 - \sqrt{\rho_g/\rho_f}\right)\left(1 - e^{-18\langle\alpha_d\rangle}\right) \tag{14-92}$$

which may be used both for a fully developed flow and for a developing flow.

For a turbulent-annular-flow regime, we have, from Eqs.(14-87) and (14-88), $C_{vm} \cong 1.02$. For all practical purposes, this may be further approximated by

$$C_{vm} \cong 1. \tag{14-93}$$

In reality, the transition from the value given by Eq.(14-92) to that given by Eq.(14-93) is a gradual one through the churn-annular (or slug-annular)-flow regime in which characteristics of churn and annular flows alternate. If a single correlation for C_{vm} is preferred, regardless of the flow-regime transitions, then Eq.(14-92) may be safely extrapolated into higher-void-fraction regime by a simple modification given by

$$C_{vm} \cong 1 + 0.3\left(1 - \sqrt{\rho_g/\rho_f}\right)\left\{1 - e^{-18\langle\alpha_d\rangle\left(1-\langle\alpha_d\rangle\right)}\right\}. \tag{14-94}$$

A similar analysis can be carried out for the enthalpy-covariance term by assuming the void, velocity, and enthalpy profiles. In general,

$$\begin{aligned}\sum_{k=1}^{2} COV\left(\alpha_k \rho_k h_k v_k\right) &= \left(C_{hm} - 1\right)\langle\rho_m\rangle \overline{\overline{h}}_m \overline{\overline{v}}_m \\ &+ \frac{\rho_c \rho_d \langle\alpha_d\rangle}{\langle\rho_m\rangle}\left[-\left(C_{hc} - 1\right)\langle\langle h_c\rangle\rangle + \left(C_{hd} - 1\right)\langle\langle h_d\rangle\rangle\right]\overline{\overline{V}}_{dj}\end{aligned} \tag{14-95}$$

where

$$\begin{cases} C_{hm} \equiv \sum_{k=1}^{2} C_{hk}\rho_k \langle\alpha_k\rangle\langle\langle h_k\rangle\rangle \big/ \left(\langle\rho_m\rangle \overline{h_m}\right) \\ C_{hk} \equiv \langle\alpha_k h_k v_k\rangle \big/ \left(\langle\alpha_k\rangle\langle\langle h_k\rangle\rangle\langle\langle v_k\rangle\rangle\right). \end{cases} \tag{14-96}$$

For a thermal-equilibrium flow, $h_g = h_{gs}$ and $h_f = h_{fs}$, where h_{gs} and h_{fs} are the saturation enthalpies of vapor and liquid. Since, in this case, the enthalpy profile is completely that for each phase, the distribution parameters become unity; namely, $C_{hg} = C_{hf} = C_{hm} = 1$. It is also evident that if one of the phases is in the saturated condition, then C_{hk} for that phase becomes unity.

In the single-phase region, the distribution parameter can be calculated from the assumed profiles for the velocity and enthalpy. Using the standard power-law profiles for a turbulent flow, namely, $v/v_0 = \left(y/R_w\right)^{1/n}$ and $\left(h - h_w\right)/\left(h_0 - h_w\right) = \left(y/R_w\right)^{1/m}$, where y is the distance from the wall, we can show that the covariance term is negligibly small both for developing and fully developed flows.

From the above two limiting cases, we can conclude that the enthalpy covariant term may become important only in highly non-equilibrium flow. Even in that case, the energy associated with phase change is considerably larger than that associated with changes in transverse temperature profiles. Therefore, except for highly transient cases, the enthalpy covariance can be neglected. Hence,

$$\frac{\partial}{\partial z}\sum_{k=1}^{2} COV\left(\alpha_k \rho_k h_k v_k\right) \cong 0 \tag{14-97}$$

1.4 One-dimensional drift-flux correlations for various flow conditions

In this section, constitutive equations of the one-dimensional drift-flux model for various flow conditions, which are of practically importance, are summarized.

1.4.1 Constitutive equations for upward bubbly flow

The constitutive equation of the distribution parameter of upward adiabatic bubbly flow in a round tube, Eq.(14-32), has been improved by considering the bubble lateral migration characteristics as (Hibiki and Ishii, 2002b; 2003b)

$$C_0 = 2.0e^{-0.000584 Re_f} + 1.2\left(1 - e^{-22\langle D_{Sm}\rangle/D}\right) \times\left(1 - e^{-0.000584 Re_f}\right) - \left[2.0e^{-0.000584 Re_f} + 1.2\left(1 - e^{-22\langle D_{Sm}\rangle/D}\right)\left(1 - e^{-0.000584 Re_f}\right) - 1\right]\sqrt{\frac{\rho_g}{\rho_f}} \tag{14-98}$$

where Re_f is defined by $\langle j_f\rangle D/\nu_f$. The bubble diameter D_{Sm} in Eq.(14-98) can be predicted by the bubble diameter correlation (Hibiki and Ishii, 2002a). As can be seen from Eq.(14-98), as the liquid Reynolds number increases, the distribution parameter predicted by Eq.(14-98) asymptotically approaches Eq.(14-32). The constitutive equation of the drift-velocity in gas-liquid bubbly flow is given by

$$\langle\langle V_{gj}\rangle\rangle = \sqrt{2}\left(\frac{\Delta\rho g\sigma}{\rho_f^2}\right)^{1/4}\left(1 - \langle\alpha_g\rangle\right)^{1.75} \quad \left(\mu_g << \mu_f\right). \tag{14-99}$$

1.4.2 Constitutive equations for upward adiabatic annulus and internally heated annulus

The applicability of Eqs.(14-98) and (14-99) has been confirmed by upward air-water turbulent bubbly flow data taken in a vertical concentric annulus at atmospheric pressure (Hibiki et al., 2003b). The constitutive equation of the distribution parameter of boiling bubbly flow in an internally heated annulus has been derived from Eq.(14-35) by considering the channel geometry difference as (Hibiki et al., 2003a)

$$C_0 = \left(1.2 - 0.2\sqrt{\frac{\rho_g}{\rho_f}}\right)\left(1 - e^{-3.12\langle\alpha_g\rangle^{0.212}}\right). \tag{14-100}$$

The drift-flux model with Eqs.(14-99) and (14-100) can predict the data taken in an internally heated annulus well (Hibiki et al., 2003a).

1.4.3 Constitutive equations for downward two-phase flow

The constitutive equation of the distribution parameter of downward two-phase flow for all flow regimes is given by (Goda et al., 2003)

$$\begin{aligned} &C_0 = \left(-0.0214\left\langle j^*\right\rangle + 0.772\right) + \left(0.0214\left\langle j^*\right\rangle + 0.228\right)\sqrt{\frac{\rho_g}{\rho_f}} \\ &\text{for } -20 \le \left\langle j^*\right\rangle < 0, \\ &C_0 = \left(0.2e^{0.00848\left(\left\langle j^*\right\rangle + 20\right)} + 1.0\right) - 0.2e^{0.00848\left(\left\langle j^*\right\rangle + 20\right)}\sqrt{\frac{\rho_g}{\rho_f}} \\ &\text{for } \left\langle j^*\right\rangle < -20 \end{aligned} \tag{14-101}$$

where $\left\langle j^*\right\rangle \equiv \langle j\rangle / \left\langle\left\langle V_{gj}\right\rangle\right\rangle$. The constitutive equation of the drift velocity of downward two-phase flow for all flow regimes is approximated by

$$\left\langle\left\langle V_{gj}\right\rangle\right\rangle = \sqrt{2}\left(\frac{\Delta\rho g\sigma}{\rho_f^2}\right)^{1/4}. \tag{14-102}$$

These constitutive equations for distribution parameter and drift velocity were developed by one-dimensional data, and they have not been validated separately by detailed local flow data. Thus, they should not be used individually.

1.4.4 Constitutive equations for bubbling or boiling pool systems

In bubbling or pool boiling systems, the ratio of the vessel diameter to the length is often large in comparison with forced convection systems. It is noted that a recirculation flow pattern may develop in a large vessel at low flow. This may significantly affect the transverse velocity and void fraction profiles. The constitutive equation of the drift velocity for bubbling or boiling pool systems ($\left\langle j_f\right\rangle = 0$) is given by (Kataoka and Ishii, 1987)

Low viscous case: $N_{\mu f} \leq 2.25\times10^{-3}$

$$\left\langle\left\langle V_{gj}^{+}\right\rangle\right\rangle = 0.0019 D_{H}^{*\,0.809}\left(\frac{\rho_g}{\rho_f}\right)^{-0.157} N_{\mu f}^{-0.562} \text{ for } D_H^{*} \leq 30$$
$$\left\langle\left\langle V_{gj}^{+}\right\rangle\right\rangle = 0.030\left(\frac{\rho_g}{\rho_f}\right)^{-0.157} N_{\mu f}^{-0.562} \text{ for } D_H^{*} \geq 30 \qquad (14\text{-}103)$$

Higher viscous case: $N_{\mu f} > 2.25\times10^{-3}$

$$\left\langle\left\langle V_{gj}^{+}\right\rangle\right\rangle = 0.92\left(\frac{\rho_g}{\rho_f}\right)^{-0.157} \quad \text{for } D_H^{*} \geq 30 \qquad (14\text{-}104)$$

where $\left\langle\left\langle V_{gj}^{+}\right\rangle\right\rangle = \left\langle\left\langle V_{gj}\right\rangle\right\rangle \Big/ \left(\sigma g \Delta\rho/\rho_f^2\right)^{1/4}$ and $D_H^{*} = D_H \Big/ \sqrt{\sigma/(g\Delta\rho)}$. The constitutive equation of the distribution parameter in bubbling or pool boiling system is given in terms of channel geometry as Eq.(14-32).

1.4.5 Constitutive equations for large diameter pipe systems

In a large diameter channel ($D_H \geq 40\sqrt{\sigma/(g\Delta\rho)}$), slug bubbles cannot be sustained due to the interfacial instability and they disintegrate to cap bubbles. A recirculation flow pattern may develop in a large diameter channel at a low-flow rate. A flow regime at a test section inlet and a flow regime transition in a developing flow may also have an influence on the liquid recirculation pattern. The liquid recirculation, inlet flow regime and flow regime transition may affect the transverse velocity and the void fraction profile significantly. The constitutive equation of the drift velocity for upward bubbly flow in large diameter pipe systems is approximated by (Hibiki and Ishii, 2003a)

$$\left\langle\left\langle V_{gj}^{+}\right\rangle\right\rangle = \left\langle\left\langle V_{gj,B}^{+}\right\rangle\right\rangle e^{-1.39\left\langle j_g^{+}\right\rangle} + \left\langle\left\langle V_{gj,P}^{+}\right\rangle\right\rangle\left(1 - e^{-1.39\left\langle j_g^{+}\right\rangle}\right) \qquad (14\text{-}105)$$

where $\left\langle\left\langle V_{gj,B}^{+}\right\rangle\right\rangle$ and $\left\langle\left\langle V_{gj,P}^{+}\right\rangle\right\rangle$ are, respectively, given by Eq.(14-102) and Eqs.(14-103) and (14-104), and $\left\langle j_g^{+}\right\rangle \equiv \left\langle j_g\right\rangle \Big/ \left(\sigma g\Delta\rho/\rho_f^2\right)^{1/4}$. The constitutive equation of the distribution parameter for upward bubbly flow in large diameter pipe systems is given by

Case for inlet flow regime as uniformly distributed bubbly flow

$$C_0 = e^{0.475\left(\langle j_g^+ \rangle / \langle j^+ \rangle\right)^{1.69}} - \left[e^{0.475\left(\langle j_g^+ \rangle / \langle j^+ \rangle\right)^{1.69}} - 1 \right] \sqrt{\frac{\rho_g}{\rho_f}}$$

for $0 \le \langle j_g^+ \rangle / \langle j^+ \rangle \le 0.9$

$$C_0 = \left\{ -2.88 \left(\frac{\langle j_g^+ \rangle}{\langle j^+ \rangle} \right) + 4.08 \right\} - \left\{ -2.88 \left(\frac{\langle j_g^+ \rangle}{\langle j^+ \rangle} \right) + 3.08 \right\} \sqrt{\frac{\rho_g}{\rho_f}} \qquad (4\text{-}106)$$

for $0.9 \le \langle j_g^+ \rangle / \langle j^+ \rangle \le 1$

where $\langle j^+ \rangle \equiv \langle j \rangle / \left(\sigma g \Delta\rho / \rho_f^2 \right)^{1/4}$.

Case for inlet flow regime as cap bubbly or slug flow

$$C_0 = 1.2 e^{0.110\langle j^+ \rangle^{2.22}} - \left\{ 1.2 e^{0.110\langle j^+ \rangle^{2.22}} - 1 \right\} \sqrt{\frac{\rho_g}{\rho_f}}$$

for $0 \le \langle j^+ \rangle \le 1.8$

$$C_0 = \left[0.6 e^{-1.2\left(\langle j^+ \rangle - 1.8\right)} + 1.2 \right] - \left[0.6 e^{-1.2\left(\langle j^+ \rangle - 1.8\right)} + 0.2 \right] \sqrt{\frac{\rho_g}{\rho_f}} \qquad (4\text{-}107)$$

for $1.8 \le \langle j^+ \rangle$

These constitutive equations for distribution parameter and drift velocity were developed by one-dimensional data, and they have not been validated separately by detailed local flow data. Thus, they should not be used individually. In slug, churn, and annular flow regime, the distribution parameter effect is dominant over the local slip effect, namely, $V_{gj}^+ << C_0 \langle j^+ \rangle$. Thus, the constitutive equations detailed in the Section 1.2 of Chapter 14 can be applicable to such flow regimes.

1.4.6 Constitutive equations at reduced gravity conditions

To extend the applicability range of the drift-flux model to reduced gravity conditions, the constitutive equations of the drift velocity detailed in the Section 1.2 of Chapter 14 have been reformulated by considering the frictional pressure loss in addition to the gravitational pressure loss as (Hibiki et al., 2004)

Bubbly Flow Regime (Distorted-fluid-particle Regime)

$$\left\langle\left\langle V_{gj}\right\rangle\right\rangle = \sqrt{2}\left\{\frac{\left(\Delta\rho g_z + M_{F\infty}\right)\sigma}{\rho_f^2}\right\}^{1/4}\left(1-\left\langle\alpha_g\right\rangle\right)\left\{F\left(\left\langle\alpha_g\right\rangle\right)\right\}^2 \times\frac{\mu_m}{\mu_f}\frac{18.67}{1+17.67\left\{F\left(\left\langle\alpha_g\right\rangle\right)\right\}^{6/7}}. \tag{14-108}$$

The frictional pressure gradients in single-particle system, $M_{F\infty}$, and in multi-particle system, M_F, are defined by

$$M_{F\infty} \equiv \frac{f}{2D}\rho_f\left\langle v_f\right\rangle^2 = \frac{f}{2D}\rho_f\left\langle j_f\right\rangle^2 \text{ and } M_F \equiv \left(-\frac{dp}{dz}\right)_F \tag{14-109}$$

where f is the wall friction factor. The function, $F\left(\left\langle\alpha_g\right\rangle\right)$, is defined by

$$F\left(\left\langle\alpha_g\right\rangle\right) \equiv \left\{\frac{\Delta\rho g_z\left(1-\left\langle\alpha_g\right\rangle\right)+M_F}{\Delta\rho g_z + M_{F\infty}}\right\}^{1/2}\left(\frac{\mu_f}{\mu_m}\right). \tag{14-110}$$

Equation (14-108) holds for $N_{\mu f} \geq 0.11\left\{1+\psi\left(r_b^*\right)\right\}/\left\{\psi\left(r_b^*\right)\right\}^{8/3}$. The parameter, $\psi\left(r_b^*\right)$, is given by

$$\psi\left(r_b^*\right) = 0.55\left\{\left(1+0.08 r_b^{*}\right)^{4/7}-1\right\}^{3/4} \tag{14-111}$$

where $r_b^* \equiv r_b\left\{\rho_f\left(\Delta\rho g_z + M_{F\infty}\right)/\mu_f^2\right\}^{1/3}$.

Slug Flow Regime

$$\left\langle\left\langle V_{gj}\right\rangle\right\rangle = 0.35\left\{\frac{\left(\Delta\rho g_z + M_{F\infty}\right)D}{\rho_f}\right\}^{1/2} \times\left[\frac{\left\{\Delta\rho g_z\left(1-\left\langle\alpha_g\right\rangle\right)+M_F\right\}}{\left(\Delta\rho g_z + M_{F\infty}\right)\left(1-\left\langle\alpha_g\right\rangle\right)}\right]^{1/2} \tag{14-112}$$

Churn Flow Regime

$$\langle\langle V_{gj}\rangle\rangle = \sqrt{2}\left\{\frac{(\Delta\rho g_z + M_{F\infty})\sigma}{\rho_f^2}\right\}^{1/4} \times\left\{\frac{\Delta\rho g_z\left(1-\langle\alpha_g\rangle\right)+M_F}{\Delta\rho g_z + M_{F\infty}}\right\}^{1/4}. \tag{14-113}$$

Annular Flow Regime

In separated flows, local relative velocity between two phases cannot be defined (Hibiki and Ishii, 2003b). If small liquid droplets are entrained in the gas core or small gas bubbles are entrained in the liquid film, local relative velocity may be approximated to be zero due to large gas and liquid velocity, resulting in $\langle\langle V_{gj}\rangle\rangle \approx 0$. This approximation may be acceptable in annular flows where the entrainment of liquid from the film to the gas-core flow is negligibly small.

The constitutive equations of the distribution parameter have been improved by considering the gravity effect on the void distribution (Hibiki et al., 2004) as

Bubbly Flow Regime

$$\begin{aligned} C_0 &= 2.0e^{-0.000584 Re_f} + \left\{1.2e^{-5.55\left(\frac{g}{g_N}\right)^3} + 1.2\left(1-e^{-22\frac{\langle D_{Sm}\rangle}{D}}\right)\left(1-e^{-5.55\left(\frac{g}{g_N}\right)^3}\right)\right\}\left(1-e^{-0.000584 Re_f}\right) \\ &- \left[2.0e^{-0.000584 Re_f} + \left\{1.2e^{-5.55\left(\frac{g}{g_N}\right)^3} + 1.2\left(1-e^{-22\frac{\langle D_{Sm}\rangle}{D}}\right) \times\left(1-e^{-5.55\left(\frac{g}{g_N}\right)^3}\right)\right\}\left(1-e^{-0.000584 Re_f}\right) - 1\right]\sqrt{\frac{\rho_g}{\rho_f}}. \end{aligned} \tag{14-114}$$

where g_N is the normal gravitational acceleration (=9.8 m/s^2).

Slug Flow Regime

$$C_0 = 1.2 - 0.2\sqrt{\frac{\rho_g}{\rho_f}} \tag{14-115}$$

Churn Flow Regime

$$C_0 = 1.2 - 0.2\sqrt{\frac{\rho_g}{\rho_f}} \tag{14-116}$$

Annular Flow Regime

$$C_0 \simeq \frac{\overline{V_{gj}}}{\langle j \rangle} + 1 = \frac{1-\langle \alpha_g \rangle}{\langle \alpha_g \rangle + \left\{ \frac{1+75\left(1-\langle \alpha_g \rangle\right)}{\sqrt{\langle \alpha_g \rangle}} \frac{\rho_g}{\rho_f} \right\}^{1/2}} \times \left(1 + \frac{\sqrt{\frac{\Delta\rho g_z D\left(1-\langle \alpha_g \rangle\right)}{0.015\rho_f}}}{\langle j \rangle} \right) + 1 \tag{14-117}$$

Chapter 15

ONE-DIMENSIONAL TWO-FLUID MODEL

The two-fluid model is the most detailed and accurate macroscopic formulation of the thermo-fluid dynamics of two-phase systems. In the two-fluid model, the field equations are expressed by the six conservation equations consisting of mass, momentum and energy equations for each phase. Since these field equations are obtained from an appropriate averaging of local instantaneous balance equations, the phasic interaction term appears in each of the averaged balance equations. These terms represent the mass, momentum and energy transfers through the interface between the phases. The existence of the interfacial transfer terms is one of the most important characteristics of the two-fluid model formulation. These terms determine the rate of phase changes and the degree of mechanical and thermal non-equilibrium between phases, thus they are the essential closure relations that should be modeled accurately. However, because of considerable difficulties in terms of measurements and modeling, reliable and accurate closure relations for the interfacial transfer terms are not fully developed. In spite of these shortcomings of two-fluid models, there is, however, no substitute available for modeling accurately two-phase phenomena where two phases are weakly coupled. Examples of these are:

- Sudden mixing of two phases;
- Transient flooding and flow reversal;
- Transient countercurrent flow;
- Two-phase flow with sudden acceleration.

A three-dimensional, two-fluid model has been obtained by using temporal or statistical averaging. In view of practical engineering problems, a one-dimensional, two-fluid model obtained by averaging local two-fluid formulation over the cross-sectional area is useful for complicated engineering problems involving fluid flow and heat transfer. This is due to the fact that field equations can be reduced to quasi-one-dimensional forms. By area averaging, the information on changes of variables in the direction normal to the main flow within a channel is basically lost. Therefore, the transfer of momentum and energy between the wall and the fluid should be expressed by empirical correlations or by simplified models. In this chapter, we develop a general one-dimensional formulation of the two-fluid model, and discuss various special cases that are important in practical applications. For simplicity, mathematical symbols of time-averaging in one-dimensional formulation are dropped in the formulation in this chapter.

1.1 Area average of three-dimensional two-fluid model

The three-dimensional form of the two-fluid model has been obtained by the temporal or statistical averaging method. For most practical applications, the model developed by Ishii (1975) can be simplified to the following forms:

Continuity equation

$$\frac{\partial \alpha_k \overline{\overline{\rho_k}}}{\partial t} + \nabla \cdot \left(\alpha_k \overline{\overline{\rho_k}} \widehat{\boldsymbol{v}_k} \right) = \Gamma_k \tag{15-1}$$

Momentum equation

$$\begin{aligned} &\frac{\partial \alpha_k \overline{\overline{\rho_k}} \widehat{\boldsymbol{v}_k}}{\partial t} + \nabla \cdot \left(\alpha_k \overline{\overline{\rho_k}} \widehat{\boldsymbol{v}_k} \widehat{\boldsymbol{v}_k} \right) = -\alpha_k \nabla \overline{\overline{p_k}} \\ &+\nabla \cdot \left[\alpha_k \left(\overline{\overline{\mathcal{T}_k}} + \mathcal{T}_k^T \right) \right] + \alpha_k \overline{\overline{\rho_k}} \widehat{\boldsymbol{g}_k} + \widehat{\boldsymbol{v}_{ki}} \Gamma_k + \boldsymbol{M}_{ik} \\ &-\nabla \alpha_k \cdot \overline{\overline{\mathcal{T}_{ki}}} + \left(\overline{\overline{p_{ki}}} - \overline{\overline{p_k}} \right) \nabla \alpha_k \end{aligned} \tag{15-2}$$

Enthalpy energy equation

$$\begin{aligned} &\frac{\partial \alpha_k \overline{\overline{\rho_k}} \widehat{h_k}}{\partial t} + \nabla \cdot \left(\alpha_k \overline{\overline{\rho_k}} \widehat{h_k} \widehat{\boldsymbol{v}_k} \right) = -\nabla \cdot \alpha_k \left(\overline{\overline{\boldsymbol{q}_k}} + \boldsymbol{q}_k^T \right) \\ &+\alpha_k \frac{D_k}{Dt} \overline{\overline{p_k}} + \widehat{h_{ki}} \Gamma_k + a_i q_{ki}'' + \Phi_k \end{aligned} \tag{15-3}$$

Here Γ_k, $\boldsymbol{M}_{ik}$, $\overline{\overline{\mathcal{T}_{ki}}}$, q_{ki}'' and Φ_k are the mass generation, generalized interfacial drag, interfacial shear stress, interfacial heat flux and dissipation, respectively. The subscript k denotes k-phase and i stands for the value at the interface. $1/a_i$ denotes the length scale at the interface and a_i has the physical meaning of the interfacial area per unit volume (Ishii, 1975; Ishii and Mishima, 1981). Thus,

$$a_i = \frac{\text{interfacial area}}{\text{mixture volume}}. \tag{15-4}$$

From the above field equations it can be seen that several interfacial transfer terms appear on the right-hand side of the equations. Since these interfacial transfer terms also should obey the balance laws at the interface, interfacial transfer conditions could be obtained from an average of the local jump conditions (Ishii, 1975). They are given by

$$\begin{cases} \sum_{k=1}^{2} \Gamma_k = 0 \\ \sum_{k=1}^{2} \boldsymbol{M}_{ik} = 0 \\ \sum_{k=1}^{2} \left(\Gamma_k \widehat{h_{ki}} + a_i q''_{ki} \right) = 0. \end{cases} \tag{15-5}$$

Therefore, constitutive equations for $\boldsymbol{M}_{ik}$, $a_i q''_{fi}$ and $a_i q''_{gi}$ are necessary for the interfacial transfer terms. The enthalpy interfacial transfer condition indicates that specifying the heat flux at the interface for both phases is equivalent to the constitutive relation for Γ_k if the mechanical energy transfer terms can be neglected (Ishii, 1975). This aspect greatly simplifies the development of the constitutive relations for interfacial transfer terms.

The rational approach to obtain a one-dimensional model is to integrate the three-dimensional model over a cross-sectional area and then to introduce proper mean values. A simple area average over the cross-sectional area A is defined by Eq.(14-5) and the void-fraction-weighted mean value is given by Eq.(14-6). In the subsequent analysis, the density of each phase within any cross-sectional area is considered to be uniform so that $\rho_k = \langle\langle \rho_k \rangle\rangle$. For most practical two-phase flow problems this assumption is valid since the transverse pressure gradient within a channel is relatively small. The axial component of the weighted mean velocity of phase k is

$$\langle\langle v_k \rangle\rangle = \frac{\langle \alpha_k v_k \rangle}{\langle \alpha_k \rangle} = \frac{\langle j_k \rangle}{\langle \alpha_k \rangle} \tag{15-6}$$

where the scalar expression of the velocity corresponds to the axial component of the vector. By area averaging Eqs.(15-1) to (15-3), and making some simplifications which are applicable to most practical problems, the following field equations can be obtained

Continuity equation

$$\frac{\partial \langle \alpha_k \rangle \rho_k}{\partial t} + \frac{\partial}{\partial z}\langle \alpha_k \rangle \rho_k \langle\langle v_k \rangle\rangle = \langle \Gamma_k \rangle \tag{15-7}$$

Momentum equation

$$\begin{aligned}
&\frac{\partial}{\partial t}\langle \alpha_k \rangle \rho_k \langle\langle v_k \rangle\rangle + \frac{\partial}{\partial z} C_{vk} \langle \alpha_k \rangle \rho_k \langle\langle v_k \rangle\rangle^2 \\
&= -\langle \alpha_k \rangle \frac{\partial}{\partial z}\langle\langle p_k \rangle\rangle + \frac{\partial}{\partial z}\langle \alpha_k \rangle \langle\langle \tau_{kzz} + \tau_{kzz}^T \rangle\rangle \\
&-\frac{4\alpha_{kw}\tau_{kw}}{D} - \langle \alpha_k \rangle \rho_k g_z + \langle \Gamma_k \rangle \langle\langle v_{ki} \rangle\rangle + \langle M_k^d \rangle \\
&+\left\langle (p_{ki} - p_k)\frac{\partial \alpha_k}{\partial z} \right\rangle
\end{aligned} \tag{15-8}$$

where α_{kw} and τ_{kw} are the mean void fraction at the wall and wall shear stress, respectively. The pressure difference and void gradient term can be important for horizontal stratified flow. Except for this case, this term may be neglected. $\langle M_k^d \rangle$ is the total interfacial shear force given by

$$\langle M_k^d \rangle = \langle \boldsymbol{M}_{ik} - \nabla \alpha_k \cdot \tau_i \rangle_z . \tag{15-9}$$

The first term on the right-hand side is the generalized particle drag which is important for a dispersed flow. The second term is the effect of the interfacial shear and the void gradient. This term is particularly important for a separated flow. In the convective term, the distribution parameter for the k-phase momentum, C_{vk}, appears due to the difference between the average of a product of variables and the product of averaged variables.

Enthalpy energy equation

$$\begin{aligned}&\frac{\partial}{\partial t}\langle\alpha_k\rangle\rho_k\langle\langle h_k\rangle\rangle+\frac{\partial}{\partial z}C_{hk}\langle\alpha_k\rangle\rho_k\langle\langle h_k\rangle\rangle\langle\langle v_k\rangle\rangle\\&=-\frac{\partial}{\partial z}\langle\alpha_k\rangle\left\langle\left\langle q_k+q_k^T\right\rangle\right\rangle_z+\langle\alpha_k\rangle\frac{D_k}{Dt}\langle\langle p_k\rangle\rangle\\&+\frac{\xi_h}{A}\alpha_{kw}q''_{kw}+\langle\Gamma_k\rangle\langle\langle h_{ki}\rangle\rangle+\langle a_i q''_{ki}\rangle+\langle\Phi_k\rangle\end{aligned}\tag{15-10}$$

where ξ_h and q''_{kw} are the heated perimeter and wall heat flux, respectively. C_{hk} is the distribution parameter for the k-phase enthalpy. From the macroscopic jump conditions at the interface the following relations between the interfacial transfer terms hold,

$$\begin{cases}\sum_{k=1}^{2}\langle\Gamma_k\rangle=0\\ \sum_{k=1}^{2}\langle M_k^d\rangle=\sum_{k=1}^{2}\langle \boldsymbol{M}_{ik}-\nabla\alpha_k\cdot\tau_i\rangle_z=0\\ \sum_{k=1}^{2}\left\{\langle\langle\Gamma_k\rangle\rangle\langle\langle h_{ki}\rangle\rangle+\langle a_i q''_{ki}\rangle\right\}=0.\end{cases}\tag{15-11}$$

1.2 Special consideration for one-dimensional constitutive relations

1.2.1 Covariance effect in field equations

In a one-dimensional model, a very careful examination of transverse distributions of various variables and their effects on the balance and constitutive equations is essential. If this is not done properly, the resulting two-phase flow formulation can be inconsistent. Improper modeling, or disregard of the distribution effects, may lead not only to a grossly inaccurate model, but also to various numerical instabilities. The distribution effects can be divided into two groups. The first one is the covariance effect which directly affects the form of the convective term in the field equation. The second effect appears in the averaging of the various local constitutive relations. These two effects are discussed separately below.

The covariance of the convective terms is defined by

$$COV\left(\rho_k\alpha_k\psi_k v_k\right) \equiv \left\langle \alpha_k\rho_k\psi_k\left(v_k - \langle\langle v_k\rangle\rangle\right)\right\rangle \tag{15-12}$$

To close the set of the governing equations we must specify relations for these covariance terms. This can be done by introducing distribution parameters for the momentum and energy fluxes. If we define a distribution parameter for a flux as

$$C_{\psi k} \equiv \frac{\langle \alpha_k\psi_k v_k\rangle}{\langle\alpha_k\rangle\langle\langle\psi_k\rangle\rangle\langle\langle v_k\rangle\rangle} \tag{15-13}$$

the covariance term becomes

$$\begin{aligned} COV\left(\rho_k\alpha_k\psi_k v_k\right) &= \rho_k\left\langle \alpha_k\psi_k\left(v_k - \langle\langle v_k\rangle\rangle\right)\right\rangle \\ &= \left(C_{\psi k} - 1\right)\rho_k\langle\alpha_k\rangle\langle\langle\psi_k\rangle\rangle\langle\langle v_k\rangle\rangle. \end{aligned} \tag{15-14}$$

For the momentum flux, the distribution parameter is defined by

$$C_{vk} \equiv \frac{\langle \alpha_k v_k^2\rangle}{\langle\alpha_k\rangle\langle\langle v_k\rangle\rangle^2}. \tag{15-15}$$

Physically, C_{vk} represents the effect of the void and momentum-flux profiles on the cross-sectional-area-averaged momentum flux of k-phase. A quantitative study of C_{vk} can be made by considering a symmetric flow in a circular duct and introducing the power-law expressions in parallel with the analysis of the drift-flux modeling (Zuber and Findlay, 1965; Ishii, 1977). The following is the summary obtained by Ishii (1977).

Hence, for bubbly, slug and churn-turbulent flow, it is postulated that

$$\frac{\alpha_k - \alpha_{kw}}{\alpha_{k0} - \alpha_{kw}} = 1 - \left(\frac{r}{R_w}\right)^n \tag{15-16}$$

and

$$\frac{v_k}{v_{k0}} = 1 - \left(\frac{r}{R_w}\right)^m \tag{15-17}$$

where the subscripts 0 and w refer to the value at the centerline and at the wall of a tube. For simplicity it is assumed that the void and velocity profiles are similar; namely, $n = m$. This assumption is widely used in mass-transfer problems, and it may not be unreasonable for fully developed two-phase flows if one considers that the vapor flux and, hence, the void concentration greatly influence the velocity distributions. Under this assumption, it can be shown that

$$C_{vk} = \frac{\frac{n+2}{n+1}\left(\alpha_{kw} + \Delta\alpha_k \frac{3n}{3n+2}\right)\left(\alpha_{kw} + \Delta\alpha_k \frac{n}{n+2}\right)}{\left(\alpha_{kw} + \Delta\alpha_k \frac{n}{n+1}\right)^2} \tag{15-18}$$

where $\Delta\alpha_k = \alpha_{k0} - \alpha_{kw}$. The volumetric-flux-distribution parameter C_0 of the drift-flux model is given by

$$C_0 \simeq \frac{(n+2)}{(n+1)} \tag{15-19}$$

where C_0 can be given by the following empirical correlation (Ishii, 1977)

$$C_0 = 1.2 - 0.2\sqrt{\rho_g/\rho_f} \tag{15-20}$$

for a fully developed flow in a round tube. For a subcooled boiling or flow in a rectangular channel, see Eq.(14-35) or Eq.(14-32), respectively. Therefore, in the standard range of n, the parameter C_{vg} can be given approximately by

$$C_{vg} \simeq 1 + 0.5(C_0 - 1). \tag{15-21}$$

For a liquid phase in a vapor-dispersed-flow regime, $\alpha_{fw} \simeq 1$ and $\alpha_{f0} < 1$. Then from Eq.(15-18) it can be shown that, for a standard range of α_{f0} in the bubbly- and churn-flow regimes, C_{vf} can be approximated by

$$C_{vf} = 1 + 1.5(C_0 - 1). \tag{15-22}$$

For an annular flow, the momentum covariance term can also be calculated by using the standard velocity profiles for the vapor and liquid flows. Thus we obtain

$$C_{vk} \cong \begin{cases} 1.02 & \text{(turbulent flow)} \\ 1.33 & \text{(laminar flow).} \end{cases} \tag{15-23}$$

Similarly, the distribution parameter for the enthalpy flux can be defined by

$$C_{hk} \equiv \frac{\langle \alpha_k h_k v_k \rangle}{\langle \alpha_k \rangle \langle\langle h_k \rangle\rangle \langle\langle v_k \rangle\rangle}. \tag{15-24}$$

For a thermal-equilibrium flow, $h_g = h_{gs}$ and $h_f = h_{fs}$, where h_{gs} and H_{fs} are the saturation enthalpies of vapor and liquid. Since, in this case, the enthalpy profile is completely flat for each phase, the distribution parameters become unity; namely, $C_{hg} = C_{hf} = 1$. It is also evident that if one of the phases is in the saturated condition, then C_{hk} for that phase becomes unity.

In the single-phase region, the distribution parameter can be calculated from the assumed profiles for the velocity and enthalpy. Using the standard power-law profiles for a turbulent flow; namely, $v/v_0 = (y/R)^{1/n}$ and $(h - h_w)/(h_0 - h_w) = (y/R)^{1/m}$, where y is the distance from the wall, we can show that the covariance term is negligibly small both for developing and fully developed flows under normal conditions. Then

$$C_{hk} \simeq 1.0. \tag{15-25}$$

Therefore, except for highly transient cases, the enthalpy covariance may be neglected.

1.2.2 Effect of phase distribution on constitutive relations

The greatest shortcoming of the conventional two-fluid model is in the modeling of the constitutive equation for the interfacial shear $\langle M_k^d \rangle$ defined by Eq.(15-9). This is particularly true when the two-fluid model was applied to other than a separated flow. The problem is twofold:

1. Modeling of the averaged drag $\langle \mathbf{M}_{ik} \rangle_z$;
2. Modeling of the effect of interfacial shear $\langle -\nabla \alpha_k \cdot \tau_i \rangle_z$.

These will be discussed separately below.

For a dispersed two-phase flow the averaged interfacial drag term could be given approximately by

$$\langle \boldsymbol{M}_{id} \rangle_z \doteq -\frac{3}{8} \frac{C_D}{r_d} \langle \alpha_d \rangle \rho_c \langle v_r \rangle |\langle v_r \rangle|. \tag{15-26}$$

Here, only the steady-state drag force part of $\boldsymbol{M}_{ik}$ is considered because it is the most important term. The above approximate form is obtained based on the experimental observation that the local relative velocity v_r is comparatively uniform across a flow channel (Serizawa et al., 1975; Hibiki and Ishii, 1999; Hibiki et al., 2001a) and the fact that the local relative velocity is much smaller than the phase velocities in most two-phase flow.

The important point, however, is that the averaged drag force should be related to the averaged local relative velocity $\langle v_r \rangle$ given by

$$\langle v_r \rangle \equiv \frac{1}{A} \int v_r dA \tag{15-27}$$

and not to the difference between the area averaged mean velocities of phases given by

$$\overline{v_r} \equiv \langle\langle v_d \rangle\rangle - \langle\langle v_c \rangle\rangle. \tag{15-28}$$

In general,

$$\langle v_r \rangle \neq \overline{v_r}. \tag{15-29}$$

The difference between these two relative velocities can be very large. The reason is that in one-dimensional formulation, the slip, $\overline{v_r}$, between two phases is caused by two completely different effects; namely, the local relative motion and integral effect of the phase and velocity distributions. The existence of these two effects is already well-known (Zuber and Findlay, 1965; Ishii, 1977; Bankoff, 1960). The first effect is the true relative motion between two phases at a local point and does not require any further explanation. The second effect of the distribution arises due to the area averaging. For example, if the dispersed phase is more concentrated in the high velocity core region, then the mean velocity of the dispersed phase should be much higher than that of the continuous phase which is concentrated near the low velocity wall region. This is true even when the two phases are locally moving with the same velocity.

Based on the drift-flux model formulation it can be shown that the approximate expression for $\langle v_r \rangle$ is given by

$$\left\langle v_r \right\rangle \simeq \frac{1 - C_0 \left\langle \alpha_g \right\rangle}{1 - \left\langle \alpha_g \right\rangle} \left\langle \left\langle v_g \right\rangle \right\rangle - C_0 \left\langle \left\langle v_r \right\rangle \right\rangle \tag{15-30}$$

for bubbly, slug and churn turbulent flow. Therefore, from the flow regime criterion (Ishii and Mishima, 1981; Ishii, 1977), it is applicable under the following conditions.

$$\begin{cases} \left\langle \alpha_g \right\rangle < \dfrac{1}{C_0}, \text{ and} \\ \left\langle j_g \right\rangle \sqrt{\dfrac{\rho_g}{\Delta \rho g D}} > \left\langle \alpha_g \right\rangle - 0.1 \end{cases} \tag{15-31}$$

This criterion is valid when the tube diameter is relatively small. For more general conditions, see Ishii and Mishima (1981). The constitutive equation for C_0 for a simple case is given by Eq.(15-20).

The expression for the drag force given by Eq.(15-26) with Eq.(15-30) compensates for the slip due to the distributions of phases and velocities. This difference between $\overline{v_r}$ and $\left\langle v_r \right\rangle$ has never been taken into account in the conventional two-fluid model. In most two-phase flow systems, the slip due to the distribution of phases is much greater than the local slip between phases. Therefore, neglecting the above-mentioned effect will lead to large errors in predictions of the void fraction and velocities in bubbly, slug and churn turbulent flow regimes. As a result, even the steady-state predictions from two-fluid model were not as good as those from a drift-flux model in these flow regimes. This was one of the most significant shortcomings of the conventional two-fluid model and it should be corrected in all future analyses.

1.2.3 Interfacial shear term

The total interfacial shear force denoted by $\left\langle M_k^d \right\rangle$ has two sources; namely, the generalized drag $\left\langle \boldsymbol{M}_{ik} \right\rangle_z$ and the contribution of the interfacial shear and void gradient $\left\langle -\nabla \alpha_k \cdot \tau_i \right\rangle_z$ as shown in Eq.(15-9). In a separated flow, the second term is the dominant one. For example, for an annular flow in a tube it can be shown that

$$\begin{aligned}\left\langle -\nabla\alpha_k\cdot\tau_i\right\rangle_z &= -\frac{1}{A}\int_A \frac{\partial \alpha_g}{\partial r}\left(\tau_{gi}\right)2\pi r dr\\ &= -\frac{1}{A}\lim_{\delta\to 0}\int_\delta \frac{\partial \alpha_g}{\partial r}\tau_{gi}2\pi r dr = -\frac{\xi_i}{A}\tau_{gi}\end{aligned} \tag{15-32}$$

where ξ_i is the wetted perimeter of the gas core.

The constitutive relation for τ_{gi} in this case can be given in terms of the standard interfacial friction factor as

$$\tau_{gi} = \frac{f_i}{2}\rho_g\overline{v_r}\left|\overline{v_r}\right| \tag{15-33}$$

where $\overline{v_r} = \langle\langle v_g\rangle\rangle - \langle\langle v_f\rangle\rangle$. There are a number of correlations for the interfacial friction factor f_i. The Wallis correlation is given by

$$f_i = 0.005\left[1+75\left(1-\langle\alpha_g\rangle\right)\right] \tag{15-34}$$

which is applicable to the case with rough wavy films.

For annular flow, this interfacial shear term has been correctly taken into account in the conventional two-fluid model. However, the effect of this term in the bubbly, slug and churn turbulent flows has been generally neglected. The inclusion of this term is important for the proper modeling of the interfacial momentum coupling between phases. In order to obtain a constitutive relation for this interfacial shear term, several assumptions are necessary since it requires information on the void and shear stress distributions. For this purpose the following power-law distribution is assumed

$$\tau_i \sim \tau_w\left(\frac{r}{R_w}\right)^m. \tag{15-35}$$

From this and the void profile of Eq.(15-16), it can be shown that

$$-\left\langle\nabla\alpha_g\cdot\tau_i\right\rangle_z = -\frac{4\tau_w}{D}\left\langle\alpha_g\right\rangle\frac{n+2}{n+1+m} \tag{15-36}$$

where α_g is the void fraction of dispersed phase. By introducing the distribution parameter C_τ given by

$$C_{\tau} = \frac{n+2}{n+1+m} \tag{15-37}$$

the interfacial shear term for a dispersed two-phase flow becomes

$$-\left\langle \nabla\alpha_{g} \cdot \tau_{i} \right\rangle_{z} = -\frac{4\tau_{w}}{D}\left\langle \alpha_{g} \right\rangle C_{\tau} \tag{15-38}$$

where C_{τ} is expected to be very close to one. In a horizontal channel this term will contribute to the slip between phases even under a steady state condition. The inclusion of this term does not alter the overall momentum balance of a two-phase mixture because of the macroscopic momentum jump condition. However, it indicates that the momentum interaction between phases is affected by the wall shear stress through the interfacial shear and void gradient distributions.

References

Adorni, N., Peterlongo, G., Ravetta, R. and Tacconi, F. A., 1964, Large Scale Experiments on Heat Transfer and Hydrodynamics with Steam-water Mixtures, *CISE Report* R-91, Italy.

Akiyama, M. and Aritomi, M., 2002, *Advanced Numerical Analysis of Two-phase Flow Dynamics –Multi-dimensional Flow Analysis-*, Corona Publishing Co. Ltd., Tokyo, Japan

Alia, P., Cravarolo, L., Hassid, A. and Pedrocchi, E., 1965, Liquid Volume Fraction in Adiabatic Two-phase Vertical Upflow-round Conduit, *CISE Report*-105, Italy.

Antal, S. P., Lahey Jr, R. T. and Flaherty, J. E., 1991, Analysis of Phase Distribution in Fully Developed Laminar Bubbly Two-phase Flow, *Int. J. Multiphase Flow* **17:** 635-652.

Aris, R., 1962, *Vectors, Tensors and the Basic Equations of Fluid Mechanics*, Prentice-Hall, Englewood Cliffs, N.J.

Arnold, G. S., Drew, D. A. and Lahey Jr., R. T., 1989, Derivation of Constitutive Equations for Interfacial Force and Reynolds Stress for a Suspension of Spheres Using Ensemble Cell Averaging, *Chem. Eng. Comm.* **86:** 43-54.

Auton, T. R., 1987, The Lift Force on a Spherical Body in a Rotational Flow, *J. Fluid Mech.* **183:** 199-218.

Azbel, D., 1981, *Two-phase Flows in Chemical Engineering*, Cambridge University Press, Cambridge, UK.

Azbel, D. and Athanasios, I. L., 1983, A Mechanism of Liquid Entrainment. *Handbook of Fluids in Motion*, Ann Arbor Sci. Pub., Ann Arbor, MI.

Baker, J. L. L., 1965, Flow-regime Transitions at Elevated Pressure in Vertical Two-phase Flow, *Argonne National Lab. Report*, ANL-7093.

Bankoff, S. G., 1960, A Variable Density Single-fluid Model for Two-phase Flow with Particular Reference to Steam Water Flow, *J. Heat Transfer* **82:** 265-272.

Bello, J. K., 1968, *Turbulent Flow in Channel with Parallel Walls,* Moskva, Mir, in Russian.

Bergonzoli, F. and Halfen, F. J., 1964, Heat Transfer and Void Formation during Forced Circulation Boiling of Organic Coolants, NAA-SR-8906, Atomics International.

Bilicki, A. and Kestin, J., 1987, Transition Criteria for Two-phase Flow Patterns in Vertical Upward Flow, *Int. J. Multiphase Flow* **13:** 283-294.

Bird, R. B., Stewart, W. E. and Lightfoot, E. N., 1960, *Transport Phenomena*, John Wiley and Sons, Inc., New York.

Bornhorst, W. J. and Hatsopoulos, G. N., 1967, Analysis of a Liquid Vapor Phase Change by the Methods of Irreversible Thermodynamics, *J. Applied Mech.* **89:** 847-853.

Boure, J. and Réocreux, M., 1972, General Equations of Two-phase Flows: Application to Critical Flows and to Non Steady Flows, *4th All Union Heat and Mass Transfer Conference*, Minsk.

Boure, J., 1973, Dynamique des Écoulements Diphasiques: Propagation des Petites Perturbations, *CEA-R-4456*.

Bowen, R. M., 1967, Toward a Thermodynamics and Mechanics of Mixtures, *Arch. Rational Mech. Anal.* **24:** 370-403.

Bridge, A. G., Lapidus, L and Elgin, J. C., 1964, The Mechanics of Vertical Gas-Liquid Fluidized System I: Counter-current Flow, *AIChE J.* **10:** 819-827.

Brinkman, H., 1952, The Viscosity of Concentrated Suspensions and Solutions, *J. Chem. Phys.* **20:** 571.

Brodkey, R. S., 1967, *The Phenomena of Fluid Motion*, Addison-Wesley.

Brodkey, R. S., 1971, Transport Phenomena at the Liquid-Vapor Interface of Mercury Using a Radioactive Tracer, *International Symposium on Two-phase Systems*, Haifa.

Burgers, J. M., 1941, On the Influence of the Concentration of a Suspension upon the Sedimentation Velocity (in Particular for a Suspension of Spherical Particles) *Proc. K. Med. Akad. Wet.* **44:** 1045-1051 (1941); **45:** 9-16 (1942)

Buevich, I., 1969, A Hydrodynamic Model of Disperse Systems, *J. Applied Math. Mech.* **33:** 466-479.

Buyevich, Y., 1972, Statistical Hydrodynamics of Disperse Systems. Part.I. Physical Background and General Equations, *J. Fluid Mech.* **49:** 489-507.

Carrier, G. F., 1958, Shock Waves in Dusty Gas, *J. Fluid Mech.* **4:** 376-382.

Callen, H. B., 1960, *Thermodynamics*, Wiley.

Chao, B., 1962, Motion of Spherical Gas Bubbles in a Viscous Liquid at Large Reynolds Numbers, *Phys. Fluids* **5:** 69-79.

Clift, R., Grace, J. R. and Weber, M. E., 1978, *Bubbles, Drops, and Particles*, Academic Press, New York.

Coleman, B. and Noll, W., 1960, An Approximate Theorem for Functionals with Applications in Continuum Mechanics, *Arch. Rational Mech. Anal.* **6:** 355-370.

Coleman, B. D., 1964, Thermodynamics of Materials with Memory, *Arch. Rational Mech. Anal.* **17:** 1-46.

Collier, J., 1972, *Convective Boiling and Condensation*, McGraw Hill, London.

Coulaloglou, C. A. and Tavlarides, L. L., 1976, Drop Size Distributions and Coalescence Frequencies of Liquid-liquid Dispersion in Flow Vessels, *AIChE J.* **22:** 289-297.

Coulaloglou, C. A. and Tavlarides, L. L., 1977, Description of Interaction Processes in Agitated Liquid-liquid Dispersions, *Chem. Eng. Sci.* **32:** 1289-1297.

Cravarolo, L., Giorgini, A., Hassid, A. and Pedrocchi, E., 1964, A Device for the Measurement of Shear Stress on the Wall of a Conduit; Its Application in the Mean Density Determination in Two-phase Flow; Shear Stress Data in Two-phase Adiabatic Vertical Flow, *CISE Report*-82, Italy.

Culick, F., 1964, Boltzman Equation Applied to a Problem of Two-phase Flow, *Phys. Fluid* **7:** 1898-1904.

De Groot, S. B. and Mazur, P., 1962, *Non-equilibrium Thermodynamics*, North Holland.

De Jarlais, G., Ishii and M., Linehan, J., 1986, Hydrodynamic Stability of Inverted Annular Flow in an Adiabatic Simulation, *J. Heat Transfer* **108**: 84-91.

Delhaye, J. M., 1968, Equations of Fondamentales des Écoulements Diphasiques, Part 1 and 2, *CEA-R-3429*.

Delhaye, J. M., 1969, General Equations of Two-phase Systems and their Application to Air-water Bubble Flow and to Steam-water Flashing Flow, *ASME Paper 69-HT-63, 11th Heat Transfer Conference*, Minneapolis.

Delhaye, J. M., 1970, Contrubution à L'étude des Écoulements Diphasiques Eau-air et Eau-vapeur, *Ph.D. Thesis*, University of Grenoble.

Delhaye, J. M., 1974, Jump Conditions and Entropy Sources in Two-phase Systems. Local Instant Formulation, *Int. J. Multiphase Flow* **1:** 395-409.

Dinh, T. N., Li, G. J. and Theofanous, T. G., 2003, An Investigation of Droplet Breakup in High Mach, Low Weber Number Regime, *Proc. 41st Aerospace Sci. Mtg and Exh.*, Paper AIAA 2003-317, Reno, Nevada.

Diunin, A. K., 1963, On the Mechanics of Snow Storms, *Siberian Branch, Akademii Nauk SSSR*, Novosibirsk.

Drew, D. A., 1971, Averaged Field Equations for Two-phase Media, *Studies Appl. Math.* **1:** 133-166.

Dumitrescu, D. T., 1943, Stomung an einer Luftblase in Senkrechten Rohr, *Z. Angew, Math. Mech.* **23:** 139-149.

Eilers, H., 1941, The Viscosity of the Emulsion of Highly Viscous Substances as Function of Cencentration, *Kolloid Z* **97:** 313-321.

Ervin, E. A. and Tryggvason, G., 1997, The Rise of Bubbles in a Vertical Shear Flow, *J. Fluids Eng.* **119:** 443-449.

Fauske, H., 1962, Critical Two-phase, Steam Water Flow, *Proc. Heat Transfer and Fluid Mechanics Institute*, Stanford Univ. Press, pp.78.

Fick, A., 1855, Über Diffusion, *Ann. Der Phys.* **94:** 59-86.

Frankl, F. I., 1953, On the Theory of Motion of Sediment Suspensions, *Soviet Physics Doklady, Academii Nauk SSSR*, **92:** 247-250.

Frankel, N. A. and Acrivos, A., 1967, On the Viscosity of a Concentrated Suspension of Solid Spheres, *Chem. Eng. Sci.* **22:** 847-853.

Friedlander, S. K., 1977, *Smoke, Dust and Haze*, Wiley, New York.

Fu, X. Y. and Ishii, M., 2002a, Two-group Interfacial Area Transport in Vertical Air-water Flow I. Mechanistic Model, *Nucl. Eng. Des.* **219:** 143-168.

Fu, X. Y. and Ishii, M., 2002b, Two-group Interfacial Area Transport in Vertical Air-water Flow II. Model Evaluation, *Nucl. Eng. Des.* **219:** 169-190.

Goda, H., Hibiki, T. Kim, S., Ishii, M. and Uhle, J., 2003, Drift-flux Model for Downward Two-phase Flow, *Int. J. Heat Mass Transfer* **46:** 4835-4844.

Gibbs, J. W., 1948, *Collected Work of J. W. Gibbs*, Yale University Press, Vol.1, New York.

Goldstein, S., 1938, *Modern Developments in Fluid Dynamics*, Oxford University Press, London.

Govier G. W. and Aziz, K, 1972, *The Flow of Complex Mixtures in Pipes*, Van Nostrant-Reinhold Co., New York.

Grace, J. R., Wairegi, T. and Brophy, J., 1978, Break-up of Drops and Bubbles in Stagnant Media, *Can. J. Chem. Eng.* **556:** 3-8.

Hadamard, J., 1911, Mouvement Permanent Lent d'une Sphere Liquide Visqueuse dans un Liquid Visqueux, *C. R. Acad. Sci. Paris Sér A-B* **152:** 1735-1739.

Hancox, W. T., and Nicoll, W. B., 1972, Prediction of Time-dependent Diabatic Two-phase Water Flows, *Prog. Heat Mass Transfer* **6:** 119-135.

Happel, J. and Brfnner, H., 1965, *Low Reynolds Number Hydrodynamics*, Prentice-Hall.

Harmathy, T. Z., 1960, Velocity of Large Drops and Bubbles in Media of Infinite and Restricted Extent, *AIChE J.* **6:** 281-288.

Hawksley, P. G. W., 1951, The Effect of Concentration on the Settling of Suspensions and Flow through Porous Media, *Some Aspect of Fluid Flow*, pp.114, Edward Arnold, London.

Hayes, W. D., 1970, Kinematic Wave Theory, Proc. *Royal Soc. London Ser. A Math. Phys. Sci.* **320:** 209-226.

Helmhotz, H., 1868, Uber Discontinuirliche Flüssigkeitsbewegungen, *Monatsber. Dtsch. Akad. Wiss. Berlin* pp.215-228.

Hewitt, G. and Hall Taylor, N. S., 1970, *Annular Two-phase Flow*, Pergamon Press, Oxford.

Hibiki, T. and Ishii, M., 1999, Experimental Study on Interfacial Area Transport in Bubbly Two-phase Flows, *Int. J. Heat Mass Transfer* **42:** 3019-3035.

Hibiki, T. and Ishii, M., 2000a, One-group Interfacial Area Transport of Bubbly Flows in Vertical Round Tubes, *Int. J. Heat Mass Transfer* **43:** 2711-2726.

Hibiki, T. and Ishii, M., 2000b, Two-group Interfacial Area Transport Equations at Bubbly-to-slug Flow Transition, *Nucl. Eng. Des.* **202:** 39-76.

Hibiki, T., Ishii, M. and Xiao, Z., 2001a, Axial Interfacial Area Transport of Vertical Bubbly Flows, *Int. J. Heat Mass Transfer* **44:** 1869-1888.

Hibiki, T., Takamasa, T. and Ishii, M., 2001b, Interfacial Area Transport of Bubbly Flow in a Small Diameter Pipe, *J. Nucl. Sci. Technol.* **38:** 614-620.

Hibiki, T. and Ishii, M., 2002a, Interfacial Area Concentration of Bubbly Flow Systems, *Chem. Eng. Sci.* **57:** 3967-3977.

Hibiki, T. and Ishii, M., 2002b, Distribution Parameter and Drift Velocity of Drift-flux Model in Bubbly Flow, *Int. J. Heat Mass Transfer* **45:** 707-721.

Hibiki, T. and Ishii, M., 2002c, Development of One-group Interfacial Area Transport Equation in Bubbly Flow Systems, *Int. J. Heat Mass Transfer* **45:** 2351-2372.

Hibiki, T. and Ishii, M., 2003a, One-dimensional Drift-flux Model for Two-phase Flow in a Large Diameter Pipe, *Int. J. Heat Mass Transfer* **46:** 1773-1790.

Hibiki, T. and Ishii, M., 2003b, One-dimensional Drift-flux Model and Constitutive Equations for Relative Motion between Phases in Various Two-phase Flow Regimes, *Int. J. Heat Mass Transfer* **46:** 4935-4948; Erratum, **48:** 1222-1223 (2005).

Hibiki, T., Situ, R., Mi, Y. and Ishii, M., 2003a, Modeling of Bubble-layer Thickness for Formulation of One-dimensional Interfacial Area Transport Equation in Subcooled Boiling Two-phase Flow, *Int. J. Heat Mass Transfer* **46:** 1409-1423; Erratum, **46:** 3549-3550 (2003).

Hibiki, T., Situ, R., Mi, Y. and Ishii, M., 2003b, Local Flow Measurements of Vertical Upward Bubbly Flow in an Annulus, *Int. J. Heat Mass Transfer* **46:** 1479-1496.

Hibiki, T., Takamasa, T. and Ishii, M., 2004, One-dimensional Drift-flux Model and Constitutive Equations for Relative Motion between Phases in Various Two-phase Flow Regimes at Microgravity Conditions, *Proc. 12th Int. Conf. Nucl. Eng.*, Arlington, VA, USA, ICONE12-49037.

Hirschfelder, J. V., Curtiss, C. F. and Bird, R. B., 1954, *Molecular Theory of Gases and Liquids*, John Wiley and Sons, Inc., New York.

Hinze, J. W., 1959, *Turbulence*, McGraw-Hill, New York.

Hughes, T. A., 1958, Steam-water Mixture Density Studies in a Natural Circulation High Pressure System, *Babcock and Wilcox, G. Report* No. 5435.

Ishii, M. and Zuber, N., 1970, Thermally Induced Flow Instabilities in Two-phase Mixtures, *Proc. 4th International Heat Transfer Conference*, Paris.

Ishii, M., 1971, Thermally Induced Flow Instabilities in Two-phase Mixture in Thermal Equilibrium, *Ph.D. Thesis*, Georgia Institute of Technology.

Ishii, M., 1975, Thermo-fluid Dynamic Theory of Two-phase Flow, Collection de la Direction des Etudes et Researches d'Electricite de France, Eyrolles, Paris, France, 22.

Ishii, M. and Grolmes, M. A., 1975, Inception Criteria for Droplet Entrainment in Two-phase Concurrent Film Flow, *AIChE J.* **21:** 308-318.

Ishii, M., Jones, O. C. and Zuber, N., 1975, Thermal Non-equilibrium Effects in Drift Flux Model of Two-phase Flow, *Trans. Am. Nucl. Soc.* **22:** 263-264.

Ishii, M., 1976, One-dimensional Drift-flux Modeling: One-dimensional Drift Velocity of Dispersed Flow in Confined Channel, *Argonne National Lab. Report*, ANL-76-49.

Ishii, M., Chawla, T. C. and Zuber, N., 1976, Constitutive Equation for Vapor Drift Velocity in Two-phase Annular Flow, *AIChE J.* **22:** 283-289.

Ishii, M. and Chawla, T. C., 1979, Local Drag Laws in Dispersed Two-phase Flow, *Argonne National Lab. Report*, ANL-79-105.

Ishii, M., 1977, One-dimensional Drift-flux Model and Constitutive Equations for Relative Motion between Phases in Various Two-phase Flow Regimes, *Argonne National Lab. Report*, ANL-77-47.

Ishii, M. and Zuber, N., 1979, Drag Coefficient and Relative Velocity in Bubbly, Droplet or Particulate Flows, *AIChE J.* **25:** 843-855.

Ishii, M. and Mishima, K., 1981, Study of Two-fluid Model and Interfacial Area, *Argonne National Lab Report* ANL-80-111.

Ishii, M. and Mishima, K., 1984, Two-fluid Model and Hydrodynamic Constitutive Relations, *Nucl. Eng. Des.* **82:** 107-126.

Ishii, M. and De Jarlais, G., 1987, Flow Visualization Study of Inverted Annular Flow of Post Dryout Heat Transfer Region, *Nucl. Eng. Des.* **99:** 187-199.

Ishii, M., Kim, S. and Uhle, J., 2002, Interfacial Area Transport Equation: Model Development and Benchmark Experiments, *Int. J. Heat Mass Transfer* **45:** 3111-3123.

Ishii, M. and Kim, S., 2004, Development of One-group and Two-group Interfacial Area Transport Equation, *Nucl. Sci. Eng.* **146:** 257-273.

Kalinin, A. V., 1970, Derivation of Fluid Mechanics Equations for a Two-phase Medium with Phase Changes, *Heat Transfer Soviet Res.* **2:** 83-96.

Kataoka, I. and Ishii, M. 1987, Drift Flux Model for Large Diameter Pipe and New Correlation for Pool Void Fraction, *Int. J. Heat Mass Transfer* **30:** 1927-1939.

Kataoka, I. and Serizawa, A., 1995, Modeling and Prediction of Turbulence in Bubbly Two-phase Flow, *Proc. 2nd Int. Conf. Multiphase Flow '95 – Kyoto*, pp. MO2-11-MO2-16.

Kelly, F. D., 1964, A Reacting Continuum, *Int. J. Engng. Sci.* **2:** 129-153.

Kelvin, W., 1871, Hydrokinetic Solutions and Observations, *London, Edinburgh and Dublin Philosophical Magazine and Journal of Science* Ser.4 **42:** 362-377.

Kim, W. K. and Lee, K. L., 1987, Coalescence Behavior of two Bubbles in Stagnant Liquids, *J. Chem. Eng. Jpn.* **20:** 449-453.

Kirkpatrick, R. D. and Lockett, M. J., 1974, The Influence of Approach Velocity on Bubble Coalescence, *Chem. Eng. Sci.* **29:** 2363-2373.

Kocamustafaogullari, G., 1971, Thermo-fluid Dynamics of Separated Two-phase Flow, *Ph.D. Thesis*, Georgia Institute of Technology.

Kocamustafaogullari, G., Chen, I. Y. and Ishii, M., 1984, Unified Theory for Predicting Maximum Fluid Particle Size for Drops and Bubbles, *Argonne National Lab. Report*, ANL-84-67.

Kocamustafaogullari, G. and Ishii, M., 1995, Foundation of the Interfacial Area Transport Equation and its Closure Relations, *Int. J. Heat Mass Transfer* **38:** 481-493.

Kolev, N., 2002, Multiphase Flow Dynamics 1 Fundamentals, 2: Mechanical and Thermal Interactions, Springer-Verlag.

Kordyban, E., 1977, Some Characteristics of High Waves in Closed Channels Approaching Kelvin-Helmholtz Instability, *J. Fluids Eng.* **99:** 389-346.

Kotchine, N. E., 1926, Sur la Théorie des Ondes De-choc dans un Fluide, *Bend. Circ. Mat. Palermo* **50:** 305-344.

Kutateladze, S. S., 1952, Heat Transfer in Condensation and Boiling, *Moscow, AEC-TR-3770*, USAEC Technical Information Service.

Kynch, G. J., 1952, A Theory of Sedimentation, *Trans. Faraday Soc.* **48:** 166-176.

Lackme, C., 1973, Two Regimes of a Spray Column in Counter-current Flow, *AIChE Symp. Heat Transfer R. D.* **70:** 59-63.

Lahey Jr., R. T., Cheng, L. Y., Drew, D. A. and Flaherty, J. E., 1978, The Effect of Virtual Mass on the Numerical Stability of Accelerating Two-phase Flows, *AIChE 71st Annual Meeting*, Miami Beach, Florida.

Lahey Jr, R. T., Lopez de Bertodano, M. and Jones Jr., O. C., 1993, Phase Distribution in Complex Geometry Conduits, *Nucl. Eng. Des.* **141:** 117-201.

Lamb, H., 1945, *Hydrodynamics*, Dover, New York.

Landau, L. D., 1941, Theory of Super Fluidity of Helium II, *Physical Review* **60:** 166-176.

Landel, R. F., Moser, B. G. and Bauman, A. J., 1965, Rheology of Concentrated Suspensions: Effects of a Surfactant, *Proc. 4th Int. Congress on Rheology*, Brown University, Part 2, pp.663.

Letan, R. and Kehat, E., 1967, Mechanics of a Spray Column, *AIChE J.* **13:** 443-449.

Levich, V. G., 1962, *Physicochemical Hydrodynamics*, Prentice-Hall.

Levy, S., 1960, Steam Slip-theoretical Prediction from Momentum Model, *J. Heat Transfer* **82:** 113-124.

Lighthill, M. J. and Whitham, G. B., 1955, On the Kinematic Waves I. Flood Movement in Long Rivers, *Proc. Royal Soc. London* **229**: 281-316.

Liu, T. J., 1993, Bubble Size and Entrance Length Effects on Void Development in a Vertical Channel, *Int. J. Multiphase Flow* **19:** 99-113.

Loeb, L. B., 1927, *The Kinetic Theory of Gases*, Dover, New York.

Lopez de Bertodano, M., Lahey Jr., R. T. and Jones, O. C., 1994, Development of a k-ε Model for Bubbly Two-phase Flow, *J. Fluids Eng.* **116:** 128-134.

Loth, E., Taeibi-Rahni, M. and Tryggvason, G., 1997, Deformable Bubbles in a Free Shear Layer, *Int. J. Multiphase Flow* **23:** 977-1001.

Lumley, J., 1970, Toward a Turbulent Constitutive Relation, *J. Fluid Mech.* **41:** 413-434.

Marchaterre, J. F., 1956, The Effect of Pressure on Boiling Density in Multiple Rectangular Channels, *Argonne National Lab. Report*, ANL-5522.

Martinelli, R. C. and Nelson, D. B., 1948, Prediction of Pressure Drop during Forced Circulation Boiling of Water, *Trans. ASME* **70:** 695-702.

Maurer, G. W., 1956, A Method of Predicting Steady State Boiling Vapour Fractions in Reactor Coolant Channels, WAPD-BT-19.

Maxwell, J., 1867, On the Dynamical Theory of Gases, *Phil. Trans. Roy. Soc. London* **157:** 49-88.

McConnell, A. S., 1957, *Application of Tensor Analysis*, Dover.

Mei, R. and Klausner, J. F., 1994, Shear Lift Force on Spherical Bubbles, *Int. J. Heat Fluid Flow* **15:** 62-65.

Meyer, J. E., 1960, Conservation Laws in One-dimensional Hydrodynamics, *Bettis Technical Review* **61:** WAPD-BT-20.

Miles, J. W., 1957, On the Generation of Surface Waves by Shear Flows, I-IV, *J. Fluid Mech.* **3:** 185-204 (1957); **6:** 568-582 (1959); **6:** 583-598 (1959); **13:** 433-448 (1962).

Mishima, K. and Ishii, M., 1980, Theoretical Prediction of Onset of Slug Flow, *J. Fluid Eng.* **102:** 441-445.

Mishima, K. and Ishii, M., 1984, Flow Regime Transition Criteria for Upward Two-phase Flow in Vertical Tubes, *Int. J. Heat Mass Transfer* **27:** 723-737.

Miller, J. W., 1993, An Experimental Analysis of Large Spherical Cap Bubbles Rising in an Extended Liquid, *M. S. Thesis*, Purdue University.

Mokeyev, Yu. G., 1977, Effect of Particle Concentration on their Drag and Induced Mass, *Fluid Mech.-Soviet Res.* **6:** 161-168.

Muller, I., 1968, A Thermodynamics Theory of Mixtures of Fluids, *Arch. Rational Mech. Anal.* **28:** 1-39.

Murray, S. O., 1954, On the Mathematics of Fluidization I, Fundamental Equations and Wave Propagation, *J. Fluid Mech.* **21:** 465-493.

Neal, L. G., 1963, An Analysis of Slip in Gas-liquid Flow Applicable to the Bubble and Slug Flow Regimes, *Kjeller Research Establishment Report, Norway* KR-2.

Nicklin, D. J., Wilkes, J. O. and Davidson, J. F., 1962, Two-phase Flow in Vertical Tubes, *Trans. Inst. Chem. Eng.* **40:** 61-68.

Nikuradse, J., 1932, Gesetzmäsigkeit der Turbulenten Strömung in Glatten Rohre, *Forsch. Arb. Ing.-Wes.* pp.356.

Oolman, T. O. and Blanch, H. W., 1986a, Bubble Coalescence in Air-Sparged Bioreactors, *Biotech. Bioeng.* **28:** 578-584.

Oolman, T. O. and Blanch, H. W., 1986b, Bubble Coalescence in Stagnant Liquids, *Chem. Eng. Commun.* **43:** 237-261.

Otake, T., Tone, S., Nakao, K., and Mitsuhashi, Y., 1977, Coalescence and Breakup of Bubbles in Liquids, *Chem. Eng. Sci.* **32:** 377-383.

Pai, S. I., 1962, *Magnetogasdynamics and Plasma Dynamics*, Wien Springer-Verlag.

Pai, S. I., 1971, Fundamental Equations of a Mixture of Gas and Small Spherical Solid Particles from Simple Kinetic Theory, *Int. Sym. on Two-phase Systems*, Paper 6-6, Haifa, Israel.

Pai, S. I., 1972, A New Classification of Two-phase Flows, *J. Mach. Phys. Sci.* **6:** 137-161.

Panton, R., 1968, Flow Properties for the Continuum View-point of a Non-Equilibrium Gas-Particle Mixture, *J. Fluid Mech.* **31:** 273-303.

Peebles, F. N. and Garber, H. J., 1953, Studies on the Motion of Gas Bubbles in Liquid, *Chem. Eng. Prog.* **49:** 88-97.

Petrick, M., 1962, A Study of Vapor Carryunder and Associated Problems, *Argonne National Lab. Report*, ANL-65-81.

Phillips, M. C. and Riddiford, A. C., 1972, Dynamic Contact Angles 2. Velocity and Relaxation Effects for Various Liquids, *J. Colloid Interface Sci.* **41**: 77-85.

Pierre, C. C. St., 1965, Frequency-response Analysis of Steam Voids to Sinusoidal Power Modulation in a Thin-walled Boiling Water Coolant Channel, *Argonne National Lab. Report*, ANL-7041.

Prigogine, I. and Mazur, P., 1951, On Two-phase Hydrodynamic Formulations and the Problem of Liquid Helium II, *Physica* **17:** 661-679.

Prince, M. J. and Blanch, H. W., 1990, Bubble Coalescence and Break-up in Air-Sparged Bubble Columns, *AIChE J.* **36:** 1485-1497.

Prosperetti, A., 1999, Some Considerations on the Modeling of Disperse Multiphase Flows by Averaged Equations, *JSME Intl. J.*, Ser. B, **42:** 573-585.

Réocreux, M., Barriere, G. and Vernay, B., 1973, Etude Expérimentale des Débits Critiques en Écoulement Diphasique Eau-vapeur à Faible Titre sur un Canal à Divergent de 7 Degrés, *Cen. G, Rapport* TT 115.

Réocreux, M., 1974, Contribution à l'étude des Débits Critiques en Éulement Diphasique eau Vapeur, *Ph. D. Thesis*, University of Grenoble.

Richardson, J. F. and Zaki, W. N., 1954, Sedimentation and Fluidization: Part 1, *Trans. Inst. Chem. Eng.* **32:** 35-53.

Roscoe, R., 1952, The Viscosity of Suspensions of Rigid Spheres, *Br. J. Appl. Phys.* **3:** 267-269.

Rose, S. C. and Griffith, P., 1965, Flow Properties of Bubbly Mixtures, *ASME Paper* 65-HT-8.

Rotta, J. C., 1972, *Turbulence Stromungen*, B. G. Teubner, Stuttgart, Germany.

Rouhani, S. Z. and Becker, K. M., 1963, Measurement of Void Fractions for Flow of Boiling Heavy Water in a Vertical Round Duct, AE-106, Aktiebolaget Atomenergi, Sweden.

Rybczynski, W., 1911, Über die Fortschreitende Bewegung einer Flüssigen Kugel in einem Zähen Medium, *Bull. Int. Acad. Sci. Cracov.* **1911A:** 40-46.

Saffman, P. G., 1965, The Lift on a Small Sphere in a Slow Shear Flow, *J. Fluid Mech.* **22:** 385-400.

Sato, Y., Sadatomi, M. and Sekoguchi, K., 1981, Momentum and Heat Transfer in Two-phase Bubble Flow - 1. Theory, *Int. J. Multiphase Flow* **7:** 167-177.

Schlichting, H., 1979, *Boundary Layer Theory*, McGraw-Hill Book Co.

Schwartz, K., 1954, Investigation of Distribution of Density, Water and Steam Velocity and of the Pressure Drop in Vertical Horizontal Tubes, *VDI Forschungsh.* 20, Series B, 445.

Schwartz, A. M. and Tejada, S. B., 1972, Studies of Dynamic Contact Angles on Solids, *J. Colloid Interface Sci.* **38:** 359-375.

Scriven, L. E., 1960, Dynamics of Fluid Interface, Equation of Motion for Newtonian Surface Fluids, *Chem. Eng. Sci.* **2:** 98-108.

Serizawa, A., Kataoka, I. and Michiyoshi, I., 1975, Turbulence Structure of Air-water Bubbly Flow, I, II and III, Int. J. Multiphase Flow **2:** 221-259.

Serizawa, A. and Kataoka, I., 1988, Phase Distribution in Two-phase Flow, *Transient Phenomena in Multiphase Flow*, Hemisphere, Washington DC, pp.179-224.

Serizawa, A. and Kataoka, I., 1994, Dispersed Flow-I., *Multiphase Science and Technology*, Vol. 8, Begell House Inc., New York, pp.125-194.

Serrin, J., 1959, *Handbuch der Physik*, Vol.8/I, Springer-Verlag.

Sevik, M. and Park, S. H., 1973, The Splitting of Drops and Bubbles by Turbulent Fluid Flow, *J. Fluids Eng.* **95:** 53-60.

Slattery, J. C., 1964, Surface – I. Momentum and Moment-of-momentum Balance for Moving Surfaces, *Chem. Eng. Sci.* **19:** 379-385.

Slattery, J. C., 1972, *Momentum, Energy and Mass Transfer in Continua*, McGraw-Hill Book Co.

Smissaert, G. E., 1963, Two-component Two-phase Flow Parameters for Low Circulation Rages, *Argonne National Lab. Report*, ANL-67-55.

Soo, S. L., 1967, *Fluid Dynamics of Multiphase Systems*, Ginn Blaisdell.

Sridhar, G. and Katz, J., 1995, Drag and Lift Forces on Microscopic Bubbles Entrained by a Vortex, *Phys. Fluids* **7:** 389-399.

Standart, G., 1964, The Mass, Momentum and Energy Equations for Heterogeneous Flow Systems, *Chem. Eng. Sci.* **19:** 227-236.

Standart, G., 1968, The Second Law of Thermodynamics for Heterogeneous Flow Systems III. Effect of Conditions of Mechanical Equilibrium and Electroneutrally on Simulataneous Heat and Mass Transfer and Prigogine Theorem, *Chem. Eng. Sci.* **23:** 279-285.

Stefan, J., 1871, Uber das Gleichgewicht und die Bewegung, Insbesondere die Diffusion von Gasmengen, *Sitzgsber, Akad. Wiss. Wien* **63:** 63-124.

Stewart, C. W., 1995, Bubble Interaction in Low-viscosity Liquids, *Int. J. Multiphase Flow* **21:** 1037-1046.

Stokes, G. G., 1851, On the Effect of Internal Friction of Fluids on the Motion of Pendulums, *Trans. Cambr. Phil. Soc.* **9, Part II:** 8-106 or Coll. Papers **III:** 55.

St. Pierre, C. C., 1965, Frequency-response Analysis of Steam Voids to Sinusoidal Power Modulation in a Thin-walled Boiling Water Coolant Channel, *Argonne National Lab. Report*, ANL-7041.

Sun, X., Ishii, M. and Kelly, J. M., 2003, Modified Two-fluid Model for the Two-group Interfacial Area Transport Equation, *Annals Nucl. Energy* **30:** 1601-1622.

Sun, X., Kim, S., Ishii, M. and Beus, S. G., 2004a, Modeling of Bubble Coalescence and Disintegration in Confined Upward Two-phase Flow, *Nucl. Eng. Des.* **230:** 3-26.

Sun, X., Kim, S., Ishii, M. and Beus, S. G., 2004b, Model Evaluation of Two-group Interfacial Area Transport Equation for Confined Upward Flow, *Nucl. Eng. Des.* **230:** 27-47.

Taylor, G. I., 1932, The Viscosity of a Fluid Containing Small Drops of Another Fluid, *Proc. R. Soc.* **A138:** 41-48.

Taylor, G. I., 1934, The Formation of Emulsions in Definable Fields of Flow, *Proc. Royal Soc. London* **A146:** 501-523.

Teletov, S. G., 1945, Fluid Dynamic Equations for Two-phase Fluids, *Soviet Physics Doklady, Akademii Nauk SSSR* **50:** 99-102.

Teletov, S. G., 1957, On the Problem of Fluid Dynamics of Two-phase Mixtures, I. Hydrodynamic and Energy Equations, *Bull. the Moscow University* **2:** 15.

Theofanous, T. G., Li, G. J. and Dinh, T. N., 2004, Aerobreakup in Rarefied Supersonic Gas Flows, *J. Fluid Eng.* **126:** 516-527.

Thomas, D. G., 1965, Transport Characteristics of Suspension: VIII A Note on Viscosity of Newtonian Suspensions of Uniform Spherical Particles, *J. Colloid Sci.* **20:** 267-277.

Thome, R. J., 1964, Effect of a Transverse Magnetic Field and Vertical Two-phase Flow through a Rectangular Channel, *Argonne National Lab. Report*, ANL-6854.

Tomiyama, A., Zun, I., Sou, A. and Sakaguchi, T., 1993, Numerical Analysis of Bubble Motion with the VOF method, *Nucl. Eng. Des.* **141:** 69-82.

Tomiyama, A., Sou, A., Zun, I., Kanami, N. and Sakaguchi, T., 1995, Effects of Eötvös Number and Dimensionless Liquid Volumetric Flux on Lateral Motion of a Bubble in a Laminar Duct Flow, *Advances in Multiphase Flow*, Elsevier, pp.3-15.

Tomiyama, A., Tamai, H., Zun, I. and Hosokawa, S., 2002, Transverse Migration of Single Bubbles in Simple Shear Flows, *Chem. Eng. Sci.* **57:** 1849-1858.

Tong, L. S., 1965, *Boiling Heat Transfer and Two-phase Flow*, John Wiley and Sons, Inc., New York.

Truesdell, C. and Toupin, R., 1960, The Classical Field Theories, *Handbuch der Physic*, Vol.3/I, Springer-Verlag.

Truesdell, C., 1969, *Rational Thermodynamics*, McGraw-Hill Book Co.

Tsouris, C. and Tavlarides, L. L., 1994, Breakage and Coalescence Models for Drops in Turbulent Dispersions, *AIChE J.* **40:** 395-406.

Tsuchiya, K., Miyahara, T. and Fan, L. S., 1989, Visualization of Bubble-wake Interactions for a Stream of Bubbles in a Two-dimensional Liquid-solid Fluidized Bed, *Int. J. Multiphase Flow* **15:** 35-49.

Vernier, P. and Delhaye, J. M., 1968, General Two-phase Flow Equations Applied to the Thermohydrodynamics of Boiling Nuclear Reactors, *Energie Primaire* **4:** No.1.

Von Karman, 1950, Unpublished Lectures (1950-1951) at Sorbonne and Published by Nachbar et al. in *Quart. Appl. Math.* **7:** 43 (1959).

Wallis, G. B., Steen, D. A., Brenner, S. N. and Turner T. M., 1964, Joint U. S. – Euratom Research and Development Program, *Quarterly Progress Report*, January, Dartmouth College.

Wallis, G. B., 1969, *One-dimensional Two-phase Flow*, McGraw-Hill Book Co.

Wallis, G. B., 1974, The Terminal Speed of Single Drops or Bubbles in an Infinite Medium, *Int. J. Multiphase Flow* **1:** 491-511.

Wang, S. K., Lee, S. J., Jones Jr., O. C. and Lahey Jr. R. T., 1987, 3-D Turbulence Structure and Phase Distribution Measurements in Bubbly Two-phase Flows, *Int. J. Heat Mass Transfer* **13:** 327-343.

Weatherburn, C. E., 1927, *Differential Geometry of Three Dimensions*, Cambridge University Press.

Werther, J., 1974, Influence of the Bed Diameter on the Hydrodynamics of Gas Fluidized Beds, *AIChE Symp. Ser.* No. 141 70: 53.

Whitaker, S., 1968, *Introduction to Fluid Mechanics*, Prentice-Hall, Inc.

White, E. T. and Beardmore, R. H., 1962, The Velocity of Rise of Single Cylindrical Air Bubbles through Liquid Contained in Vertical Tubes, *Chem. Eng. Sci.* **17:** 351-361.

Wu, Q. and Ishii, M., 1996, Interfacial Wave Instability of Co-current Two-phase Flow in Horizontal Channel, *Int. J. Heat Mass Transfer* **39:** 2067-2075.

Wu, Q., Kim, S., Ishii, M. and Beus, S. G., 1998, One-group Interfacial Area Transport in Vertical Bubbly Flow, *Int. J. Heat Mass Transfer* **41:** 1103-1112.

Wundt, H., 1967, Basic Relationships in n-components Diabatic Flow, *EUR 3459e*.

Yoshida, F. and Akita, K., 1965, Performance of Gas Bubble Columns: Volumetric Liquid-phase Mass Transfer Coefficient and Gas Holdup, *AIChE J.* **11:** 9-13.

Zhang, D. Z., 1993, Ensemble Phase Averaged Equations for Multiphase Flows, *Ph D. Thesis*, Johns Hopkins University.

Zhang, D. Z., Prosperetti, A., 1994a, Averaged Equations for Inviscid Disperse Two-phase Flow, *J. Fluid Mech.* **267:** 185-219.

Zhang, D. Z., Prosperetti, A., 1994b, Ensemble Phase-averaged Equations for Bubbly Flows, *Phys. Fluids* **6:** 2956-2970.

Zuber, N., 1964a, On the Dispersed Flow in the Laminar Flow Regime, *Chem. Eng. Sci.* **19:** 897-917.

Zuber, N., 1964b, On the Problem of Hydrodynamic Diffusion in Two-phase Flow Media, *Proc. 2nd All Union Conference on Heat and Mass Transfer*, Minsk, USSR, **3:** 351.

Zuber, N., Staub, F. W. and Bijwaard, G., 1964, Steady state and Transient Void Fraction in Two-phase Flow Systems, Vol.1, *GEAP 5417*.

Zuber, N. and Findlay, J. A., 1965, Average Volumetric Concentration in Two-phase Flow Systems, *J. Heat Transfer* **87:** 453-468.

Zuber, N. and Staub, F. W., 1966, Propagation and the Wave Form of the Volumetric Concentration in Boiling Water Forced Convection Systems under Oscillatory Conditions, *Int. J. Heat Mass Transfer* **9:** 871-895.

Zuber, N., 1967, Flow Excursions and Oscillations in Boiling, Two-phase Flow Systems with Heat Addition, *Proc. Symp. Two-phase Flow Dynamics*, **1:** 1071.

Zuber, N. and Dougherty, D. E., 1967, Liquid Metals Challenge to the Traditional Methods in Two-phase Flow Investigations, *Proc. EURATOM Symposium on Two-phase Flow Dynamics*, pp.1085.

Zuber, N., Staub, F. W., Bijwaard, G. and Kroeger, P. G., 1967, Steady State and Transient Void Fraction in Two-phase Flow Systems, *General Electric Co.* Report GEAP-5417, vol.1

Zuber, N., 1971, Personal Communication at Georgia Institute of Technology.

Zun, I., 1988, Transition from Wall Void Peaking to Core Void Peaking in Turbulent Bubbly Flow, *Transient Phenomena in Multiphase Flow*, Hemisphere, Washington DC, pp.225-245.

Nomenclature

Latin

A	surface of a volume
A	frontal area of bubble
$A^{\alpha\beta}$	surface metric tensor (Aris, 1962)
$\mathbb{A}$	turbulence anisotropy tensor
A_d	projected area of a typical particle
A_i	mathematical surface between A_1 and A_2
A_i	surface area
A_k	surface bounding the interfacial region and adjacent to phase *k*
A_m	surface of fixed mass volume
A_p	projected area of a particle
a	cross sectional radius of cap or slug bubble
a_c^i	mobility of the fluid at the interface
a_i	interfacial area concentration
a_{sk}, a_{tk}	isentropic and isothermal sound velocities based on the average thermodynamic properties
B_d	volume of a typical particle
B_S	balance at an interface
B_V	balance in each phase
b_k^{Γ}, b_k^M, b_k^E	Transport coefficients associated with interfacial

	transfer of mass, momentum and energy
C	wave velocity
C	constant
C_D	drag coefficient
$C_{D\infty}$	ideal drag coefficient
C_g	variable defined by $\sqrt{2g\Delta\rho/\rho_f}$
C_{hk}	distribution parameter
C_{hm}	mixture-enthalpy-distribution parameter
C_i	closed curve on an interface
$\boldsymbol{C}_K$	kinematic wave velocity
C_{LW}	coefficient of lift force caused by slanted wake
C_M	virtual mass constant
C_T	adjustable parameter
C_{vk}	distribution parameter
C_{vm}	virtual volume coefficient
C_{vm}	mixture-momentum-distribution parameter
C_τ	distribution parameter
$C_{\psi k}$	distribution parameter for flux
C^i	shape factor
C_0	distribution parameter
C_∞	propagation velocity
C_∞	asymptotic value of distribution parameter
c_k	mass concentration of phase k
c_{pk}, c_{vk}	specific heat at constant pressure and density based on averaged properties
D	hydraulic-equivalent diameter
D^*	length scale ratio
D_b	bubble diameter
D_{bc}	critical bubble size
D^*_{c1}	ratio of D_{crit} to D_{Sm1}
$D_{c,max}$	maximum diameter of stable bubble
D_{crit}	volume-equivalent diameter of a bubble at boundary between groups 1 and 2
$D_{d,max}$	maximum distorted bubble limit

D_d^*	ratio of bubble diameter to bubble diameter at distorted bubble limit
D_e	volume-equivalent diameter of a fluid particle
D_e	eddy diameter
D_E	effective diameter of mixture volume that contains one bubble
D_H	hydraulic-equivalent diameter
D_H^*	non-dimensional hydraulic-equivalent diameter
D_k	diffusion coefficient
$\mathbb{D}_k$	total deformation tensor of phase *k*
$\mathbb{D}_{kb}$	bulk deformation tensor
$\mathbb{D}_{ki}$	interfacial extra deformation tensor
D_k^α	drift coefficient
D_{Sm}	Sauter mean diameter
D_s	surface-equivalent diameter of a fluid particle
d_B	bubble diameter
$\widehat{d_B}$	cross-sectional mean diameter of bubbles
E_B	average energy required for bubble breakup
E_d	area fraction of liquid entrained in gas core from total liquid area at any cross section
E_e	average energy of a single eddy
E_k	total energy gain through interfaces for phase *k*
E_m	mixture total energy source from interfaces
E_m^H	mixture energy gain due to changes in mean curvature
Eo	Eötvös number
$\widehat{e_k}$, $\widehat{e_{ki}}$	weighted mean virtual internal energy (with turbulent kinetic energy included) at the bulk phase and at the interfaces
$F(\boldsymbol{x},t)$	general function
$\boldsymbol{F}^B$	Basset force
$\boldsymbol{F}^D$	standard drag force
$\boldsymbol{F}^L$	lift force
$\boldsymbol{F}^T$	turbulent dispersion force
$\boldsymbol{F}^V$	virtual mass force

$\boldsymbol{F}^W$	wall lift force
F_D	drag force
F_k, $\mathbb{F}_k$	general function associated with phase k
$f(\boldsymbol{x},t)$	function for interface position
$f(\boldsymbol{x},t,\boldsymbol{\xi})$	molecular density function
f	collision frequency
f	friction factor
f^*	correction factor for drag coefficient
f_i	interfacial friction factor
f_k	Helmholtz potential
$f_{kn}(\boldsymbol{x},t,\boldsymbol{\xi})$	particle density function of the nth-kind particles
f_{TW}	two-phase friction factor
G	mass velocity
G	cap bubble thickness
G_s	non-dimensional velocity gradient
$\boldsymbol{g}$	gravity field
$\boldsymbol{g}_k$	body force field
g_k, $\widehat{g_k}$, $\widehat{g_{ki}}$	Gibbs free energy: local instant, bulk mean and interfacial mean values
g_{ln}	space metric tensor (Aris, 1962)
g_N	normal gravitational acceleration
H_{21}, $\overline{\overline{H_{21}}}$	local instant and averaged mean curvature ($\overline{\overline{H_{21}}} > 0$ if phase 2 is the dispersed phase)
h	bubble height
h_1, h_2	average thickness of upper (1) and lower (2) fluid layers
$\widehat{h_k}$, $\widehat{h_{ki}}$	weighted mean virtual enthalpy (with turbulent kinetic energy included) at the bulk phase and at the interfaces
h_m	mixture virtual enthalpy
$\mathbb{I}$	unit tensor
I_k	interfacial source term in the balance equations for phase *k*
I_m	interfacial source term for mixture balance equations
I_{ka}, I_{ma}	interfacial source terms in the shock conditions for

	phase *k* and for mixture
$i_k, \widehat{i_k}$	local instant and mean enthalpies
$\widehat{i_{ki}}$	mean enthalpy of phase k at interfaces
i_m	mixture enthalpy
i_a	local instant surface enthalpy per area
$\mathcal{J}$	flux
$\mathcal{J}^D$	drift flux
$\mathcal{J}_a$	line flux for interface
$\mathcal{J}_k$	surface flux for phase *k*
$\mathcal{J}_k^T, \mathcal{J}^T$	turbulent fluxes
J_k, J_m	Jacobians based on macroscopic field
$\boldsymbol{j}_k, \boldsymbol{j}$	volumetric fluxes of phase k and mixture
j^*	non-dimensional mixture volumetric flux
j^+	non-dimensional mixture volumetric flux
K	constant
K_k	thermal conductivity
$\mathbb{K}_k$	thermal conductivity tensor
K_k^T	turbulent conductivity
K_k^{T*}	thermal mixing length coefficient
k	wave number
k^{SI}	turbulent kinetic energy due to shear-induce turublence
k_e	wave number of eddy
L	pitch of slug unit
L_b	cylindrical bubble length
$1/L_j$	area concentration of jth-interface
$1/L_s$	total area concentration
L_T	mean traveling distance between two bubbles for one collision
L_W	effective wake length
l	mixing length
l_B	mixing length due to bubble-induced turbulence
l_{SP}	mixing length of single-phase flow
l_{TP}	mixing length of two-phase flow
m_e	mass per a single eddy

$\dot{m}_k, \overline{\overline{\dot{m}_k}}$	local instant and mean mass transfer rates per unit area (mass loss)
M	Morton number
M_F	frictional pressure gradient in multi-particle system
$M_{F\infty}$	frictional pressure gradient in single particle system
$\boldsymbol{M}_{ik}$	generalized interfacial drag
M_k, M_s	state density functions for phase k and interface
$\boldsymbol{M}_k, \boldsymbol{M}_m$	momentum sources for phase k and mixture
$\boldsymbol{M}_m^H$	force due to changes in mean curvature
$\boldsymbol{M}_k^n, \boldsymbol{M}_k^t, \boldsymbol{M}_k^d$	form, skin and total drag forces
$\boldsymbol{M}_{\tau m}$	force associated with mixture transverse stress gradient
$\boldsymbol{N}$	unit normal vector to a curve on an interface
N	number of samples
N_b	number of bubbles
N_D	drift number
N_{drag}	drag number
N_e	number of eddies of wave number k_e per volume of fluid
N_{Ec}	Eckert number
N_{Eu}	Euler number
N_{Fr}	Froude number
N_i	converted enthalpy ratio
N_{Jk}	Jakob number
N_{pch}	phase change number
N_{pch}^i	interfacial phase change effect number
N_{Pe}	Peclet number
$N_{Pr\,k}^T$	turbulent Prandtl number
N_q	interface heating number
N_{Re}	Reynolds number
N_{Re}^i	interfacial Reynolds number
N_{Sl}	Strouhal number
N_{We}	Weber number
N_σ	surface tension number

N_{Pr}	Prandtl number
N_W	number of bubbles inside effective volume
N_μ	viscosity number
N_ρ	density ratio
n	fluid particle number per unit mixture volume
$\boldsymbol{n}$	unit normal vector
n_b	bubble number density
n_e	number of eddies of wave number per volume of two-phase mixture
$\boldsymbol{n}_k$	outward unit normal vector for phase k
P^{SI}	production of shear-induced turbulence
P_C	probability for a bubble to move toward neighboring bubble
$\overline{\mathbb{P}_k}$	partial pressure tensor
P_i	interfacial wetted perimeter
P_{wf}	wall wetted perimeter
p	pressure
p_c	critical pressure
$p_k, \overline{\overline{p_k}}, \overline{\overline{p_{ki}}}$	partial, bulk mean and interfacial mean pressure
p_m	mixture pressure
$\boldsymbol{q}$	heat flux
$\boldsymbol{q}^D$	diffusion (drift) heat flux
$\overline{\overline{\boldsymbol{q}_k}}, \boldsymbol{q}_k^T$	mean conduction and turbulent heat fluxes
$\overline{\boldsymbol{q}}, \boldsymbol{q}_k^T$	mixture conduction and turbulent heat fluxes
$\dot{q}_k$	local instant body heating
$\overline{\overline{q_k''}}$	average heat transfer pert interfacial area (energy gain)
$\boldsymbol{q}_k^C$	mean conduction heat flux
R	ideal gas constant
R	radius of a pipe
R	radius of curvature
R^+	variable defined by Rv_f^*/ν_f
$\overline{\overline{R_d}}$	mean radius of fluid particles
R_j	particle number source and sink rate

R_w	tube radius
Re	Reynolds number
$(Re)_d$	particle Reynolds number
r	radial coordinate
r_d^*	non-dimensional radius
S_B, S_C	surface available to collision
S_j	particle source and sink rates per unit mixture volume due to j-th particle interactions such as disintegration or coalescence
S_{ph}	particle source and sink rates per unit mixture volume due to phase change
s	entropy
s_a	surface entropy per area
$\widehat{s_k}$, $\widehat{s_{ki}}$	weighted mean entropy at bulk phase and at interfaces
s_m	mixture entropy
T	temperature
$T_i, \overline{\overline{T_i}}$	instant and mean interface temperature
$\overline{\overline{T_k}}$, $\overline{\overline{T_{ki}}}$	mean temperature at bulk phase and at interface
$\mathcal{T}_k$	stress tensor
t	time
t_C	time required for bubble coalescence
t_j	time when the jth-interface passes the point
t_α^m (or t_α)	hybrid tensor of interface, see Aris (1962)
$\boldsymbol{U}$	velocity of shock in mixture
U_0	velocity of stream
U_B, U_C	volume available to collision
u	internal energy
u_a	surface energy per area
u_b	mean fluctuation velocity
u_B, u_C	bubble velocity
u_e	eddy velocity
$\widehat{u_k}$, $\widehat{u_{ki}}$	weighted mean internal energy at bulk phase and at interfaces
u_m	mixture internal energy

u_{rW}	averaged relative velocity between leading bubble and bubble in wake region
u_t	root-mean-square approaching velocity of two bubbles
$u_{t,crit}$	critical fluctuation velocity
V	volume
$\dot{V}$	time derivative of volume V
V_c	critical bubble volume
V_{gj}^{+}	non-dimensional drift velocity
V_i	interfacial region
$\boldsymbol{V}_{kj}$	drift velocity
$\boldsymbol{V}_{km}$	diffusion velocity
V_m	fixed mass volume
V_s^*	ratio of $V_{s,min}$ to $V_{s,max}$
V_W	effective wake volume
V_{1p}	peak bubble volume in group 1
$\boldsymbol{v}$	velocity
v_f'	liquid velocity fluctuation independent of bubble agitation
v_f''	liquid velocity fluctuation dependent on bubble agitation
v_f^*	friction velocity
$\boldsymbol{v}_g$	average center-of-volume velocity of dispersed phase
$\boldsymbol{v}_i$	velocity of interface
$\widehat{\boldsymbol{v}_k}$, $\widehat{\boldsymbol{v}_{ki}}$	weighted mean velocity at bulk phase and at interfaces
$\widehat{\left(v_k'\right)^2}/2$	mean turbulent kinetic energy
$\boldsymbol{v}_m$	mixture center of mass velocity
$\boldsymbol{v}_{pm}$	average local particle velocity weighted by particle number
$\boldsymbol{v}_r$	relative velocity
$\overline{v_r}$	difference between area averaged mean velocities of phases
$v_{r\infty}$	relative velocity of a single particle in an infinite medium

$\boldsymbol{v}_s$	velocity of interfacial particles
W_{ki}^T	work due to fluctuations in drag forces
We	Weber number
We_{crit}	critical Weber number
$\boldsymbol{X}$	convective coordinates
$\boldsymbol{x}$	spatial coordinates
x	spatial coordinate
y	spatial coordinate
y^+	variable defined by yv_f^* / ν_f
z	spatial coordinate

Greek

α_b	void fraction in slug bubble section
α_{core}	ratio of liquid-film cross-sectional area to total cross-sectional area
α_d	average overall void fraction
α_{drop}	ratio of cross-sectional area of drops to cross-sectional area of core
$\alpha_{g,crit}$	critical void fraction when center bubble cannot pass through free space among neighboring bubbles
$\alpha_{g,max}$	maximum void fraction
α_k	time (void) fraction of phase *k*
β	ratio of mixing length and width of wake
β_C	variable to take account of overlap of excluded volume
β_k	thermal expansivity based on averaged properties
Γ	constant
Γ_k	mass generation for phase *k*
γ	constant
γ_k	ratio of specific heats
Δ_a	interfacial entropy generation per area
Δ_k	entropy generation for phase *k*
$\Delta\dot{m}_{12}$	inter-group mass transfer rates from group 1 to group 2
Δt	time interval of averaging
Δt_B	time interval to drive daughter bubble apart with

	characteristic length of D_b
Δt_C	time interval for one collision
Δt_k, Δt_s	time intervals associated with phase *k* and interfaces
Δt_W	average time interval for a bubble in wake region to catch up with preceding bubble
δ	thickness of interface
δ	film thickness
δ'	collective parameter
δ_{crit}	critical film thickness where rapture occurs
δ_{init}	initial film thickness
δp_k	pressure deviation from saturation pressure
$\delta\mu$	volume element in μ space
ε	energy dissipation rate per unit mass
2ε (or $2\varepsilon_j$)	time associated with the jth-interface
ε^{SI}	dissipation of shear-induced turbulence
ε', ε''	eddy diffusivity
η_{ph}	rate of volume generated by nucleation source per unit mixture volume
η_0	amplitude
Θ	contact angle
θ	angle in cylindrical coordinates
θ_w	wake angle
κ_{fr}	variable defined by $1-exp\left(-C_{fr}V_s^{*1/2}/D^{1/2}\right)$
κ_{Sk}, κ_{Tk}	isentropic and isothermal compressibilities of phase *k*
Λ_k	interfacial thermal energy transfer term in the averaged equation
λ	wavelength
λ	constant
λ_B	breakup efficiency
λ_C	coalescence efficiency
λ_c	critical wavelength
λ_k	bulk viscosity
μ	viscosity
$\overline{\overline{\mu_k}}$, μ_k^T	mean molecular and turbulent viscosities

μ_k^{T*}	mixing length coefficient
μ_m	mixture viscosity
ν	kinematic viscosity
ν_t	turbulent kinematic viscosity
$\boldsymbol{\xi}$	particle (phase) velocity in Boltzmann statistical average
ξ	ratio of V_{1p} to V_c
ξ	variable defined by $2\left(1-0.2894D_{c1}^{*\,3}\right)^2$
ξ	variable defined by P_i/P_{wf}
ρ	density
ρ_a	surface mass per area
$\overline{\rho_k}$, $\overline{\overline{\rho_k}}$	partial and mean densities
ρ_k'	modified density defined by $\rho_k \coth\left(kh_k\right)$
ρ_m	mixture density
σ	surface tension
$\mathcal{T}$	viscous stress tensor
$\mathcal{T}^D$	diffusion (or drift) stress tensor
$\mathcal{T}_f^{BI}$	bubble-induced turbulent stress tensor
$\mathcal{T}_f^{SI}$	shear-induced turbulent stress tensor
$\overline{\mathcal{T}}$, $\mathcal{T}^T$	mixture viscous and turbulent stress tensors
$\overline{\overline{\mathcal{T}_k}}$, $\mathcal{T}_k^T$	average viscous and turbulent stress tensor
$\mathcal{T}_k^\mu$	average viscous stress
$\overline{\overline{\mathcal{T}_{ki}}}$, $\mathcal{T}_{ki}$	interfacial shear stress
τ_C	contact time for two bubbles
τ_i	interfacial shear stress
τ_o	reference time constant
$\boldsymbol{\tau}_{tk}$, $\boldsymbol{\tau}_{nk}$	tangential and normal stresses at interface
τ_{wf}	wall shear
Φ	velocity potential
Φ_k^T	turbulent work effect in enthalpy energy equation
Φ_m^i	interfacial mechanical energy exchange effect in the mixture thermal energy equation
Φ_k^μ	viscous dissipation

Φ_m^μ	mixture viscous dissipation
Φ_m^σ	surface tension effect in the mixture thermal energy equation
ϕ	source term
ϕ_a	interfacial source per area
ϕ_j	source and sink rate for interfacial area concentration
ϕ_k	velocity potential
χ	coefficient accounting for contribution from inter-group transfer
ψ	property of extensive characteristics
ψ	shape factor
$\widehat{\psi}, \widehat{\psi_k}$	mass weighted mean values for mixture and phase k
ψ_a	property per interfacial area
Ω	potential function

Subscripts and Superscripts

a	surface (property per area)
c	continuous phase
d	dispersed phase
f	liquid phase
g	vapor phase
i	interface
j	j^{th}-interface
k	each phase : (k=1 & 2), (k=c & d), (k=f & g)
ki	k^{th}-phase at interfaces
m	$\begin{cases}\text{mixture (in macroscopic formulation)}\\ \text{fixed mass (in local instant formulation)}\end{cases}$
n	normal to interface
o	reference
RC	random collision
r, θ, z	cylindrical coordinate
sat	saturation
s	$\begin{cases}\text{surface (surface property per mass)}\\ \text{solid phase}\end{cases}$

SI	surface instability
SO	shearing off
TI	turbulent impact
WE	wake entrainment
t	tangential to interface
w	wall
x, y, z	rectangular coordinate
$+, -$	+ and – side of shock in macroscopic field
$1, 2$	phase 1 and phase 2

Symbols and Operators

$\mathbb{A}$	tensor
$\boldsymbol{A}$	vector
A	scalar
$\boldsymbol{A} \cdot \boldsymbol{B}$	dot product
$\boldsymbol{AB}$	dyadic product of two vectors (=tensor)
$\mathbb{A}:\mathbb{B}$	double dot product of two tensors (=scalar)
$\nabla \cdot$	divergence operator
∇	gradient operator
$\nabla_s \cdot$	surface divergence operator (Aris, 1962)
$(\mathbb{A})^+$	transposed tensor
$\dfrac{D_k}{Dt}$	$= \dfrac{\partial}{\partial t} + \widehat{\boldsymbol{v}_k} \cdot \nabla$
$\dfrac{D}{Dt}$	$= \dfrac{\partial}{\partial t} + \boldsymbol{v}_m \cdot \nabla$
$\dfrac{D_c}{Dt}$	$= \dfrac{\partial}{\partial t} + \boldsymbol{C}_k \cdot \nabla$
$\dfrac{D_i}{Dt}$	$= \dfrac{\partial}{\partial t} + \widehat{\boldsymbol{v}_i} \cdot \nabla$
$\dfrac{d_s}{dt}$	surface convective derivative with $\widehat{\boldsymbol{v}_s}$ (Aris, 1962)
$\overline{F}$	time average
$\overline{F}^{w}$	weighted mean value
$\overline{F}^{w_k}$	k^{th}-phase weighted mean value
$\overline{\overline{F}}$	phase average

$\widehat{\psi_k}$	k^{th}-phase mass weighted mean value
$\widehat{\psi}$	mixture mass weighted mean value
F_k'	fluctuating component with respect to mean value
F_{ki}'	fluctuation component with respect to surface mean value
$\overline{\overline{F_{(i)}}}, \overline{\overline{F_{ki}}}$	surface average
$\widehat{F_{ki}}$	mass flux weighted mean value at interfaces
$(\ )_{,\beta}$	surface covariant derivative (Aris, 1962)
$[\Delta t]_k$	with (k=T,S,1,2) ; sets of time intervals
$\sum_k$	summation on both phases
$\sum_j$	summation on the interfaces passing in Δt at $\boldsymbol{x}$

Index

A

B

C